Advanced Physics

Advanced Physics

Edited by
Aydan Bailey

Willford Press
www.willfordpress.com

Published by Willford Press,
118-35 Queens Blvd., Suite 400,
Forest Hills, NY 11375, USA

ISBN: 978-1-68285-598-0

Cataloging-in-Publication Data

Advanced physics / edited by Aydan Bailey.
p. cm.
Includes bibliographical references and index.
ISBN 978-1-68285-598-0
1. Physics. I. Bailey, Aydan.
QC21.3 .A38 2019
530--dc23

For information on all Willford Press publications visit our website at www.willfordpress.com

WILLFORD PRESS

Printed in China.

Contents

Preface

The main aim of this book is to educate learners and enhance their research focus by presenting diverse topics covering this vast field. This is an advanced book which compiles significant studies by distinguished experts in the area of analysis. This book addresses successive solutions to the challenges arising in the area of application, along with it; the book provides scope for future developments.

Physics is a fundamental science concerned with the study of matter and its motion in space-time as well as its interaction with force and energy. Some of the classical branches of physics are optics, solid-state physics, classical mechanics, electromagnetism and nuclear physics. These have diverse applications in engineering, optical devices, forensic investigations, etc. The chief theories of modern physics include quantum theory and the theory of relativity, which describes physical phenomena in an approach entirely different from that of classical physics. Quantum theory explores the understanding of phenomena at the atomic and sub-atomic level. General relativity explores accelerated motions in a gravitational framework. Research in fundamental particles, supersymmetry, dark matter and gravitational waves is being widely studied. This book elucidates the concepts and innovative models around prospective developments in advanced physics. It presents researches and studies performed by experts across the globe. This book will assist researchers and students in this field.

It was a great honour to edit this book, though there were challenges, as it involved a lot of communication and networking between me and the editorial team. However, the end result was this all-inclusive book covering diverse themes in the field.

Finally, it is important to acknowledge the efforts of the contributors for their excellent chapters, through which a wide variety of issues have been addressed. I would also like to thank my colleagues for their valuable feedback during the making of this book.

Editor

1

Accelerating Bianchi type-VI_0 bulk viscous cosmological models in Lyra geometry

A. Asgar · M. Ansari

Abstract The present study deals with spatially homogeneous and totally anisotropic Bianchi type-VI_0 bulk viscous cosmological models in Lyra geometry. The Einstein's field equations have been solved exactly by taking the shear (σ) in the model proportional to expansion scalar (θ) which leads to $A = B^n$, where A and B are metric functions and n is a positive constant ($n > 1$). We also adopt a condition $\zeta\theta = L$ (constant) where ζ is the coefficient of bulk viscosity. It has been found that the displacement vector (β) is a decreasing function of time and it approaches to a small positive value at late time which is supported by recent observations. It is also found that the distance modulus curve of derived model matches with observations perfectly.

Keywords Bianchi type-VI_0 universe · Accelerated expansion · Bulk viscosity · Lyra geometry

Introduction

Several modifications of Riemannian geometry have been proposed so far in an attempt to unify gravitation, electromagnetic field and many other effects in the universe. Weyl [1] tries to unify gravitation and electromagnetism in single space–time geometry. But Weyl's theory was not taken seriously because it was based on the non-integrability of length transfer. Later on, Lyra [2] proposed a further modification of Riemannian geometry and removed non-integrability of length transfer by introducing a gauge function into the structure-less manifold as a result of which a displacement vector arise naturally. In consecutive investigations Sen [3], Sen and Dunn [4] proposed a new scalar–tensor theory of gravitation and constructed an analog of the Einstein field equations based on Lyra's geometry. Halford [5] has pointed out that the constant vector displacement field ϕ_i in Lyra's geometry plays the role of cosmological constant Λ in the normal general relativistic treatment. It is shown by Halford [6] that the scalar–tensor treatment based on Lyra's geometry predicts the same effects within observational limits as the Einstein's theory.

Cosmological observations on expansion history of the universe indicate that current universe is not only expanding but also accelerating. This late time accelerated expansion of the universe has been confirmed by high redshift supernovae experiments (Riess et al. [7], Perlmutter et al. [8], Bennett et al. [9]). Also, observations such as cosmic background radiation [10, 11] and large-scale structure [12] provide an indirect evidence for late time acceleration.

The simplest model of the observed universe is well represented by Friedmann–Robertson–Walker (FRW) models, which are both spatially homogeneous and isotropic. These models in some sense are good global approximation of the present day universe. But on smaller scales, the universe is neither homogeneous nor isotropic. There are theoretical arguments [13, 14] and recent experimental data regarding cosmic background radiation anisotropies which support the existence of an anisotropic phase that approaches an isotropic one [15]. Bianchi types I–IX cosmological models are important in the sense that these are homogeneous and anisotropic, from which the

A. Asgar (✉) · M. Ansari
Department of Mathematics, North-Eastern Hill University, Shillong 793022, Meghalaya, India
e-mail: aasgar77@yahoo.com

M. Ansari
e-mail: mansarinehu@gmail.com

process of isotropization of the universe is studied through the passage of time. Bianchi type-VI$_0$ space–time is of special interest in anisotropic cosmology. Barrow [16] pointed out that Bianchi type-VI$_0$ models of the universe give a better explanation of some of the cosmological problems like primordial helium abundance and they also isotropize in a special sense.

Astronomical observations of the large-scale distribution of galaxies in the universe show that the distribution of matter can be satisfactorily described by perfect fluid. However, bulk viscosity is expected to play an important role at certain stages of expanding universe. Various authors [14, 17, 18] have shown that bulk viscosity leads to inflationary-like solution and acts like a negative energy field in an expanding universe. At an early stage of the universe, when neutrino decoupling occurs during radiation era and decoupling of radiation with matter takes place during recombination era, the matter behaves like a viscous fluid. The coefficient of viscosity is known to decrease as the universe expands. Gron [19] has reviewed viscous cosmological models and deduced that viscosity plays an important role in the process of isotropization of the universe. Apart from these qualitative discussions, suitable viscous fluid cosmological models have been discussed in different contexts by several authors such as Pavon et al. [20], Burd and Coley [21], Fabris et al. [22], Johri and Sudharsan [23], Murphy [24], Heller and Klimek [25] and Bali et al. [26, 27].

Motivated by the situation discussed above, in this paper we have obtained accelerating Bianchi type-VI$_0$ bulk viscous cosmological models with a time-dependent displacement field within the framework of Lyra's geometry. To get the deterministic model of the universe, we have assumed two conditions: (1) $\zeta\theta = L$ (constant) and (2) $\sigma \propto \theta$. The physical and geometrical aspects of the models are also discussed.

Metric and field equations

We consider the spatially homogeneous and anisotropic Bianchi type-VI$_0$ space–time in the form

$$ds^2 = -dt^2 + A^2 dx^2 + B^2 e^{2x} dy^2 + C^2 e^{-2x} dz^2 \tag{1}$$

where the metric potentials A, B and C are functions of the cosmic time t.

Einstein's modified filed equation in normal gauge for Lyra's manifold obtained by Sen [3] is given by (in geometrized unit where $8\pi G = 1$, $c = 1$)

$$R_i^j - \frac{1}{2} g_i^j R + \frac{3}{2} \phi_i \phi^j - \frac{3}{4} \phi_k \phi^k g_i^j = -T_i^j, \tag{2}$$

where ϕ_i is the displacement vector defined as $\phi_i = (0, 0, 0, \beta(t))$ and other symbols have their usual meaning as in Riemannian geometry. We assume the cosmic matter consisting of bulk viscous fluid given by the energy momentum tensor

$$T_i^j = (p + \rho) v_i v^j + p g_i^j - \zeta\theta (g_i^j + v_i v^j), \tag{3}$$

where $v_i = (0, 0, 0, -1)$, $v^i v_i = -1$, $v_4 = -1$, $v^4 = 1$, p is the isotropic pressure, ρ the matter density, v^i the fluid flow vector and β the gauge function.

The equation of state for the fluid is taken as

$$p = \omega\rho, \tag{4}$$

where $0 \le \omega \le 1$ is a constant. For the metric (1), the field Eqs. (2) together with (3) lead to

$$\frac{\ddot{B}}{B} + \frac{\ddot{C}}{C} + \frac{\dot{B}\dot{C}}{BC} + \frac{1}{A^2} = -(p - \zeta\theta) - \frac{3}{4}\beta^2, \tag{5}$$

$$\frac{\ddot{A}}{A} + \frac{\ddot{C}}{C} + \frac{\dot{A}\dot{C}}{AC} - \frac{1}{A^2} = -(p - \zeta\theta) - \frac{3}{4}\beta^2, \tag{6}$$

$$\frac{\ddot{A}}{A} + \frac{\ddot{B}}{B} + \frac{\dot{A}\dot{B}}{AB} - \frac{1}{A^2} = -(p - \zeta\theta) - \frac{3}{4}\beta^2, \tag{7}$$

$$\frac{\dot{A}\dot{B}}{AB} + \frac{\dot{B}\dot{C}}{BC} + \frac{\dot{A}\dot{C}}{AC} - \frac{1}{A^2} = \rho + \frac{3}{4}\beta^2, \tag{8}$$

$$\frac{\dot{B}}{B} - \frac{\dot{C}}{C} = 0. \tag{9}$$

Here and in what follows, an overhead dot denotes ordinary differentiation with respect to t. The energy conservation equation $T^j_{i;j} = 0$ leads to,

$$\dot{\rho} + (\rho + p - \zeta\theta)\left(\frac{\dot{A}}{A} + \frac{\dot{B}}{B} + \frac{\dot{C}}{C}\right) = 0 \tag{10}$$

and conservation of L.H.S of (2) leads to

$$\left(R_i^j - \frac{1}{2} g_i^j R\right)_{;j} + \frac{3}{2} (\phi_i \phi^j)_{;j} - \frac{3}{4} (\phi_k \phi^k g_i^j)_{;j} = 0. \tag{11}$$

Equation (11) leads to

$$\frac{3}{2}\phi_i \left[\frac{\partial \phi^j}{\partial x^j} + \phi^l \Gamma^j_{lj}\right] + \frac{3}{2}\phi^j \left[\frac{\partial \phi_i}{\partial x^j} - \phi_l \Gamma^l_{ij}\right] - \frac{3}{4} g_i^j \phi_k \left[\frac{\partial \phi^k}{\partial x^j} + \phi^l \Gamma^k_{lj}\right] - \frac{3}{4} g_i^j \phi^k \left[\frac{\partial \phi_k}{\partial x^j} - \phi_l \Gamma^l_{kj}\right] = 0. \tag{12}$$

Equation (12) is automatically satisfied for $i = 1, 2, 3$. For $i = 4$, Eq. (12) leads to

$$\frac{3}{2}\beta \left[\frac{\partial}{\partial x^4}(g^{44}\phi_4) + \phi^4 \Gamma^4_{44}\right] + \frac{3}{2} g^{44}\phi_4 \left[\frac{\partial \phi_4}{\partial t} - \phi_4 \Gamma^4_{44}\right] - \frac{3}{4} g_4^4 \phi_4 \left[\frac{\partial \phi_4}{\partial x^4} + \phi^4 \Gamma^4_{44}\right] - \frac{3}{4} g_4^4 g^{44} \phi_4 \left[\frac{\partial \phi_4}{\partial t} - \phi^4 \Gamma^4_{44}\right] = 0, \tag{13}$$

which leads to

$$\frac{3}{2}\beta\dot{\beta}+\frac{3}{2}\beta^2\left(\frac{\dot{A}}{A}+\frac{\dot{B}}{B}+\frac{\dot{C}}{C}\right)=0. \tag{14}$$

Integrating Eq. (9), we obtain

$$B=lC. \tag{15}$$

Here, l is a constant of integration which can be taken as unity without any loss of generality so that

$$B=C. \tag{16}$$

The average scale factor a for the metric (1) is defined by

$$a=\left(AB^2\right)^{\frac{1}{3}}. \tag{17}$$

The spatial volume V is given by

$$V=a^3=AB^2. \tag{18}$$

The generalized mean Hubble parameter H is given by

$$H=\frac{\dot{a}}{a}=\frac{1}{3}(H_1+H_2+H_3) \tag{19}$$

where $H_1=\frac{\dot{A}}{A}$, $H_2=H_3=\frac{\dot{B}}{B}$ are the directional Hubble parameters in the directions of x, y and z axes, respectively.

The expansion scalar θ and shear scalar σ are given by

$$\theta=\frac{\dot{A}}{A}+2\frac{\dot{B}}{B}, \tag{20}$$

$$\sigma^2=\frac{1}{2}\left[\frac{\dot{A}^2}{A^2}+2\frac{\dot{B}^2}{B^2}\right]-\frac{1}{6}\theta^2. \tag{21}$$

The mean anisotropy parameter A_m is given by

$$A_m=\frac{1}{3}\sum_{i=1}^{3}\left(\frac{\Delta H_i}{H}\right)^2, \tag{22}$$

where $\Delta H_i=H_i-H(i=1,\,2,\,3)$. An important observational quantity in cosmology is the deceleration parameter q which is defined as

$$q=-\frac{a\ddot{a}}{\dot{a}^2}=-\left(\frac{\dot{H}+H^2}{H^2}\right). \tag{23}$$

Solutions of the Field Equations

To solve the field equations completely, we constrain, the system of equations with proportionality relation of shear scalar (σ) and expansion scalar (θ) [28]. This condition leads to the following relation between metric potentials

$$A=B^n, \tag{24}$$

where n is a positive constant ($n>1$).

The reasons for consideration of Eq. (24) can be explained by the work of Thorne [29]. The observations of the velocity–redshift relation for extragalactic sources suggest that the Hubble expansion of the universe is isotropic today within approximately 30 % [30, 31]. More precisely, the redshift studies place the limit $\sigma/H\le 0.30$ where σ is shear and H the Hubble constant. Collins et al. [32] have pointed out that for spatially homogeneous metric, the normal congruence to the homogeneous hypersurface satisfies the condition σ/θ as constant which leads to the assumption $A=B^n$.

We also assume that coefficient of bulk viscosity ζ is inversely proportional to the expansion θ, i.e.,

$$\zeta\theta=L\,(\text{constant}). \tag{25}$$

The motive behind assuming this condition is explained in Ref. [33–35]. Using Eq. (24) in Eqs. (5) and (6), we obtain

$$2\ddot{B}+2\alpha\dot{B}^2B^{-1}=2\gamma B^{1-2n} \tag{26}$$

where $\alpha=n+1$, $\gamma=\frac{2}{n-1}$. Putting $\dot{B}=\eta(B)$, $\ddot{B}=\eta\frac{d\eta}{dB}$ in Eq. (26) and then integrating, we obtain

$$dt=\left[\frac{\gamma}{2}B^{4-2\alpha}+k_1B^{-2\alpha}\right]^{\frac{-1}{2}}dB, \tag{27}$$

where k_1 is a constant of integration.

If we put $n=1$, in Eq. (26) (since $\gamma=\frac{2}{n-1}$) then there arises a singularity. So, we cannot consider $n=1$ in the present model to explain the feature of the universe.

With the help of Eq. (27), the line element (1) reduces to

$$ds^2=-\frac{dB^2}{B^{-2\alpha}\left[\frac{\gamma}{2}B^4+k_1\right]}+B^{2n}dx^2+B^2\left(e^{2x}dy^2+e^{-2x}dz^2\right). \tag{28}$$

After using the suitable transformation of coordinates $B=T$, the above model (28) transforms to

$$ds^2=-\frac{dT^2}{T^{-2(n+1)}\left[\left(\frac{1}{n-1}\right)T^4+k_1\right]}+T^{2n}dx^2+T^2\left(e^{2x}dy^2+e^{-2x}dz^2\right). \tag{29}$$

Some physical and geometrical features

The displacement vector (β), energy density (ρ) and pressure (p) for the model (29) are found to be

$$\beta=k_2\left[\frac{1}{n-1}T^{-4n-2}+k_1T^{-4n-6}\right]^{\frac{1}{2}}, \tag{30}$$

$$\rho=\left(\frac{n+2}{n-1}\right)T^{-2n}+k_1(2n+1)T^{-2n-4}-\frac{3k_2^2}{4(n-1)}T^{-4n-2}-\frac{3}{4}k_1k_2^2T^{-4n-6}, \tag{31}$$

$$p=\omega\left[\left(\frac{n+2}{n-1}\right)T^{-2n}+k_1(2n+1)T^{-2n-4}-\frac{3k_2^2}{4(n-1)}T^{-4n-2}-\frac{3}{4}k_1k_2^2T^{-4n-6}\right], \tag{32}$$

where k_2 is a constant of integration. It is observed that the displacement vector (β) was large in the beginning, but decreases fast with the evolution of the model analogous to cosmological constant (Λ). The energy density (ρ) and pressure (p) are also decreasing function of time T. The energy density $\rho \to \infty$ when $T \to 0$ and $\rho \to 0$ when $T \to \infty$.

The mean Hubble parameter (H), expansion scalar (θ), shear scalar (σ) are given by

$$H = \left(\frac{n+2}{3}\right)\left[\left(\frac{1}{n-1}\right)T^{-2n} + k_1 T^{-2n-4}\right]^{\frac{1}{2}}, \tag{33}$$

$$\theta = (n+2)\left[\left(\frac{1}{n-1}\right)T^{-2n} + k_1 T^{-2n-4}\right]^{\frac{1}{2}}, \tag{34}$$

$$\sigma^2 = \frac{1}{3}(n-1)^2\left[\left(\frac{1}{n-1}\right)T^{-2n} + k_1 T^{-2n-4}\right]. \tag{35}$$

The Hubble parameter (H), expansion scalar (θ) and shear scalar (σ) are tend to zero when $T \to \infty$ and they become infinite when $T \to 0$.

Average scale factor (a) and spatial volume (V) are given by

$$a = \left[T^{-(2n+3)}\sqrt{\left(\frac{1}{n-1}\right)T^4 + k_1}\right]^{\frac{-1}{3}}, \tag{36}$$

$$V = T^{n+2}. \tag{37}$$

The spatial volume $V \to 0$ when $T \to 0$ and $V \to \infty$ when $T \to \infty$.

Also,

$$\lim_{T\to\infty}\frac{\sigma^2}{\theta^2} = \frac{1}{3}\left(\frac{n-1}{n+2}\right)^2 \neq 0. \tag{38}$$

Therefore, the model does not isotropic for large values of T.

Coefficient of bulk viscosity (ζ) and mean anisotropy parameter (A_m) are found to be

$$\zeta = \left(\frac{L}{n+2}\right)\left[\frac{1}{n-1}T^{-2n} + k_1 T^{-2n-4}\right]^{\frac{-1}{2}}. \tag{39}$$

$$A_m = \frac{1}{3}\left[\left(1 - \frac{3n}{n+2}\right)^2 + 2\left(1 - \frac{3}{n+2}\right)^2\right]. \tag{40}$$

Therefore, the model has constant anisotropy parameter throughout the evolution of the universe except $n = 1$.

The deceleration parameter q is found to be

$$q = -1 + \left(\frac{3}{n+2}\right)\left(\frac{1}{n-1}T^4 + k_1\right)^{-1} \times \left((n+2)T^{n+1}\sqrt{\frac{1}{n-1}T^4 + k_1} - \frac{\frac{2}{n-1}T^{n+5}}{\sqrt{\frac{1}{n-1}T^4 + k_1}}\right). \tag{41}$$

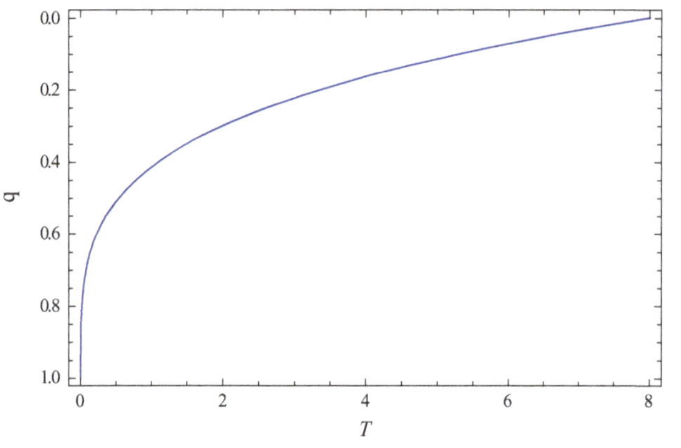

Fig. 1 The variation of q vs. T for the model (29) with parameters $n = 1.25$, $k_1 = 0.001$

Recent observations (Perlmutter et al. [8, 36, 37]; Riess et al. [7, 38]; Tonry et al. [39]; John [40]; Knop et al. [41]) reveal that the value of deceleration parameter q is confined in the range $-1 \leq q < 0$ and the present day universe is undergoing an accelerated expansion. We have seen in Fig. 1 that the value of q lies in the range $-1 \leq q < 0$ which is consistent with recent observations. The negative value of deceleration parameter implies that our proposed model (29) of the universe is accelerating.

Distance modulus curve

The distance modulus is given by

$$\mu = 5\log d_L + 25, \tag{42}$$

where the luminosity distance d_L is defined as

$$d_L = r_1(1+z)a_0. \tag{43}$$

Here, z and a_0 represent redshift parameter and present scale factor, respectively. For the determination r_1, we assume that photon is emitted by a source with co-ordinate (r, t) and received at a time t_0 by an observer located at $r = 0$. Then, we determine r_1 from

$$r_1 = \int_t^{t_0}\frac{dt}{a}. \tag{44}$$

Using Eqs. (27) and (36) (considering $k_1 = 0$) and solving Eqs. (42)–(44), we can obtain the expression for distance modulus (μ) in terms of redshift parameter (z) as

$$\mu = 5\log\left[\frac{3}{\sqrt{n-1}}\left(\frac{2n+1}{3H_0}\right)^{\frac{n+2}{3}}(1+z)\left(1-(1+z)^{\frac{1-n}{2n+1}}\right)\right] + 25. \tag{45}$$

In this paper, we have analyzed 18 data set out of recently released 38 data set of SN Ia in the range

Table 1 Comparison of the results of present model with the observational data

Redshift (z)	Supernovae Ia (μ)	Our model (μ)
0.0132	$30.35^{+0.26}_{-0.26}$	30.0531
0.0133	$32.54^{+0.22}_{-0.22}$	32.1711
0.0134	$33.10^{+0.22}_{-0.22}$	32.1875
0.0264	$33.93^{+0.2}_{-0.2}$	33.6684
0.0274	$34.33^{+0.1}_{-0.1}$	33.7498
0.0284	$34.61^{+0.23}_{-0.23}$	33.8283
0.0295	$34.82^{+0.17}_{-0.17}$	33.9115
0.0398	$35.18^{+0.17}_{-0.17}$	34.5684
0.0399	$35.91^{+0.17}_{-0.17}$	34.5739
0.0409	$35.84^{+0.16}_{-0.16}$	34.6283
0.0519	$36.09^{+0.16}_{-0.16}$	35.1524
0.0619	$36.35^{+0.17}_{-0.17}$	35.5412
0.0721	$36.33^{+0.17}_{-0.17}$	35.8787
0.0813	$36.75^{+0.16}_{-0.16}$	36.1451
0.0854	$36.71^{+0.16}_{-0.16}$	36.2544
0.0951	$36.86^{+0.16}_{-0.16}$	36.4938
0.0967	$36.76^{+0.16}_{-0.16}$	36.531
0.0979	$37.5^{+0.19}_{-0.19}$	37.54

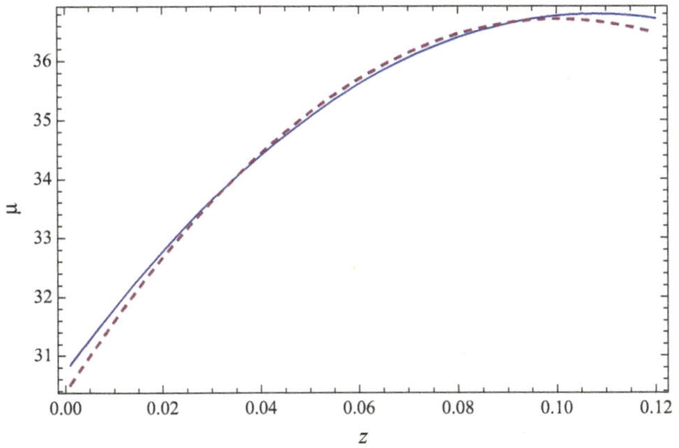

Fig. 2 Plot of distance modulus (μ) vs. redshift (z) for Supernovae Ia data (*solid line*) from Amanullah et al. [42] and for our model (*dotted line*)

$0.0015 \leq z < 0.12$ as reported by Amanullah et al. [42] which are shown in Table 1. The comparison between calculated $\mu(z)$ and observed $\mu(z)$ SN Ia data as reported by Amanullah et al. [42] can be seen in Fig. 2. We observed that derived model is fit well with SN Ia observations.

Conclusion

In this paper, we have obtained accelerating Bianchi type-VI_0 bulk viscous cosmological models in the framework of Lyra's geometry for time-dependent displacement field. To get the deterministic model of the universe, we have assumed two conditions: (1) $\zeta\theta = L$ (constant) and (2) $\sigma \propto \theta$, where ζ is the coefficient of bulk viscosity, θ the expansion in the model and σ the coefficient of shear viscosity. It is observed that the spatial volume $V \to 0$ as $T \to 0$ and $V \to \infty$ as $T \to \infty$. This shows that the universe starts expanding with zero volume at $T = 0$ and expands with cosmic time T. When $T \to 0$, the expansion scalar $\theta \to \infty$ and energy density $\rho \to \infty$. Again, when $T \to \infty$ then $\theta \to 0$ and $\rho \to 0$. Therefore, the model represents expanding, shearing and non-rotating universe with big bang start. Since $\lim_{T\to\infty} \frac{\sigma^2}{\theta^2} \neq 0$, therefore, the model is not isotropic for large values of T. At $T = 0$, Hubble parameter H, expansion scalar θ, shear scalar σ, pressure p and energy density ρ are infinite and at late times they become zero. The role of bulk viscosity is to retard expansion in the model. We can see from the above discussion that the bulk viscosity plays a significant role in the evolution of the universe. We also observe that the displacement vector β is large in the beginning of the universe and reduces fast during its evolution so that its behavior coincides with the nature of the cosmological constant Λ. The distance modulus of the derived model fits well with the observational $\mu(z)$ values (see Fig. 2 and Table 1).

Acknowledgments Authors are thankful to referee for valuable comments and constructive suggestions to improve the manuscript.

References

1. Weyl, H: Vorlesungen uker Allgemeine relativitotstheorie. Sber. Preuss. Akad. der Wiss. 465 Sitzunbgsberichte, zu Berlin (1918)
2. Lyra, G.: Über eine Modifikation der Riemannschen Geometrie. Math. Z. **54**, 52–64 (1951)
3. Sen, D.K.: A static cosmological model. Z. Phys. **149**, 311–323 (1957)
4. Sen, D.K., Dunn, K.A.: A scalar-tensor theory of gravitation in a modified Riemannian manifold. J. Math. Phys. **12**, 578–586 (1971)
5. Halford, W.D.: Cosmological theory based on Lyra's geometry. Aust. J. Phys. **23**, 863–869 (1970)
6. Halford, W.D.: Scalar-tensor theory of gravitation in a Lyra manifold. J. Math. Phys. **13**, 1699–1703 (1972)
7. Riess, A.G., et al.: Observational evidence from supernovae for an accelerating universe and a cosmological constant. Astron. J. **116**, 1009–1038 (1998)
8. Perlmutter, S., et al.: Measurements of Ω and Λ from 42 high-redshift supernovae. Astrophys. J. **517**, 565–585 (1999)
9. Bennett, C.L., et al.: First-year Wilkinson microwave anisotropy probe (WMAP) observations: preliminary maps and basic results. Astrophys. J. Suppl. Ser. **148**, 1–27 (2003)
10. Spergel, D.N., et al.: First-year Wilkinson microwave anisotropy probe (WMAP) observations: determination of cosmological parameters. Astrophys. J. Suppl. Ser. **148**, 175–194 (2003)
11. Oli, S.: Interacting and non-interacting two fluid cosmological

models in a bianchi type-VI_0 space times. Ind. J. Phys. **86**, 755–761 (2012)
12. Tegmark, M., et al.: Cosmological parameters from SDSS and WMAP. Phys. Rev. D **69**, 103501–103526 (2004)
13. Chimento, L.P.: Extended tachyon field, chaplygin gas and solvable k-essence cosmologies. Phys. Rev. D **69**, 123517-10 (2004)
14. Misner, C.W.: The isotropy of the universe. Astrophys. J. **151**, 431–457 (1968)
15. Land, K., Magueijo, J.: Examination of evidence for a preferred axis in the cosmic radiation anisotropy. Phys. Rev. Lett. **95**, 071301–071304 (2005)
16. Barrow, J.D.: Helium formation in cosmologies with anisotropic curvature. Mon. Not. R. Astr. Soc. **211**, 221–227 (1984)
17. Padmanabhan, T., Chitre, S.M.: Viscous universes. Phys. Lett. A **120**, 433–436 (1987)
18. Johri, V.B., Sudarshan, R.: FRW cosmological models with bulk viscosity and entropy production. Proceedings of International Conference on Mathematical Modeling in Science and Technology. vol. 2, Tata McGraw-Hill (1988)
19. Gron, O.: Viscous inflationary universe models. Astrophys. Space Sci. **173**, 191–225 (1990)
20. Pavon, D., Bafaluy, J., Jou, D.: Causal Friedmann–Robertson–Walker cosmology. Class. Quantum Grav. **8**, 347–360 (1991)
21. Burd, A., Coley, A.: Viscous fluid cosmology. Class. Quantum Grav. **11**, 83–105 (1994)
22. Fabris, J.C., Goncalves, S.V.B., Ribeiro, R.: Bulk viscosity driving acceleration of the universe. Gen. Relativ. Gravit. **38**, 495–506 (2006)
23. Johri, V.B., Sudharsan, R.: Friedmann universes with bulk viscosity. Phys. Lett. A **132**, 316–320 (1988)
24. Murphy, G.L.: Big-bang model without singularities. Phys. Rev. D **8**, 4231–4233 (1973)
25. Heller, M., Klimek, Z.: Viscous universes without initial singularity. Astrophys. Space Sci. **33**, L37–L39 (1975)
26. Bali, R., Pradhan, A., Amirhashchi, H.: Bianchi type-VI_0 magnetized barotropic bulk viscous fluid massive string universe in general relativity. Int. J. Theor. Phys. **47**, 2594–2604 (2008)
27. Bali, R., Banerjee, R., Banerjee, S.K.: Bianchi type-VI_0 magnetized bulk viscous massive string cosmological model in general relativity. Astrophys. Space Sci. **317**, 21–26 (2008)
28. Tiwari, R.K., Dwivedi, U.: Kantowski-Sachs cosmological models with time-varying G and Λ. FIZIKA B (Zagreb). **19**, 1–8 (2010)
29. Thorne, K.S.: Primordial element formation, primordial magnetic fields and the isotropy of the universe. Astrophys. J. **148**, 51–68 (1967)
30. Kantowski, R., Sachs, R.K.: Some spatially homogeneous anisotropic relativistic cosmological models. J. Math. Phys. **7**, 443–446 (1966)
31. Kristian, J., Sachs, R.K.: Observations in cosmology. Astrophys. J. **143**, 379–399 (1966)
32. Collins, C.B., Glass, E.N., Wilkinson, D.A.: Exact spatially homogeneous cosmologies. Gen. Rel. Grav. **12**, 805–823 (1980)
33. Bali, R., Pradhan, A.: Bianchi type-III string cosmological models with time dependent bulk viscosity. Chin. Phys. Lett. **24**, 585–588 (2007)
34. Saha, B.: Bianchi type-I universe with viscous fluid. Mod. Phys. Lett. A **20**, 2127–2143 (2005)
35. Yadav, V.K., Yadav, L.: Some bulk viscous magnetized LRS bianchi type-I string cosmological models in Lyra's geometry. Rom. J. Phys. **58**, 64–74 (2013)
36. Perlmutter, S., et al.: Discovery of a supernova explosion at half the age of the universe. Nature **391**, 51–54 (1998)
37. Perlmutter, S., et al.: Measurements of the cosmological parameters Ω and Λ from the first seven supernova at $z \geq 0.35$. Astrophys. J. **483**, 565–581 (1997)
38. Riess, A.G., et al.: Type Ia supernova discoveries at $z > 1$ from the Hubble space telescope: evidence for past deceleration and constraints on dark energy evolution. Astrophys. J. **607**, 665–687 (2004)
39. Tonry, J.L., et al.: Cosmological results from high-z supernovae. Astrophys. J. **594**, 1–24 (2003)
40. John, M.V.: Cosmographic evaluation of the deceleration parameter using type Ia supernova data. Astrophys. J. **614**, 1–5 (2004)
41. Knop, R.A., et al.: New constraints on Ω_M, Ω_Λ and ω from an independent set of 11 high-redshift supernovae observed with the Hubble space telescope. Astrophys. J. **598**, 102–137 (2003)
42. Amanullah, R., et al.: Spectra and Hubble space telescope light curves of six type Ia supernovae at $0.511 < z < 1.12$ and the union 2 compilation. Astrophys. J. **716**, 712–738 (2010)

2

Fabrication of graded helical square tower-like Mn sculptured thin films and investigation of their electrical properties: comparison with perturbation theory

Mahsa Fakharpour[1] **· Hadi Savaloni**[2]

Abstract Mn sculptured thin films were fabricated in form of graded helical square tower-like terraced sculptured Mn thin films (GHSTTS) using oblique angle deposition together with rotation of substrate about its surface normal with fixed rotation angle (90°) and a shadowing block which was fixed at the centre of the substrate holder. The anisotropy of the samples was examined by resistivity measurements at two orthogonal angles. Direct relationship is obtained between resistivity and the anisotropy of the produced samples which showed that both of these parameters increase with decreasing distance from the edge of the shadowing block. Simulation work using the perturbation theory produced results consistent with the experimental observations.

Keywords Mn sculptured thin films · Anisotropy · Electrical property · Perturbation theory · Shadowing effect · Helical

Introduction

The size and shape of nano-particle as well as the environment (medium) are important parameters in determining the properties of the nano-particle. The different properties of the nano-particle or an assembly of nano-particles in form of a thin film may be adjusted by controlling the size and shape of the nano-particle(s). Hence, it is of utmost importance to have control on the fabrication process of such structures. Two of such techniques are proposed and implemented during the last few years namely oblique angle deposition (OAD) (vapor incident angle less than 85°) of thin films and glancing angle deposition (GLAD) (vapor incident angle greater than 85°) as physical vapor deposition methods have provided facilities for production of variety of nano-structures with structural anisotropy which can be controlled by pre-design of the structure [1]. Once these techniques are combined with rotation of substrate in two different directions, two-dimensional and three-dimensional structures can be fabricated [2, 3] which is known as thin film nano-engineering. The main phenomenon in oblique angle is the shadowing effect. The self shadowing effect prohibits areas at the rear of the formed nuclei where the growth is extensively reduced [4, 5], and further deposition of atoms cause growth of these nuclei which ultimately tilted columnar structure with voids form on the substrate surface [6–8]. Sculptured thin films consist of columnar structures that their growth direction is being controlled/changed by movement of substrate. Complex structures such as chiral nano-flower shaped sculptured thin films are produced and their optical properties are investigated both experimentally [9] and theoretically [10–12].

Sculptured nano-structures fabricated by these methods have a vast area of applications such as optical filters [13], reflectors [14], polarizers [15], sensors [16, 17], bioscaffolds [18], and microchannels [19].

In this work graded helical square tower-like terrace shaped sculptured thin films (GHSTTS) were produced by oblique angle deposition together with the rotation of the substrate sequentially by 90°, while a shadowing block was fixed at the centre of the substrate holder having substrates

✉ Hadi Savaloni
Savaloni@khayam.ut.ac.ir

1 Department of Physics, Science and Research Branch, Islamic Azad University, Tehran, Iran

2 Department of Physics, University of Tehran, North-Kargar Street, Tehran, Iran

mounted on it at different distances from the edge of the shadowing block. In addition, the length of the arms of the second and third pitches of the GHSTTS were reduced by a third and two thirds relative to the length of the arms of the first pitch; hence, terraced structures as well as slanted shapes in which the tilt angle changes with distance from the edge of the shadowing block were formed.

Theory

Perturbation method for calculation of electrical resistance

Hori and Yonezawa [20] in 1977 introduced a formalism in which a relationship is made between the electric field of a conductive material with voids (inhomogeneous region) and the electric field of that material with homogenized structure. The relationship between localized current density, conductivity and localized electric field in the introductory electromagnetism is given as follows:

$$J_i(r) = \sigma(r)E_i(r), \tag{1}$$

If the region under examination consists of voids (as a perturbing medium), then one may introduce $\delta\sigma(r)$ as the local perturbation in the conductivity $\sigma(r)$;

$$\sigma(r) = \sigma_0 + \delta\sigma(r), \tag{2}$$

where σ_0 is conductivity of the material. Using the electrostatic potential $\Phi(r)$ and the continuity relationship, we have

$$\Phi(r_1) = \Phi_0(r_1) + \int dr_2 g(r_{12}) \frac{\partial}{\partial x_{2,i}} \left(\delta\sigma(r_2) \frac{\partial \Phi(r_2)}{\partial x_{2,i}} \right), \tag{3}$$

where $r_{12} = r_2 - r_1$ and the integral is taken over whole space v or s. $\Phi_0(r)$ only depends on σ_0 and the boundary condition. The electric field is calculated as

$$E_i(r_1) = E_i^0(r_1) + \int dr_2 G_{ij}(r_{12}) \delta\sigma(r_2) E_j(r_2), \tag{4}$$

$E_i^0(r_1)$ is independent of $\delta\sigma(r_1)$ because it is the electric field of homogenized medium with σ_0 conductivity at r_1, $E_i(r_1)$ is the electric field of the ith component at r_1 and $G_{ij}(r_{12})$ is the Green function defined as

$$G_{ij}(r_{12}) = -\frac{\partial^2 g(r_{12})}{\partial x_{1,i} \partial x_{2,j}} = \begin{cases} -\dfrac{\delta_{ij}\delta(r_{12})}{3\sigma_0} + p\dfrac{1}{4\pi\sigma_0 r_{12}^3}\left(3\dfrac{x_{12,i}}{r_{12}}\dfrac{x_{12,j}}{r_{12}} - \delta_{ij}\right) 3D \\ -\dfrac{\delta_{ij}\delta(r_{12})}{2\sigma_0} + p\dfrac{1}{2\pi\sigma_0 r_{12}^2}\left(2\dfrac{x_{12,i}}{r_{12}}\dfrac{x_{12,j}}{r_{12}} - \delta_{ij}\right) 2D \end{cases}. \tag{5}$$

In order to solve Eq. (4), we considered the thin film as a large rectangular cube consisting of many small cubes (see Fig. 2 in [21]). It may also be assumed that the electric field and conductivity in each of these small cubes is fixed (i.e. variation of these two quantities with respect to the small size of the cube is negligible). With these assumptions and some calculations, Eq. (4) becomes

$$E_i^0(r_k) = E_i(r_k)\left(1 + \frac{\delta\sigma(r_k)}{3\sigma_0}\right) - \sum_{j=1}^{3}\sum_{l \neq k} v_l \frac{1}{4\pi\sigma_0 r_{kl}^3}\left(3\frac{x_{kl,i}}{r_{kl}}\frac{x_{kl,j}}{r_{kl}} - \delta_{ij}\right)\delta\sigma(r_l)E_j(r_l), \tag{6}$$

where v_l is the volume of each small cube, r_{kl} is the distance between kth cube and lth cube and $x_{kl,i}$ is the ith component of vector connecting r_k to r_l whose value is r_{kl}. Since this value under no circumstance becomes zero, there is no need to use the principle value [21].

In addition, one should also consider some of these small cubes as voids (void fraction) and some of them as the film material (fraction of material inclusions). In order to implement this idea according to the film structure, one should consider some of the columns made of small cubes as the film material with σ_0 conductivity and the columns between them as voids with zero conductivity. In order to perform the calculations in this work, the dimensions of the cubes were considered as large enough that classical concepts can be applied [21]. At this stage, after considering the above operations, Eq. (6) transfers to 3N equations with 3N unknowns, where N is the number of small cubes and the components of local electric fields are the unknowns of the problem. After solving these 3N equations the local current density can be obtained from multiplication of the electric field of each cube with its conductivity. The summation of local current density of all cubes will provide the total current density. Finally, multiplying the total current density with cross-section area of the thin film in a certain direction yields the resistance of the film. This resistance is the relative resistance of the film (i.e., the ratio of the resistance of the porous film to a homogeneous film from the same material). Hence in the perturbation theory the type of the material is not included and only the geometry of the structure is involved.

Exprimental details

The graded helical square tower-like (terraced) sculptured (GHSTTS) Mn thin films were deposited on glass (microscope slide) substrates by electron beam evaporation from a graphite crucible at room temperature. The purity of Mn was 99.99%. An Edwards (Edwards E19 A3) coating plant with a base pressure of 2×10^{-7} mbar was used. The deposition angle was fixed at 85° and the substrate was rotated clock-wise for 90° azimuthally for production of each arm of the GHSTTS. The substrate holder system is controlled by two stepping motors which can rotate the substrate holder by the two angles, α and φ, with 0.01 °/step accuracy and with controlled speed. The movement of the stepper motor for rotation of substrate about its surface normal (φ) and its speed of revolution as well as facility for dividing each revolution to different sectors are controlled through interface to a computer in which the related software (in the LABVIEW format) is written and installed. All these are domestic made. The substrate holder was a stainless steel disc of 12 cm diameter. At the centre of this disc a cylindrical block (2 cm in diameter and 2 cm in height) was fixed as a shadowing block. The substrates were fixed at 1, 3 and 5 cm distance from this shadowing block along four mutually normal radii of the substrate holder disc. Hence, in each run four sets of three samples were produced for use in different analyses and reproducibility check of the samples.

Three different GHSTTS Mn thin films consisting of different number of arms, namely eight, nine and ten arms were produced. Each arm of the first pitch was deposited for 110 nm while each arm of the second and third pitches was 72 and 36 nm, respectively.

The deposition rate was fixed at 1.0 Ås^{-1}. The deposition rate was measured by a quartz crystal deposition rate controller (Sigma Instruments, SQM-160, USA) positioned close to the substrate holder and at almost the same azimuthal angle as that of the substrate. This was calibrated after obtaining the film thickness using a field emission electron microscope (FESEM) and dividing by the deposition time. The distance between the evaporation source and the substrate was 30 cm. In this arrangement, the vapor source (6 mm in diameter) behaves like a point source with a cosine distribution, and because of the 30 cm distance between the vapor source and the substrate it is expected that the vapor has straight trajectories (i.e., no appreciable scattering due to the large mean free path (~103–104 cm [22]) occurs).

Prior to deposition, all glass substrates were ultrasonically cleaned in heated acetone and then ethanol. The surface roughness of the substrates was measured by a Talysurf profilometer and a AFM and the rms (route mean square) substrate roughness (Rq) obtained using these methods were 0.3 and 0.9 nm, respectively.

The deposition process was repeated a few times and the reproducibility of the results was confirmed. The film thicknesses and column shapes and sizes were measured by field emission electron microscope (FESEM) (Hitachi S-4100 SEM, Japan). The FESEM samples were coated with a very thin layer of gold to prevent the charging effect.

The surface physical morphology and surface roughness of the samples were obtained by means of AFM (NT-MDT, SOLVER, Nova Tech) analysis with a Si tip of 10 nm diameter and in non-contact mode. The electrical measurements of these thin films were obtained by four point probe.

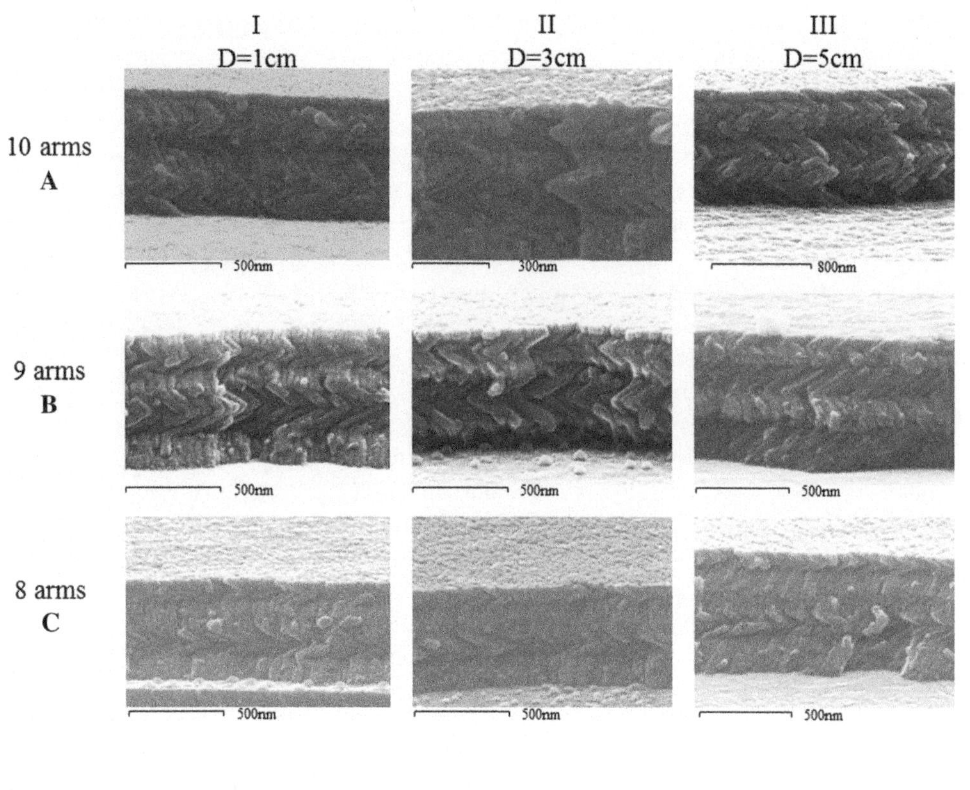

Fig. 1 FESEM images of graded helical square tower-like (terraced) Mn sculptured thin films produced at different distances from the edge of the shadowing block with different number of arms; column (I) 1 cm distance; column (II) 3 cm distance; column (III) 5 cm distance. Row (**a**) 10 arms; Row (**b**) 9 arms; Row (**c**) 8 arms

Table 1 Details of graded helical sculptured Mn thin film samples and their different parameters

Sample	Distance from the shadowing block (cm)	Film thickness (nm)	Helical column tilt angle (αt) (°)	D_{AFM} (nm)	R_{ave} (nm)	R_{rms} (nm)	VF_{surf} (%)
A1	1	465.5	14	122.0	1.75	2.24	13.32
A3	3	504.1	8	112.0	1.33	1.68	13.19
A5	5	550.8	4	96.0	1.20	1.50	9.60
B1	1	436.7	12	120.7	2.33	2.87	19.04
B3	3	450.0	8	109.6	2.20	2.86	16.44
B5	5	544.8	4	81.0	1.96	2.58	12.87
C1	1	417.4	12	132.3	2.60	3.42	22.92
C3	3	441.0	5	128.9	2.57	3.28	29.32
C5	5	525.0	3	91.9	2.03	2.60	19.55

D_{AFM} grain size obtained from AFM analysis. R_{ave} and R_{rms} average and root mean square surface roughnesses, respectively, VF_{surf} surface void fraction

Results and discussion

Structure and surface morphology of GHSTTS thin films

In Fig. 1 (columns I to III) the FESEM images of the cross sections of the GHSTTS Mn thin films with eight, nine and ten arms produced at three different distances of 1, 3 and 5 cm from the edge of the shadowing block are shown. The helical structure of the grown nano-columns and the dense bottom layer of the films [23, 24] can clearly be distinguished in these images, while the grown arms of the square pitches and their angle as well as the growth angle of the columns (tilt angle) are also seen. In addition, one can also see the reduction of the length of the arms from bottom pitch to upper pitches consistent with the deposition procedure (110, 72 and 36 nm lengths of arms for the bottom, middle and top pitches, respectively). The tilt angle of the helical columns is measured from the FESEM images and is given in Table 1 for GHSTTSs with eight, nine and ten arms as mentioned above. In order to clearly identify the produced samples with three different number of arms and that each set contains of three different samples which were positioned at three different distances from the edge of the shadowing block (i.e., 1, 3 and 5 cm) we have assigned the following names/symbols to them; the set with 10 arms are assigned as "A" group followed by distance from the shadowing block in centimeters: A1, A3 and A5. Similarly for the set with 9 arms we assigned them as B1, B3 and B5 and for the group with 8 arms we named them as C1, C3 and C5. Results of growth/tilt angle of columns and the thickness of the films measured from the FESEM images are given in Table 1 and these are consistent with the calculated results.

Figure 2i–iii shows the 2D and 3D AFM images (columns I and II) as well as surface void/inclusion fractions (column III), while in Table 1 the results of grain size distribution, percentage of surface void fraction (measured using JMicroVision software) and also average and root mean square surface roughness for all samples produced in this work obtained from the AFM results are given.

Electrical resistivity

The electrical resistivity of the Mn GHSTTS films produced in this work was measured using a four point probe instrument to obtain their I–V curves. Measurements were carried out at four different directions of x (vapour incident direction) and y axes of the sample as well as two diagonal directions. These measurements provide us with the required data to evaluate the degree of anisotropy in the structure of the produced samples due to the structural differences (tilt angle, number of arms) and the bundling effect in these obliquely deposited thin films [25]. Resistivity of the films was calculated using [26]:

$$\rho = \frac{2\pi s V}{I}, \tag{7}$$

The uncertainty of these results was obtained as [27]:

$$\Delta\rho = \frac{\pi t}{\ln 2}\left(\frac{V}{I}\right)\sqrt{\left(\frac{\Delta v}{v}\right)^2 + \left(\frac{\Delta I}{I}\right)^2}. \tag{8}$$

As mentioned earlier due to anisotropic structure of these films their anisotropy was achieved using

$$Anisotropy = \frac{\rho_x - \rho_y}{\rho_x + \rho_y}. \tag{9}$$

In Fig. 3 results obtained for the I–V curves of the Mn GHSTTS thin films with 8, 9 and 10 arms deposited at

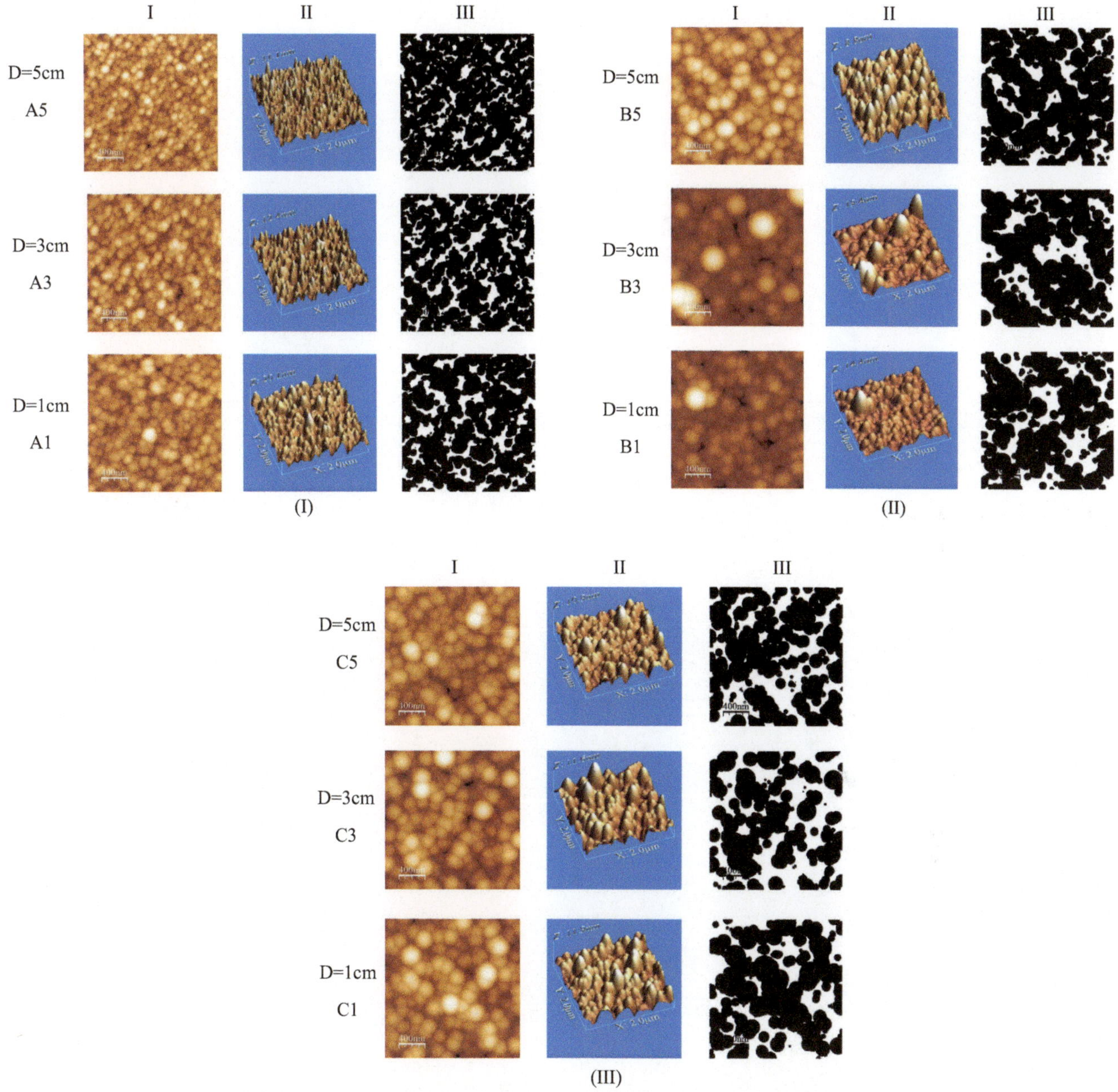

Fig. 2 (i) AFM images of graded helical square tower-like (terraced) Mn sculptured thin films produced at different distances from the edge of the shadowing block with 10 arms; column (I) 2D images; column (II) 3D images; column (III) void fraction. Row (**a**) 5 cm distance; Row (**b**) 3 cm distance; Row (**c**) 1 cm distance. (ii) AFM images of graded helical square tower-like (terraced) Mn sculptured thin films produced at different distances from the edge of the shadowing block with 9 arms; column (I) 2D images; column (II) 3D images; column (III) void fraction. Row (**a**) 5 cm distance; Row (**b**) 3 cm distance; Row (**c**) 1 cm distance. (iii) AFM images of graded helical square tower-like(terraced) Mn sculptured thin films produced at different distances from the edge of the shadowing block with 8 arms; column (I) 2D images; column (II) 3D images; column(III) void fraction. Row (**a**) 5 cm distance; Row (**b**) 3 cm distance; Row (**c**) 1 cm distance

different distances from the edge of the shadowing block in both x and y directions and two diagonal directions obtained from Eq. 7 are given. The influential parameters on the variation of resistivity can be surface porosity, surface roughness, film thickness or the length of the arms in the structures produced in this work, geometry of the structure and grain sizes. Now we may examine our results of the resistivity in Fig. 3. Considering the aforementioned parameters in relationship to the structure of different samples produced in this work, one can perform two types of comparison (analyses): (1) comparison of the resistivity of each structure (i.e., GHSTTS with 10, or 9 or 8 arms), with increasing distance from the edge of the shadowing block; (2) comparison of the three different GHSTTS with

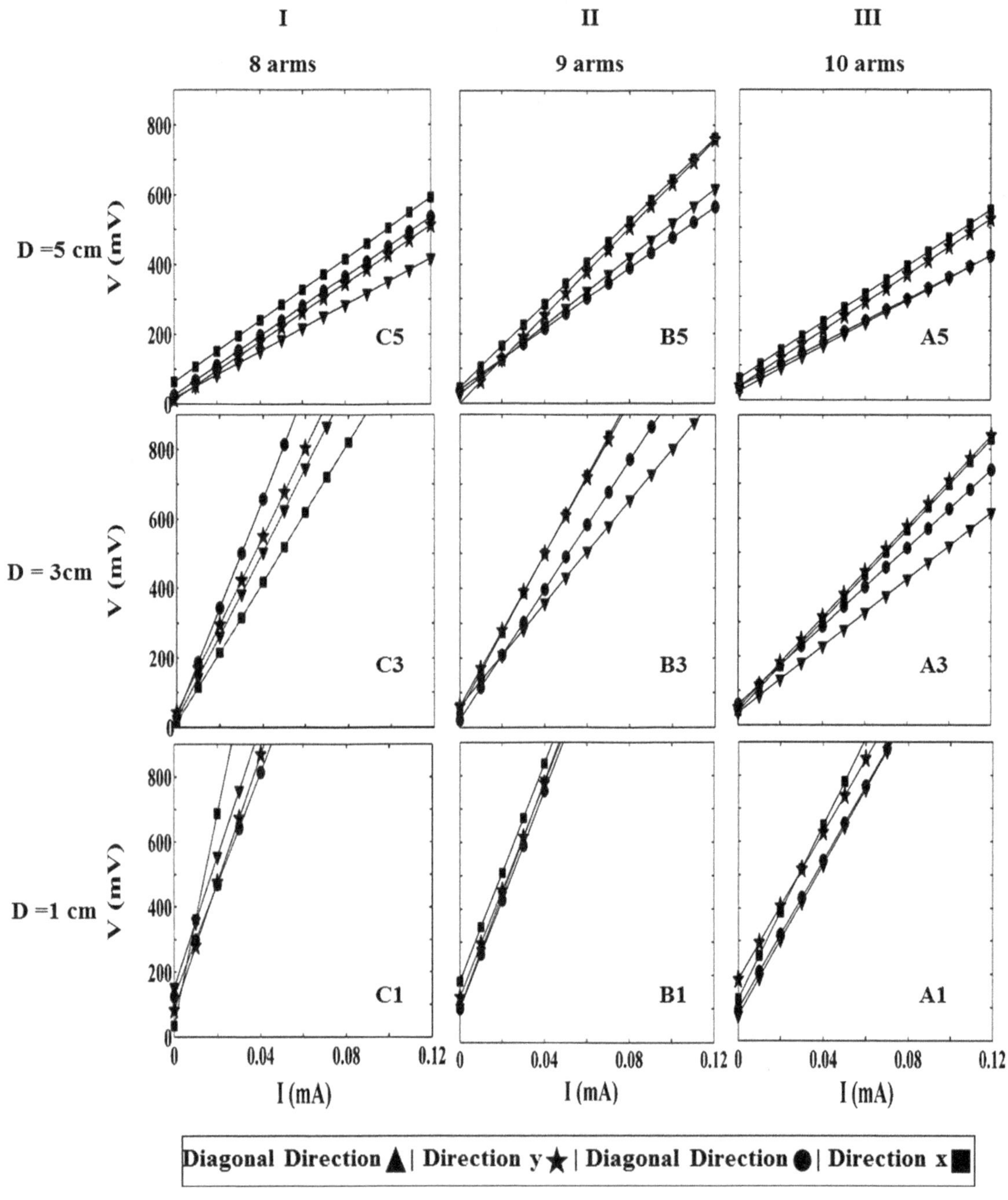

Fig. 3 I–V *curves* for GHSTTS structures with 8, 9 and 10 arms

the same distance from the edge of the shadowing block, but with different number of arms.

Comparison of samples of each structure with increasing distance from the edge of the shadowing block

The following observations may be made:

(a) The slop of the I–V curves is different for different measurement directions; hence the film structure is inhomogeneous and anisotropic which will be presented in the following paragraphs.

(b) The slop of I–V curves for the samples closer to the edge of the shadowing block is higher than those at further distance which reduces with increasing this distance. This shows that samples grown at close distance to the edge of the shadowing block have higher anisotropy/resistance. This in fact can be due to several factors as follows: the higher tilt angle of columns in this type of samples (i.e., A1, B1 and C1) relative to samples deposited at further distance

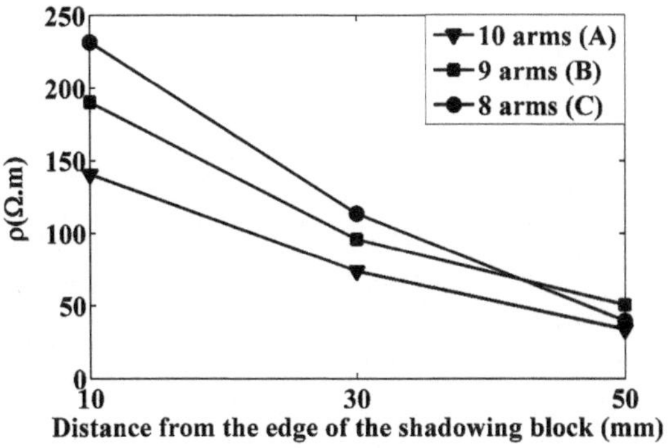

Fig. 4 Resistivity of GHSTTS structures with different number of arms versus distance from the edge of the *shadowing block*

(Table 1); higher void fraction of this type of samples obtained from the analysis of the AFM results (Table 1); higher surface roughness of these samples (Table 1), and the lower film thickness (Table 1).

Comparison of the resistivity of GHSTTS with different number of arms at the same distance from the edge of the shadowing block

In Fig. 4 the resistivity of A, B and C samples positioned at different distances from the edge of the shadowing block are compared. It can be observed that the resistivity of these samples varies as follows: $\rho A5 \prec \rho C5 \prec \rho B5, \rho A3 \prec \rho B3 \prec \rho C3, \rho A1 \prec \rho B1 \prec \rho C1$. The order/trend of variation of the resistivity for the samples positioned at 5 cm distance from the edge of the shadowing block differs from the other two sets of samples though this difference is very small and may be ignored as it could be either due to the uncertainty in the data or due to the lower tilt angle of the helical columns for the C5 sample (see Table 1).

The general trend of variation seen in Fig. 4 can be explained on the bases (with reference to the data presented in Table 1) of the thickness of the films (the thicker films should have lower resistivity), surface roughness of the samples (surface roughness is decreased with increasing the number of arms, hence less electron scattering from surface roughness may occur which in turn reduces the resistivity of the sample) and void fraction of the samples (void fraction is increased for the samples closer to the edge of the shadowing block; hence it can be assumed that large potential barriers are produced between the grains which obstructs the movement of electrons along the length of the sample (or the surface).

Electrical anisotropy of GHSTTS thin films

In Fig. 5 results obtained for the resistivity of the GHSTTS nano-structures with 8, 9 and 10 arms in both x and y

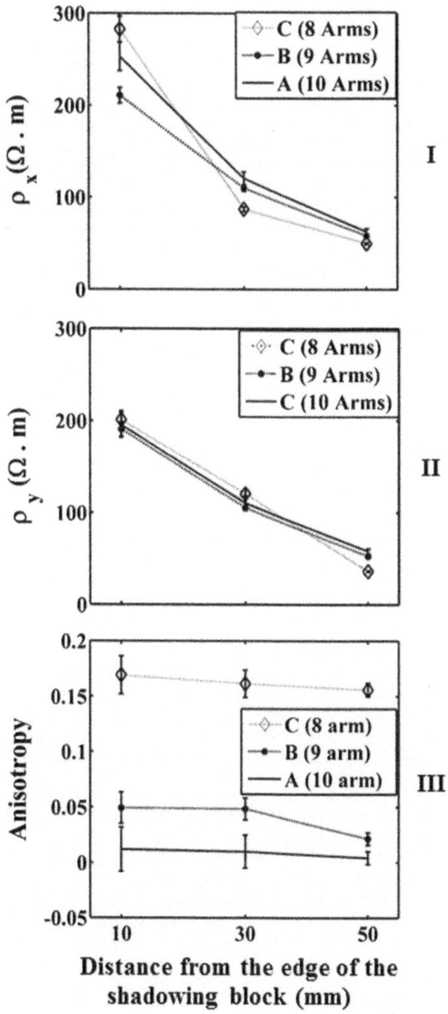

Fig. 5 Resistivity of GHSTTS structures with 8, 9 and 10 arms measured at two *x* and *y* directions and the obtained anisotropy

directions and the anisotropy of the samples obtained from Eq. (9) are given. It can be seen that there is a decreasing behaviour for all of these plots with increasing the distance from the edge of the shadowing block. Hence, the electrical measurements give us a proof and degree of anisotropy in the structure of the GHSTTS films produced in this work.

Estimating the resistance of GHSTTS thin films

As mentioned before, in order to calculate the relative resistivity of the samples produced in this work, the thin film was assumed in shape of a rectangle with $a \times b \times h$ dimensions. Then this rectangle was divided into many small cubes with a side length of d. It should be mentioned that dimensions of the rectangle are multiplication of the small cubes side length d. Hence, this procedure provides the required condition for working with dimensionless parameters/lengths. Then the geometry of

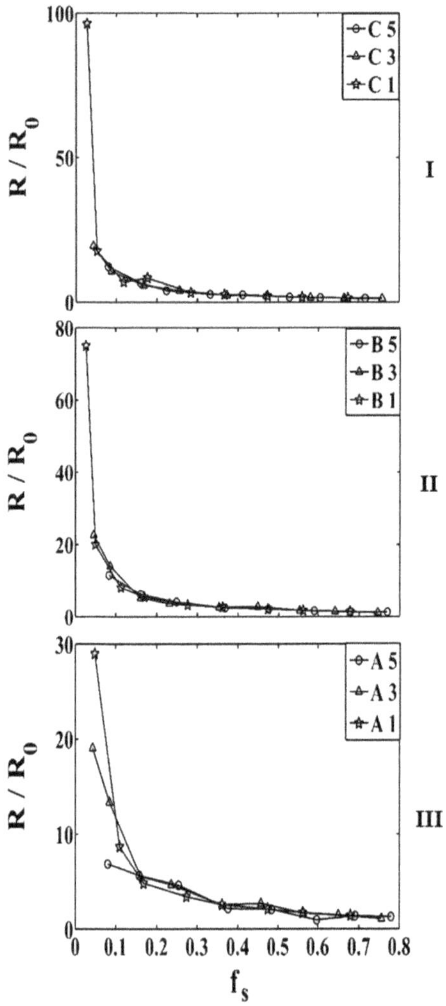

Fig. 6 Relative resistivity of GHSTTS structures with 8, 9 and 10 arms as a function of void fraction for different distances from the edge of the *shadowing block*

the structure is considered (i.e., some cubes were chosen as the material inclusion while they make the helical structure of GHSTTS with σ_0 conductivity. The remaining cubes produce the voids between the material inclusions with zero conductivity). The GHSTTS structure was made with different arm lengths so that in the first pitch the length of arms was *2 l* and in the second pitch it was reduced to two third (i.e., *2l/3*) and in the third pitch was reduced to one-third of the length of arm in the first pitch (i.e., *l/3*). Hence in the perturbation simulation work, the same procedure was assumed as in the first pitch the length of arm was taken as *6d,* and in the second pitch it was *4d* and in the last (top) pitch it was *2d.* Hence a cube is chosen with $6d \times 6d \times h$ dimensions. The structure of GHSTTS with 8, 9 and 10 arms is the same and the only difference between them is the number of arms; hence, the thickness/height of the cube in these samples is different which is included in the *h* parameter. The thickness of the film was obtained by considering the tilt angle of the helical columns and the deposition angle and the length of the arms. Therefore, it is expected that the relative resistivity obtained from the perturbation theory/program does not show a large variation as the result of this program does not depend on the thin film dimensions and only depends on the geometry of the structure and the void fraction. Hence, as discussed in the preceding section in this calculation, for each value of material inclusion, $6d \times 6d \times h$ cubes are involved and a system of $3 \times 6d \times 6d \times h$ equations with the same number of unknowns is solved. We emphasise again that resistance calculations using this procedure give us the resistance of a columnar structure relative to a homogeneous structure as a function of material inclusion and the growth/rise angle of columns, while effects due to size/dimensions must be considered separately.

Results of relative resistivity as a function of void fraction obtained from calculations are given in Fig. 6. It can be seen that in the GHSTTS structures resistance increases with decreasing material inclusion. This is consistent with Siabi-Garjan et al.s' results for columnar structures [28]. In addition, from comparison of sub-figures in Fig. 6 it can be deduced that the relative resistivity decreases with increasing the number of arms in the GHSTTS structure. This is in agreement with the experimental measurements discussed in the preceding section (Fig. 4). Hence, the simulation work of the perturbation method clearly shows the influence of the geometrical shape of the structure on the electrical resistivity.

Conclusion

Graded helical square tower-like (terraced) sculptured (GHSTTS) Mn thin films were produced by deposition at oblique angle, rotation of substrate about its surface normal with fixed rotation angle (90°) and use of a shadowing block at the centre of the substrate holder. The produced films are anisotropic due to use of shadowing block which also leads to formation of graded structures. The variation of this anisotropy influences the electrical properties of the GHSTTS. The anisotropy of the samples was obtained by measuring the electrical resistivity of the samples at two orthogonal *x* and *y* directions. Results showed that the electrical resistivity and the anisotropy of the structure increase with decreasing distance from the edge of the shadowing block. This observation was confirmed by use of perturbation theory.

Acknowledgements This work was carried out with the support of the University of Tehran and the Islamic Azad University. HS is grateful to the Centre of Excellence for Physics of Structure and Microscopic Properties of Matter, Department of Physics, University of Tehran, for partial support of this work. Partial support of this work by Iran National Science Foundation (INSF) is greatly appreciated.

References

1. Hawkeye, M., Brett, M.J.: Glancing angle deposition: fabrication, properties, and applications of micro-and nanostructured thin films. J. Vac. Sci. Technol. A **25**(5), 1317–1335 (2007)
2. Jensen, M.O., Brett, M.J.: Periodically structured glancing angle deposition thin films. IEEE Trans. Nanotechnol. **4**(2), 269–277 (2005)
3. Li, H.F., Kar, A.K., Parker, T., Wang, G.C., Lu, T.M.: The morphology and texture of Cu nanorod films grown by controlling the directional flux in physical vapor deposition. Nanotechnology **19**(33), 335708 (2008)
4. Wang, F., Lakhtakia, A.: Response of slanted chiral sculptured thin films to dipolar sources. Opt. Commun. **235**(1), 133–151 (2004)
5. Wang, F., Lakhtakia, A.: Lateral shifts of optical beams on reflection by slanted chiral sculptured thin films. Opt. Commun. **235**(1), 107–132 (2004)
6. Nieuwenhuizen, J.M., Haanstra, H.B.: Microfractography of thin films. Philips Tech. Rev. **27**(3), 87–91 (1966)
7. Messier, R., Yehoda, J.E.: Geometry of thin-film morphology. J. Appl. Phys. **58**(10), 3739–3746 (1985)
8. Motohiro, T., Taga, Y.: Thin film retardation plate by oblique deposition. Appl. Opt. **28**(13), 2466–2482 (1989)
9. Savaloni, H., Haydari-Nasab, F., Malmir, M.: Nano-structural characteristics and optical properties of silver chiral nano-flower sculptured thin films. Appl. Surf. Sci. **257**(21), 9044–9055 (2011)
10. Siabi-Garjan, A., Savaloni, H.: Extinction spectra and electric field enhancement of silver chiral nano-flower shaped nanoparticle; comparison of discrete dipole approximation results with experimental results. Eur. Phys. J. B **86**(6), 257 (2013)
11. Abdi, F., Siabi-Garjan, A., Savaloni, H.: Investigation on the dependence of optical spectra of silver chiral nanostructures on shape, dimensions and incident light by discrete dipole approximation. J. Theor. Appl. Phys. **6**(1), 1–7 (2012)
12. Abdi, F., Siabi-Gerjan, A., Savaloni, H.: On the discrete dipole approximation investigation of the extinction spectra of Ag/glass nano-flower thin film with threefold symmetry. J. Theor. Appl. Phys. **6**(1), 4 (2012)
13. Lakhtakia, A., Messier, R.: Sculptured Thin Films: Nanoengineered Morphologyand Optics, Messier, p. 233. SPIE, Bellingham (2005)
14. Robbie, K., Brett, M.J.: Chiral sculptured thin films. Nature **38**, 4616 (1996)
15. Hodgkinson, I.J., Wu, Q., Arnold, M., McCall, W., Lakhtakia, A.: Chiral mirror and optical resonator designs for circularly polarized light: suppression of cross-polarized reflectances and transmittances. Opt. Commun. **210**(3), 201–211 (2002)
16. Steele, J.J., Van popta, A.C., Hawkeye, M.M., Sit, J.C., Brett, M.J.: Nanostructured gradient index optical filter for high-speed humidity sensing. Sens. Actuators **B120**(1), 213–219 (2006)
17. Kesapragada, S.V., Victor, O.N.P., Gall, D.: Nanospring pressure sensors grown by glancing angle deposition. Nano Lett. **6**(4), 854–857 (2006)
18. Harris, K.D., Brett, M.J., Smy, T.J., Backhouse, C.: Microchannel surface area enhancement using porous thin films. J. Electrochem. Soc. **147**(5), 2002–2006 (2000)
19. Lakhtakia, A., Demirel, M., Horn, M., Xu, J.: Six emerging directions in sculptured-thin-film research. Adv. Solid State Phys. **46**, 295–307 (2007)
20. Hori, M., Yonezawa, F.: Theoretical approaches to inhomogeneous transport in disordered media. J. Phys. C **10**(2), 229 (1977)
21. Siabi-Garjan, A., Savaloni, H., Beik-Mohammadi, J., Grayeli-Korpi, A.-R.: On the characterization of sculptured columnar Mn thin films; correlation between nanostructure and the optical and electrical properties using experimental and simulation results. Philos. Mag. **93**(26), 3527–3546 (2013)
22. Eckertova, L.: Physics of Thin Films, 2nd edn. Plenum Press, Berlin (1986)
23. Savaloni, H., Babaei, F., Song, S., Placido, F.: Characteristics of sculptured Cu thin films and their optical properties as a function of deposition rate. Appl. Surf. Sci. **255**, 8041–8047 (2009)
24. Alouach, H., Mankey, G.J.: Texture orientation of glancing angle deposited copper nanowire arrays. J. Vac. Sci. Technol. A **22**(4), 1379–1382 (2004)
25. Abelmann, L., Lodder, C.: Oblique evaporation and surface diffusion. Thin Solid Films **305**(1–2), 1–21 (1997)
26. Taylor, J.R.: An Introduction to Error Analysis, 2nd edn. University Science Books, Sausalito (1997)
27. Savaloni, H., Babaei, F., Song, S., Placido, F.: Influence of substrate rotation speed on the nanostructure of sculptured Cu thin films. Vacuum **85**(7), 776–781 (2011)
28. Siabi-Garjan, A., Savaloni, H.: Extinction spectra and electric near-field distribution of Mn nano-rod based sculptured thin films: experimental and discrete dipole approximation results. Plasmonics **10**(4), 861–872 (2015)

Study of virtual annihilation and retardation effects in relativistic two-, four-, and six-body wave equations in scalar QFT

Mohsen Emami-Razavi[1] · Somayeh Asgary[2]

Abstract The variational method within the Hamiltonian formalism of QFT using Darewych reformulated model has been used for scalar particles and antiparticles interacting through a scalar mediating field. We have investigated the relativistic effects such as virtual annihilation interactions and retardation effects for relativistic two-, four-, and six-body wave equations in scalar QFT. Approximate ground-state solutions have been studied for different strengths of coupling, for both massive and massless mediating fields where the virtual annihilation terms or retardation effects in the wave equations have been included or eliminated.

Keywords Relativistic · Wave equations · Bound state · Virtual annihilation interactions · Retardation effects

Introduction

The study of few- or many-body systems in relativistic quantum field theory (QFT) is of fundamental interest and has played a major role in the progress of modern physics. It contributes to the development of our idea about the nature of particle interactions. For example, we can mention that the thorough study of the spectrum of the hydrogen atom was an early test of the theory of quantum electrodynamics (QED). The Bethe–Salpeter (BS) formalism [1, 2] is the traditional tool for analyzing relativistic bound state, but this method is hard to implement and is all but intractable for systems that contain more than two particles. The study of two-body bound states using BS equation in the ladder approximation has been known as the Wick–Cutkosky model (WC) [3, 4].

The relativistic quantum-mechanical treatment of bound states of a particle in a given field is principally based on the Dirac and Klein–Gordon equations. The Dirac equation is suitable for particles, like electrons, with spin 1/2 and the Klein–Gordon equation is used to solve problems of particles without spin like π^+ or π^-. The Klein–Gordon equation is also called the scalar equation of relativistic quantum mechanics.

An important problem in quantum field theory has been the description of relativistic n-body scalar particles wave equations or n-body scalar particles and antiparticles wave equations (spin-0 bosons), including the interactions, from the underlying Lagrangian. In previous works by Emami-Razavi [5, 6], it has been demonstrated that, by employing a reformulated model in QFT proposed by Darewych [7, 8], formulas concerning relativistic n-body wave equations for scalar particles and/or antiparticles can be obtained (see also [9]). It has also been proven that using Darewych's formalism [7, 8] relativistic n-fermion wave equations (particles and antiparticles; spin-1/2) in quantum electrodynamics can be obtained [10, 11]. One of the appealing features of having n-body wave equations formula is that we can explicitly write the relativistic wave equations for an arbitrary number of n including the interactions. Therefore, one can easily write, for example four- or six-body wave equations, and thereupon one can perform different calculations (employing various

✉ Mohsen Emami-Razavi
mrazavi@srbiau.ac.ir

1 Plasma Physics Research Center, Science and Research Branch, Islamic Azad University, Tehran, Iran

2 Department of Physics, West Tehran Branch, Islamic Azad University, Tehran, Iran

numerical methods) to study relativistic effects such as retardation effects or virtual annihilation interaction terms.

In this paper we study the effects of inclusion or exclusion of retardation terms or virtual annihilation interactions for relativistic two-, four- and six-body scalar particles and antiparticles which interact through a scalar mediating field which can be massless ($\mu = 0$) or massive ($\mu/m = 0.15$). The case of two-body system has been investigated before [12, 13]. However, we remind it again here to better follow the four- and six-body cases. In this work the coupling constant that shows the strength of the interaction is presented by the symbol α which is related to another constant g by the relation $\alpha = g^2/16\pi m^2$. α can vary from a relatively small coupling (such as $\alpha \sim 0.1$) to strong coupling numbers such as $\alpha \sim 1$.

Our calculations show that even though the retardation effects or virtual annihilation interactions can be small effects relative to Yukawa interactions at low coupling for the two-, four-, and six-body systems, for large values of α ($0.4 < \alpha \leq 1$) the magnitude of the contribution of virtual annihilation terms and retardation effects augments for the ground-state energy solutions of the systems under study.

The presentation of the manuscript is as follows. The first section is the introduction. The second section is about variational methods and the reformulated scalar model. The relativistic wave equations, virtual annihilation interactions and retardation effects are reminded in the following section. In the subsequent section, the results of approximate ground-state energy solutions for various cases have been discussed. Concluding remarks have been presented in the last section. Appendix contains some formulas and examples concerning relativistic wave equations.

Variational methods and the reformulated model

From a practical point of view, the variational method is independent of the magnitude of the coupling constant strength, contrary to perturbation theory which usually depends on expansion of some small parameters. The variational methods in Hamiltonian QFT is similar to Schrödinger type of representation of few-body systems, and can easily be applied to systems of more than two bodies [14].

The variational methods have been moderately used in QFT since the construction of realistic trial states can be a difficult task. The variational approach to the treatment of relativistic bound states has received increasing attention since the early papers on the variational methods in QFT (see, for examples, [15, 16]). We should note that the variational approach is basically as good as the trial states we use [17–19].

As pointed out before, the variational method is only as good as the trial states that we use. In the functional formulation we are restricted mainly to wave functionals of Gaussian type, due to the difficulty of handling analytically non-Gaussian functional integrals. Another possible approach that we can employ is to expand the trial state on a Fock-space basis. For example, the Tamm's work [20] is along these lines, though it was not formulated variationally. In this paper, we have applied the approach given by Darewych [7] to particle–antiparticle bound states in the scalar Yukawa model. In this model, we have assumed that particles or antiparticles have no spin. This means that they are scalar particles or antiparticles. Moreover, the mediating fields which act between particles–antiparticles are also scalar mediating fields.

In Ref. [6] simple trial states have been employed to obtain the n-body wave equations in the scalar QFT by the variation method. Thereupon, one can easily write the relativistic wave equations for the two-, four-, and six-body systems. The equations have the Schrödinger non-relativistic limit, with Yukawa potentials in the event of a massive mediating scalar field or Coulombic interparticle potentials on condition of having a massless mediating scalar field. Ground-state energy solutions of the equations have been approximately calculated for various strengths of the coupling. A comparison of various cases have been considered where virtual annihilation terms and/or retardation effects have been included or excluded for our different systems under study.

For the sake of completeness and having a self-contained paper it is better to recall some of the main points from previous works [12, 13]. The Lagrangian density for scalar particles which interact via a mediating scalar field is

$$\mathcal{L} = \partial^\nu \phi^*(x)\partial_\nu \phi(x) - m^2\phi^*(x)\phi(x) + \frac{1}{2}\partial^\nu \chi(x)\partial_\nu \chi(x) - \frac{1}{2}\mu^2\chi^2(x) - g\phi^*(x)\phi(x)\chi(x), \tag{2.1}$$

where g indicates the coupling constant. We are working in the unit $\hbar = c = 1$. m is the mass of particle or antiparticle. ϕ and ϕ^* present the field associated with the particle and antiparticle, respectively. $\chi(x)$ designates the mediating field which can be massless (i.e., $\mu = 0$), (cf. [3]), or massive ($\mu \neq 0$).

The Lagrangian density obtained in the reformulated model is (see [7] or [17])

$$\mathcal{L} = \partial^\nu \phi^*(x)\partial_\nu \phi(x) - m^2\phi^*(x)\phi(x) + \frac{1}{2}\partial^\nu \chi_0(x)\partial_\nu \chi_0(x) - \frac{1}{2}\mu^2\chi_0^2(x) - g\phi^*(x)\phi(x)\,\chi_0(x) + \frac{1}{2}\int \mathrm{d}x'\rho(x)D(x-x')\rho(x'), \tag{2.2}$$

where $\mathrm{d}x' = \mathrm{d}t\ \mathrm{d}^3r$, $\rho(x) = -g\phi^*(x)\phi(x)$; $\chi_0(x)$ denotes the free field and $D(x-x')$ indicates a covariant Green function in a way that we have

$$(\partial^\nu\partial_\nu + \mu^2)D(x-x') = \delta^{3+1}(x-x'). \tag{2.3}$$

As it has been mentioned, for example, in [7] or [12, 13] the theories based on (2.1) and (2.2) are equivalent in the sense that they lead to the same invariant matrix elements in different orders of perturbation theory. The Lagrangian density (2.2) can be written as:

$$\mathcal{L} = \partial^\nu \phi^*(x)\partial_\nu \phi(x) - m^2\phi^*(x)\phi(x) - g\phi^*(x)\phi(x)\,\chi_0(x) + \frac{1}{2}\int \mathrm{d}x'\rho(x)D(x-x')\rho(x'), \tag{2.4}$$

where we have suppressed the free part of the Lagrangian in the above equation. As a matter of fact, one should note that the trial states, $|\psi_n\rangle$, that we employ (see Eq. (3.1)) are insensitive to $\mathcal{H}_\chi$ (Eq. 2.7) and $\mathcal{H}_{I_1}$ (Eq. 2.8) in the sense that $\langle\psi_n| : \hat{H}_\chi : |\psi_n\rangle = 0$ and $\langle\psi_n| : \hat{H}_{I_1} : |\psi_n\rangle = 0$ for the current choice of trial states. The total Hamiltonian of our scalar systems under study is $\hat{H}(t) = \int \mathrm{d}^N x \hat{\mathcal{H}}(x)$, where the Hamiltonian density, $\hat{\mathcal{H}}(x)$, has been provided in Eqs. (2.5)–(2.10).

We can write down the Hamiltonian density [7], [14] as follows:

$$\mathcal{H}(x) = \mathcal{H}_\phi(x) + \mathcal{H}_\chi(x) + \mathcal{H}_{I_1}(x) + \mathcal{H}_{I_2}(x), \tag{2.5}$$

where

$$\mathcal{H}_\phi(x) = p_{\phi^*}(x)p_\phi(x) + \nabla\phi^*(x)\cdot\nabla\phi(x) + m^2\,\phi^*(x)\phi(x), \tag{2.6}$$

$$\mathcal{H}_\chi(x) = \frac{1}{2}p_{\chi_0}^2 + \frac{1}{2}(\nabla\chi_0)^2 + \frac{1}{2}\mu^2\chi_0^2, \tag{2.7}$$

$p_{\phi^*} = \frac{\partial\mathcal{L}}{\partial\dot\phi} = \dot\phi^*$, $p_\phi = \frac{\partial\mathcal{L}}{\partial\dot\phi^*} = \dot\phi$ and $p_{\chi_0} = \frac{\partial\mathcal{L}}{\partial\dot\chi_0} = \dot\chi_0$ are conjugate momenta,

$$\mathcal{H}_{I_1}(x) = g\,\phi^*(x)\phi(x)\chi_0(x), \tag{2.8}$$

$$\mathcal{H}_{I_2}(x) = -\frac{g^2}{2}\int \mathrm{d}x'\,\phi^*(x)\phi(x)D(x-x')\phi^*(x')\phi(x'), \tag{2.9}$$

where we indicate $\mathrm{d}x = \mathrm{d}^N x\,\mathrm{d}t$ in N spatial plus time dimensions and

$$D(x-x') = \int \frac{\mathrm{d}k}{(2\pi)^{N+1}}\,\mathrm{e}^{-ik\cdot(x-x')}\,\frac{1}{\mu^2 - k^2 + i\epsilon}, \tag{2.10}$$

where $\mathrm{d}k = \mathrm{d}^{N+1}k$ and $k^2 = k^\nu k_\nu$.

To indicate our notation, we quote the usual decomposition of the fields in $N+1$ dimensions:

$$\phi(x) = \int \mathrm{d}^N q\,\frac{1}{[(2\pi)^N 2\omega_q]^{\frac{1}{2}}}\,[A(\mathbf{q})\mathrm{e}^{-iq\cdot x} + B^\dagger(\mathbf{q})\mathrm{e}^{iq\cdot x}] \tag{2.11}$$

$$\chi_0(x) = \int \mathrm{d}^N p\,\frac{1}{[(2\pi)^N 2\Omega_p]^{\frac{1}{2}}}\,[d(\mathbf{p})\mathrm{e}^{-ip\cdot x} + d^\dagger(\mathbf{p})\mathrm{e}^{ip\cdot x}], \tag{2.12}$$

where $\omega_q = (\mathbf{q}^2 + m^2)^{\frac{1}{2}}$, $\Omega_p = (\mathbf{p}^2 + \mu^2)^{\frac{1}{2}}$, $q\cdot x = q^\nu x_\nu$ and $q^\nu = (q^0 = \omega_q, \mathbf{q})$.

The momentum-space operators, $A^\dagger, A, B^\dagger, B$ obey the usual commutation relations, the non-vanishing ones are

$$[A(\mathbf{p}), A^\dagger(\mathbf{q})] = [B(\mathbf{p}), B^\dagger(\mathbf{q})] = \delta^N(\mathbf{p}-\mathbf{q}), \tag{2.13}$$

$$[d(\mathbf{p}), d^\dagger(\mathbf{q})] = \delta^N(\mathbf{p}-\mathbf{q}). \tag{2.14}$$

The vacuum state $|0\rangle$ is designated by $A_\mathbf{p}|0\rangle = B_\mathbf{p}|0\rangle = d_\mathbf{k}|0\rangle = 0$. The Hamiltonian is normal ordered (specified by $:\hat{H}:$) since we are not interested in vacuum-energy matters in this work.

In the Hamiltonian formalism we wish to find solutions of the equation

$$\hat{P}^\beta|\Psi\rangle = Q^\beta|\Psi\rangle, \tag{2.15}$$

where $\hat{P}^\beta = (\hat{H}, \hat{\mathbf{P}})$ indicates the energy–momentum operator and $Q^\beta = (E, \mathbf{Q})$ presents the energy–momentum eigenvalue. The above equation usually cannot be solved analytically. Hence, we should find some approximation solutions. Thereupon, we apply the following variational principle

$$\delta\langle\Psi_{\mathrm{trial}}| : \hat{H} - E : |\Psi_{\mathrm{trial}}\rangle_{t=0} = 0, \tag{2.16}$$

where $|\Psi_{\mathrm{trial}}\rangle$ is a suitable trial state with adjustable parameters or functions. We set the time $t = 0$ for the calculations.

Relativistic wave equations, virtual annihilation terms and retardation effects

For the sake of completeness and to have a self-contained manuscript it is better to recall some main points of the n-body scalar particle and antiparticle relativistic wave equations in Ref. [6]. In the reformulated scalar QFT variational model simple trial states have been used (see Eq. 3.1) where $n = n_1 + n_2$ (n_1: number of particles and n_2: number of antiparticles, $n_1 = n_2$ or $n_2 = n_1 - 1$). One should note that since we are studying here the two-, four-, and six-body systems the total number of particles and antiparticles in the present paper is even. We have for (n/2) particles and (n/2) antiparticles the following simple trial state [6]:

$$|\psi_n\rangle = \int d^N p_1 d^N p_2 d^N p_3 d^N p_4 \ldots d^N p_{n-1} d^N p_n \; G(\mathbf{p}_1, \mathbf{p}_2, \mathbf{p}_3, \mathbf{p}_4, \ldots, \mathbf{p}_{n-1}, \mathbf{p}_n) \times A^\dagger(\mathbf{p}_1) B^\dagger(\mathbf{p}_2) A^\dagger(\mathbf{p}_3) B^\dagger(\mathbf{p}_4) \cdots A^\dagger(\mathbf{p}_{n-1}) B^\dagger(\mathbf{p}_n)|0\rangle, \tag{3.1}$$

where G is normalizable to unity and it should be a well-behaved function with adjustable parameters. The matrix element for the rest-plus-kinetic energy of the n-body system is

$$\langle\psi_n| : \hat{H}_\phi - E : |\psi_n\rangle = \int d^N p_1 d^N p_2 \cdots d^N p_n \, G^*(\mathbf{p}_1, \mathbf{p}_2, \ldots, \mathbf{p}_n) \times G(\mathbf{p}_1, \mathbf{p}_2, \ldots, \mathbf{p}_n)[\omega_{p_1} + \omega_{p_2} + \cdots + \omega_{p_n} - E] \tag{3.2}$$

The matrix elements related to the interactions of our system have $n(n-1)/2$ terms (all of them present attractive nature, i.e., gravity-like). We have also $n^2/4$ terms (when n is an even number) analogous to virtual annihilation terms (interaction among the particle and antiparticle pairs).

The matrix element for the interaction of the system under study has the following form [6]:

$$\langle\psi_n| : \hat{H}_I : |\psi_n\rangle = [G_1] + [G_2] + \cdots [G_{n(n-1)/2}] + [A_1] + [A_2] + \cdots [A_{n^2/4}]. \tag{3.3}$$

As an example we can quote the following. In case we consider three particles and three antiparticles ($n = 6$), we will have ten interparticle interactions (labeled $G_1, G_2 \ldots, G_{15}$, all of them attractive, i.e., gravity-like) and we will have six terms that correspond to virtual annihilation terms. These interactions occur among particle and antiparticle pairs (labeled $A_1, A_2, \ldots, A_9$). The virtual annihilation interactions are repulsive contact potentials if $\mu < 2m$.

As it can be seen, for example in [21], the description of interactions of two or more than two bodies with equal or comparable masses is much more demanding than the case of particles with non-identical masses. We cannot generally differentiate between the test particles or antiparticles and the source of the field for the case of equal or comparable masses. Furthermore, we should consider all particles on an equal footing. Moreover, the interactions among particles or antiparticles do not propagate with infinite velocity. Therefore, there are further complications for the treatment of interactions among particles or antiparticles. These phenomena lead to the terms corresponding to "retardation effects".

The matrix element for the interaction related to our n-body wave equations (retardation effects are included) is [6] (for $j < k$):

$$\begin{aligned}
\langle\psi_n| : \hat{H}_I : |\psi_n\rangle &= \langle\psi_n| : \hat{H}_{I_2} : |\psi_n\rangle \\
&= -\frac{g^2}{4(2\pi)^N} \int d^N p_1 \ldots d^N p_n d^N p'_1 \ldots d^N p'_n \\
&\quad G^*(\mathbf{p}'_1, \mathbf{p}'_2, \ldots, \mathbf{p}'_n) G(\mathbf{p}_1, \mathbf{p}_2, \ldots, \mathbf{p}_n) \\
&\quad \times \Bigg\{ \sum_{j=1}^{n-1} \sum_{k=j+1}^{n} \prod_{i=1\ldots n}^{(j,k)} \delta^N(\mathbf{p}'_i - \mathbf{p}_i) \\
&\quad \frac{\delta(p'_j + p'_k - p_j - p_k)}{\sqrt{\omega_{p'_j}\omega_{p'_k}\omega_{p_j}\omega_{p_k}}} \frac{1}{\mu^2 - (p_j - p'_j)^2} \\
&\quad + \sum_{j=1}^{n-1} {\sum_{k=j+1}^{n}}' \prod_{i=1\ldots n}^{(j,k)} \delta^N(\mathbf{p}'_i - \mathbf{p}_i) \\
&\quad \frac{\delta(p'_j + p'_k - p_j - p_k)}{\sqrt{\omega_{p'_j}\omega_{p'_k}\omega_{p_j}\omega_{p_k}}} \frac{1}{\mu^2 - (p_j + p_k)^2} \Bigg\},
\end{aligned} \tag{3.4}$$

where we can write $\delta(p'_j + p'_k - p_j - p_k) = \delta(\omega_{p'_j} + \omega_{p'_k} - \omega_{p_j} - \omega_{p_k})\delta(\mathbf{p}'_j + \mathbf{p}'_k - \mathbf{p}_j - \mathbf{p}_k)$, and $\Sigma_{k=a}{}' u_k$ signifies $u_a + u_{a+2} + u_{a+4} + \cdots$.

The superscript notation (j, k) in $\prod_{i=1\ldots n}^{(j,k)} \delta^N(\mathbf{p}'_i - \mathbf{p}_i)$ in Eq. (3.4) indicates that the terms with indices j and k are left out. We can write down:

$$\begin{aligned}
\prod_{i=1\ldots n}^{(j,k)} \delta^N(\mathbf{p}'_i - \mathbf{p}_i) &= \prod_{i=1}^{j-1} \delta^N(\mathbf{p}'_i - \mathbf{p}_i) \\
&\quad \prod_{i=j+1}^{k-1} \delta^N(\mathbf{p}'_i - \mathbf{p}_i) \prod_{i=k+1}^{n} \delta^N(\mathbf{p}'_i - \mathbf{p}_i) \\
&= \prod_{i=1}^{n} \delta^N(\mathbf{p}'_i - \mathbf{p}_i) \times [\delta^N(\mathbf{p}'_j - \mathbf{p}_j)\delta^N(\mathbf{p}'_k - \mathbf{p}_k)]^{-1}.
\end{aligned}$$

One should note that if $n = 2$, we will have $\prod_{i=1\ldots n}^{(j,k)} \delta^N(\mathbf{p}'_i - \mathbf{p}_i) = 1$.

We should also consider the statistical factors [22] in our systems under study due to the fact we have identical particles or antiparticles. This means that we should not double count the states of identical particles [22]. For the case where n is an even number, the statistical factor is

$1/(n/2)! \times 1/(n/2)!$. For examples, if we have four- and six-body systems consisting of two particles–two antiparticles and three particles–three antiparticles, respectively, we will have the following statistical factors for four-body and six-body systems, respectively: $1/2! \times 1/2!$ and $1/3! \times 1/3!$.

In Eq. (3.4), *i*, *j* and *k* indicate indices. Even indices are considered for antiparticles and odd indices are considered for particles. The relativistic *n*-body wave equation for scalar particles–antiparticles can be obtained from the matrix elements as below [6] (for $j<k$):

$$
\begin{aligned}
&G(\mathbf{p}_1,\mathbf{p}_2,\ldots,\mathbf{p}_n)[\omega_{\mathbf{p}_1}+\omega_{\mathbf{p}_2}+\cdots+\omega_{\mathbf{p}_n}-E] \\
&= \frac{g^2}{4(2\pi)^N}\int \mathrm{d}^N p_1' \mathrm{d}^N p_2' \ldots \mathrm{d}^N p_n' G^*(\mathbf{p}_1',\mathbf{p}_2',\ldots,\mathbf{p}_n') \\
&\times \Bigg\{ \sum_{j=1}^{n-1}\sum_{k=j+1}^{n}\prod_{i=1\ldots n}^{(j,k)} \delta^N(\mathbf{p}_i'-\mathbf{p}_i) \\
&\quad \frac{\delta(p_j'+p_k'-p_j-p_k)}{\sqrt{\omega_{p_j'}\omega_{p_k'}\omega_{p_j}\omega_{p_k}}}\frac{1}{\mu^2-(p_j-p_j')^2} \\
&+ \sum_{j=1}^{n-1}{\sum_{k=j+1}^{n}}' \prod_{i=1\ldots n}^{(j,k)} \delta^N(\mathbf{p}_i'-\mathbf{p}_i) \\
&\quad \frac{\delta(p_j'+p_k'-p_j-p_k)}{\sqrt{\omega_{p_j'}\omega_{p_k'}\omega_{p_j}\omega_{p_k}}}\frac{1}{\mu^2-(p_j+p_k)^2} \Bigg\}.
\end{aligned} \tag{3.5}
$$

The first terms on the right-hand side of the above equation are related to attractive interactions among the *n* bodies. The second terms correspond to virtual annihilation interactions among odd–even or even–odd indices (particles–antiparticles pairs). The Eq. (3.5) has the Schrödinger-like structure, with positive-energy solutions only. This can easily be seen by putting the right-hand side of this equation to zero (i.e., no interactions). In Appendix we explicitly quote some examples of the two-, four-, and six-body wave equations.

We should mention that if the terms corresponding to retardation effects are eliminated we should substitute in the kernel of the equations the terms like $1/[\mu^2-(p_1-p_2)^2]$ by $1/[\mu^2+(\mathbf{p}_1-\mathbf{p}_2)^2]$, and the terms like $1/[\mu^2-(p_1+p_2)^2]$ by $1/[\mu^2-4m^2]$.

The non-relativistic limit of the relativistic *n*-body wave equations can be obtained (which corresponds to $\mathbf{p}^2/m^2 << 1$) by the Fourier transformation (from momentum space to coordinate space),

$$
\begin{aligned}
\Psi(\mathbf{r}_1,\mathbf{r}_2,\ldots,\mathbf{r}_n) &= \frac{1}{(2\pi)^{nN/2}}\int \mathrm{d}^N p_1 \mathrm{d}^N p_2 \ldots \mathrm{d}^N p_n \\
&\times G(\mathbf{p}_1,\mathbf{p}_2,\ldots,\mathbf{p}_n) \\
&\times \exp(i(\mathbf{p}_1\cdot\mathbf{r}_1+\mathbf{p}_2\cdot\mathbf{r}_2+\cdots+\mathbf{p}_n\cdot\mathbf{r}_n)).
\end{aligned} \tag{3.6}
$$

Hence, the resulting *n*-body Schrödinger equation is

$$
\begin{aligned}
&-\frac{1}{2m}\left(\sum_{i=1}^{n}\nabla_i^2\right)\Psi(\mathbf{r}_1,\mathbf{r}_2,\ldots,\mathbf{r}_n) \\
&+(V(\mathbf{r}_1,\mathbf{r}_2,\ldots,\mathbf{r}_n)-\epsilon)\Psi(\mathbf{r}_1,\mathbf{r}_2,\ldots,\mathbf{r}_n)=0,
\end{aligned} \tag{3.7}
$$

where $\epsilon = E - nm$.

The non-relativistic potential $V(\mathbf{r}_1,\mathbf{r}_2,\ldots,\mathbf{r}_n)$ can be explained as follows. The first terms of Eq. (3.8) correspond to sum of attractive Coulomb potentials for the case of massless mediating field ($\mu = 0$) or Yukawa potentials for the case of massive mediating field ($\mu \neq 0$) among the *n* bodies, and repulsive (if $\mu < 2m$) contact potentials, due to virtual annihilation interactions among the particle and antiparticle pairs (among odd–even or even–odd indices of *j* and *k*). The *n*-body system potential in coordinate space in the non-relativistic limit can be written as below:

$$
\begin{aligned}
V(\mathbf{r}_1,\mathbf{r}_2,\ldots,\mathbf{r}_n) &= -\alpha\sum_{j=1}^{n-1}\sum_{k=j+1}^{n}\frac{\mathrm{e}^{-\mu|r_j-r_k|}}{|\mathbf{r}_j-\mathbf{r}_k|} \\
&+\sum_{j=1}^{n-1}{\sum_{k=j+1}^{n}}'\alpha\frac{4\pi}{4m^2-\mu^2}\delta(\mathbf{r}_j-\mathbf{r}_k), \quad \text{for } j<k.
\end{aligned} \tag{3.8}
$$

Note again that we have $\alpha = g^2/(16\pi m^2)$ and $\Sigma_{k=a}{}' u_k$ means $u_a + u_{a+2} + u_{a+4} + \cdots$.

Approximate ground-state energy solutions for two-, four-, and six-body systems with and/or without virtual annihilations and retardation effects

The simplest case for the systems under study is the two-body case. As mentioned before, the case of two-body system has been investigated before [12, 13]. However, it is good to recall some key points of the two-body particle and antiparticle system since it will help to better understand the four- and six-body cases.

As pointed out before, the scalar Yukawa [or Wick–Cutkosky (WC)] model [3, 4] has been investigated by several researchers in different formalisms and approximations. Figure 1 of Ref. [12, 13] (which is plotted here again as Fig. 1) contains various results of WC or BS formalism as well as other calculations. Moreover, the results of Di Leo and Darewych [23] correspond with the solutions obtained in [12, 13] (the two-body system in the present formalism) with $\omega_{p_1} = \omega_{p_1'}$ (i.e., no retardation). In other words, the retardation term is not included in [23]. Furthermore, Refs. [3, 4] and [23] do not contain the virtual annihilation interaction terms which are repulsive interactions. The effect of this repulsive interaction is to increase the ground-state energy solutions of the two-body system

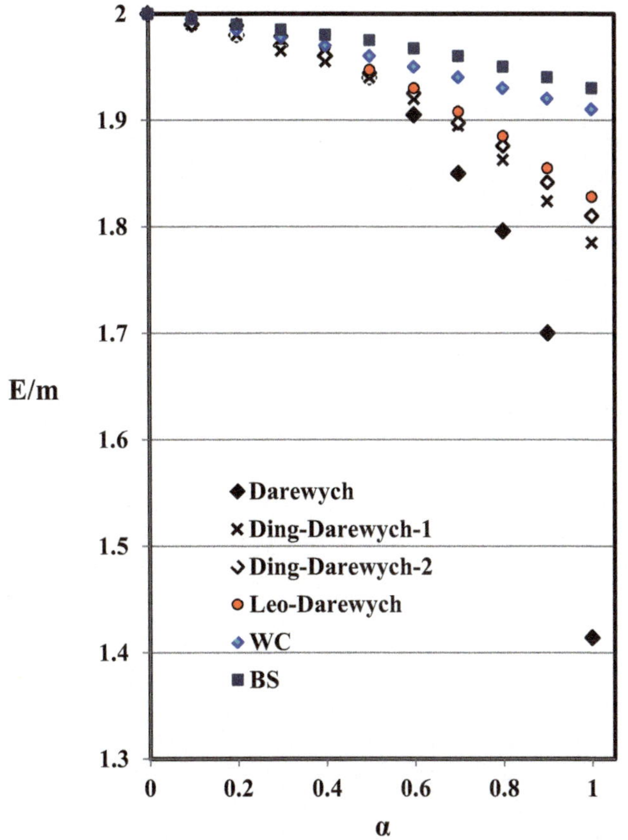

Fig. 1 Ground-state energies in units of m of two-body system. Curves from lowest to highest: Feshbach–Villar formalism [24]; with retardation and without virtual annihilation [12, 13]; with retardation and with virtual annihilation [12, 13]; without virtual annihilation interactions and retardation terms [23]; Wick–Cutkosky formalism [3, 4]; Bethe–Salpeter formalism [1]

by a small amount or to decrease the binding energy of the two-body system [$\mathbf{B}_2 = E_2 - 2$ in units of m (i.e., $m = 1$)] by a small amount. One should note that the magnitude of contribution of virtual annihilation interactions augments with increase of the coupling constant α. This means that at very high coupling ($\alpha \sim 1$) the contribution of this repulsive interaction becomes more important.

Figure 1 of Refs. [12, 13] (also plotted here as Fig. 1) shows the retardation effects for the two-body case. The inclusion of this effect causes to decrease the ground-state energy solutions of the two-body system by a small but not negligible amount. One should note that the magnitude of contribution of retardation effects augments with increasing of the coupling constant α. A comparison of the present two-body equation [12, 13] with BS [1, 2] or WC results [3, 4] proves that for the present case we have stronger binding energy solutions particularly at strong coupling. One should keep in mind that the obtained results compared to BS or WC model are from different sets of equations. One of the differences between variational method Eq. (3.5) in this manuscript and BS equation is clear even in the no interaction case ($\alpha = 0$), that is, the former equation has no negative energy solutions. Thereupon, perhaps it is not surprising that there are differences between the variational method equations here and BS and WC models, particularly at strong coupling α.

Figure 1 also contains the results obtained in [24] showing that the two-body eigenstates can be written down for Hamiltonian of the present model without free fields, provided that an empty vacuum state $|0\rangle$, annihilated by both negative and positive-energy components of the field operator $\phi(x)$, is employed. The use of such an empty vacuum state did give both negative and positive ground-state energy solutions. Moreover, The two-body system in Ref. [24] could be solved analytically for the massless mediating field case and also the results suggest a critical value of the coupling strength around $\alpha \sim 1$.

All the results related to different formalisms and approaches for the two-body system mentioned above have been presented in Fig. 1 (Fig. 1 of Refs. [12, 13]).

The four-body system (consisting of two particles and two antiparticles) and six-body system (consisting of three particles and three antiparticles) cannot be solved analytically. Therefore some approximate solutions should be sought. In the Appendix, four-, and six-body wave equations have been written explicitly using the Eq. (3.5) for $n = 4$ and $n = 6$, respectively. One can obtain variational expression for the energy E_n of the n-body system (for our case: $n = 4$ or $n = 6$) of the systems under study by substituting $G(\mathbf{p}_1, \mathbf{p}_2, \ldots, \mathbf{p}_n)$ with analytical functions that contain adjustable constants or parameters to compute the energy of four- or six-body system (see Eqs. 3.2 and 3.4).

$$E_n = \frac{1}{\langle \psi_n | \psi_n \rangle} \langle \psi_n | : \hat{H} : | \psi_n \rangle. \tag{4.1}$$

We consider ground-state energy solutions and using the variational principle (Eq. 2.16) and taking simple forms for the wave function G we can obtain some approximate solutions. We have, namely

$$G(\mathbf{p}_1, \mathbf{p}_2, \ldots, \mathbf{p}_n) = f(\mathbf{p}_1) \cdots f(\mathbf{p}_n) \delta(\mathbf{p}_1 + \cdots + \mathbf{p}_n), \tag{4.2}$$

where

$$f(\mathbf{p}_i) = \frac{1}{(\mathbf{p}_i^2 + p_0^2)^{\nu}}, \tag{4.3}$$

where p_0 and ν describe adjustable parameters. The calculations have been performed in the rest frame (with zero total momentum).

For the four- or six-body wave equations, the multidimensional integrals that we have cannot be solved analytically, hence we should seek some numerical solutions using the Monte Carlo method [25]. The parameter p_0 indicates the scale of the approximate wave functions. We locate the minimum of the variational trial energy as a function of p_0. Moreover, varying the parameter v does not change the numerical solutions sensibly (see, for example, [12, 13]); therefore, we leave it at the value $v = 2$ which is the non-relativistic, hydrogenic ground-state value.

For the four-body system the numerical results have been presented for different cases, "without" and "with" retardation effects and virtual annihilation interactions. The virtual annihilation interactions are mostly not included in other works. One should note that the solutions corresponding to four-body relativistic wave equations when the retardation effects are included and virtual annihilation interactions are eliminated have been done previously in [14] and they are added here for the comparison purposes with other cases in Tables 1 and 2. The numerical solutions for the variationally obtained trial wave functions are given in Table 1 for the massive-exchange case with $\mu = 0.15\,m$ and for the massless-exchange case $\mu = 0$ in Table 2. These solutions have been also plotted in Fig. 2a, b, respectively. Energies, E_4 and p_0 have been provided in units of m (i.e., $m = 1$) for given coupling α.

Binding energy $\mathbf{B}_4$ is: $\mathbf{B}_4 = E_4 - 4$ in units of m (i.e., $m = 1$); the binding energy augments when we increase the coupling α. This is because of the fact that for present scalar mediating field, the dominant one-quantum exchange interactions are attractive. The parameter p_0 corresponding to our wave functions also augments when we increase the coupling strength α. This means that the wave function becomes more peaked in coordinate space at higher values of the coupling. Similar results have been presented in Table 2 and Fig. 2b for the massless-exchange case ($\mu = 0$). In this case, Yukawa potentials transform to Coulombic potentials in the non-relativistic limits. Although the qualitative behaviors of $E(\alpha)$ for $\mu = 0$ are very close to that for $\mu = 0.15\,m$, the binding energy, for a specific α is stronger for $\mu = 0$ than for $\mu = 0.15\,m$, and it sets in at $\alpha_0 \to 0$.

The numerical results in Tables 1 and 2 generally show that the retardation effects and virtual annihilation interactions are somehow minor effects but not negligible in contrast to Yukawa interactions for the scalar four-body particle–antiparticle systems. Moreover, the contribution of these retardation effects and virtual annihilation terms increases at high coupling and they become more noticeable for the range of the values of coupling strength $0.5 \leq \alpha \leq 1$. Furthermore, for a given value of coupling α we can observe (from Tables or plotted figures) that for the case where virtual annihilation terms are included and retardation effects are excluded we have the highest values of the energy solutions (the top curves in the Fig. 2a, b, respectively). In addition, for the case where virtual annihilation terms are excluded and retardation effects are included we have the lowest values of the energy solutions (the lowest curves in the Fig. 2a, b, respectively). We can observe these differences of the ground-state energy values with respect to inclusion and/or exclusion of virtual annihilation terms and retardation effects particularly at higher values of the coupling strength $0.5 < \alpha \leq 1$.

We should mention that the four-body system similar to the case of two-body system in [24] (that was solved analytically for the massless mediating field case) may suggest a critical value of the coupling constant around $\alpha \sim 1$. This means that the values of energy solutions obtained for $\alpha \geq 1$ may not be reliable or no energy solutions exist beyond a critical value of the coupling $\alpha \sim 1$. Hence, we did not consider the energy results for α larger than 1. Note also that the error estimates in the numerical

Table 1 Four-body ground-state energies E_4 and wave function parameter, p_0, for the massive-exchange case $\mu = 0.15m$

α	E_4 (p_0) with annihilation and without retardation	E_4 (p_0) without annihilation and without retardation	E_4 (p_0) with annihilation and with retardation	E_4 [14] (p_0) without annihilation and with retardation
0.3	3.995 ± 0.020 (0.080)	3.992 ± 0.020 (0.085)	3.989 ± 0.020 (0.090)	3.985 ± 0.025 (0.095)
0.4	3.968 ± 0.030 (0.200)	3.959 ± 0.030 (0.210)	3.952 ± 0.035 (0.220)	3.946 ± 0.035 (0.225)
0.5	3.916 ± 0.045 (0.270)	3.902 ± 0.050 (0.275)	3.894 ± 0.050 (0.285)	3.882 ± 0.050 (0.290)
0.6	3.851 ± 0.060 (0.330)	3.843 ± 0.065 (0.340)	3.819 ± 0.070 (0.345)	3.798 ± 0.070 (0.350)
0.7	3.781 ± 0.075 (0.365)	3.758 ± 0.075 (0.370)	3.739 ± 0.075 (0.380)	3.706 ± 0.075 (0.385)
0.8	3.694 ± 0.080 (0.410)	3.663 ± 0.080 (0.420)	3.638 ± 0.085 (0.430)	3.594 ± 0.085 (0.430)
0.9	3.603 ± 0.090 (0.440)	3.561 ± 0.090 (0.450)	3.525 ± 0.095 (0.455)	3.477 ± 0.095 (0.465)
1	3.504 ± 0.100 (0.455)	3.458 ± 0.100 (0.465)	3.415 ± 0.100 (0.475)	3.356 ± 0.100 (0.480)

Table 2 Four-body E_4 ground-state energies and wave function parameter, p_0, for the massless-exchange case, $\mu = 0$

α	E_4 with annihilation and (p_0) without retardation	E_4 without annihilation and (p_0) without retardation	E_4 with annihilation and (p_0) with retardation	E_4 [14] without annihilation and (p_0) with retardation
0.1	3.986 ± 0.040 (0.020)	3.980 ± 0.040 (0.025)	3.975 ± 0.050 (0.030)	3.971 ± 0.050 (0.030)
0.2	3.938 ± 0.050 (0.130)	3.930 ± 0.050 (0.135)	3.925 ± 0.060 (0.140)	3.919 ± 0.060 (0.145)
0.3	3.854 ± 0.060 (0.185)	3.835 ± 0.065 (0.195)	3.829 ± 0.070 (0.200)	3.821 ± 0.070 (0.205)
0.4	3.758 ± 0.070 (0.265)	3.743 ± 0.075 (0.275)	3.733 ± 0.080 (0.280)	3.721 ± 0.085 (0.290)
0.5	3.643 ± 0.080 (0.385)	3.625 ± 0.085 (0.390)	3.613 ± 0.090 (0.400)	3.598 ± 0.090 (0.405)
0.6	3.512 ± 0.090 (0.410)	3.484 ± 0.095 (0.420)	3.470 ± 0.095 (0.430)	3.454 ± 0.100 (0.430)
0.7	3.379 ± 0.100 (0.445)	3.342 ± 0.100 (0.450)	3.324 ± 0.105 (0.460)	3.304 ± 0.105 (0.465)
0.8	3.237 ± 0.105 (0.450)	3.185 ± 0.105 (0.460)	3.155 ± 0.105 (0.470)	3.131 ± 0.105 (0.470)
0.9	3.095 ± 0.110 (0.475)	3.042 ± 0.110 (0.480)	3.010 ± 0.110 (0.490)	2.970 ± 0.110 (0.500)
1	2.935 ± 0.110 (0.515)	2.875 ± 0.110 (0.525)	2.830 ± 0.110 (0.530)	2.775 ± 0.110 (0.535)

results obtained by Monte Carlo method grow significantly for higher values of the coupling strength.

Similar to the case of four-body system (two particles and two antiparticles) we have the case of six-body problem (system consisting of three particles and three antiparticles). Tables 3 and 4 show the numerical results corresponding to the effects of inclusion or exclusion of virtual annihilation interactions and/or retardation effects for the six-body system for the massive-exchange case with $\mu = 0.15\,m$ and for the massless-exchange case $\mu = 0$, respectively. The wave function parameters p_0 are also available in Tables 3 and 4 for each case at a given coupling strength α.

Note that as for the previous cases of two-, and four-body systems the virtual annihilation terms have small contributions for the ground-state energy results. The virtual annihilation term is a "repulsive" interaction that increases the energy of the system by a small amount, particularly for coupling values around $0.1 \leq \alpha \leq 0.4$. However, the contribution of this interaction becomes more important (not negligible) at higher coupling strength values such as $0.5 < \alpha \leq 1$.

Moreover, the contribution of retardation effects become more significant at higher values of coupling α. In addition, for a given value of coupling α if we consider that virtual annihilation interactions are included in the calculations, we will have two different cases. The first case is when retardation term is added and the second one is when retardation term is neglected. Similar to the case of two- and four-body systems the inclusion of retardation terms do decrease the energy solutions of the six-body system. This means that the binding energy of the systems in the example mentioned above will increase for both cases: the massive-exchange case with $\mu = 0.15\,m$ and for the massless exchange case $\mu = 0$, respectively. Note that binding energy $\mathbf{B}_6$ is: $\mathbf{B}_6 = E_6 - 6$ in units of m (i.e., $m = 1$).

The numerical solutions have been presented in Tables 3 and 4. These values have been also plotted in Fig 3a, b for the massive-exchange case with $\mu = 0.15\,m$ and for the massless-exchange case $\mu = 0$, respectively.

The case of six-body system, for particles only and no virtual annihilation interaction [5], was done before and it has been included in the Tables 3 and 4 as well as the calculation of the case of combined three particles and three antiparticles with virtual annihilations and with retardation effects [9]. The corresponding graphs are also put in Fig. 3a, b for comparison purposes. However, the other two cases when the retardation terms are included or excluded have not done previously and correspond to the present work.

We should note that for the six-body system under study, as for the two- and four-body cases mentioned before, we may have a critical value of the coupling around $\alpha \sim 1$. Note also that the error estimates in the numerical results obtained by Monte Carlo method is growing with increase of the coupling strength values. Furthermore, the error estimates augment notably with the increasing number of particles and antiparticles. This means that, for example for $\alpha = 1$, the error bars in the graphs related to

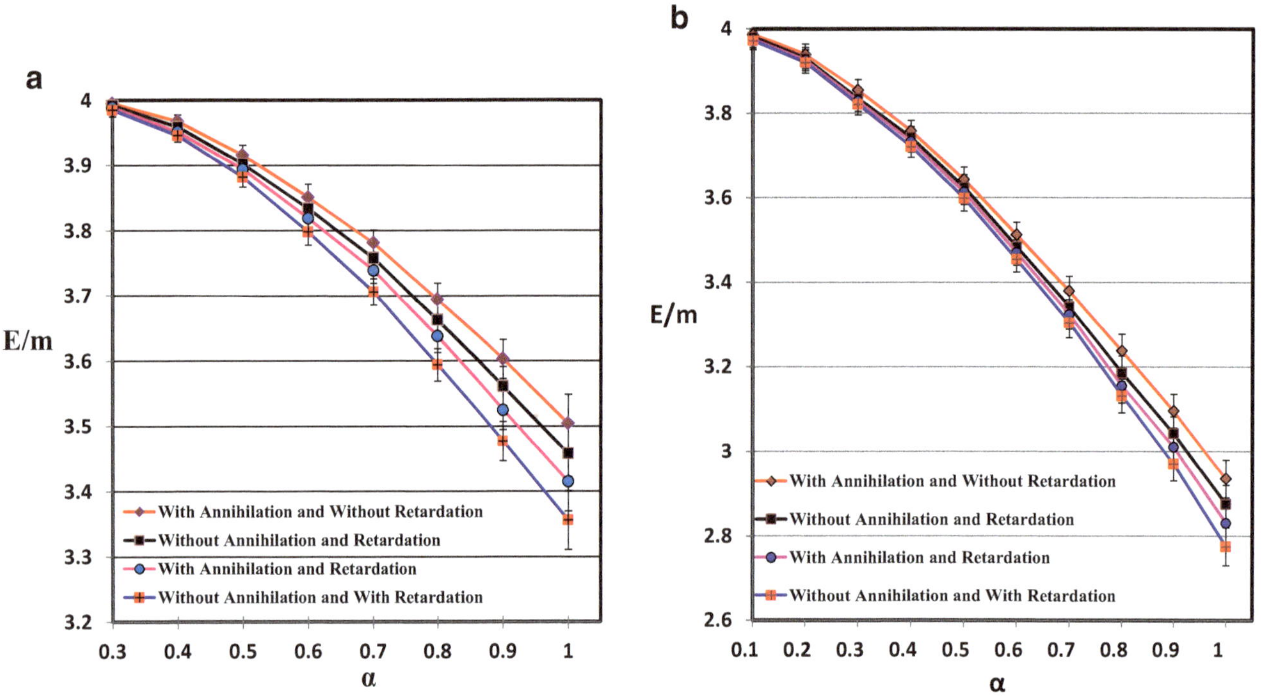

Fig. 2 **a** Relativistic ground-state energies in units of m of four-body systems, "with" and/or "without" retardation and/or virtual annihilation terms, for the massive-exchange case $\mu = 0.15m$ and different coupling α. **b** Relativistic ground-state energies in units of m of four-body systems, "with" and/or "without" retardation and/or virtual annihilation terms, for the massless-exchange case $\mu = 0$ and different coupling α

the six-body system are much larger than the error bars corresponding to the four-body system investigated earlier in this section.

Binding energy of the six-body system $\mathbf{B}_6$ is: $\mathbf{B}_6 = E_6 - 6$ in units of m (i.e., $m = 1$). As one can observe from the results of six-body system the binding energy of the system augments with increase of coupling strength α similar to the previous cases of two- and four-body systems. Although the qualitative behaviors of $E(\alpha)$ for $\mu = 0$ are very close to that for $\mu = 0.15\,m$, the binding energy for a particular value of α is stronger for $\mu = 0$ than for $\mu = 0.15\,m$, and it sets in at $\alpha_0 \rightarrow 0$.

Table 3 Six-body ground-state energies E_6 and wave function parameter, p_0, for massive-exchange case $\mu = 0.15m$

α	E_6 (p_0) with annihilation and without retardation	E_6 (p_0) without annihilation and without retardation	E_6 [9] (p_0) with annihilation and with retardation	E_6 [5] (p_0) without annihilation and with retardation
0.3	5.935 ± 0.0065 (0.085)	5.922 ± 0.065 (0.225)	5.915 ± 0.065 (0.235)	5.891 ± 0.060 (0.245)
0.4	5.820 ± 0.100 (0.300)	5.777 ± 0.100 (0.310)	5.765 ± 0.100 (0.315)	5.715 ± 0.095 (0.330)
0.5	5.651 ± 0.110 (0.350)	5.596 ± 0.110 (0.360)	5.570 ± 0.110 (0.370)	5.505 ± 0.105 (0.380)
0.6	5.477 ± 0.120 (0.400)	5.401 ± 0.120 (0.410)	5.345 ± 0.120 (0.420)	5.274 ± 0.110 (0.435)
0.7	5.252 ± 0.155 (0.435)	5.161 ± 0.155 (0.445)	5.101 ± 0.155 (0.450)	5.021 ± 0.145 (0.470)
0.8	5.008 ± 0.200 (0.505)	4.910 ± 0.200 (0.515)	4.842 ± 0.200 (0.520)	4.742 ± 0.195 (0.535)
0.9	4.725 ± 0.260 (0.540)	4.620 ± 0.260 (0.550)	4.535 ± 0.260 (0.570)	4.425 ± 0.250 (0.580)
1	4.380 ± 0.290 (0.615)	4.260 ± 0.290 (0.625)	4.155 ± 0.290 (0.635)	4.105 ± 0.280 (0.645)

Table 4 Six-body E_6 ground-state energies and wave function parameter, p_0, for the massless-exchange case, $\mu = 0$

α	E_6 with annihilation and (p_0) without retardation	E_6 without annihilation and (p_0) without retardation	E_6 [9] with annihilation and (p_0) with retardation	E_6 [5] without annihilation and (p_0) with retardation
0.1	5.967 ± 0.070 (0.090)	5.945 ± 0.070 (0.100)	5.935 ± 0.055 (0.105)	5.915 ± 0.060 (0.105)
0.2	5.881 ± 0.090 (0.205)	5.852 ± 0.090 (0.220)	5.840 ± 0.090 (0.230)	5.812 ± 0.085 (0.230)
0.3	5.665 ± 0.090 (0.285)	5.632 ± 0.110 (0.290)	5.617 ± 0.110 (0.300)	5.587 ± 0.100 (0.310)
0.4	5.412 ± 0.165 (0.360)	5.367 ± 0.165 (0.375)	5.340 ± 0.165 (0.380)	5.291 ± 0.155 (0.390)
0.5	5.102 ± 0.205 (0.405)	5.051 ± 0.205 (0.410)	5.025 ± 0.205 (0.415)	4.971 ± 0.195 (0.425)
0.6	4.822 ± 0.260 (0.440)	4.745 ± 0.260 (0.450)	4.710 ± 0.260 (0.455)	4.638 ± 0.250 (0.465)
0.7	4.498 ± 0.310 (0.495)	4.410 ± 0.310 (0.500)	4.360 ± 0.310 (0.505)	4.278 ± 0.295 (0.515)
0.8	4.171 ± 0.330 (0.525)	4.061 ± 0.330 (0.535)	3.990 ± 0.335 (0.540)	3.885 ± 0.325 (0.550)
0.9	3.770 ± 0.360 (0.570)	3.655 ± 0.360 (0.580)	3.585 ± 0.360 (0.590)	3.461 ± 0.345 (0.610)
1	3.405 ± 0.385 (0.640)	3.260 ± 0.385 (0.650)	3.170 ± 0.385 (0.655)	3.015 ± 0.375 (0.675)

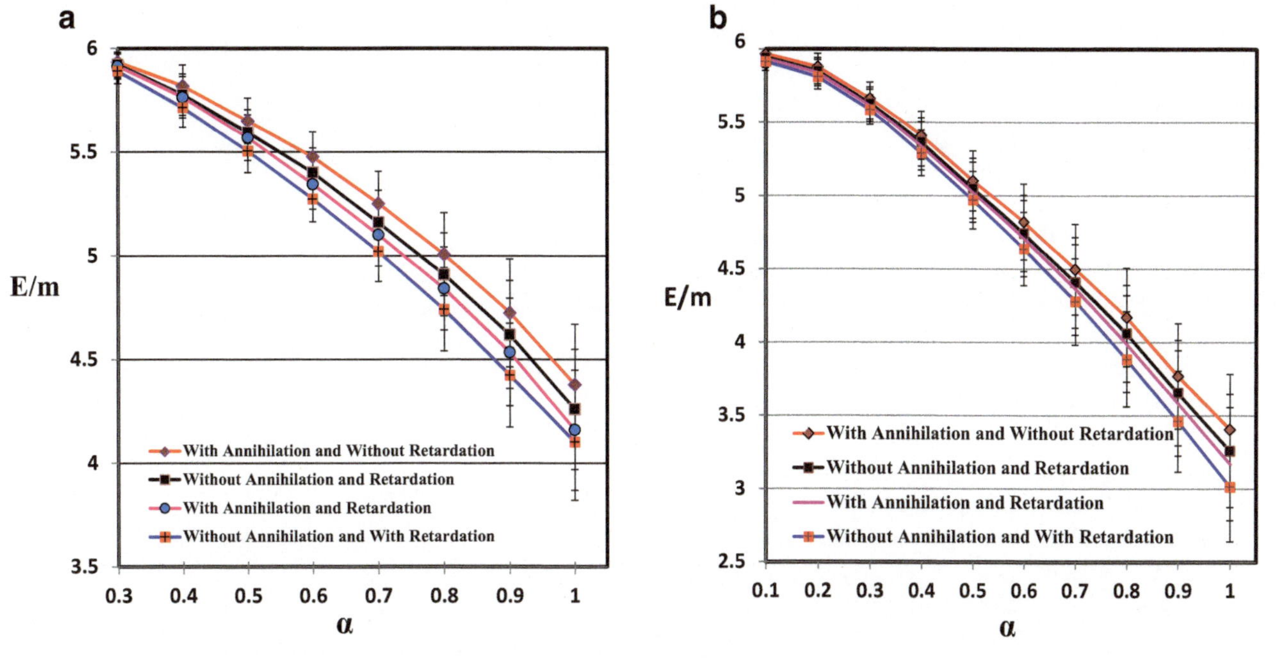

Fig. 3 **a** Relativistic ground-state energies in units of m of six-body systems, "with" and/or "without" retardation and/or virtual annihilation terms, for the massive-exchange case $\mu = 0.15m$ and different coupling α. **b** Relativistic ground-state energies in units of m of six-body systems, "with" and/or "without" retardation and/or virtual annihilation terms, for the massless-exchange case $\mu = 0$ and different coupling α

Concluding remarks

Approximate ground-state solutions of the two-, four-, and six-body systems have been presented for different strength of coupling, for both massive and massless scalar mediating field to compare the effects of virtual annihilation interaction terms and retardation effects in the relativistic equations. The numerical solutions of the two-, four-, and six-body systems have been discussed considering that "the virtual annihilation interactions" and "the retardation effects" are neglected or included.

The results for our various systems under study show that at low coupling the retardation effects and virtual annihilation interactions are minor effects compared to Yukawa interactions in our particle–antiparticle systems. In addition, at high coupling the results show that we have a noticeable increase in the effects of virtual annihilation interactions or retardation terms in the two-, four-, and six-body systems.

As mentioned before, the effect of retardation terms is as follows. The binding energies of our systems under study have been increased by a small amount for low values of the coupling and by a noticeable amount for high values of the coupling. However, as we can see, for example in the work of Gross [26], the inclusion of retardation effects reduces the binding energy of the two-body system. The argument that we can use to express the difference between these results with respect to the inclusion of the retardation effects is as follows. In a general point of view, we should consider the types of interactions that have been used. As a matter of fact, retardation terms that did arise in the equations are relativistic effects. The relativistic equations in the current model have only positive-energy solutions. In other words, the present relativistic equations are Salpeter-like and not Klein–Gordon like. The relativistic results lie above non-relativistic solutions, as we can see, for example, in [5] or [12, 13]. This fact, that the relativistic solutions are above the non-relativistic results, is the characteristic of formalisms, like the present model, which has no negative energy answers. One should also note that for the two-body case, the relativistic solutions which are obtained by employing an "empty" vacuum (Klein–Gordon-like equations) lie below the non-relativistic solutions. As one can see from eq. (50) of Darewych [24] for the massless-exchange case (the same as Todorov equation [27] using a quasipotential approach), at small values of the coupling α, the energy solutions are: $E_2 = 2 - \frac{1}{4}\alpha^2 - \frac{5}{64}\alpha^4$. To sum up, the retardation effect augments the binding energy of the systems mostly due to the characteristic of the present model or formalism.

From the practical point of view, it is of interest to remind what follows. Wheeler [28] speculated that two positronium atoms can combine to form the four-body system, namely positronium molecule (Ps_2: $e^-e^+e^-e^+$), in 1946. He also postulated the possible existence of larger atoms such as the six-body systems Ps_3 ($e^-e^+e^-e^+e^-e^+$). The first calculation of the binding energy of Ps_2 was performed [29] later on. In 2007 the positronium molecule was observed by Cassidy and Mills [30]. Similarly, the existence of a scalar "fundamental" particle, named Higgs boson, was theorized in 1960 by Higgs [31] and some other scientists. Observation of this new particle was confirmed in 2012 [32]. Hence, it would be also of interest to investigate the existence of bound states and related relativistic effects of the few-body Higgs bosons such as four-body Higgs systems or even larger structures. For example, the existence of two-Higgs-boson bound states has been studied in references [33, 34].

Therefore, from experimental and theoretical points of view it will be of fundamental interest to study the existence of bound states of scalar particles (or other exotic systems [35]) and the corresponding relativistic effects. Moreover, since the bound-state solutions, for example, of the scalar Higgs bosons are mainly found in the very strong coupling regime [34] the use of variational method seems to be more appropriate for the computations of relativistic effects of such "scalar" particles and/or antiparticles (the perturbation theory does depend on the coupling constant and the calculations may not be reliable) particularly at strong coupling.

Appendix

For the two-body system (one particle and one antiparticle), choosing the trial state

$$| \psi_2\rangle = \int \mathrm{d}^N p_1 \mathrm{d}^N p_2\, G(\mathbf{p}_1, \mathbf{p}_2) A^\dagger(\mathbf{p}_1) B^\dagger(\mathbf{p}_2)|0\rangle, \tag{6.1}$$

we obtain the following wave equation:

$$\begin{aligned} G(\mathbf{p}_1, \mathbf{p}_2)[\omega_{p_1} + \omega_{p_2} - E] &= \frac{g^2}{4(2\pi)^N} \int \mathrm{d}^N p'_1 \mathrm{d}^N p'_2 G(\mathbf{p}'_1, \mathbf{p}'_2) \\ &\times \delta^N(\mathbf{p}'_1 + \mathbf{p}'_2 - \mathbf{p}_1 - \mathbf{p}_2) \frac{1}{\sqrt{\omega_{p'_1}\omega_{p'_2}\omega_{p_1}\omega_{p_2}}} \\ &\times \left[\frac{1}{\mu^2 - (p'_1 - p_1)^2} + \frac{1}{\mu^2 - (p_1 + p_2)^2} \right] \end{aligned} \tag{6.2}$$

For the four-body system (two particles and two antiparticles), choosing the trial state

$$\begin{aligned} | \psi_4\rangle = \int \mathrm{d}^N p_1 \mathrm{d}^N p_2 \mathrm{d}^N p_3\, \mathrm{d}^N p_4 G(\mathbf{p}_1, \mathbf{p}_2, \mathbf{p}_3, \mathbf{p}_4) \\ A^\dagger(\mathbf{p}_1) B^\dagger(\mathbf{p}_2) A^\dagger(\mathbf{p}_3) B^\dagger(\mathbf{p}_4)|0\rangle, \end{aligned} \tag{6.3}$$

we have the following wave equation:

$$G(\mathbf{p}_1,\mathbf{p}_2,\mathbf{p}_3,\mathbf{p}_4)[\omega_{\mathbf{p}_1}+\omega_{\mathbf{p}_2}+\omega_{\mathbf{p}_3}+\omega_{\mathbf{p}_4}-E]$$
$$=\frac{g^2}{4(2\pi)^N}\int d^Np_1'd^Np_2'd^Np_3'd^Np_4'G(\mathbf{p}_1',\mathbf{p}_2',\mathbf{p}_3',\mathbf{p}_4')$$
$$\times\Bigg\{\Bigg[\delta^N(\mathbf{p}_2'-\mathbf{p}_2)\delta^N(\mathbf{p}_4'-\mathbf{p}_4)\frac{\delta^N(\mathbf{p}_1'+\mathbf{p}_3'-\mathbf{p}_1-\mathbf{p}_3)}{\sqrt{\omega_{p_1'}\omega_{p_3'}\omega_{p_1}\omega_{p_3}}}\frac{1}{\mu^2-(p_1-p_1')^2}\Bigg]_{G_1}$$
$$+\Bigg[\delta^N(\mathbf{p}_1'-\mathbf{p}_1)\delta^N(\mathbf{p}_4'-\mathbf{p}_4)\frac{\delta^N(\mathbf{p}_2'+\mathbf{p}_3'-\mathbf{p}_2-\mathbf{p}_3)}{\sqrt{\omega_{p_2'}\omega_{p_3'}\omega_{p_2}\omega_{p_3}}}\frac{1}{\mu^2-(p_2-p_2')^2}\Bigg]_{G_2}$$
$$+\Bigg[\delta^N(\mathbf{p}_4'-\mathbf{p}_4)\delta^N(\mathbf{p}_3'-\mathbf{p}_3)\frac{\delta^N(\mathbf{p}_1'+\mathbf{p}_2'-\mathbf{p}_1-\mathbf{p}_2)}{\sqrt{\omega_{p_1'}\omega_{p_2'}\omega_{p_1}\omega_{p_2}}}\frac{1}{\mu^2-(p_2-p_2')^2}\Bigg]_{G_3}$$
$$+\Bigg[\delta^N(\mathbf{p}_1'-\mathbf{p}_1)\delta^N(\mathbf{p}_2'-\mathbf{p}_2)\frac{\delta^N(\mathbf{p}_4'+\mathbf{p}_3'-\mathbf{p}_3-\mathbf{p}_4)}{\sqrt{\omega_{p_4'}\omega_{p_3'}\omega_{p_3}\omega_{p_4}}}\frac{1}{\mu^2-(p_3-p_3')^2}\Bigg]_{G_4}$$
$$+\Bigg[\delta^N(\mathbf{p}_3'-\mathbf{p}_3)\delta^N(\mathbf{p}_2'-\mathbf{p}_2)\frac{\delta^N(\mathbf{p}_4'+\mathbf{p}_1'-\mathbf{p}_4-\mathbf{p}_1)}{\sqrt{\omega_{p_4'}\omega_{p_1'}\omega_{p_4}\omega_{p_1}}}\frac{1}{\mu^2-(p_1-p_1')^2}\Bigg]_{G_5}$$
$$+\Bigg[\delta^N(\mathbf{p}_3'-\mathbf{p}_3)\delta^N(\mathbf{p}_1'-\mathbf{p}_1)\frac{\delta^N(\mathbf{p}_4'+\mathbf{p}_2'-\mathbf{p}_4-\mathbf{p}_2)}{\sqrt{\omega_{p_4'}\omega_{p_2'}\omega_{p_4}\omega_{p_2}}}\frac{1}{\mu^2-(p_4-p_4')^2}\Bigg]_{G_6}$$
$$+\Bigg[\delta^N(\mathbf{p}_1'-\mathbf{p}_1)\delta^N(\mathbf{p}_2'-\mathbf{p}_2)\frac{\delta^N(\mathbf{p}_4'+\mathbf{p}_3'-\mathbf{p}_3-\mathbf{p}_4)}{\sqrt{\omega_{p_4'}\omega_{p_3'}\omega_{p_3}\omega_{p_4}}}\frac{1}{\mu^2-(p_4+p_3)^2}\Bigg]_{A_1}$$
$$+\Bigg[\delta^N(\mathbf{p}_3'-\mathbf{p}_3)\delta^N(\mathbf{p}_2'-\mathbf{p}_2)\frac{\delta^N(\mathbf{p}_4'+\mathbf{p}_1'-\mathbf{p}_4-\mathbf{p}_1)}{\sqrt{\omega_{p_4'}\omega_{p_1'}\omega_{p_4}\omega_{p_1}}}\frac{1}{\mu^2-(p_4+p_1)^2}\Bigg]_{A_2}$$
$$+\Bigg[\delta^N(\mathbf{p}_1'-\mathbf{p}_1)\delta^N(\mathbf{p}_4'-\mathbf{p}_4)\frac{\delta^N(\mathbf{p}_2'+\mathbf{p}_3'-\mathbf{p}_2-\mathbf{p}_3)}{\sqrt{\omega_{p_2'}\omega_{p_3'}\omega_{p_2}\omega_{p_3}}}\frac{1}{\mu^2-(p_3'+p_2')^2}\Bigg]_{A_3}$$
$$+\Bigg[\delta^N(\mathbf{p}_4'-\mathbf{p}_4)\delta^N(\mathbf{p}_3'-\mathbf{p}_3)\frac{\delta^N(\mathbf{p}_1'+\mathbf{p}_2'-\mathbf{p}_1-\mathbf{p}_2)}{\sqrt{\omega_{p_1'}\omega_{p_2'}\omega_{p_1}\omega_{p_2}}}\frac{1}{\mu^2-(p_1'+p_2')^2}\Bigg]_{A_4}\Bigg\} \tag{6.4}$$

For the six-body system (consisting of three particles and three antiparticles: $n=6$), the simplest trial state for a system of three particles and three antiparticles is

$$|\psi_6\rangle=\int d^Np_1d^Np_2d^Np_3d^Np_4d^Np_5d^Np_6G(\mathbf{p}_1,\mathbf{p}_2,\mathbf{p}_3,\mathbf{p}_4,\mathbf{p}_5,\mathbf{p}_6)$$
$$\times A^\dagger(\mathbf{p}_1)B^\dagger(\mathbf{p}_2)A^\dagger(\mathbf{p}_3)B^\dagger(\mathbf{p}_4)A^\dagger(\mathbf{p}_5)B^\dagger(\mathbf{p}_6)|0\rangle \tag{6.5}$$

The resulting six-body wave equation is

$$G(\mathbf{p}_1,\mathbf{p}_2,\mathbf{p}_3,\mathbf{p}_4,\mathbf{p}_5,\mathbf{p}_6)[\omega_{\mathbf{p}_1}+\omega_{\mathbf{p}_2}+\omega_{\mathbf{p}_3}+\omega_{\mathbf{p}_4}+\omega_{\mathbf{p}_5}+\omega_{\mathbf{p}_6}-E]$$
$$=\frac{g^2}{4(2\pi)^N}\int d^Np_1'd^Np_2'd^Np_3'd^Np_4'd^Np_5'd^Np_6'$$
$$G^*(\mathbf{p}_1',\mathbf{p}_2',\mathbf{p}_3',\mathbf{p}_4',\mathbf{p}_5',\mathbf{p}_6')$$
$$\times\Bigg\{\Bigg[\delta^N(\mathbf{p}_3'-\mathbf{p}_3)\delta^N(\mathbf{p}_4'-\mathbf{p}_4)\frac{\delta^N(\mathbf{p}_1'+\mathbf{p}_2'-\mathbf{p}_1-\mathbf{p}_2)}{\sqrt{\omega_{p_1'}\omega_{p_2'}\omega_{p_1}\omega_{p_2}}}\frac{\delta^N(\mathbf{p}_5'-\mathbf{p}_5)\delta^N(\mathbf{p}_6'-\mathbf{p}_6)}{\mu^2-(p_1-p_1')^2}\Bigg]_{G_1}$$
$$+\Bigg[\delta^N(\mathbf{p}_2'-\mathbf{p}_2)\delta^N(\mathbf{p}_4'-\mathbf{p}_4)\frac{\delta^N(\mathbf{p}_1'+\mathbf{p}_3'-\mathbf{p}_1-\mathbf{p}_3)}{\sqrt{\omega_{p_1'}\omega_{p_3'}\omega_{p_1}\omega_{p_3}}}\frac{\delta^N(\mathbf{p}_5'-\mathbf{p}_5)\delta^N(\mathbf{p}_6'-\mathbf{p}_6)}{\mu^2-(p_1-p_1')^2}\Bigg]_{G_2}$$
$$+\Bigg[\delta^N(\mathbf{p}_2'-\mathbf{p}_2)\delta^N(\mathbf{p}_3'-\mathbf{p}_3)\frac{\delta^N(\mathbf{p}_1'+\mathbf{p}_4'-\mathbf{p}_1-\mathbf{p}_4)}{\sqrt{\omega_{p_1'}\omega_{p_4'}\omega_{p_1}\omega_{p_4}}}\frac{\delta^N(\mathbf{p}_5'-\mathbf{p}_5)\delta^N(\mathbf{p}_6'-\mathbf{p}_6)}{\mu^2-(p_4-p_4')^2}\Bigg]_{G_3}$$
$$+\Bigg[\delta^N(\mathbf{p}_2'-\mathbf{p}_2)\delta^N(\mathbf{p}_3'-\mathbf{p}_3)\frac{\delta^N(\mathbf{p}_1'+\mathbf{p}_5'-\mathbf{p}_1-\mathbf{p}_5)}{\sqrt{\omega_{p_1'}\omega_{p_5'}\omega_{p_1}\omega_{p_5}}}\frac{\delta^N(\mathbf{p}_4'-\mathbf{p}_4)\delta^N(\mathbf{p}_6'-\mathbf{p}_6)}{\mu^2-(p_5-p_5')^2}\Bigg]_{G_4}$$
$$+\Bigg[\delta^N(\mathbf{p}_2'-\mathbf{p}_2)\delta^N(\mathbf{p}_3'-\mathbf{p}_3)\frac{\delta^N(\mathbf{p}_1'+\mathbf{p}_6'-\mathbf{p}_1-\mathbf{p}_6)}{\sqrt{\omega_{p_1'}\omega_{p_6'}\omega_{p_1}\omega_{p_6}}}\frac{\delta^N(\mathbf{p}_4'-\mathbf{p}_4)\delta^N(\mathbf{p}_5'-\mathbf{p}_5)}{\mu^2-(p_6-p_6')^2}\Bigg]_{G_5}$$
$$+\Bigg[\delta^N(\mathbf{p}_1'-\mathbf{p}_1)\delta^N(\mathbf{p}_4'-\mathbf{p}_4)\frac{\delta^N(\mathbf{p}_2'+\mathbf{p}_3'-\mathbf{p}_2-\mathbf{p}_3)}{\sqrt{\omega_{p_2'}\omega_{p_3'}\omega_{p_2}\omega_{p_3}}}\frac{\delta^N(\mathbf{p}_5'-\mathbf{p}_5)\delta^N(\mathbf{p}_6'-\mathbf{p}_6)}{\mu^2-(p_3-p_3')^2}\Bigg]_{G_6}$$
$$+\Bigg[\delta^N(\mathbf{p}_1'-\mathbf{p}_1)\delta^N(\mathbf{p}_3'-\mathbf{p}_3)\frac{\delta^N(\mathbf{p}_2'+\mathbf{p}_4'-\mathbf{p}_2-\mathbf{p}_4)}{\sqrt{\omega_{p_2'}\omega_{p_4'}\omega_{p_2}\omega_{p_4}}}\frac{\delta^N(\mathbf{p}_5'-\mathbf{p}_5)\delta^N(\mathbf{p}_6'-\mathbf{p}_6)}{\mu^2-(p_2-p_2')^2}\Bigg]_{G_7}$$

$$+\left[\delta^N(\mathbf{p}_1'-\mathbf{p}_1)\delta^N(\mathbf{p}_3'-\mathbf{p}_3)\frac{\delta^N(\mathbf{p}_2'+\mathbf{p}_5'-\mathbf{p}_2-\mathbf{p}_5)}{\sqrt{\omega_{p_2'}\omega_{p_5'}\omega_{p_2}\omega_{p_5}}}\frac{\delta^N(\mathbf{p}_4'-\mathbf{p}_4)\delta^N(\mathbf{p}_6'-\mathbf{p}_6)}{\mu^2-(p_2-p_2')^2}\right]_{G_8}$$

$$+\left[\delta^N(\mathbf{p}_1'-\mathbf{p}_1)\delta^N(\mathbf{p}_3'-\mathbf{p}_3)\frac{\delta^N(\mathbf{p}_2'+\mathbf{p}_6'-\mathbf{p}_2-\mathbf{p}_6)}{\sqrt{\omega_{p_2'}\omega_{p_6'}\omega_{p_2}\omega_{p_6}}}\frac{\delta^N(\mathbf{p}_4'-\mathbf{p}_4)\delta^N(\mathbf{p}_5'-\mathbf{p}_5)}{\mu^2-(p_6-p_6')^2}\right]_{G_9}$$

$$+\left[\delta^N(\mathbf{p}_1'-\mathbf{p}_1)\delta^N(\mathbf{p}_2'-\mathbf{p}_2)\frac{\delta^N(\mathbf{p}_3'+\mathbf{p}_4'-\mathbf{p}_3-\mathbf{p}_4)}{\sqrt{\omega_{p_3'}\omega_{p_4'}\omega_{p_3}\omega_{p_4}}}\frac{\delta^N(\mathbf{p}_5'-\mathbf{p}_5)\delta^N(\mathbf{p}_6'-\mathbf{p}_6)}{\mu^2-(p_3-p_3')^2}\right]_{G_{10}}$$

$$+\left[\delta^N(\mathbf{p}_1'-\mathbf{p}_1)\delta^N(\mathbf{p}_2'-\mathbf{p}_2)\frac{\delta^N(\mathbf{p}_3'+\mathbf{p}_5'-\mathbf{p}_3-\mathbf{p}_5)}{\sqrt{\omega_{p_3'}\omega_{p_5'}\omega_{p_3}\omega_{p_5}}}\frac{\delta^N(\mathbf{p}_4'-\mathbf{p}_4)\delta^N(\mathbf{p}_6'-\mathbf{p}_6)}{\mu^2-(p_5-p_5')^2}\right]_{G_{11}}$$

$$+\left[\delta^N(\mathbf{p}_1'-\mathbf{p}_1)\delta^N(\mathbf{p}_2'-\mathbf{p}_2)\frac{\delta^N(\mathbf{p}_3'+\mathbf{p}_6'-\mathbf{p}_3-\mathbf{p}_6)}{\sqrt{\omega_{p_3'}\omega_{p_6'}\omega_{p_3}\omega_{p_6}}}\frac{\delta^N(\mathbf{p}_4'-\mathbf{p}_4)\delta^N(\mathbf{p}_5'-\mathbf{p}_5)}{\mu^2-(p_6-p_6')^2}\right]_{G_{12}}$$

$$+\left[\delta^N(\mathbf{p}_1'-\mathbf{p}_1)\delta^N(\mathbf{p}_2'-\mathbf{p}_2)\frac{\delta^N(\mathbf{p}_4'+\mathbf{p}_5'-\mathbf{p}_4-\mathbf{p}_5)}{\sqrt{\omega_{p_4'}\omega_{p_5'}\omega_{p_4}\omega_{p_5}}}\frac{\delta^N(\mathbf{p}_3'-\mathbf{p}_3)\delta^N(\mathbf{p}_6'-\mathbf{p}_6)}{\mu^2-(p_4-p_4')^2}\right]_{G_{13}}$$

$$+\left[\delta^N(\mathbf{p}_1'-\mathbf{p}_1)\delta^N(\mathbf{p}_2'-\mathbf{p}_2)\frac{\delta^N(\mathbf{p}_4'+\mathbf{p}_6'-\mathbf{p}_4-\mathbf{p}_6)}{\sqrt{\omega_{p_4'}\omega_{p_6'}\omega_{p_4}\omega_{p_6}}}\frac{\delta^N(\mathbf{p}_3'-\mathbf{p}_3)\delta^N(\mathbf{p}_5'-\mathbf{p}_5)}{\mu^2-(p_4-p_4')^2}\right]_{G_{14}}$$

$$+\left[\delta^N(\mathbf{p}_1'-\mathbf{p}_1)\delta^N(\mathbf{p}_2'-\mathbf{p}_2)\frac{\delta^N(\mathbf{p}_5'+\mathbf{p}_6'-\mathbf{p}_5-\mathbf{p}_6)}{\sqrt{\omega_{p_5'}\omega_{p_6'}\omega_{p_5}\omega_{p_6}}}\frac{\delta^N(\mathbf{p}_3'-\mathbf{p}_3)\delta^N(\mathbf{p}_4'-\mathbf{p}_4)}{\mu^2-(p_5-p_5')^2}\right]_{G_{15}}$$

$$+\left[\delta^N(\mathbf{p}_3'-\mathbf{p}_3)\delta^N(\mathbf{p}_4'-\mathbf{p}_4)\frac{\delta^N(\mathbf{p}_1'+\mathbf{p}_2'-\mathbf{p}_1-\mathbf{p}_2)}{\sqrt{\omega_{p_1'}\omega_{p_2'}\omega_{p_1}\omega_{p_2}}}\frac{\delta^N(\mathbf{p}_5'-\mathbf{p}_5)\delta^N(\mathbf{p}_6'-\mathbf{p}_6)}{\mu^2-(p_1+p_2)^2}\right]_{A_1}$$

$$+\left[\delta^N(\mathbf{p}_2'-\mathbf{p}_2)\delta^N(\mathbf{p}_3'-\mathbf{p}_3)\frac{\delta^N(\mathbf{p}_1'+\mathbf{p}_4'-\mathbf{p}_1-\mathbf{p}_4)}{\sqrt{\omega_{p_1'}\omega_{p_4'}\omega_{p_1}\omega_{p_4}}}\frac{\delta^N(\mathbf{p}_5'-\mathbf{p}_5)\delta^N(\mathbf{p}_6'-\mathbf{p}_6)}{\mu^2-(p_1+p_4)^2}\right]_{A_2}$$

$$+\left[\delta^N(\mathbf{p}_2'-\mathbf{p}_2)\delta^N(\mathbf{p}_3'-\mathbf{p}_3)\frac{\delta^N(\mathbf{p}_1'+\mathbf{p}_6'-\mathbf{p}_1-\mathbf{p}_6)}{\sqrt{\omega_{p_1'}\omega_{p_6'}\omega_{p_1}\omega_{p_6}}}\frac{\delta^N(\mathbf{p}_4'-\mathbf{p}_4)\delta^N(\mathbf{p}_5'-\mathbf{p}_5)}{\mu^2-(p_1'+p_6')^2}\right]_{A_3}$$

$$+\left[\delta^N(\mathbf{p}_1'-\mathbf{p}_1)\delta^N(\mathbf{p}_4'-\mathbf{p}_4)\frac{\delta^N(\mathbf{p}_2'+\mathbf{p}_3'-\mathbf{p}_2-\mathbf{p}_3)}{\sqrt{\omega_{p_2'}\omega_{p_3'}\omega_{p_2}\omega_{p_3}}}\frac{\delta^N(\mathbf{p}_5'-\mathbf{p}_5)\delta^N(\mathbf{p}_6'-\mathbf{p}_6)}{\mu^2-(p_2'+p_3')^2}\right]_{A_4}$$

$$+\left[\delta^N(\mathbf{p}_1'-\mathbf{p}_1)\delta^N(\mathbf{p}_3'-\mathbf{p}_3)\frac{\delta^N(\mathbf{p}_2'+\mathbf{p}_5'-\mathbf{p}_2-\mathbf{p}_5)}{\sqrt{\omega_{p_2'}\omega_{p_5'}\omega_{p_2}\omega_{p_5}}}\frac{\delta^N(\mathbf{p}_4'-\mathbf{p}_4)\delta^N(\mathbf{p}_6'-\mathbf{p}_6)}{\mu^2-(p_2+p_5)^2}\right]_{A_5}$$

$$+\left[\delta^N(\mathbf{p}_1'-\mathbf{p}_1)\delta^N(\mathbf{p}_2'-\mathbf{p}_2)\frac{\delta^N(\mathbf{p}_4'+\mathbf{p}_3'-\mathbf{p}_3-\mathbf{p}_4)}{\sqrt{\omega_{p_4'}\omega_{p_3'}\omega_{p_3}\omega_{p_4}}}\frac{\delta^N(\mathbf{p}_5'-\mathbf{p}_5)\delta^N(\mathbf{p}_6'-\mathbf{p}_6)}{\mu^2-(p_4'+p_3')^2}\right]_{A_6}$$

$$+\left[\delta^N(\mathbf{p}_1'-\mathbf{p}_1)\delta^N(\mathbf{p}_2'-\mathbf{p}_2)\frac{\delta^N(\mathbf{p}_3'+\mathbf{p}_6'-\mathbf{p}_3-\mathbf{p}_6)}{\sqrt{\omega_{p_3'}\omega_{p_6'}\omega_{p_3}\omega_{p_6}}}\frac{\delta^N(\mathbf{p}_4'-\mathbf{p}_4)\delta^N(\mathbf{p}_5'-\mathbf{p}_5)}{\mu^2-(p_4+p_3)^2}\right]_{A_7}$$

$$+\left[\delta^N(\mathbf{p}_1'-\mathbf{p}_1)\delta^N(\mathbf{p}_2'-\mathbf{p}_2)\frac{\delta^N(\mathbf{p}_4'+\mathbf{p}_5'-\mathbf{p}_4-\mathbf{p}_5)}{\sqrt{\omega_{p_4'}\omega_{p_5'}\omega_{p_4}\omega_{p_5}}}\frac{\delta^N(\mathbf{p}_3'-\mathbf{p}_3)\delta^N(\mathbf{p}_6'-\mathbf{p}_6)}{\mu^2-(p_4'+p_5')^2}\right]_{A_8}$$

$$+\left[\delta^N(\mathbf{p}_1'-\mathbf{p}_1)\delta^N(\mathbf{p}_2'-\mathbf{p}_2)\frac{\delta^N(\mathbf{p}_5'+\mathbf{p}_6'-\mathbf{p}_5-\mathbf{p}_6)}{\sqrt{\omega_{p_5'}\omega_{p_6'}\omega_{p_5}\omega_{p_6}}}\frac{\delta^N(\mathbf{p}_3'-\mathbf{p}_3)\delta^N(\mathbf{p}_4'-\mathbf{p}_4)}{\mu^2-(p_5'+p_6')^2}\right]_{A_9}\Bigg\}. \tag{6.6}$$

As an example for the non-relativistic potential of six-body system, namely $V(\mathbf{r}_1, \mathbf{r}_2, \ldots, \mathbf{r}_6)$, we can write it as below:

$$
\begin{aligned}
V(\mathbf{r}_1,\mathbf{r}_2,\mathbf{r}_3,\mathbf{r}_4,\mathbf{r}_5,\mathbf{r}_6) &= -\alpha\frac{e^{-\mu|\mathbf{r}_1-\mathbf{r}_2|}}{|\mathbf{r}_1-\mathbf{r}_2|} - \alpha\frac{e^{-\mu|\mathbf{r}_1-\mathbf{r}_3|}}{|\mathbf{r}_1-\mathbf{r}_3|} \\
&- \alpha\frac{e^{-\mu|\mathbf{r}_1-\mathbf{r}_4|}}{|\mathbf{r}_1-\mathbf{r}_4|} - \alpha\frac{e^{-\mu|\mathbf{r}_1-\mathbf{r}_5|}}{|\mathbf{r}_1-\mathbf{r}_5|} \\
&- \alpha\frac{e^{-\mu|\mathbf{r}_1-\mathbf{r}_6|}}{|\mathbf{r}_1-\mathbf{r}_6|} - \alpha\frac{e^{-\mu|\mathbf{r}_2-\mathbf{r}_3|}}{|\mathbf{r}_2-\mathbf{r}_3|} - \alpha\frac{e^{-\mu|\mathbf{r}_2-\mathbf{r}_4|}}{|\mathbf{r}_2-\mathbf{r}_4|} - \alpha\frac{e^{-\mu|\mathbf{r}_2-\mathbf{r}_5|}}{|\mathbf{r}_2-\mathbf{r}_5|} \\
&- \alpha\frac{e^{-\mu|\mathbf{r}_2-\mathbf{r}_6|}}{|\mathbf{r}_2-\mathbf{r}_6|} - \alpha\frac{e^{-\mu|\mathbf{r}_3-\mathbf{r}_4|}}{|\mathbf{r}_3-\mathbf{r}_4|} - \alpha\frac{e^{-\mu|\mathbf{r}_3-\mathbf{r}_5|}}{|\mathbf{r}_3-\mathbf{r}_5|} - \alpha\frac{e^{-\mu|\mathbf{r}_3-\mathbf{r}_6|}}{|\mathbf{r}_3-\mathbf{r}_6|} \\
&- \alpha\frac{e^{-\mu|\mathbf{r}_4-\mathbf{r}_5|}}{|\mathbf{r}_4-\mathbf{r}_5|} - \alpha\frac{e^{-\mu|\mathbf{r}_4-\mathbf{r}_6|}}{|\mathbf{r}_4-\mathbf{r}_6|} - \alpha\frac{e^{-\mu|\mathbf{r}_5-\mathbf{r}_6|}}{|\mathbf{r}_5-\mathbf{r}_6|} \\
&+ \alpha\frac{4\pi}{4m^2-\mu^2}\,\delta^3(\mathbf{r}_1-\mathbf{r}_2) + \alpha\frac{4\pi}{4m^2-\mu^2}\,\delta^3(\mathbf{r}_1-\mathbf{r}_4) \\
&+ \alpha\frac{4\pi}{4m^2-\mu^2}\,\delta^3(\mathbf{r}_1-\mathbf{r}_6) + \alpha\frac{4\pi}{4m^2-\mu^2}\,\delta^3(\mathbf{r}_2-\mathbf{r}_3) \\
&+ \alpha\frac{4\pi}{4m^2-\mu^2}\,\delta^3(\mathbf{r}_2-\mathbf{r}_5) + \alpha\frac{4\pi}{4m^2-\mu^2}\,\delta^3(\mathbf{r}_3-\mathbf{r}_4) \\
&+ \alpha\frac{4\pi}{4m^2-\mu^2}\,\delta^3(\mathbf{r}_3-\mathbf{r}_6) + \alpha\frac{4\pi}{4m^2-\mu^2}\,\delta^3(\mathbf{r}_4-\mathbf{r}_5) \\
&+ \alpha\frac{4\pi}{4m^2-\mu^2}\,\delta^3(\mathbf{r}_5-\mathbf{r}_6),
\end{aligned}
\tag{6.7}
$$

References

1. Salpeter, E.E., Bethe, H.A.: A relativistic equation for bound-state problems. Phys. Rev. **84**, 1232 (1951)
2. Bethe, H.A., Salpeter, E.E.: Quantum Mechanics of One- and Two-Electron Atoms. Springer, Berlin (1957)
3. Wick, G.C.: Properties of Bethe–Salpeter wave functions. Phys. Rev. **96**, 1124 (1954)
4. Cutkosky, R.E.: Solutions of a Bethe–Salpeter equation. Phys. Rev. **96**, 1135 (1954)
5. Emami-Razavi, M.: Relativistic *n*-body wave equations in scalar quantum field theory. Phys. Lett. B **640**, 285–291 (2006)
6. Emami-Razavi, M.: Relativistic wave equations of combined particles and antiparticles in scalar quantum field theory. Phys. Rev. D **77**, 045025 (2008)
7. Darewych, J.W.: Partial decoupling of electrodynamics. Annales F. Louis de Broglie (Paris) **23**, 15 (1998)
8. Darewych, J.W.: Interparticle interactions and nonlocality in QFT. In: Hunter, G., Jeffers, S., Vigier, J.-P. (eds.) Causality and Locality in Modern Physics. Proceedings of a Symposium in Honour of Jean-Pierre Vigier. Kluwer Academic Publishers, Dordrecht, pp. 333–344 (1998)
9. Emami-Razavi, M., Bergeron, N., Darewych, J.W.: Relativistic wave equations of *n*-body systems of particles and antiparticles of various masses in scalar quantum field theory. J. Phys. G Nucl. Part. Phys. **38**, 065004 (2011)
10. Emami-Razavi, M.: Relativistic *n*-fermion wave equations in quantum electrodynamics. Phys. Rev. A **77**, 042104 (2008)
11. Emami-Razavi, M., Bergeron, N., Darewych, J.W.: Relativistic wave equations of *n*-body systems of fermions and antifermions of various masses in quantum electrodynamics. Int J. Mod. Phys. E **21**, 1250091 (2012)
12. Ding, B., Darewych, J.W.: A variational calculation of particle-antiparticle bound states in the scalar Yukawa model. J. Phys. G **26**, 907 (2000)
13. Ding, B., Darewych, J.W.: A variational calculation of particle-antiparticle bound states in the scalar Yukawa model. J. Phys. G **27**, 253 (2001)
14. Emami-Razavi, M., Darewych, J.W.: Relativistic two-, three- and four-body wave equations in scalar QFT. J. Phys. G Nucl. Part. Phys. **31**, 1095–1109 (2005)
15. Schiff, L.I.: Application of the variation method to field quantization. Phys. Rev. **130**, 458 (1963)
16. Coleman, S., Weinberg, E.: Radiative corrections as the origin of spontaneous symmetry breaking. Phys. Rev. D **7**, 1888 (1973)
17. Emami-Razavi, M., Darewych, J.W.: Relativistic two-body bound states in scalar QFT: variational basis-state approach. J. Phys. G Nucl. Part. Phys. **32**, 1171–1191 (2006)
18. Emami-Razavi, M., Kowalski, M.: Effect of virtual pairs for relativistic bound-state problems. Phys. Rev. D **76**, 045006 (2007)
19. Emami-Razavi, M., Bergeron, N., Darewych, J.W., Kowalski, M.: Relativistic three-body bound states in scalar QFT: variational basis-state approach. Phys. Rev. D **80**, 085006 (2009)
20. Tamm, I.: Relativistic interaction of elementary particles. J. Phys. USSR **9**, 449 (1945)
21. Greiner, W., Reinhard, J.: Quantum Electrodynamics, section 6, 3rd edn. Springer, Berlin (2003)
22. Gross, Franz: Relativistic Quantum Mechanics and Field Theory, pp. 259–261. Wiley InterScience Publication, New York (1992)
23. Di Leo, L., Darewych, J.W.: Bound and resonant relativistic two-particle states in scalar quantum field theory. Can. J. Phys. **70**, 412 (1992)
24. Darewych, J.W.: Some exact solutions of reduced scalar Yukawa theory. Can. J. Phys. **76**, 523 (1998)
25. Press, W.H., Teukolsky, S.A., Vetterling, W.T., Flannery, B.P.: Numerical Recipes, 2nd edn. Cambridge University Press, Cambridge (1992)
26. Gross, F.: Relativistic few-body problem. I. Two-body equations. Phys. Rev. C **26**, 2203 (1982)
27. Todorov, I.T.: Quasipotential equation corresponding to the relativistic Eikonal approximation. Phys. Rev. D **3**, 2351 (1971)
28. Wheeler, J.A.: Polyelectrons. Ann. N.Y. Acad. Sci. **48**, 219 (1946)
29. Hylleraas, E.A., Ore, A.: Binding energy of the positronium molecule. Phys. Rev. **71**, 493 (1947)
30. Cassidy, D.B., Mills, A.P.: The production of molecular positronium. Nature **449**, 195 (2007)
31. Higgs, P.W.: Broken symmetries and the masses of gauge bosons. Phys. Lett. **12**, 132 (1964)
32. ATLAS Collaboration: Observation of a new particle in the search for the Standard Model Higgs boson with the ATLAS detector at the LHC. Phys. Lett. B **716**, 1 (2012)
33. Di Leo, L., Darewych, J.W.: Bound states in the Higgs model. Phys. Rev. D **49**, 1659 (1994)
34. Siringo, F.: Variational study of bound states in the Higgs model. Phys. Rev. D **62**, 116009 (2000)
35. Emami-Razavi, M.: A short review about some exotic systems containing electrons, muons, and tauons. J. Theor. Appl. Phys. **9**, 1–6 (2015)

4

Multiple-layered structure of obelisk-shaped crystalline nano-ZnO prepared by sol–gel route

Saeideh Jurablu[1] · Majid Farahmandjou[1] · Tahereh Pormirjaafari Firoozabadi[1]

Abstract Zinc oxide nanopowders were synthesized by the simple sol–gel method from an ethanol solution of zinc nitrate hexahydrate. Structural and surface morphological investigations were carried out using X-ray diffraction (XRD), high-resolution transmission electron microscopy (HRTEM), scanning electron microscopy, Fourier transform infrared spectroscopy (FTIR) and ultraviolet–visible (UV–Vis) spectrophotometry analyses. XRD pattern showed that the zinc oxide nanoparticles exhibited hexagonal wurtzite structure. A multiple-layered structure of obelisk-shaped ZnO nanoparticles was achieved after calcinations. The average particle size of ZnO was around 20 nm as estimated by direct HRTEM observation. The size of sphere-like shaped ZnO nanoparticles was measured in the range of 20–80 nm and the size of pyramid-like shaped annealed samples was achieved in the range of 40–100 nm with less agglomeration. The energy dispersive spectroscopy spectrum showed peaks of zinc and oxygen. The sharp peaks in FTIR spectrum determined the Zn–O stretching and absorbance peak of UV–Vis spectrum showed the wide bandgap energy of 3.35 eV.

Keywords ZnO nanoparticles · Obelisk-shaped · Synthesis · Sol–gel

✉ Majid Farahmandjou
farahmandjou@iauvaramin.ac.ir

1 Department of Physics, Varamin Pishva Branch, Islamic Azad University, Varamin, Iran

Introduction

ZnO has received lot of attention as a nanostructured material because of its unique properties rendering it suitable for various applications. One-dimensional nanostructures exhibit interesting electronic and optical properties due to their low dimensionality leading to quantum confinement effects. The novel properties of nanoscale zinc oxide particles have found applications in a variety of applications such as luminescence [1–3], varistors [4, 5], solar cells [6], gas sensors [7, 8]. Many of the synthetic approaches such as sol–gel method [13], co-precipitation [12], hydrothermal method [14], microwave synthesis [9–11], and thermal evaporation method [15] have been used for the preparation of ZnO powders. Amongst the different methods of synthesis of ZnO nanostructures, the sol–gel method is attractive for its simplicity and environment-friendly conditions. Sol–gel preparation of solid catalysts has been reported by many research groups [16–20]. Using this method, the crystal grain can develop completely and the particle size is uniform. The soft-chemistry routes especially sol–gel procedures offer unique advantages such as the possibility of obtaining metastable materials, achieving superior purity and compositional homogeneity of the products at moderate temperatures with simple laboratory equipment [21]. Semiconductor photo-catalysts offer huge potential for elimination of toxic chemicals [22]. ZnO, with bang gap = 3.37 eV, has become promising in the past few years because of its distinctive optoelectronic, catalytic, and photochemical properties [24, 25]. It crystallizes in a hexagonal wurtzite structure (zincite). The ZnO-mediated photocatalytic process has been successfully used to degrade organic pollutants [23, 26]. The aim of this study was to synthesize zinc oxide of low dimension and investigate the morphological properties and surfactant

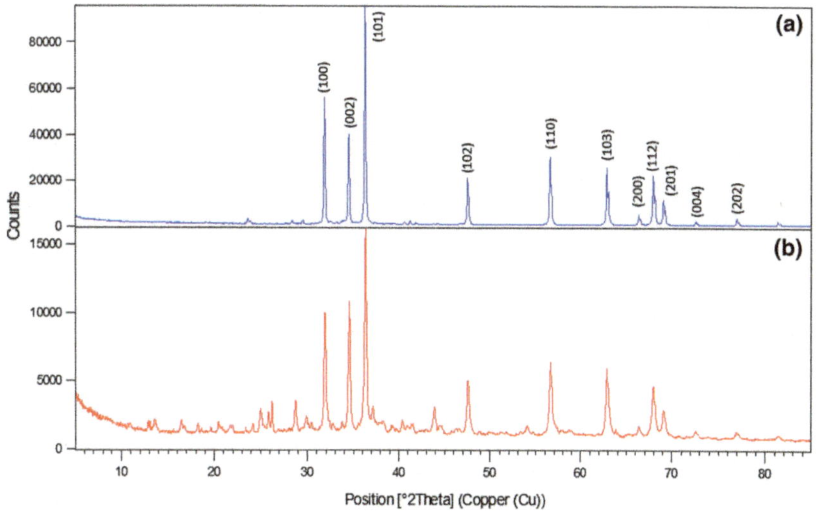

Fig. 1 XRD pattern of ZnO nanoparticles: **a** as prepared, **b** annealed sample

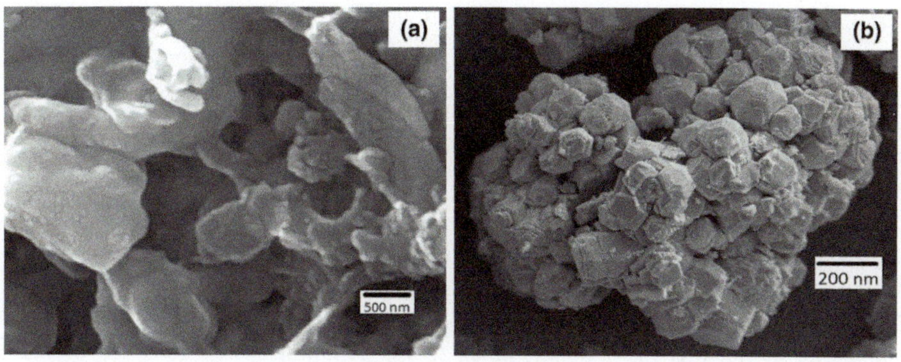

Fig. 2 SEM images of the **a** as-prepared, **b** annealed ZnO nanoparticles at 500 °C

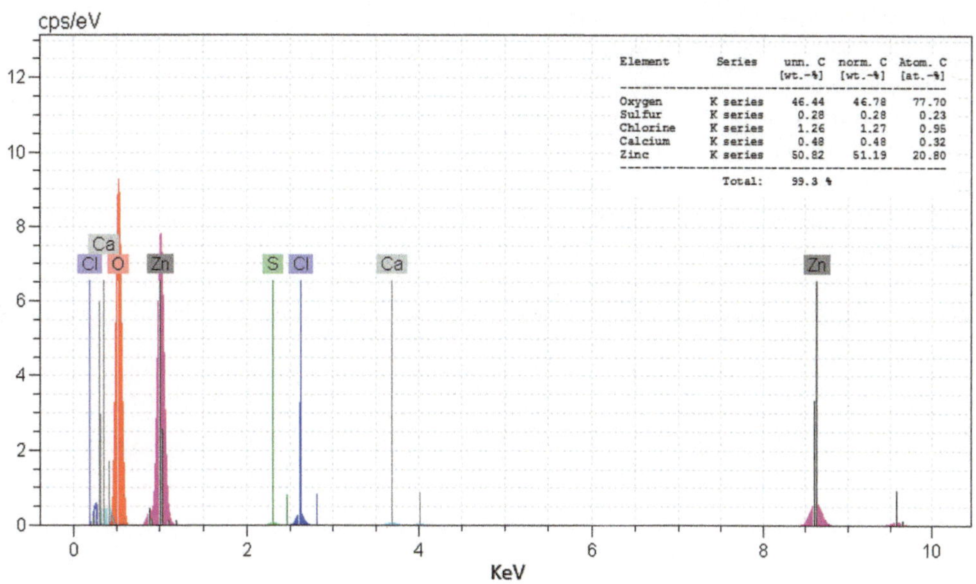

Fig. 3 EDS spectra of the as-synthesized ZnO prepared by wet synthesis

effect on the particle size. This method has features, which are of considerable interest due to its low cost, easy preparation and industrial viability. The novelty of this method is that the samples were not purified and centrifuged in the fabrication stage because of oxidation. In many synthesis methods, the researchers purified samples to have uniform particles which oxidize ZnO nanoparticles by purification and centrifugation process [2, 8, 10–15]. Synthesis of ZnO nanoparticles by sol–gel technique is reported by $Zn(NO_3)_2{\cdot}6H_2O$ precursor and calcined at 500 °C. The structural and morphological properties of ZnO have been studied by X-ray diffractometry (XRD), Fourier transform infrared spectroscopy (FTIR), ultraviolet–visible spectrophotometry (UV–Vis), spectroscopy, high-resolution transmission electron microscopy and scanning electron microscopy (SEM) analyses.

Experimental details

ZnO nanoparticles were synthesized by a new simple sol–gel approach in the following manner: In the first 30 g, $ZnSO_4{\cdot}7H_2O$ was dissolved in 150 mL deionized water and then 14 mL ethanol (99.7 %) was added to the solution. The mixed solution was stirred with a magnetic stirrer at 80 °C for 3 h to obtain the gel. The obtained gel was dried at 220 °C for 1 h and then ground into fine particles. The temperature of the dried precursor powder was increased at the rate of 1 °C/min to attain the required temperature and then the sample was allowed to stay at 500 °C for 3 h to obtain the ZnO nanoparticles.

The morphology of the as-synthesized and annealed ZnO nanoparticles was carried out. XRD was used to identify the crystalline phase and to estimate the crystalline size. The XRD patterns were recorded with 2θ in the range 4°–85° with Cu-Kα: $\lambda = 1.54$ Å model X – PRt PR$_\theta$MPD. The morphology was characterized by an SEM model 3200 Ky-EM and transmission electron microscope (TEM) 900 ZEISSEM. All the measurements were carried out at room temperature. The optical properties of absorption were measured by UV–Vis model 300-sp and FTIR model 510-WQF.

Results and discussion

XRD at 40 kV was used to identify crystalline phases and to estimate the crystalline sizes. Figure 1a shows the XRD morphology of the as-prepared ZnO nanoparticles and Fig. 1b shows the sample annealed at 500 °C for 3 h. In our case, all the diffraction peaks at angles (2θ) of 31.36°, 34.03°, 35.8°, 47.16°, 56.26°, 62.54°, 67.64°, 68.79°, 69.45°, 72.82° and 77.33° correspond to the reflection from (100), (002), (101), (102), (110), (103), (200), (112), (201), (004) and (202) crystal planes of the hexagonal wurtzite zinc oxide structure. The mean size of the ordered ZnO nanoparticles has been estimated from full width at half maximum (FWHM) and Debye–Scherrer formula according to the following equation: [27]:

$$D = \frac{0.89\lambda}{B\cos\theta} \tag{1}$$

where, 0.89 is the shape factor, λ is the X-ray wavelength, B is the line broadening at half the maximum intensity (FWHM) in radians, and θ is the Bragg angle. The mean size of as-prepared ZnO nanoparticles was in the range 20–50 nm from this Debye–Scherrer equation.

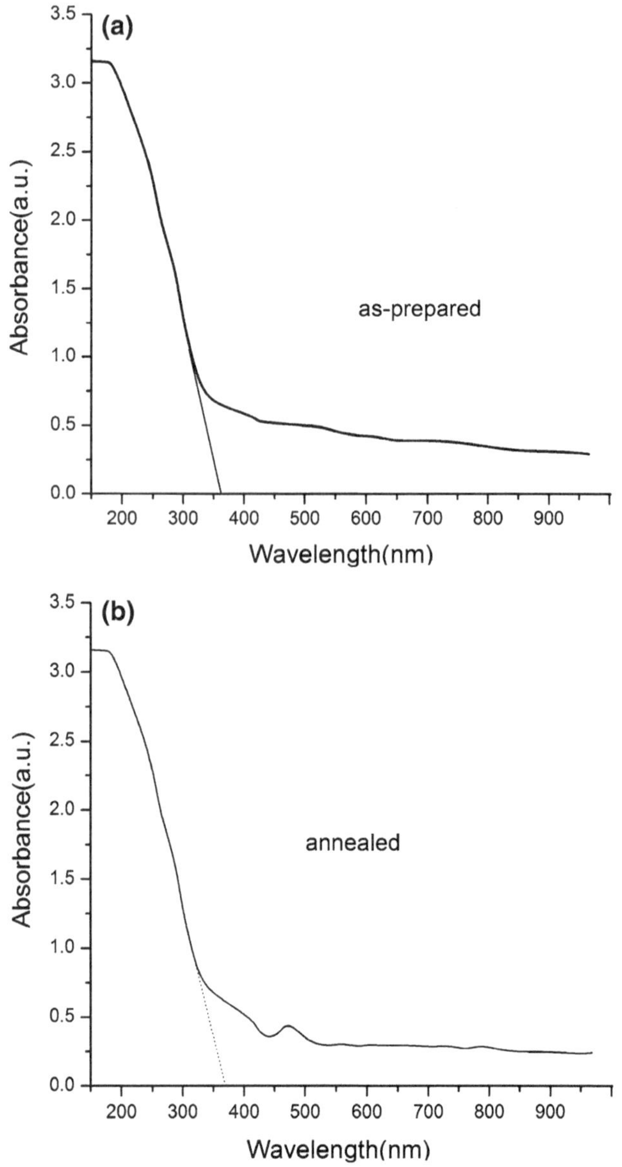

Fig. 4 UV–Vis absorption spectra of ZnO: **a** as-prepared, **b** annealed sample

SEM analysis was used for the morphological study of ZnO nanoparticles. These analyses show that the multiple layered structure of obelisk-shaped ZnO nanoparticles are formed by increasing annealing temperature [28]. With increasing temperature the morphology of the particles changes to the pyramid shape and nanopowders were less agglomerated. Figure 2a shows the SEM image of the as-prepared ZnO nanoparticles prepared by sol–gel method. In this figure, the particles were prepared with formation of clusters. Figure 2b shows the SEM image of the ZnO nanoparticles annealed at 500 °C for 3 h. It can be seen that the multiple layered of obelisk-shaped ZnO nanoparticles were less agglomerated.

Energy dispersive spectroscopy analysis of the ZnO samples is shown in Fig. 3. It confirms the existence of Zn and O with weight percent. EDS was used to analyze the chemical composition of a material under SEM. EDS measurement shows peaks of zinc and oxygen and indicates a few of impurities in prepared samples.

UV–visible absorption spectral study was done to understanding of the optical band gap of the material. Absorption in the near-ultraviolet region arises from electronic transitions associated within the sample. UV–Vis absorption spectra of as-prepared and annealed ZnO nanoparticles are shown in Fig. 4. As you can see from the pictures, the strong absorption band at a low wavelength near 370 nm corresponds to the bandgap energy of 3.35 eV for the as-prepared sample, and the strong absorption band at the low wavelength near 379 nm corresponds to 3.27 eV for the annealed nanoparticles. The size of particles changes with increasing temperature and the bandgap energy decreases with increasing annealing temperature.

Figure 5 shows the FTIR spectra of the ZnO powders, which were acquired in the range of 4000–400 cm^{-1}. All of the spectra exhibit a strong absorption peak at 3508 cm^{-1} for stretching vibration of non-chemical bond association OH groups and at 1637 cm^{-1} for H–O–H bending vibrations. The peaks at 2390 cm^{-1} are attributed to the presence of carbon dioxide. The absorption peaks around 1392 cm^{-1} are assigned to the bending vibration of C–H stretching. The peaks at 514–442 cm^{-1} are for Zn–O. The above results are in accordance with the XRD results.

The TEM analysis was carried out to confirm the actual size of the particles, their growth pattern and the distribution of the crystallites. Figure 6 shows the TEM image of spherical ZnO nanoparticles prepared by chemical sol–gel route. Figure 6a shows the as-synthesized samples with an average diameter of 20 nm. Figure 6b indicates the annealed samples with an average diameter of 80 nm. It can be seen that the size of nanoparticles increases with increasing annealing temperature. Because the increasing temperature remove the surfactants around the particles and the particles are closed to each other because of interactions.

Conclusion

A multiple-layered structure of obelisk-shaped ZnO nanoparticles was successfully synthesized by a new and simple sol–gel method using an ethanol solution of $Zn(NO_3)_2 \cdot 6H_2O$. The XRD results show that ZnO nanoparticles are hexagonal wurtzite in the size range 20–50 nm. From SEM images, it is clear that with increasing temperature, the morphology of the particles changes to the obelisk shape and the nanopowders were less agglomerated. The TEM image exhibits the as-synthesized ZnO nanoparticles prepared the by sol–gel route

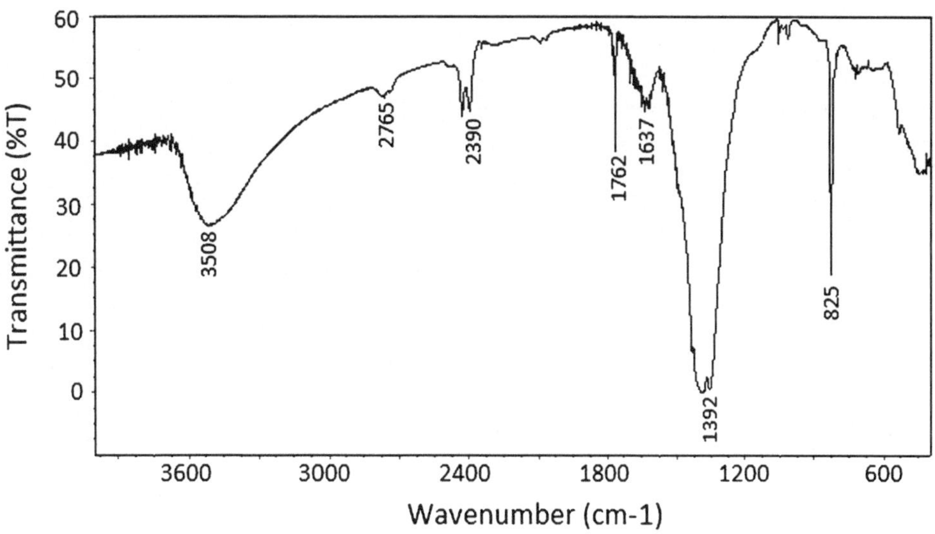

Fig. 5 The Fourier transform infrared spectroscopy pattern of the zinc oxide nanoparticles

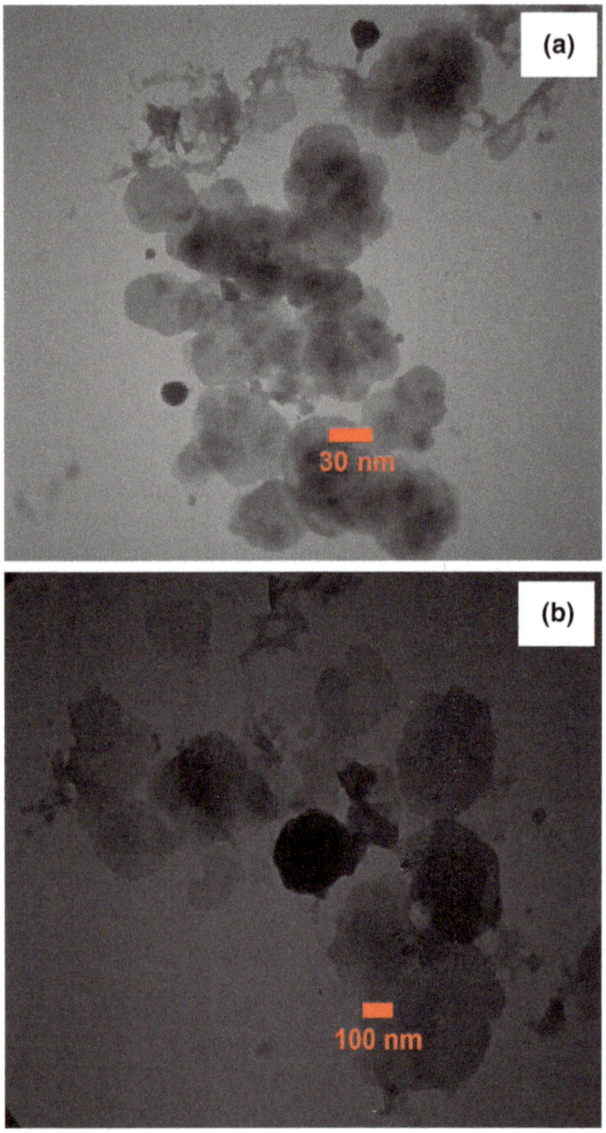

Fig. 6 TEM images of the ZnO nanoparticles: **a** as-prepared, **b** annealed samples

with average size of 20 nm. The EDS measurements showed only peaks of zinc and oxygen and indicated the absence of impurities in the prepared ZnO samples. From the FTIR data, the presence of Zn–O stretching mode of ZnO is shown. The zinc oxide nanoparticles show a strong UV–Vis absorption below 400 nm with a well-defined absorption peak at 370 nm; the direct bandgap is found to be 3.35 eV.

Acknowledgments The authors are thankful for the financial support of Varamin Pishva Branch at Islamic Azad University for analysis and the discussions on the results.

References

1. van Dijken, A., Meulenkamp, E.A., Vanmaekelbergh, D., Meijerink, A.: The luminescence of nanocrystalline ZnO particles: the mechanism of the ultraviolet and visible emission. J. Lumin. **87–89**, 454–456 (2000)
2. Guo, L., Yang, S.H., Yang, C.L., Yu, P., Wang, J.N., Ge, W.K., Wong, G.K.L.: Synthesis and characterization of poly(vinylpyrrolidone)-modified zinc oxide nanoparticles. Chem. Mater. **12**, 2268–2274 (2000)
3. Feng, Y., Zhou, Y.X., Liu, Y.Q., Zhang, G.B., Zhang, X.Y.: Photoluminescence spectra of nano-structured ZnO thin films. J. Lumin. **119–120**, 233–236 (2006)
4. Hingorani, S., Pillai, V., Kumar, P., Multani, M.S., Shah, D.O.: Microemulsion mediated synthesis of zinc-oxide nanoparticles for varistor studies. Mater. Res. Bull. **28**, 1303–1310 (1993)
5. Pillai, S.C., Kelly, J.M., McCormack, D.E., O'Brien, P., Ramesh, R.: The effect of processing conditions on varistors prepared from nanocrystalline ZnO. J. Mater. Chem. **13**, 2586–2590 (2003)
6. Beek, W.J.E., Wienk, M.M., Janssen, R.A.J.: Efficient hybrid solar cells from zinc oxide nanoparticles and a conjugated polymer. Adv. Mater. **16**, 1009–1013 (2004)
7. Epifani, M., Diaz, R., Arbiol, J., Comini, E., Sergent, N., Pagnier, T., Siciliano, P., Taglia, G., Morante, J.R.: Nanocrystalline metal oxides from the injection of metal oxide sols in coordinating solutions: synthesis, characterization, thermal stabilization, device processing, and gas-sensing properties. Adv. Funct. Mater. **16**, 1488–1498 (2006)
8. Tang, H.X., Yan, M., Ma, X.F., Zhang, H., Wang, M., Yang, D.R.: Gas sensing behavior of polyvinylpyrrolidone-modified ZnO nanoparticles for trimethylamine. Sens. Actuators B **113**, 324–328 (2006)
9. Trandafilović, L.V., Božanić, D.K., Dimitrijević-Branković, S., Luyt, A.S.: Fabrication and antibacterial properties of ZnO–alginate nanocomposites. Carbohydr. Polym. Djoković V **88**, 263–269 (2012)
10. Martins, N.C.T., Freire, C.R.S., Neto, C.P., Silvestre, A.J.D., Causio, J., Baldi, G.: Antibacterial paper based on composite coatings of nanofibrillated cellulose and ZnO. Colloids Surf. A **417**, 111–119 (2013)
11. Jalal, R., Goharshadi, E.K., Abareshi, M., Moosavi, M., Yousefi, A., Nancarrow, P.: ZnO nanofluids: green synthesis, characterization, and antibacterial activity. Mater. Chem. Phys. **121**, 198–201 (2010)
12. Nair, M.G., Nirmala, M., Rekha, K., Anukaliani, A.: Structural, optical, photo catalytic and antibacterial activity of ZnO and Co doped ZnO nanoparticles. Mater. Lett. **65**, 1797–1800 (2011)
13. Talebian, N., Nilforoushan, M.R., Zargar, E.B.: Enhanced antibacterial performance of hybrid semiconductor nanomaterials: ZnO/SnO_2 nanocomposite thin films. Appl. Surf. Sci. **258**, 547–555 (2011)
14. Phan, D.T., Chung, G.S.: Effects of defects in Ga-doped ZnO nanorods formed by a hydrothermal method on CO sensing properties. Sens. Actuators B **187**, 191–197 (2013)
15. Li, Q., Chen, Y., Luo, L., Wang, L., Yu, Y., Zhai, L.: Photoluminescence and wetting behavior of ZnO nanoparticles/nanorods array synthesized by thermal evaporation. J Alloys Compd. **560**, 156–160 (2013)
16. Azarang, M., Shuhaimi, A., Yousefi, R., Jahromi, S.P.: One-pot sol–gel synthesis of reduced graphene oxide uniformly decorated zinc oxide nanoparticles in starch environment for highly efficient photodegradation of methylene blue. RSC Adv. **5**, 21888–21896 (2015)

17. Yousefi, R., Jamali-Sheini, F., Cheraghizade, M., Khosravi-Gandomani, S., Sáaedi, A., Huang, N.M., Basirun, W.J., Azarang, M.: Enhanced visible-light photocatalytic activity of strontium-doped zinc oxide nanoparticles. Mater. Sci. Semicond. Process. **32**, 152–159 (2015)
18. Zak, Khorsand: A., Majid, W.H.A., Darroudi, M., Yousefi, R.: Synthesis and characterization of ZnO nanoparticles prepared in gelatin media. Mater. Lett. **65**, 70–73 (2011)
19. Ghandomani, S., Yousefi, R., Jamli-Sheini, F., Ming Huang, N.: Optical and electrical properties of p-type Ag-doped ZnO nanowires. Ceram. Int. **40**, 7957–7963 (2014)
20. Azarang, M., Shuhaimi, A., Yousefi, R., Golsheikh, A.M., Sookhakian, M.: Synthesis and characterization of ZnO NPs/graphene oxide nanocomposite prepared in gelatin medium as highly efficient photo—degradation of MB. Ceram. Int. **40**(2014), 10217–10221 (2014)
21. Yousefi, R., Khorsand-Zak, A., Jamali-Sheini, F.: Growth, X-ray peak broadening studies, and optical properties of Mg-doped ZnO nanoparticles. Mater. Sci. Semicond. Process. **16**, 771–776 (2013)
22. Azarang, M., Shuhaimi, A., Yousefi, R., Sookhakian, M.: Effects of graphene oxide concentration on optical properties of ZnO/RGO nanocomposites and their application to photocurrent generation. J. Appl. Phys. **116**, 084307–084312 (2014)
23. Height, M.J., Pratsinis, S.E., Mekasuwandumrong, O., Praserthdam, P.: Ag-ZnO catalysts for UV-photodegradation of methylene blue. Appl. Catal. B Environ. **63**, 305–312 (2006)
24. Chen, C.C., Lu, C.S.: Mechanistic studies of the photocatalytic degradation of methyl green: an investigation of products of the decomposition processes. Environ. Sci. Technol. **41**, 4389–4396 (2007)
25. Kuo, T.J., Lin, C.N., Kuo, C.L., Hung, M.H.: Growth of ultralong ZnO nanowires on silicon substrates by vapor transport and their use as recyclable photocatalysts. Chem. Mater. **19**, 5143–5147 (2007)
26. Akyol, A.H., Yatmaz, C., Bayramoblu, M.: Photocatalytic decolorization of Remazol Red R in aqueous ZnO suspensions. Appl. Catal. B Environ. **54**, 19–24 (2004)
27. Monshi, A., Foroughi, M.R., Monshi, M.R.: Modified Scherrer equation to estimate more accurately nano-crystallite size using XRD. World J. Nano Scie. Eng. **2**, 154–160 (2012)
28. Baruah, S., Dutta, J.: Hydrothermal growth of ZnO nanostructures. Sci. Technol. Adv. Mater. **10**, 013001–013019 (2009)

5

Electrical properties and *I–V* characteristics of 5,14-dihydro- 5,7,12,14-tetraazapentacene doped Schottky barrier diode

Hassan Ghalami Bavil Olyaee[1] · Peter J. S. Foot[2] · Vincent Montgomery[2]

Abstract The current–voltage (*I–V*) characteristics of Aluminium/5,14-dihydro-5,7,12,14-tetraazapentacenes (L5H2 or DHTAPs) doped surface-type structures were investigated in air at ambient temperature and moisture. The conventional forward bias *I–V* method, Semi-logarithm, Cheung functions and modified Norde's function were used to extract the diode parameters including ideality factor, series resistance and barrier height. The parameter values obtained from these four different methods were found in good agreement.

Keywords Schottky diodes · 5,14-dihydro-5,7,12,14-tetraazapentacenes (L5H2 or DHTAPs) doped · Series resistance · Ideality factor · Barrier height

Introduction

The large range of electronic and optical properties of organic conducting polymers makes them very good candidates for use in modern electronic devices. The fabrication of electronic devices based on conducting polymers, especially Schottky diodes, can be simple and of low cost compared with devices based on inorganic semiconductors. The photoelectrical properties of metal–semiconductor (MS)—have been studied as photovoltaic cells, photodiodes and Schottky diodes by a number of authors [1–22]. Schottky diodes are playing an important role in modern electronic devices. Although Schottky diodes have the simplest structure, they are the base of a great number of compound semiconductor electronic devices [23, 24]. Schottky diodes are less expensive to fabricate than p–n junction diodes and the device performance is better due to faster response, low resistance, very small transient reverse current during switching and the reverse saturation current of Schottky barrier diodes is higher than that of the conventional p–n junction diodes [25–27]. The Schottky diode requires lower forward bias voltage to deliver a given current compared to a p–n junction diodes. The operation of Schottky barrier diodes relies upon the device characteristics of the MS junction. Therefore, the understanding of the electronic properties of the surface between metal and the organic layer is important for device applications. Series resistance (R_s), formation of Schottky barrier, insulating layer between metal and semiconductor and also the interface states can affect the properties of the junction.

Recently, a number of authors [28] investigated Schottky diodes fabricated using non-polymeric organic and oligomer compounds have reported that the non-polymeric organic materials are suitable for electronic devices, especially Schottky diodes, mainly due to their stability [28–30]. The compound, 5,14-dihydro-5,7,12,14-tetraazapentacenes (L5H2) used in this work is a doped organic semiconductor in the form of dark blue powder. The heterojunctions of L5H2 have been investigated in Ref. [31]. These studies showed the potential of L5H2 for the application in electronic devices.

✉ Hassan Ghalami Bavil Olyaee
olyaee@gmail.com; olyaee@azad.ac.ir

Peter J. S. Foot
P.Foot@kingston.ac.uk

Vincent Montgomery
V.Montgomery@kingston.ac.uk

1 Islamic Azad University-South Tehran Branch, Iranshaher Shomali, Tehran, Iran

2 Kingston University London, Penrhyn Road, Kingston upon Thames, Surrey, UK

In this work, the doped oligomer compound L5H2 is chosen for the fabrication of the Schottky diode due to its conjugated structure richness in 20-π-electron [31]. The plan of this investigation is to extract the parameters that control the device performance such as ideality factor, barrier height and series resistance from *I–V* device characteristics by using different methods at ambient temperature and moisture.

Sample preparation

L5H2 with molecular formula $C_{18}H_{12}N_4$ was used for the fabrication of Schottky diodes. The molecular structure of the L5H2 is shown in Fig. 1. The ITO glass substrate was cleaned for 1 min using acetone and after drying. The thin films of L5H2 were deposited by evaporation from a resistive heating element in a oil-pumped vacuum system with the chamber pressure at 5.2×10^{-5} torr on to the ITO glass substrates. The thin film layers were removed from the vacuum system and then doped by exposure to acetic acid vapour for 15 min. Finally, thermal vacuum depositions of aluminium metal contacts were produced using a shadow mask in the oil-pumped vacuum system. During thermal deposition the chamber pressure was 5.8×10^{-5} torr. The thickness of the electrode was approximately 100 nm and the thickness of L5H2 film was approximately 300 nm. A diagram of a Al/L5H2-doped surface-type Schottky diode device is shown in Fig. 2. The measurements were taken using current–voltage source keithley-230 Programmable Voltage and using keithley-485 Autoranging Picoammeter.

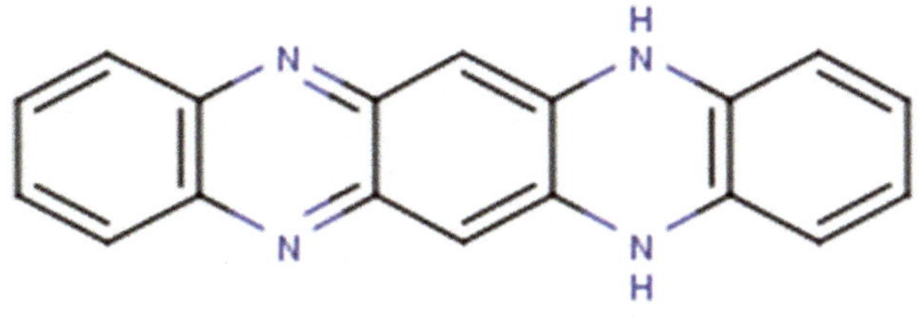

Fig. 1 Structure of L5H2

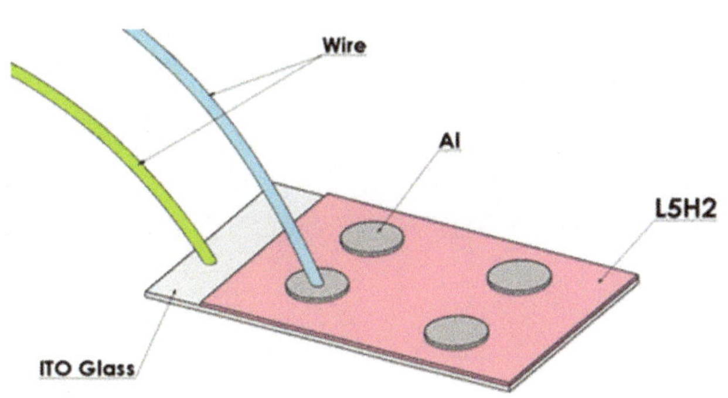

Fig. 2 The Schottky diode

Results and discussion

The forward- and reverse-biased *I–V* and semi-log *I–V* characteristics of the Al/L5H2-doped surface Schottky barrier diode at ambient temperature and moisture are shown in Figs. 3 and 4. In forward bias, the Al contact was positively biased and the ITO glass was negatively biased. The *I–V* characteristics of the Al/L5H2-doped surface Schottky barrier diode showed rectification behaviour which is limited by the magnitude of the energy barrier at the junction interface. The rectification ratio was found to be 1.92 at ±2.8 V. The rectification ratio is the ratio of forward current to reverse current at certain voltage. It is assumed that the current in the Schottky contact is due to thermionic emission. According to the thermionic emission

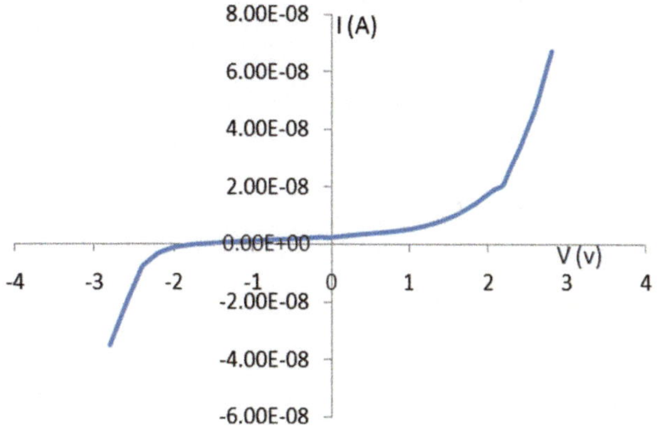

Fig. 3 The forward bias and the inverse bias *I* vs. *V* plot

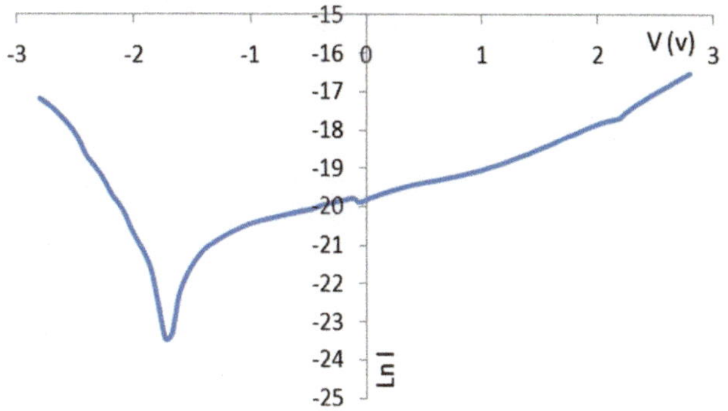

Fig. 4 Semi-log *I–V* curve of Al/L5H2-doped Schottky diode

theory, the current in Schottky barrier diodes can be expressed as [32].

$$I = I_0\left(e^{\frac{qV}{nkT}} - 1\right), \tag{1}$$

where I_o is the reverse saturation current and can be expressed as

$$I_0 = AA^*T^2e^{-\frac{q\varphi_b}{kT}}, \tag{2}$$

where V is the applied voltage, A is the effective diode area (7.85×10^{-7} m^2), A^* is the effective Richardson constant (1.02×10^6 A/m^2 K^2, taken from Richardson's law), φ_b is the effective barrier height at zero bias, k is the Boltzmann constant (1.38×10^{-23} J/K or 8.62×10^{-5} eV/K). T is the temperature in Kelvin and n is the ideality factor. The value of ideality factor n can be calculated from the slope of the linear region of the forward bias Ln I–V plot using the following relation:

$$n = \frac{q}{kT}\left(\frac{\mathrm{dV}}{\mathrm{d}(\ln I)}\right) \tag{3}$$

By using the above equation and slope of line in Ln I–V plot in Fig. 5, the value of ideality factor was calculated to be 42.9. The value of reverse saturation current obtained from forward bias semi-Ln I–V curve was 2.13×10^{-9} A.

For an ideal diode the value is unique. However, generally n has a value higher than unity. High values of ideality factor can be ascribed to the existence of local oxide layer on electrodes and non-homogenous barrier [23, 33] and [34]. Also, the high values of ideality factor may be attributed to other effects, for example organic layer effect, non-homogeneous thickness of organic film, etc. The value of zero bias barrier height was found to be 0.78 eV, which was calculated using Eq. (2). At low voltage, forward bias is linear in semi-log scale, but at higher voltages the device characteristics deviate from linear behaviour. It is well known that the curvature of the forward biased I–V plots at higher voltages is due to the R_S, apart from the effect of interface traps distribution, which are in equilibrium with the semiconductor materials [35]. If the affect of series resistance is more, then the nonlinear region of forward bias I–V curve will be large [36]. From the non linearity of semi-log I–V curve of Al/L5H2-doped Schottky diode at higher voltage, the value of series resistance was found to be 1.02×10^7 Ω. The graph of $\mathrm{d}V/\mathrm{d}(\ln I)$ vs. I is shown in Fig. 6. To find out the electronic parameter of the Schottky diode, the third method, which is known as Cheung and Cheung method, was used [37]. According to this method the forward bias I–V characteristics due to the thermionic emission with the effect of series resistance can be written as

$$I = I_0\mathrm{e}^{\frac{q(V-IR_S)}{\mathrm{nkT}}}, \tag{4}$$

where $\mathrm{IR_S}$ is the voltage drop across the series resistance. According to Cheung and Cheung the values of series resistance, ideality factor and barrier can be determined from following the functions and by using Eq. (4):

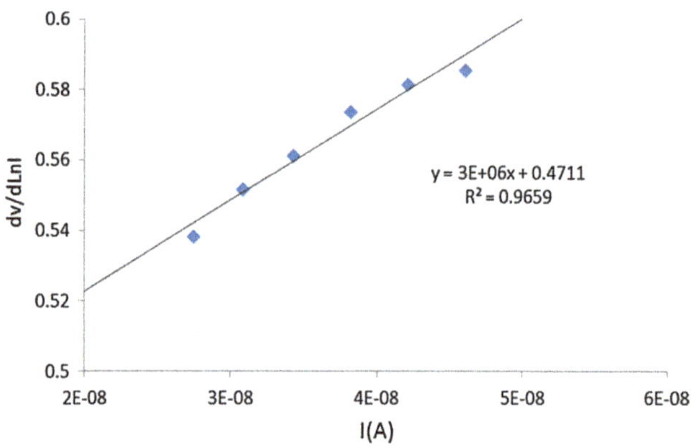

Fig. 6 The forward bias dV/(d(LnI) vs. I) plot of Al/L5H2-doped Schottky diode

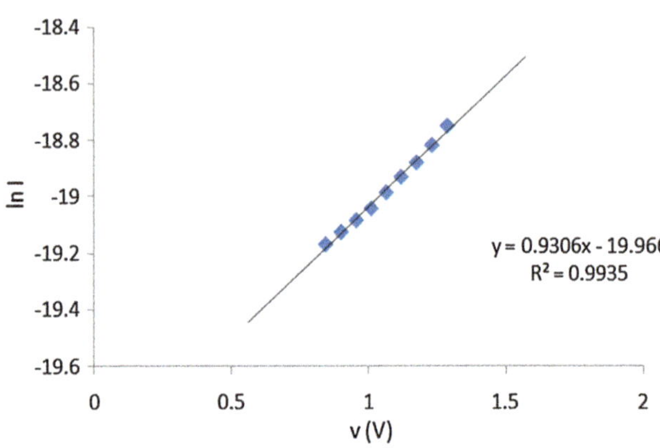

Fig. 5 The forward bias Ln(I) vs. V plot of Al/L5H2-doped Schottky diode

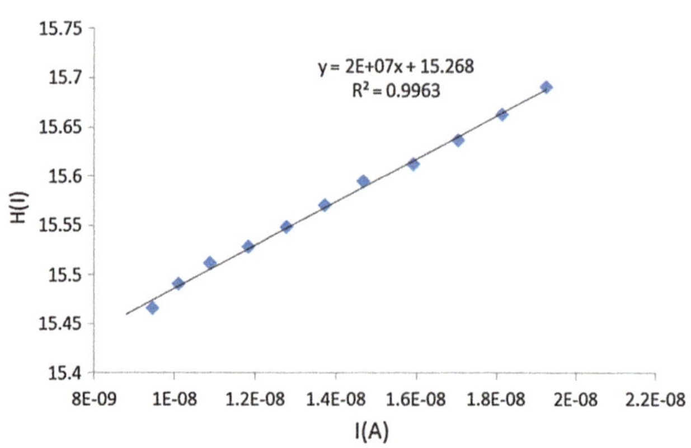

Fig. 7 The forward bias $H(I)$ vs. I plot of Al/L5H2-doped Schottky diode

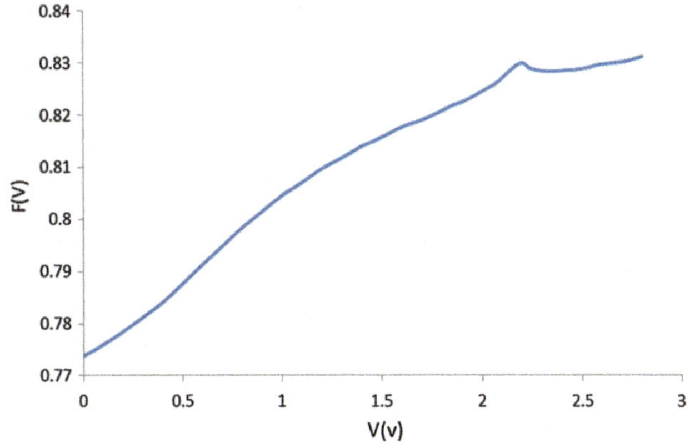

Fig. 8 The forward bias *F(V)* vs. *V* plot of Al/L5H2-doped Schottky diode

$$\frac{dV}{d(\ln I)} = n\left(\frac{kT}{q}\right) + IR_s \tag{5}$$

The values of ideality factor and series resistance are obtained from the slope and *y*-axis intercept of the graph $dV/d(\ln I)$ vs. *I*.

And

$$H(I) = V - n\left(\frac{kT}{q}\right)\ln\left(\frac{I}{AA^*T^2}\right) \tag{6}$$

$$H(I) = IR_s + n\varphi_b \tag{7}$$

Also, the plot of *H(I)* vs. *I* according to Eq. (7) gives a straight line with the *y*-axis intercept equal to $n\varphi_b$. The slope of this plot also gives the value of series resistance by which the consistency of Cheung's approach can be checked. The graph of *H(I)* vs. *I* is shown in Fig. 7. The determined values of series resistance from the plots of $dV/d(\ln I)$ vs. *I* and *H(I)* vs. *I* were found to be 3×10^6 and $2 \times 10^7\ \Omega$, respectively. The average value of series resistance for the Schottky diode was found to be $1.15 \times 10^7\ \Omega$. As it can be seen, the values of series resistance from the Cheung plots $dV/d(\ln I)$ vs. *I* and *H(I)* vs. *I* are in good agreement with each other and also with the value obtained from semi-log *I–V* curve.

Beside the conventional *I–V* method and Cheung functions, there is another method to determine the value of series resistance, which was proposed by Norde and known as Norde's Method in 1978. The following equation has been used to express the modified Norde's method [38, 39]:

$$F(V) = \frac{V}{\gamma} - \frac{1}{\beta}\ln\left(\frac{I}{AA^*T2}\right) \tag{8}$$

$$\beta = \frac{q}{kT}, \tag{9}$$

where γ is a constant number ($\gamma \geq n$), n is the ideality factor and q is the electronic charge. It is shown that by plotting the function *F(V)* a consistent value of the barrier height can be obtained even if there is a series resistance which would fetter the evaluation of the standard method. In this case, γ value is taken as 20.The plot of *F(V)* vs. *V* has a maximum, if less than 20 are chosen, while the function should be a minimum only. The *I(V)* is value of current taken from the *I–VI–V* curve. If the minimum of the *F(V)* vs. *V* plot is found then the barrier height can be calculated using the following equation:

$$\varphi_b = F(V_{min}) + \frac{V_{min}}{\gamma} - \frac{1}{\beta} \tag{10}$$

$F(V_{min})$ is the minimum value of *F(V)*, and $V_{min} = 2.61 \times 10^{-15}\ v$ is the corresponding voltage and it is close to zero. The Norde plot for Al/L5H2-doped surface-type Schottky diode is shown in Fig. 8. The value of the series resistance has been calculated from Norde's function for Al/L5H2-doped Schottky diode junction using the following relation:

$$R_s = \frac{\gamma - n}{\beta I_{min}}, \tag{11}$$

where I_{min} is current in the minimum point of *F(V)*. Using Norde's method, the values of series resistance and barrier height were obtained as $1.2 \times 10^7\ \Omega$ and 0.75 eV, respectively. We can see from Table 1 the values of Schottky diode parameters obtained from the different methods. The barrier height and Series resistance show good agreement but not the ideality factors.

Table 1 Experimentally obtained values of barrier height, series resistance and ideality factor

	Barrier height (eV)	Series resistance (Ω)	Ideality factor
Semi-log *I–V*	0.78	1.02×10^7	42.9
dV/d(ln *I*) vs. *I* plot	–	3×10^6	18.85
H(I) vs. *I* plot	0.81	2×10^7	–
Norde's function	0.75	1.2×10^7	20

Conclusion

In conclusion, we have fabricated surface-type Al/L5H2-doped Schottky barrier diode. The electronic parameters such as ideality factor, series resistance and barrier height of the diode were determined by four methods. The majority of the parameters obtained using different methods are in good accord.

According to Table 1, barrier height is acquired minimum by Norde's function method (0.75 eV) and maximum by *H(I)* vs. *I* plot method (0.81 eV), and series resistance is achieved minimum by d*V*/d(ln*I*) vs. I plot method (3 MΩ) and maximum by *H(I)* vs. *I* plot method (20 MΩ). Also ideality factor is attained minimum by d*V*/d(ln*I*) vs. *I* plot method (18.85) and maximum by semi-log *I–V* method (42.9). Then minimum values for series resistance and ideality factor is given by d*V*/d(ln*I*) vs. *I* plot method and maximum values for barrier height and series resistance by *H*(*I*) vs. *I* plot method.

Acknowledgments This research project has been supported by Islamic Azad University, South Tehran Branch, contract number B/16/601.

References

1. El-Nahass, M.M., Zeyada, H.M., Abd-El-Rahman, K.F., Darwish, A.A.A.: Energy Mater. Sol. Cells **91**, 1120 (2007)
2. Güllü, Ö., Türüt, A.: Sol. Energy Mater. Sol. Cells **92**, 1205 (2008)
3. Akkılıç, K., Yakuphanoğlu, F.: Microelectron. Eng. **85**, 1826 (2008)
4. Yakuphanoğlu, F.: Sens. Actuators A **141**, 383 (2008)
5. Ghosh, A.K., Morel, D.L., Feng, T., Shaw, R.F., Rowe, C.A.: J. Appl. Phys. **45**, 230 (1974)
6. Gupta, R.K., Singh, R.A.: J. Polym. Res. **11**, 269 (2004)
7. Forrest, S.R., Schmidt, P.H.: J. Appl. Phys. **59**, 513 (1986)
8. Forrest, S.R., Kaplan, M.L., Schmidt, P.H.: J. Appl. Phys. **56**, 543 (1984)
9. Kampen, T., Schuller, A., Zahn, D.R.T., Biel, B., Ortega, J., Perez, R., Flores, F.: Appl. Surf. Sci. **234**, 341 (2004)
10. Güllü, Ö., Aydoğan, Ş., Türüt, A.: Microelectron. Eng. **85**, 1647 (2008)
11. Çakar, M., Yıldırım, N., Karataş, Ş., Temirci, C., Türüt, A.: J. Appl. Phys. **100**, 1 (2006)
12. Güllü, Ö., Barış, Ö., Biber, M., Türüt, A.: Appl. Surf. Sci. **254**, 3039 (2008)
13. Kılıçoğlu, T., Aydın, M.E., Ocak, Y.S.: Phys. B **388**, 244 (2007)
14. Kılıçoğlu, T.: Thin Solid Films **516**, 967 (2008)
15. Kılıçoğlu, T., Aydın, M.E., Topal, G., Ebeoğlu, M.A., Sayğılı, H.: Synth. Met. **157**, 540 (2007)
16. Akkılıç, K., Ocak, Y.S., İlhan, S., Kılıçoğlu, T.: Synth. Met. **158**, 969 (2008)
17. Çakar, M., Türüt, A., Onganer, Y.: J. Mater. Sci. Mater. Electron. **15**, 47 (2004)
18. Akkılıç, K., Uzun, İ., Kılıçoğlu, T.: Synth. Met. **157**, 297 (2007)
19. Temirci, C., Çakar, M.: Phys. B **38**, 454 (2004)
20. Roberts, A.R.V., Evans, D.A.: Appl. Phys. Lett. **86**, 072105 (2005)
21. Brabec, C.J., Dyakonov, V., Parisi, J., Sariciftci, N.S.: Organic Photovoltaics. Springer, New York (2003)
22. Yuechao, Wu, Yin, Zongyou, Xiao, Jinchong, Liu, Yi, Wei, Fengxia, Tan, K.J., Kloc, C., Huang, L., Yan, Q., Hu, F., Zhang, H., Zhang, Q.: Acs Appl Mater Interfaces **4**, 1883 (2012)
23. Rhoderick, E.H., Williams, R.H.: Metal-semiconductor contacts. Clarendon Press, Oxford (1988)
24. Van Meirhaeghe, R.L., Laflere, W.H., Cardon, F.: J. Appl. Phys. **76**, 403 (1994)
25. Krishnan, S.: Department of Electrical Engineering, College of Engineering, M.S. Thesis, University of South Florida (2004)
26. Aydogan, S., Saglam, M., Turut, A.: Polymer **46**, 10982 (2005)
27. Chung, G.-S., Ahn, J.-K.: Microelectron. Eng. **85**, 1772 (2008)
28. Janardhanam, V., Yun, H.J., Jyothi, I., Lee, J., Hong, H.: Alloy. Compd. **637**, 84 (2015)
29. Forrest, S.R., Kaplan, M.L., Schmidt, P.H., Feldmann, W.L., Yanowski, E.: Appl. Phys. Lett. **41**, 90 (1982)
30. Cakar, M., Temirci, C., Turut, A.: Chem. Phys. Chem. **8**, 701 (2002)
31. Seillan, Claire, Brisset, Hugues, Siri, Olivier: Org. Lett. **10**, 4013 (2008)
32. Rhoderick, E.H.: Metal-semiconductors contacts. Oxford University Press, Oxford (1978)
33. Karatas, S., Turut, A.: Vacuum **74**, 45 (2004)
34. Cetinkara, H.A., Turut, A., Zengin, D.M., Erel, S.: Appl. Surf. Sci. **207**, 190 (2003)
35. Aydogan, S., Saglam, M., Turut, A.: Microelectron. Eng. **85**, 278 (2008)
36. Turut, A., Saglam, M., Efeoglu, H., Yalcin, N., Yildirim, M., Abay, B.: Phys. B **205**, 41 (1995)
37. Cheung, S.K., Cheung, N.W.: Appl. Phys. Lett. **49**, 85 (1986)
38. Norde, H.: J. Appl. Phys. **50**, 5052 (1979)
39. Karatas, S., Altindal, S., Turut, A., Cakar, M.: Phys. B **392**, 43 (2007)

6

Design and fabrication of highly hydrophobic Mn nano-sculptured thin films and evaluation of surface properties on hydrophobicity

Somaye Hosseini[1] · Hadi Savaloni[2] · Mehran Gholipour-Shahraki[3]

Abstract The wettability of solid surfaces is important from the aspects of both science and technology. The Mn nano-sculptured thin films were designed and fabricated by oblique angle deposition of Mn on glass substrates at room temperature. The obtained structure was characterized by field emission scanning electron microscopy and atomic force microscopy. The wettability of thin films samples was investigated by water contact angle (WCA). The 4-pointed helical star-shaped structure exhibits hydrophobicity with static WCAs of more than 133° for a 10-mg distilled water droplet. This sample also shows the rose petal effect with the additional property of high adhesion. The Mn nano-sculptured thin films also act as a sticky surface which is confirmed by hysteresis of the contact angle obtained from advancing and receding contact angles measurements. Physicochemical property of liquid phase could effectively change the contact angle, and polar solvents in contact with hydrophobic solid surfaces do not necessarily show high contact angle value.

Keywords Sculptured thin films · Hydrophobicity · Surface free energy · Cassie–Baxter state · Wenzel state · Petal effect

✉ Hadi Savaloni
savaloni@khayam.ut.ac.ir

1 Plasma Physics Research Center, Science and Research Branch, Islamic Azad University, Tehran, Iran

2 Department of Physics, University of Tehran, North-Kargar Street, Tehran, Iran

3 Physics Department, Faculty of Science, Arak University, Arak 38156-8-8349, Iran

Introduction

There is a growing interest in hydrophobic surfaces due to their extensive applications in science and technology [1–3]. Roughness and symmetry [4–7] are the most effective factors to influence the hydrophobicity. Measurement of contact angle is a common method to describe the wettability of solid surfaces and their hydrophobicity. Surfaces with contact angle between 90° and 150° and more than 150° are considered as hydrophobic and super hydrophobic, respectively [8]. Natural and most of the synthetic hydrophobic surfaces consist of micro- and nano-structures simultaneously, but nanoscale structure considered less attractive by scientific research [7, 9].

Glancing or oblique angle deposition cares about the new generation of physical techniques for preparation of thin films with predesign morphology and desirable roughness. The rotation of the substrate about two axes during the deposition procedure could prepare condition for production of three-dimensional structures with anisotropic columns in 1–100 nm scale [10]. The deposition angle, rate of substrate rotation and solid material are the effective factors which control morphology, roughness, symmetry and porosity of structures.

In this study, Mn nano-sculptured thin films with different morphologies and structural properties are designed and fabricated by oblique angle deposition technique. The effects of structural properties (i.e., deposition angle, morphology, number and diameter of arms, roughness and void fraction) on thin films hydrophobicity are investigated.

Experimental details

Mn (99.99% purity) helical star-shaped nano-sculptured thin films with different sizes and morphologies were deposited on glass substrate (microscope slide; 3 × 2 cm;

Thermo Scientific Menzel–Glaser, Soda lime glass) by electron beam evaporation from a graphite crucible of 6 mm diameter at room temperature. An Edwards (EdwardsE19 A3) coating plant with a base pressure of 2×10^{-7} mbar and a deposition rate of 1 $\text{Å}s^{-1}$ was used. In order to produce a uniform deposition on the substrates, distance between the evaporation source and the substrate was fixed at 30 cm. In this distance, evaporation source behave like a point source and due to the large mean free path ($\sim$ 103 to 104 cm) vapor has straight trajectories [11]. Reproducibility of the deposition results was confirmed by repeating the deposition a few times. A quartz crystal monitor (Sigma Instruments, SQM-160, USA) positioned close to the substrate for controlling the deposition rate. The mechanical movement of the substrate holder [rotation about its surface normal (φ), the facility for dividing each revolution into different sectors and the speed of revolution] was controlled by the stepper motor which controlled through interface to a computer in which the related software (in the LabVIEW format) is written and installed. All these are domestic made. Prior to deposition, all glass substrates were ultrasonically cleaned sequentially in heated acetone and ethanol. The surface roughness of the substrates was measured by a Talysurf profilometer and AFM, and the root-mean-square (rms) substrate roughness (Rq) obtained using these methods was 0.3 and 0.9 nm, respectively.

In oblique angle deposition, growth angle of columns which is with respect to the surface normal is "β" when vapor incident angle is "α." Experimental data of FESEM are more agreeable with Tait rule (Eq. 1) when deposition angles are larger than 60° and tangent rule when deposition angles are smaller than 60° (Eq. 2) [12].

$$\beta = \alpha - \sin^{-1}\left(\frac{1 - \cos\alpha}{2}\right) \tag{1}$$

$$\tan(\beta) = \frac{1}{2}\tan(\alpha) \tag{2}$$

Five different Mn nano-sculptured thin films, namely vertical nano-rod, inclined nano-rod, 3-pointed helical star, 4-pointed helical star and 5-pointed helical star, were produced with different deposition angles. The most influential deposition conditions are as follows:

Vertical and inclined nano-rod thin films were produced by deposition at $\alpha = 0°$ and $\alpha = 45°$ with films thickness of 500 and 200 nm, respectively, where the substrate holder was fixed (φ = constant).

The Mn helical stars nano-sculptured were deposited in the following stages:

Stage 1: substrate was fixed in an arbitrary position ($\varphi = 0°$), and 60 nm Mn film was deposited.

Stage 2: substrate was rotated by 180° ($\varphi = 180°$), and at this new position (opposite to Stage 1) 60 nm Mn film was deposited (first tip point of the star is made).

Stage 3: substrate was rotated by 180° to return to the position of Stage 1 and then rotated by 120°, 90° and 72° for deposition of 3-pointed helical star, 4-pointed helical star and 5-pointed helical star, respectively, and 60 nm Mn film was deposited.

Stage 4: substrate was rotated by 180° ($\varphi = 180°$), and at this new position (opposite to Stage 3) 60 nm Mn film was deposited (second tip point of the star is made).

Stage 5: substrate was rotated by 180° to return to the position of Stage 3 and then rotated by 120°, 90° and 72° for deposition of 3-pointed helical star, 4-pointed helical star and 5-pointed helical star, respectively, and 60 nm Mn film was deposited.

Stage 6: substrate was rotated by 180° ($\varphi = 180°$), and at this new position (opposite to Stage 5) 60 nm Mn film was deposited (third tip point of the star is made).

Note: the first pitch of 3-pointed helical star is completed and should continue from Stage 11.

Stage 7: substrate was rotated by 180° to return to the position of Stage 5 and then rotated by 90° and 72° for deposition of 4-pointed helical star and 5-pointed helical star, and 60 nm Mn film was deposited.

Stage 8: substrate was rotated by 180° ($\varphi = 180°$), and at this new position (opposite to Stage 7) 60 nm Mn film was deposited (forth tip point of the 4-pointed helical star and 5-pointed helical star is made).

Note: the first pitch of 4-pointed helical star is completed and should continue from Stage 11.

Stage 9: substrate was rotated by 180° to return to the position of Stage 7 and then rotated by 72°, and 60 nm Mn film was deposited.

Stage 10: substrate was rotated by 180° ($\varphi = 180°$), and at this new position (opposite to Stage **9**) 60 nm Mn film was deposited (fifth tip point of the star for 5-pointed helical star is made).

Stage 11: substrate was rotated by 180° to return to the position of Stage 9 (for 3-pointed helical star this position is Stage 5 and for 4-pointed helical star this position is Stage 7) and then rotated by 72° (for 3-pointed helical star rotate by 120° and for 4-pointed helical star rotate by 90°).

One pitch of the helical Mn star-like structure is completed, and the substrate holder is in the position of Stage 1; Stages 1–11 may be repeated for fabrication of second and third pitches of the helical structures.

The second pitch and the third pitch of the sculptured structures were fabricated by repeating the above stages while the lengths of arms were decreased to 30 and 15 nm, respectively. Therefore, the final structure should look like a pine tree.

The film thicknesses and column shapes and sizes were measured by field emission electron microscope (FESEM) (Hitachi S-4100 SEM, Japan). The FESEM samples were coated with a very thin layer of gold to prevent the charging effect. The surface physical morphology and roughness were obtained by means of AFM (Park scientific instruments model autoprobe) analysis with a Si tip of 10 nm in diameter and in non-contact mode.

The static and dynamic contact angles were measured by domestic made instrument with a 5-diopter lens and digital camera (model DCR-SR200E, Sony, Japan). A 10-mg distilled water droplet was placed smoothly on the solid surfaces of the samples, and the image of static contact angle is recorded by using ImageJ software code and the method of Low-Bond Axisymmetric Drop Shape Analysis (LB-ADSA) [13]. Dynamic and hysteresis contact angles were measured according to the method described by Abelmane et al. [14]. The advancing and receding contact angles were measured by pumping in and out further of liquid in initial droplet. The contact angle hysteresis is attributed to the difference between advancing and receding contact angles.

Theory: surface free energy

There are several intermolecular bonds in the bulk of a material compared with the molecules on the surface. Therefore, surface has more energy that is named surface free energy (SEF) or interfacial energy. Surface free energy is calculated by Young equation:

$$\gamma_{sv} = \gamma_{sl} + \gamma_{lv} \cos \theta_Y \quad (3)$$

where θ_Y is the contact angle between the liquid and the solid surfaces, γ_{lv} and γ_{sl} are liquid surface and surface-solid tensions, respectively. γ_{sl} was calculated by acid–base method (Eq. 4) which is described by Van Oss and Good [15, 16].

$$\gamma_l^{\text{tot}}(1 + \cos \theta) = 2\sqrt{\gamma_s^{\text{LW}}\gamma_l^{\text{LW}}} + 2\sqrt{\gamma_s^{-}\gamma_l^{+}} + 2\sqrt{\gamma_s^{+}\gamma_l^{-}} \quad (4)$$

where γ_l^{tot} is the total surface tension of the liquid drop, γ^{LW} is Lifshitz–Van der Waals, γ^+ is electron acceptor and γ^- is electron donor components. θ is the contact angle, and s and l define solid and liquid, respectively. γ_s^{LW}, γ_s^- and γ_s^+ were calculated by measuring the contact angle between three different liquids (α-bromonaphthalene, water and formamide) with known surface tensions (Table 1), and then, the values of solid surface tensions were obtained by using Eqs. 5 and 6.

$$\gamma_s^{\text{AB}} = 2\sqrt{\gamma_s^{+}\gamma_s^{-}}, \quad (5)$$

Table 1 Components of surface tension at room temperature (mJ m^{-2}) [6]

Liquid	γ^{tot}	γ^{LW}	γ^{+}	γ^{-}
α-Bromonaphthalene (apolar)	44.4	44.4	0	0
Water (polar)	72.8	21.8	25.5	25.5
Formamide (polar)	58	39	2.28	39.6

$$\gamma_s^{\text{tot}} = \gamma_s^{\text{LW}} + \gamma_s^{\text{AB}}, \quad (6)$$

where γ_s^{AB} is the polar component of Lewis acid–base interaction.

Works of adhesion and spreading that are dependent on the contact angle of water droplet on solid surfaces and water total surface tension were determined by Eqs. 7 and 8, respectively [17]:

$$W_a = \gamma_l^{\text{tot}}(\cos \theta + 1), \quad (7)$$

$$W_S = \gamma_l^{\text{tot}}(\cos \theta - 1). \quad (8)$$

Water surface tension also was used to calculate Laplace pressure [18] (Eq. 9):

$$P_{\text{L}} = \frac{2\gamma_{lv}}{R}, \quad (9)$$

where R is the water droplet radius on solid surface which is defined as:

$$R = \left[3V/\pi(1 - \cos \theta)^2(2 + \cos \theta)\right]^{1/3}, \quad (10)$$

where V is droplet volume [19].

Results and discussion

FESEM and AFM analyses

The FESEM top and cross-sectional views of samples are shown in Fig. 1. Morphology of thin films indicates that Mn layers are deposited uniformly onto the substrate and their controlled growth, produced different shapes/nano-sculptured thin films. The thickness of pitches in each layer is less than 100 nm which were put together in specific configurations to make it possible to trap air bubble in pore body of structure and form rough surfaces. Structural characteristics of thin layer films are given in Table 2. As shown in Fig. 1, column width increases by increasing the films thickness and arm length and forms cone-shaped rough structure with cauliflower-like morphology. These results are in agreement with predation of structural zone model (SZM) which is suggested by Messier et al. [20]. In this model, evolutionary growth of structures is classified

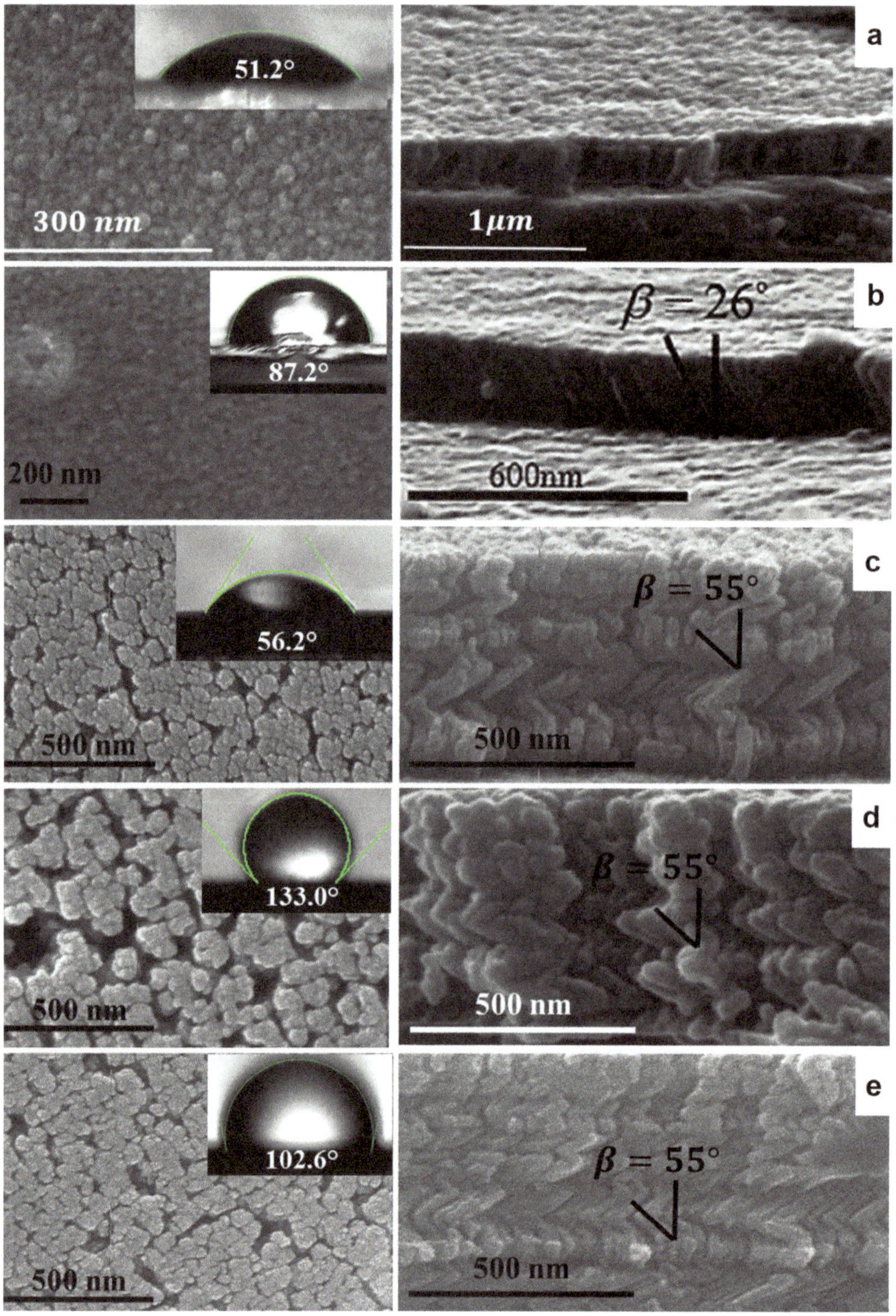

Fig. 1 FESEM images of surfaces and cross sections of nano-sculptured Mn thin films with different morphologies, vertical nano-rod (**a**); inclined nano-rod (**b**); 3-pointed helical star (**c**); 4-pointed helical star (**d**); 5-pointed helical star (**e**)

into different-size zones 1–3, 5–20, 20–40, 50–200 and 200–400 nm. The scaling of the column width will multiply 3 times per stage. The results of Messier's study showed that random ballistic aggregation of atoms leads to cluster structure in 1–3 nm scale.

During the development of nano-sculptured thin films, clustering and conical structure are formed. The low adatom mobility on Mn surfaces during low-temperature deposition causes cauliflower-like structure of about 10 nm in size [20]. Savaloni et al. reported that increasing thin film thickness increases grain size and consequently the crystal structure and morphology change [21, 22].

Figure 2 shows AFM images of the two- and three-dimensional configuration of samples with different morphologies. Number of grain, grain sizes and surface roughness were calculated by applying Spip code and 2D image of each structure. The surface void fraction can be calculated as:

Table 2 Structural characteristics of different Mn sculptured thin films designed for this work

Helical morphology	Deposition angle α (°)	Growth angle β (°)	Number of arms	Arm thickness (t nm)	Number of pitches	Film thickness (T nm)	Arm length (L nm)	Mean diameter (nm)
Vertical nano-rod	0	0	1	15	0	500	500	12.8 ± 1.1
inclined nano-rod	45	26	1	18	0	180	200	16.2 ± 2.0
3-Pointed helical star	80	55.6	18	30	3	480	60 30 15	60.4 ± 2.3
4-Pointed helical star	80	55.6	24	60	3	500	60 30 15	68.1 ± 3.1
5-Pointed helical star	80	55.6	30	20	3	710	60 30 15	58.3 ± 1.6

$$\text{Void Fraction} = \left(1 - \frac{N \times A'}{A}\right) \times 100 \quad (11)$$

where N, A' and A are number of grains, mean surface of top area of grains and total area of the 2D AFM image, respectively.

The surface void fraction is important specifics of nano-sculptured thin films which are produced by oblique angle deposition and self-shadowing effects. Figure 2a shows the AFM image of vertical nano-rod structure. As shown in Fig. 2a and first column of Table 3, in vertical deposition method void fraction is very low. For more investigation of surface properties, void fraction oblique angle deposition ($\alpha = 45°$) was used to fabricate inclined nano-rod structure (Fig. 2b). In this structure, column growth angle (β) is 26° according to the tangent equation [12]. Self-shadowing effect increases void fraction of inclined nano-rod structure (Table 3), but there is not significant difference between void fraction of two nano-rod structures. In Fig. 2c–e, 3-, 4- and 5-pointed helical star-structured films deposited at 80° oblique angle are given. The difference in morphologies and size of grooves between all the samples is clearly observed. In these Mn sculptured thin films, lengths of arms in each pitch are the same, but number of arms is different. The growth (rise) angle of all of these structures is the same and equal to 54.5° according to the Tait equation [12]. The analyses of the AFM images indicate that the shadowing effect modifies void fraction of these structures.

Solid surface wettability

Contact angle is one of the common ways to measure the wettability of a surface. The contact angles between 10 mg distilled water droplet and the thin film surfaces examined in this work are given in Table 3. The 4-pointed helical star structure thin film shows the highest static contact angle (133.1°), while contact angle of 5-pointed helical star, inclined nano-rod structure, 3-pointed helical star and the vertical nano-rod structure is 102.6°, 87.2°, 56.2° and 52.2°, respectively. Figure 1 shows the image of 10 mg distilled water droplet on the examined sample's surfaces.

Factors affecting the wettability of surfaces

The contact angle and hydrophobic property of surface depend on their structural properties (roughness, void fraction, film thickness and symmetry) and almost certainly on the water droplet constituents.

Influence of different structural parameters on the contact angle and the surface free energy

As shown in Table 3 for all samples except 3-pointed helical star structure, slight increase in surface void fraction has led to increase in contact angle more than 36° (the void fraction of the samples (4- and 5-pointed helical stars, vertical and inclined nano-rod structure samples) changes as a result of self-shadowing effects of oblique angle deposition). The increase in deposition angle from 45° to 80° has led to a drastic increase in the 3- and 4-pointed helical star thin film void fraction, although the increase in the 5-pointed helical star is low when compared with other helical star samples. On the other hand, increase in void fraction is not always the cause of enhancement of surface hydrophobicity. Table 3 shows that the void fraction is higher in the 3-pointed helical star than in 5-pointed helical

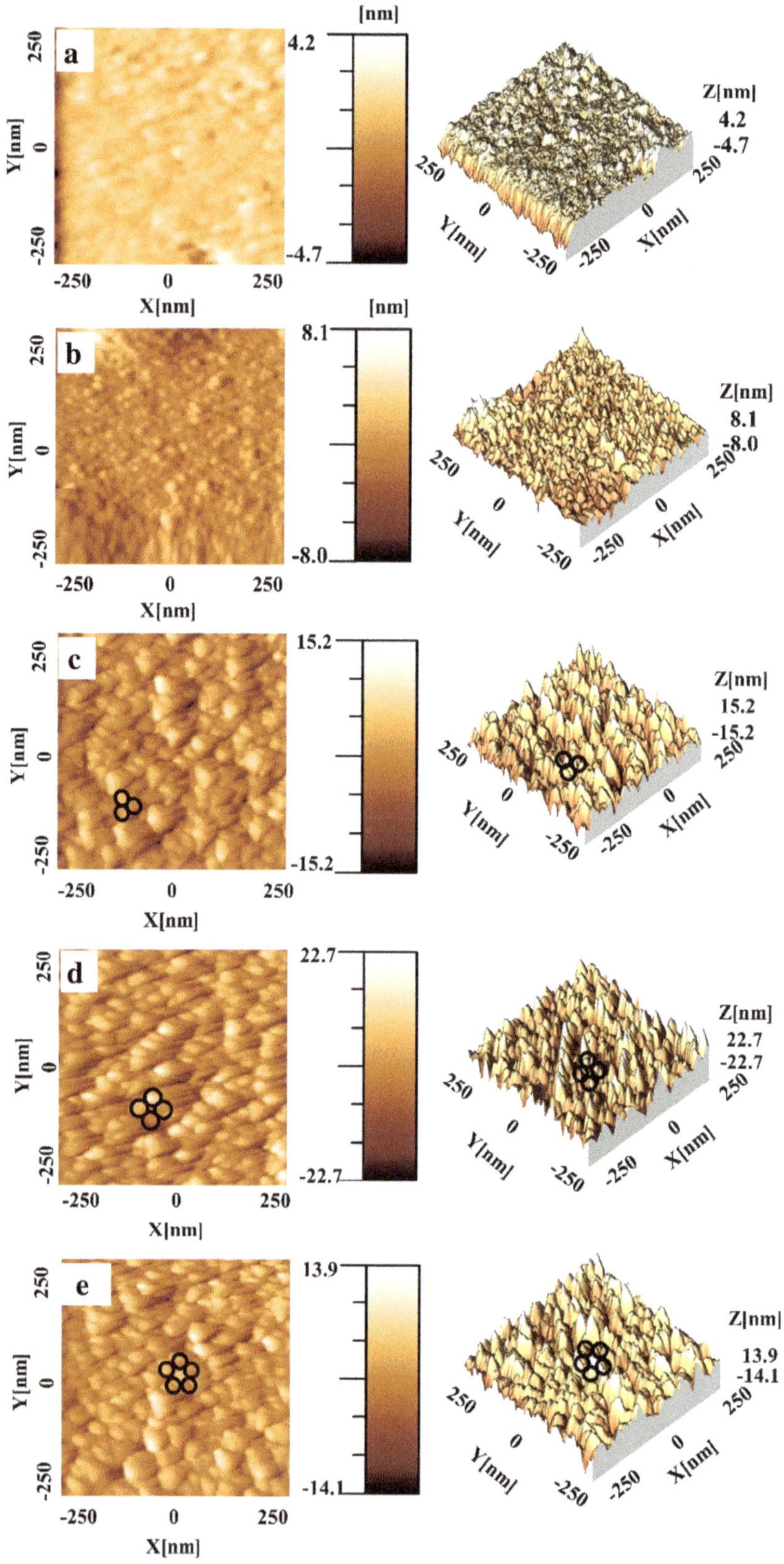

Fig. 2 2D and 3D AFM images of nano-sculptured Mn thin films with different morphologies, vertical nano-rod (**a**); inclined nano-rod (**b**); 3-pointed helical star (**c**); 4-pointed helical star (**d**); 5-pointed helical star (**e**)

star and inclined nano-rod samples but contact angle is lower. Hence, there are some other structural parameters that may influence the surface hydrophobicity.

Comparison of the void fraction and surface roughness values in Table 3 for 3-pointed helical star and inclined nano-rod samples showed that these two major parameters

Table 3 Structural characteristics of different Mn sculptured thin films designed for this work, work of adhesion (W_a) (mJ m^{-2}) and work of spreading (W_S) (mJ m^{-2}) for 3 different liquid drops on different types of sculptured structures produced in this work

Helical morphology	Void fraction (%)	R_{avg} (nm)	R_{rms} (nm)	Experimental static WCA (°)	Cassie–Baxter predicted WCA	Wenzel predicted WCA	R (mm)	Laplace pressure (Pa)	W_a	W_S
Vertical nano-rod	12.2 ± 0.2	1.4 ± 0.8	1.8 ± 0.2	51.2 ± 0.4	160.0	63.3	1.5	96.1	118.6	−27.1
Inclined nano-rod	17.0 ± 1.7	2.9 ± 0.6	3.7 ± 0.8	87.2 ± 2.2	163.7	63.2	2.6	55.9	76.5	−69.2
3-Pointed helical star	40.9 ± 3.4	5.7 ± 0.3	7.0 ± 0.3	56.2 ± 1.8	102.7	59.3	1.7	86.3	113.5	−32.3
4-Pointed helical star	51.2 ± 2.	8.6 ± 0.8	10.6 ± 0.7	133.0 ± 1.4	112.7	54.5	3.3	44.2	23.2	−122.4
5-Pointed helical star	30.6 ± 1.0	5.1 ± 0.9	6.3 ± 1.1	102.6 ± 1.5	116.9	59.3	2.9	49.6	57.0	−88.6

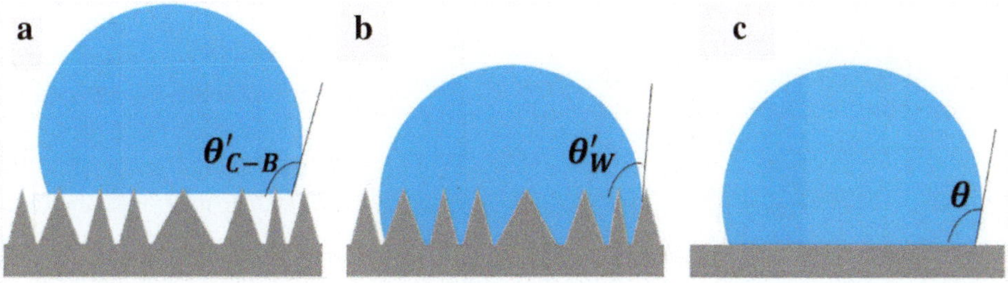

Fig. 3 Surfaces wetting models. Rough surface and Cassie–Baxter model (**a**); rough surface and Wenzel model (**b**); flat surface (**c**)

are increased by ~23% and 2.8 nm, respectively, but the contact angle of this sample is decreased by ~1.5 degrees which may be due to arm length (the arm length is 60 and 200 nm for 3-pointed helical star and inclined nano-rod samples, respectively). Abdolsalam et al. [23] also showed that higher values of contact angles (higher hydrophobicity) may be achieved with longer nano-rods (nano-structure with longer arms).

Surface roughness is also another effective parameter on the hydrophobicity of surfaces. Increasing the surface roughness considerably increases the hydrophobicity of all samples except 3-pointed helical star structure (in all samples the contact angle increases about 30° by 3 nm increase in the surface roughness), but high surface roughness and void fraction of 3-pointed helical star do not enhance their hydrophobicity properties. 3-pointed helical star in compare with other helical star structures has more asymmetric structure. With regard to these results, it could be concluded that structural symmetry effectively enhances the hydrophobicity of solid surfaces.

Design and deposition of different Mn nano-sculptured thin films indicated that structural symmetry, surface roughness and surface void fraction as result of shadowing effect (80° deposition angle) may enhance hydrophobicity. Hence, from the obtained results, one may deduce that the higher contact angle (133°) obtained for the 4-pointed helical star nano-sculptured sample could be due to higher degree of symmetry, higher surface roughness (8.6 nm) and higher void fraction (51.28%) (Fig. 1; Fig. 2). Structural symmetry with roughness and void fraction of 5-pointed star sample led to lower contact angle (102.6°) than of 4-pointed star. Asymmetric structure in 3-pointed star leads to lowest contact angle between helical star structure (56.2°) and dominates influence of the void fraction and surface roughness.

Rough surfaces wetting models

Two main approaches are proposed in the literature to explain wetting property of rough surfaces, namely Cassie–Baxter state [24] and Wenzel state [25].

According to the Cassie–Baxter model, air is trapped between surface grooves and liquid droplet sits on top of the arms (Fig. 3a) and so the hydrophobicity increases. In this case, the relationship between the contact angle on the

Table 4 Contact angles for 10 mg water droplet and surface free energy (mJ m^{-2}) obtained from contact angle of 3 different liquid drops on different types of sculptured structures produced in this work

Helical morphology	Advancing WCA (°)	receding WCA (°)	Hysteresis WCA (°)	θ (α-bromonaphthalene) (°)	θ_{water} (°)	$\theta_{formamid}$ (°)	γ_s^{LW}	γ_s^+	γ_s^-	γ_{AB}	γ_s^{tot}
Vertical nano-rod	54.4 ± 1.9	18.5 ± 2.3	35.9	23.4 ± 2.4	51.2 ± 0.4	12.2 ± 2.4	40.8	3.3	16.1	14.6	55.4
Inclined nano-rod	93.1 ± 1.7	22.1 ± 0.8	71	5.2 ± 1.6	87.2 ± 2.2	11.4 ± 1.7	44.2	8.3	2.1	8.4	52.6
3-Pointed helical star	60.3 ± 2.0	20.6 ± 1.5	39.7	8.3 ± 1.0	56.2 ± 1.8	12.9 ± 2.0	43.9	3.2	7.0	16.7	60.7
4-Pointed helical star	141.1 ± 2.6	39.2 ± 0.6	101.9	27.2 ± 1.0	133.0 ± 1.4	21.6 ± 1.5	39.6	21.0	65.9	74.4	114.0
5-Pointed helical star	109.7 ± 1.2	24.4 ± 1.3	85.3	5.2 ± 1.1	102.6 ± 1.5	27.4 ± 1.4	44.2	8.5	11.8	20.0	64.3

smooth surface θ and the contact angle on the rough surface θ' is given as [24]:

$$\cos\theta' = \varphi - 1 + \varphi\cos\theta \tag{12}$$

where φ is an area fraction of liquid–solid interface and $(1 - \varphi)$ is that of the solid–air interface, while φ can be considered as:

$$\varphi = \frac{N' \times A'}{S} \tag{13}$$

where N' is:

$$N' = \frac{N \times S}{A}, \tag{14}$$

and S can be obtained by defining the radius of the spherical drop as R and the radius of circular interface of liquid–solid as X, hence we have:

$$X = R\sin\theta, \tag{15}$$

and

$$S = \pi X^2. \tag{16}$$

According to Wenzel model, the penetration of water droplet between grooves of the structure decreases surface hydrophobicity and also its contact angle (Fig. 3b). In this case, the relationship between the contact angle formed on a smooth surface θ and the contact angle formed on the rough surface θ' is given as [25]:

$$\cos\theta' = R_1\cos\theta, \tag{17}$$

where R_1 is the ratio of the actual contact area of liquid–solid to the projected area.

Considering that the structures do not have an ordered geometrical shape, the following equation was used to obtain values of R_1:

$$R_1 = 1 + \frac{N' \times 2\pi r \times R_{rms}}{S}. \tag{18}$$

In this equation, r is the average grain radius and R_{rms} is the root-mean-square surface roughness.

Predicted contact angle values for samples obtained from Cassie–Baxter and Wenzel equations (Eqs. 12 and 17, respectively) are given in Table 3. By comparing experimental and theoretical data, it can be deduced that the vertical nano-rod, inclined nano-rod and 3-pointed helical star are in the Wenzel state and vertical 4-pointed helical star and 5-pointed helical star are in the Cassie–Baxter state.

It is worthy to note that there is slight difference between predicted and experimental contact angles which may depend on some anisotropy of sculptured thin films that is not considered in Eqs. 12 and 17.

The result of Laplace pressure in different samples is in agreement with hydrophobicity data (i.e., 4-pointed helical star structure with lowest Laplace pressure shows the largest contact angle). Increasing the Laplace pressure induced transition from Cassie–Baxter to Wenzel state.

Rose petal effect

Surface free energy values for different samples examined in this work were calculated using acid–base method (Eqs. 4–6) by considering the contact angle for three different liquids (α-bromonaphthalene, formamide and distilled water) which are given in Table 4. Except 3-pointed helical star structure, all surface free energies of samples in this work are increased with surface roughness (i.e., 4-pointed helical star structure shows highest surface free energy). Results are in agreement with theoretical study on role of surface roughness on the surface free energy [26]:

$$\gamma_s^{tot} = W = \frac{1}{N}\sum_i^N \overrightarrow{F_i} \cdot \overrightarrow{r_i} = \frac{1}{N}\overrightarrow{F_c} \cdot \sum_i^N \overrightarrow{r_i} \sim F_c R_{rms}, \quad (19)$$

where F_c is the constant adhesion force, R_{rms} is the root-mean-square surface roughness, W is the average work done by the adhesion force vector on the surface as the tip of the AFM moves the displacement vector $\overrightarrow{r_i} \cdot \overrightarrow{F_i}$ is the adhesion force which can be obtained from the spring constant of the AFM tip. N is the number of samples. According to this equation, the surface energy increases with increasing surface roughness.

The application of hydrophobic surfaces depends on droplet behavior on the surfaces and their wettability. Natural superhydrophobic surfaces classified into two groups: slippery and sticky [27, 28]. Both of these superhydrophobic surfaces exhibit a large contact angle, but water droplet shows weak and strong adhesion onto slippery and sticky superhydrophobic surfaces, respectively, according to their contact angle hysteresis [27, 28]. In sticky case, the water droplet is strongly pinned on the surface, even when the surface is upside down which is called rose petal effect [28], while in case of slippery surfaces water droplet can easily move on the surface and make self-cleaning property which is so called lotuses effect.

As shown in Table 3, 4-pointed helical star structure has highest work of adhesion and work of spreading values that means this surface is hydrophobic. Surfaces, with contact angle hysteresis (difference between advancing and receding contact angle) higher than 20°, can show large adhesion property [29]. It can be seen that all of the contact angle hysteresis values are higher than 20° with the highest value obtained for 4-pointed helical star structure. Hence, it can be considered that these samples are sticky surfaces. The result of contact angle hysteresis (Table 4) showed that this property is related to the structural property and their hydrophobicity. By increasing samples hydrophobicity, contact angle hysteresis and adhesion increase. Therefore, 4-pointed helical star hydrophobic structure with contact angle of 133° and contact angle hysteresis of 101.9° is a hydrophobic sticky surface which is acting as a rose petal surface.

The effects of physicochemical property of liquid phase on contact angle

The effect of liquid polarization

Not only substrate structural property influences contact angle between liquid and solid surfaces but also

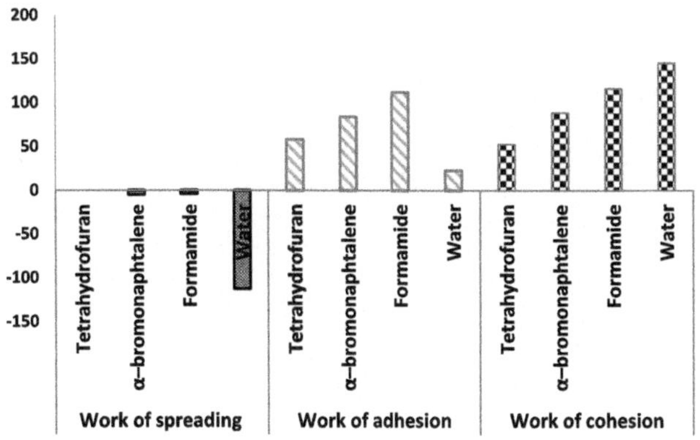

Fig. 4 Work of spreading, adhesion and cohesion of different solvent onto the 4-pointed helical star structure

physicochemical property of liquid phase could effectively change the contact angle.

In order to investigate the chemical property, solvents with different polarizations were selected and their contact angle onto the 4-pointed helical star structure was studied. As illustrated in Fig. 4, water has the highest work of cohesion and spreading when compared with other solvents. Therefore, in contact with hydrophobic surface, it shows largest contact angle. The different behaviors of formamide, tetrahydrofuran, water and α-bromonaphthalene at solid liquid interface may be due to their polarization, viscosity and surface tension. Formamide has highest dipole momentum (3.4 D) while dipole momentum of tetrahydrofuran, water and α-bromonaphthalene are 1.7, 1.8 and 0.0 D, respectively [30]. Viscosity is one of the main factors in formation of droplet which are 3.30, 0.48 and 0.89 mPa s for formamide, tetrahydrofuran and water, respectively, whereas viscosity of α-bromonaphthalene is negligible [30]. Surface tension of formamide, tetrahydrofuran, water and α-bromonaphthalene at 20 °C is 58.20, 26.40, 72.8 and 44.4 mN/m [31], respectively. These parameters are responsible for contact angle variation of liquid droplet on solid surfaces, but quantifying the impact of parameters subjects to some complications and needs some further investigations. Hence, polar solvents in contact with hydrophobic solid surfaces do not necessarily show high contact angle value.

The effect of total dissolved solids (TDS) on contact angle

Concentration of calcium carbonate in water could be an important parameter in contact angle measurements. This was studied by using four different dissolved solid concentrations in 10 mg water droplet. The results of contact angle onto the 4-pointed helical star structure, resistivity and TDS concentrations are shown in Fig. 5. It can be seen that the contact angle (hydrophobicity) increases with TDS

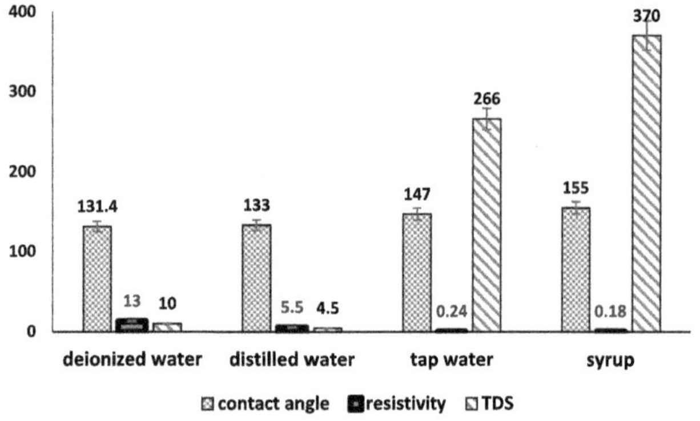

Fig. 5 Contact angle of 10 mg water with different TDS concentration onto the 4-pointed helical star structure (electrical conductivity is directly related to TDS). Sucrose syrup (370 μg/l)

which has led to increased viscosity and surface free energy. The aforementioned discussion indicates that an increase in viscosity and surface free energy could enhance hydrophobicity as a result of increased W_a, W_c and W_s. It can be seen that the distilled water has lowest TDS concentration in 10 mg droplet. Hence, in order to achieve more accurate contact angle values, distilled water is a good choice.

Conclusion

In this study, hydrophobic nano-sculptured thin films with different roughness were fabricated and the result showed that structures which trap air pocket between grooves are more hydrophobic. By increasing roughness, void fraction and structural symmetry contact angle (hydrophobicity) increased. In addition, the wetting model of 4- and 5-pointed helical star structures was described by Cassie–Baxter equation, while other structures, namely vertical nano-rod, inclined nano-rod and 3-pointed helical star, were in Wenzel state. All of the Mn nano-sculptured thin films indicate contact angle hysteresis higher than 20° that means these surfaces are acting as a sticky superhydrophobic surface with petal effect. The physicochemical property of the surface with 4-pointed star-like helical structure was examined for different liquids with different polarities (i.e., formamide, tetrahydrofuran, water and α-bromonaphthalene), and it was found that water has the highest work of cohesion and spreading as well as largest contact angle. As this issue is affected by many parameters, further investigations are needed to make a more elaborate conclusion.

Acknowledgements This work was carried out with the support of the University of Tehran. HS gratefully acknowledges the Iran National Science Foundation (INSF) and the Centre of Excellence for Physics of Structure and Microscopic Properties of Matter, Department of Physics, University of Tehran, for partial support of this work.

References

1. Lafuma, A., Quere, D.: Superhydrophobic states. Nat. Mater. **2**, 457–460 (2003)
2. Bingwei, X., Jingcheng, H.: Reversibly switchable wettability. Chem. Soc. Rev. **39**, 769–782 (2010)
3. Yang, C., Tartaglino, U., Persson, B.N.J.: Influence of surface roughness on superhydrophobicity. Phys. Rev. Lett. **97**, 1161–1163 (2006)
4. Ichimura, K., Oh, S.K., Nakagawa, M.: Light-driven motion of liquids on a photoresponsive surface. Science **288**, 1624–1629 (2000)
5. Gleiche, M., Chi, L.F., Fuchs, H.: Nanoscopic channel lattices with controlled anisotropic wetting. Nature **403**, 173 (2000)
6. Guo, Z.G., Fang, J., Hao, J.C., Liang, Y.M., Liu, W.M.: A Novel approach to stable superhydrophobic surfaces. Chem. Phys. Chem. **7**, 1674 (2006)
7. Shang, H.M., Wang, Y., Limmer, S.J., Chou, T.P., Takahashi, K., Cao, G.Z.: Optically transparent superhydrophobic silica-based films. Thin Solid Films **472**, 37–43 (2005)
8. Zhang, X., Shi, F., Niu, J., Jiang, Y., Wang, Z.: Superhydrophobic surfaces: from structural control to functional application. J. Mater. Chem. **18**, 621–633 (2008)
9. Ebert, D., Bhushan, B.: Durable lotus-effect surfaces with hierarchical structure using micro and nanosized hydrophobic silica particles. J. Colloid Interface Sci. **368**, 584–591 (2012)
10. Siabi-Garjan, A., Savaloni, H.: Extinction spectra and electric near-field distribution of Mn nano-rod based sculptured thin films: experimental and discrete dipole approximation results. Plasmonics **10**, 1007–1011 (2014)
11. Eckertova, L.: Physics of Thin Films, 2nd edn, pp. 28–49. Plenum Press. Berlin, Springer (1986)
12. Messeier, R., Giri, A.P., Rou, R.A.: Revised structure zone model for thin film physical structure. J. Vac. Sci. Technol. A **2**, 496–500 (1984)
13. Stalder, A.F., Kulik, G., Sage, D., Barbieri, L., Hoffmann, P.: A snake based approach to accurate determination of both contact points and contact angles. Colloids Surf. A **286**, 92–103 (2006)
14. Abelmane, L., Lodder, C.: oblique evaporation and surface diffusion. Thin Solid Films **305**, 1–5 (1997)
15. van Oss, C.J., Good, R.J., Chaudhury, M.K.: The role of van der Waals forces and hydrogen bonds in "hydrophobic interactions" between biopolymers and low energy surfaces. J. Colloid Sci. **111**, 378–390 (1986)
16. van Oss, C.J.: Interfacial Force in Aqueous Media, pp. 9–48. Taylor & Francis, New York (2006)
17. Israelachvili, J.N.: Intermolecular and Surface Forces, 2nd edn, pp. 223–281. Academic Press, London (1985)
18. Butt, H.J., Graf, K., Kappl, M.: Physics and Chemistry of Interfaces, 2nd edn, ScienceDirect, pp. 8–9. Wiley-VCH Verlag, Weinheim (2003)
19. Extrand, C.W.: A thermodynamic model for wetting free energies from contact angles. Langmuir **19**, 646–649 (2002)
20. Messeier, R., Giri, A.P., Rou, R.A.: Revised structure zone model for thin film physical structure. J. Vac. Sci. Tech. A **2**, 496–500 (1984)
21. Savaloni, H., Player, M.A.: Influence of deposition conditions and of substrate on the structure of uhv deposited erbium films. Vacuum **46**, 167–172 (1995)

22. Savaloni, H., Player, M.A.: Morphological changes in UHV deposited Er/a-C films: nucleation, growth and grain structure. Thin Solid Films **256**, 48–54 (1995)
23. Abdelsalam, M.E., Bartlett, P.N., Kelf, T., Baumberg, J.: Wetting of regularly structured gold surfaces. Langmuir **21**, 1753–1757 (2005)
24. Baxter, S., Cassie, A.B.D.: The water repellency of fabrics and a new water repellency test. J. Text. **36**, 67–90 (1946)
25. Wenzel, R.N.: The evaluation of textile waterproofing agents. Surf. Coat. Techol. **25**, 505–514 (1936)
26. Sun, C.C., Lee, S.C., Hwang, W.C., Hwang, J.S., Tang, I.T., Fu, Y.S.: Surface free energy effects in sputter-deposited WNx films. Metall. Trans. A **47**, 2533–2539 (2006)
27. Stark, A.Y., Badgeb, I., Wucinicha, N.A., Sullivana, T.W., Niewiarowskia, P.H., Dhinojwala, A.: Surface wettability plays a significant role in gecko adhesion underwater. PNAS **16**, 6340–6346 (2013)
28. Barthlott, W., Neinhuis, C.: Purity of the sacred lotus, or escape from contamination in biological surfaces. Planta **202**, 1–8 (1997)
29. Stanton, M.M., Ducker, R.E., MacDonald, J.C., Lambert, C.R., McGimpsey, W.G.: Super-hydrophobic, highly adhesive, polydimethylsiloxane (PDMS) surfaces. J. Colloid Interface Sci. **367**, 502–508 (2012)
30. Smith, J.W.: Some developments of Guggenheim's simplified procedure for computing electric dipole moments. Trans. Faraday Soc. **46**, 394–399 (1950)
31. Wypych, G.: Handbook of Solvents, pp. 1453–1454. ChemTec Publishing, Victoria (2001)

7

Experimental measurements of a prototype high-concentration Fresnel lens and sun-tracking method for photovoltaic panel's efficiency enhancement

Meraj Rajaee[1] · **Seyed Mohamad Bagher Ghorashi**[1]

Abstract Concentrator photovoltaic modules are a promising technology for highly efficient solar energy conversion. This system presents several advantages due to additional degrees of freedom that has been provided by the spectral separation such as cost and mass reduction, increase in the incident solar flux on PV cells and performances. This paper has proposed a unique photovoltaic solar cell system that consists of semi-Fresnel lens convergent structure and a novel two axis sun tracking module to enhance the efficiency of solar cell by using less cell area and energy losses. The grooves of this lens are calculated according to the refraction and convergent angles of the light easy for perpendicular incidence angle. The update time interval during tracking causes misalignment of the lens' optical axis versus the sunrays. Then an inventive sun-tracking method is introduced to adjust the module so that the incident rays are always perpendicular to the module's surface. As a result, all rays will be refracted with the predetermined angles. This way the focus area is reduced and smaller cells can be used. We also mentioned different module connections in order to provide compensation method during losses, for networks and power systems. Experimental results show that using semi-Fresnel lens, along with the sun-tracking method increases the efficiency of PV panel.

Keywords Concentrator · Semi-Fresnel · Photovoltaic · Efficiency · Tracking method

✉ Seyed Mohamad Bagher Ghorashi
mghorashi@kashanu.ac.ir

1 Department of Laser and Photonics, University of Kashan, Kashan, Iran

Introduction

PV concentrator systems reduce the area of photovoltaic cells required, increase the solar radiation intensity on the photovoltaic cells and can lead to a reduction in total system cost, if the costs of the concentrator and tracking assembly are less than that of the area of photovoltaic cells displaced.

Although they are mainly composed of silicon, which is a widespread material in nature, the process and thus the final product is expensive. Many investigations have been done to raise the performance of PV cells and reduce the required cell area and its final cost. The proposed methods are mainly based on concentrating the sunrays on a small cell area. There are several concentrator technologies that include parabolic trough solar concentrator, linear refractive Fresnel lens, linear reflective Fresnel lens, dish-shaped reflectors, etc. [1, 2]. Compared to other concentrator technologies, Fresnel lens is characterized by some advantages such as small volume, low weight, short focus, high efficiency, and low cost that motivate researchers to design and commercialize Fresnel-based solar concentrators [3, 4]. With the improvements of lens material since the official invention of Fresnel lens in 1822, especially polymethylmethacrylate (PMMA), which has a good transmissivity and resistance to sunlight, interest in Fresnel lenses for concentrated solar energy applications rose in the latter half of the twentieth century. The recent developments of concentrated solar energy applications using Fresnel lens systems include imaging Fresnel lens solar concentration systems and non-imaging systems [5]. Researches have revealed that it is the refractive surface of a conventional lens which defines its focusing properties. Based on this fact the Fresnel lens was invented. Later, researchers found out that using Fresnel lens as a concentrator in PV systems can effectively increase the system's efficiency. A concentrating photovoltaic (CPV) system with a steady electric output adopting an electrically tunable concentration ratio of the liquid crystals (LC) lens is demonstrated [6]. The module consists of a Fresnel lens array (including 16 Fresnel lenses, each 0.120 m × 0.120 m), a solar cell array (including 16 solar cells connected in series, each 0.025 m × 0.025 m) which was pasted onto an aluminum plate heat sink and below the Fresnel array. The authors pointed out that using a Fresnel lens array, instead of a large Fresnel lens, can reduce the thermal energy of the solar cells. Zhai et al. [1] experimentally investigated a concentrating solar collector using linear Fresnel lens. Test results revealed that the indication of lost energy was 0.578 W/m^2 K, which was much smaller than solar collectors without concentrator. In [1], authors used five small Fresnel lenses made of polymethyl, evacuated tube absorber, stepping motor, etc. as the main parts of the module. The module was equipped with a sun-tracking system that tracked the sunlight and a 20 mm light strip was produced. The advantages of this solar collector are: (1) low cost due to accessibility of polymethyl and evacuated glass tube, (2) the area of receiver solar cells is smaller than modules without concentrator, (3) the module is easy to deploy in large scale at high temperature [1]. Rai et al. [9] have investigated the Fresnel lens application for space photovoltaic. The authors proposed the optimum linear Fresnel lens concentrator in terms of concentration ratio. Shvartz and Soluyanov [8] presented space Fresnel concentrated modules with GaInp/GaAs/Ge triple-junction (TJ) solar cells. The module included 4 linear Fresnel lenses, 4 linear photo receivers containing 12 TJ solar cells, and a copper heat sink. Optical characteristics of the linear Fresnel lens were estimated accordingly [8]. Wagner et al. [9] have proposed a refractive system, which enhances solar system's efficiency and reduces its cost. The proposed model works based on separating sunlight (white light) into different wavelengths. Then using a Fresnel lens and with the help of progresses in semiconductors, each wavelength can be concentrated on a different material to be absorbed effectively. This will increase the solar system's efficiency. A switchable Fresnel zone plate lens is demonstrated using a polymer-stabilized liquid crystal. The fabrication process is relatively simple and the device can be operated below 10 V with fast response time [10]. In this paper, a novel refractive structure, named semi-Fresnel lens, is proposed. The proposed model works by calculating the groove angles of Fresnel lens based on the refraction and convergent angles of the orthogonal incident ray. To be clearer, we assume that the incoming ray is directly refracted and converged to a special point. Then using Snell's law and simple trigonometry we find these refraction and convergent angles. Finally based on these two parameters, we calculate the groove angles. This method ensures that every incoming ray will refract directly with refraction angle calculated before. As a result all rays will be converged to a predetermined point with a special convergent angle. Therefore, the needed solar cell area can be significantly reduced. It should not be left unmentioned that, since all calculations are based on the incidence angle orthogonal to the lens surface, the most efficiency will be obtained when the Sun is exactly above the module. To provide this situation in practice, we suggest a sun-tracking method which tracks the sun and adjusts the module so that the incidence angle is always 90 degrees. This way, all incoming rays are concentrated to a line resulting in producing a 2-cm light strip. This will reduce the needed solar cell area resulting in lower cost and lower loss of energy. Based on this assumption a solar panel containing four rows of 2 cm × 90 cm solar cells is designed and implemented. A Fresnel lens panel including four 90 cm × 20 cm semi-Fresnel arrays is placed on a the solar cell module at a distance equal to its focal length which is 5 cm. Test results show that the designed model effectively increases the efficiency of the PV system, reducing its cost and loss of energy [7].

Designing the model

The theory behind the proposed approach

Centuries ago, it was recognized that the contour of the refracting surface of a conventional lens defines its focusing properties. The bulk of material between the refracting surfaces has no effect (other than increasing absorption losses) on the optical properties of the lens [8].

Much research has been carried out to improve the performance of the Fresnel lens in concentrating the sunlight, which will result in a smaller focus area. As a result, smaller solar cells can be used. These methods also tried to reduce the losses of Fresnel lenses. Here we consider two main losses that result from spherical aberration and total internal reflection (TIR). For lenses made with spherical surfaces, rays which are parallel to the optical axis, but at different distances from the optical axis, fail to converge at the same point. This is called spherical aberration. When moving from a denser medium into a less dense one (i.e., $n1 > n2$), above an incidence angle known as the critical angle, all light is reflected. This phenomenon is known as TIR (later we will discuss how TIR affects our designed model). The critical angle is approximately 41.5 degrees for Plexiglas in air [9, 10].

Although some researchers have tried to use TIR to reduce the spherical aberration and reduce the focus area [11], we prefer to eliminate TIR by designing the groove angles based on the convergent angle α so that all the incoming rays can be refracted directly to the focus point. This way the loss of energy of the light is considerably reduced. α (for incidence angle orthogonal to the surface of the lens) is derived by using simple trigonometry. Then we can find $\theta 2$ (the refraction angle) and θ (the groove angle) accordingly. Therefore, we can make sure that for the orthogonal incident rays there will be no spherical aberration. Using this method for calculating the grooves' angles, the only problem is in keeping the incidence angle orthogonal to the lens which will be done by using a sun-tracking system.

Calculating the groove angles

The proposed structure for the semi-Fresnel lens is shown in Fig. 1. As it can be seen from the figure, the lens is flat on one side and has fine grooves on the other side. Our purpose is to concentrate all incoming rays to the central line in the focus of the lens. Assuming the lens focal length to be f, we will derive the angle of the grooves based on Snell's law. We will follow the derivation for the right groove. We name the angle of this groove with the horizontal direction to be θ. Assume the light beam makes the angle of θ_1 with the orthogonal line to the groove (Fig. 1) and it refracts with an angle of θ_2 with the same orthogonal line. This exit beam will finally make angle α with the focal line. Therefore, we can write Snell's law for this optical beam:

$$n_1 \sin\theta_1 = n_0 \sin\theta_2 \tag{1}$$

Again referring to Fig. 1, the θ and θ_1 angles have both axes orthogonal to each other and thus based on simple geometrical laws they are equal:

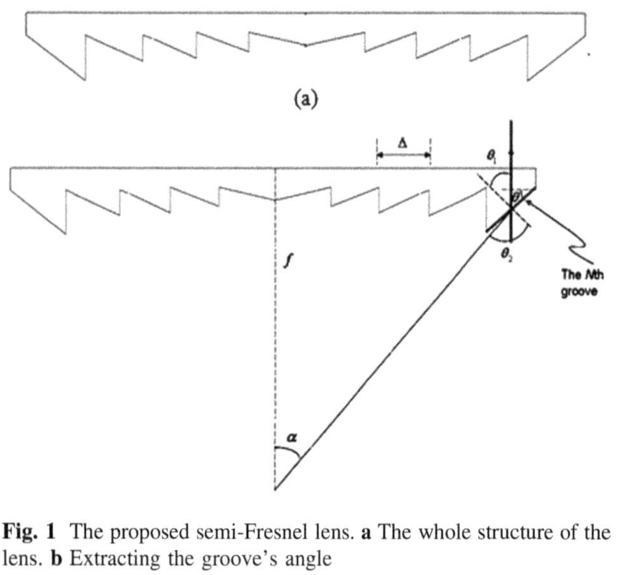

Fig. 1 The proposed semi-Fresnel lens. **a** The whole structure of the lens. **b** Extracting the groove's angle

$$\theta_1 = \theta \tag{2}$$

On the other side, by extending the incident beam we see that the angle θ_2 is divided into two angles, one equal to θ_1 (and thus θ) and the other one equal to α which is again clear from the figure. This yields the following relation:

$$\theta_2 = \theta + \alpha \tag{3}$$

Using Eqs. (2) and (3), Eq. (1) can be rewritten in the following way:

$$n_1 \sin\theta = n_0 \sin(\theta + \alpha) \tag{4}$$

Thus $n0$ in this equation is the refractive index of the air and can be set equal to 1. Parameter $n1$ is the refractive index of the medium where for Plexiglas is around 1.49. The only remaining parameter to be determined for the Eq. (4) to yield the value of the groove's angle θ, is α. This angle could be simply calculated from simple trigonometric relations. The sine of this angle is the ratio of the groove distance from the center of the lens and its distance from the focal line:

$$\sin\alpha = \frac{(N-1+N/2)\Delta}{\sqrt{((N-1+N/2)\Delta)^2 + f^2}} \Rightarrow$$

$$\alpha = \sin^{-1}\left(\frac{(N-1+N/2)\Delta}{\sqrt{((N-1+N/2)\Delta)^2 + f^2}}\right) \tag{5}$$

where Δ is the width of each groove and N is the number of grooves counting from the center part. By determining α from Eq. (4), the groove's angle θ could be extracted using Eq. (4).

Table 1 represents the angles α, $\theta 2$, and θ for the grooves on the right side of the Fresnel lens. The angles will be the same for the left side grooves.

As it is shown in Table 1, θ is calculated for each groove based on the convergent angle (α) of that groove for orthogonal incident ray. It means all rays with non-orthogonal angles will be considered as loss. To be more clear, when the incidence angle is non-orthogonal to the lens's surface one of these situations will occur: TIR or spherical aberration. The reason is that every change in incidence angle will make changes to the angles $\theta 1$, $\theta 2$ and α, which in turn enlarge the focus area of the Fresnel lens.

Table 1 Values of the refraction, convergent, and groove angles

Groove/angles	α	$\theta 2$	θ
GCR (center-right)	11.33	31.79	20.46
G1R	21.82	54.81	32.99
G2R	30.9	69.5	38.6
G3R	38.68	79.5	41
G4R	44.42	86	41.5

Since our purpose is to reduce the solar cell area, we should prevent the focus area of the Fresnel lens from being changed. To do this, we used a sun-tracking system that determines the position of the sun periodically and adjusts the module so that the incidence angle is always orthogonal to the lens's surface.

Constructing the model

In this section the results of experimental implementation is presented. Figure 4a shows the used configuration to test the performance of the proposed concentrating structure. As it is obvious in the Fig. 4a, a Fresnel lens panel including four 90 cm × 20 cm semi-Fresnel lens arrays is placed on a PV module in a distance equal to its focal length which is 5 cm. On the solar panel we have placed four narrow cell rows with a width of 2 cm, equal to the width of light strip resulting from the lens. The solar panel is shown in Fig. 2. Figure 3 shows our proposed semi-Fresnel lens made of Plexiglas.

Experimental results

In this section the results of experimental implementation are presented. Figure 4 shows the used setup to test the performance of the proposed concentrating structure. As it is obvious in Fig. 4, a Fresnel lens panel including four 90 cm × 20 cm semi-Fresnel arrays is placed on a solar panel at a distance equal to its focal length which has a value of 5 cm. On the solar panel, we have placed four narrow cell rows with a width of 2 cm, equal to the concentrating area of the lens.

Fig. 2 Solar panel

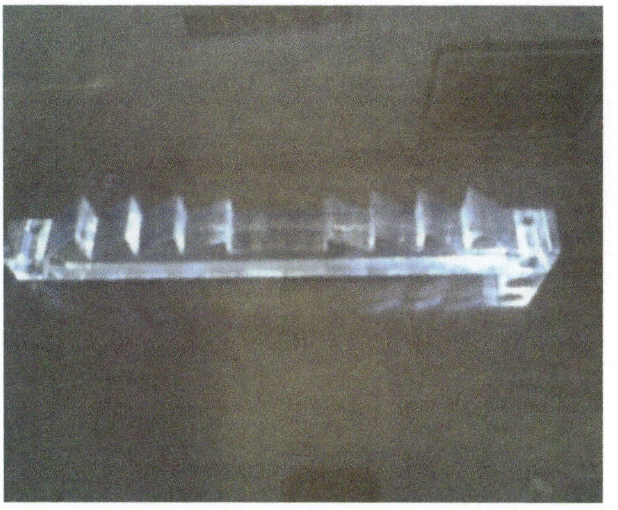

Fig. 3 A single semi-Fresnel lens

To show the superiority of the proposed lens structure in improving the efficiency of PV panel, we have done the same experiment with a conventional Fresnel lens.

The output power versus derived voltage for different amount of incident energy per unit area (in lux) is plotted in Fig. 5. As is clear from the curves, the peak power for different amount of incident energies per unit area is higher for the proposed semi-Fresnel lens compared to the conventional Fresnel lens.

Sun-tracking system method

As mentioned above, we first calculate the convergent angle of the incoming light and then the groove angles are determined accordingly. The mathematical analyses given above have been obtained for the perpendicular incidence angle. But in practice because of the sun's movement during the day, we cannot have the orthogonal incidence angle during the entire daytime. Based on this fact and in order to observe the impact of the orthogonal incidence angle on the performance of PV panel, we manually set the panel so that the orthogonal incidence angle was obtained for the entire daytime.

What we have used is a two-axis sun-tracking system. Single-axis tracking systems are considerably cheaper and easier to construct, but their efficiency is lower than that of two-axis sun-tracking systems. Two-axis sun-tracking systems can be applied in all types of solar systems to increase their efficiency [12].

The two-axis tracking system could be an azimuth-elevation tracking or polar tracking. In azimuth-elevation tracking, which is the most common method used in large solar dish systems, the dish rotates in a plane parallel to the earth (azimuth) and in another plane perpendicular to it

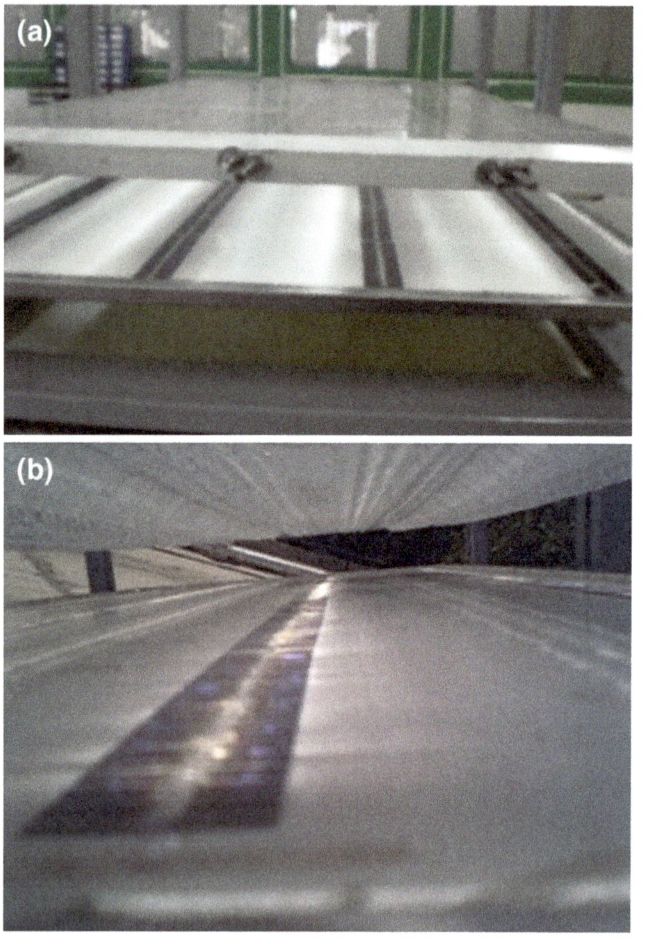

Fig. 4 Experimental setup. **a** The whole structure for testing the performance of the proposed structure. **b** Semi-Fresnel lens concentrating the incident radiation on the panel

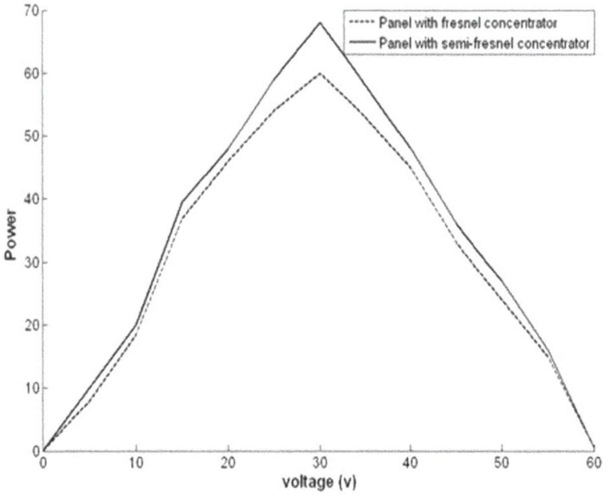

Fig. 5 Power versus voltage for the panel with Fresnel concentrator (*dashed line*) and semi-Fresnel concentrator (*solid line*)

(elevation), which gives the collector left–right and up-down rotation. Rotational rates can be easily calculated during all day. In polar tracking, which is the most common method in small solar dish systems, the collector rotates about an axis parallel to the earth's axis of rotation. The collector rotates at a constant rate of 15° per hour to match the rotational speed of the earth. The other axis of rotation, the declination axis, is perpendicular to the polar axis. Movement about this axis occurs slowly and varies by ±23.5° over a year [13].

The sun-tracking systems can be realized as closed- or open-loop control systems. The closed-loop systems use photo sensors to position the solar cell modules. Due to permanent position changes, these systems can spend more energy than they gain, especially when the weather is changing. The other possibility is to use open-loop systems where different algorithms provide pre-defined trajectories for the tracking systems. Their trajectories can be accurately determined because the relative position of the sun can be precisely calculated at any time for any location on the earth. The hybrid control systems combine the open-loop tracking trajectories based on the solar movement models and the closed-loop control, using dynamic feedback controllers [14].

Many investigations have been carried out to design and employ two-axis sun-tracking systems; however, only a few were cited in the literature that investigated the effect of using two-axis sun-tracking systems controlled by modern, computerized control systems, such as the programmable logic controller (PLC) control system. Abu-Malouh et al. [12] proposed a spherical solar cooker design and construction using two-axis sun-tracking system to control the spherical solar cooker rotation. This sun-tracking system depends on PLC and frequency control. It assures that the sunlight beam is normal to the dish at any time of the day.

Al-Mohamad [15] presented a sun-tracking design, whereby the movement of a photovoltaic module was controlled to follow the sun's radiation using a PLC unit. A PLC unit was employed to control and monitor the mechanical movement of the PV module and to collect and store data related to the sun's radiation. It also allows the connection of many PV modules in series or parallel connection. It is found that the daily output power of the PV was increased by more than 20 % in comparison with that of a fixed module and exceeded it by at least 40 % for the period of spanning morning and evening hours. The PV tracking system can be employed as a stand-alone device and it could be connected to a personal computer through the RS232 serial port to monitor the whole process on a computer screen. This technique reduces the costs of tracking systems and makes it a cost-effective technology.

Tejwany and Solanki [16] designed a novel mechanism of sun tracking with automatic cleaning of PV modules and cleaning mechanism of the PV module consists of sliding brushes, which slides over module and cleans it twice a day, wherein PV panel makes a rotation of 360° in a day. It is observed that the daily energy generation of a flat PV module (which is kept on the ground) increases the efficiency to 15 % for single axis tracking.

Seme et al. [14] proposed a new method for prediction of the time-dependent solar radiation on a tilted surface for the two-axis sun-tracking system for a PV system. The trajectories of the sun-tracking system are determined in an optimization procedure. It is applied in the new procedure for determining the tilt and azimuth angle trajectories, which assures the maximum energy production in the PV system, considering the energy consumed in the tracking system. Their method is general and gives the optimal results for the applied solar radiation prediction and the tracking system model. Generality of the proposed method remains even in the case when the applied tracking system model and the prediction of the solar radiation are replaces with the more advanced ones.

Hiramatsu et al. [17] developed a two-axis tracking apparatus, which was put into operation with non-imaging Fresnel lens. The excellent tracking ability was verified by yielding results in all three tracking performance evaluations.

Radhi Mahmood and Muhammad [18] designed and implemented a smart relay-based sun-tracking systems to keep the PV array perpendicular to the sun during the day hours. The smart relay-based controller (SR3B261BD) was used to control the motion of the two-axis sun-tracking surface. Preliminary measurements indicated that the use of the two-axis sun tracker would increase daily energy 42 % more than the stationary solar panel and 1 % more than the single axis. The designed system operated smoothly with precise positioning about 7.5° per half hour.

Chong and Wong [19] derived a novel on-axis general sun-tracking formula and integrated it into an open-loop azimuth-elevation sun-tracking system for improving the tracking accuracy. The installation defect of a 25 m^2 solar concentrator prototype had induced significant sun-tracking errors, range from 12.12 to 17.54 mrad. With the use of a CCD camera for recording the solar image, three misaligned angles from ideal azimuth-elevation axes have been determined and corrected by changing of the general formula parameters' values to better the tracking accuracy to 2.99 mrad. Further fine-tuning has been made to attain the tracking accuracy up to 0.96 mrad, by including the motor step count. The general formula has been demonstrated to be capable of achieving high tracking accuracy, cost-effectively without complicated sun-tracking solutions.

Figure 6 shows the output power of panel with and without sun-tracking method. As it is obvious in Fig. 6, tracking the sun to adjust the panel, significantly increased the output power of our PV system. Hence, we propose a two-axis open-loop automatic sun-tracking system, the flowchart of which is given in Fig. 7.

Our proposed two-axis sun-tracking algorithm works based on calculations of the incidence angle according to

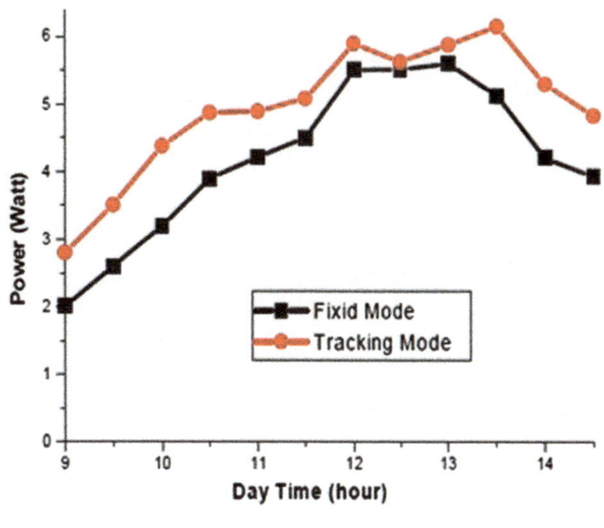

Fig. 6 Output power of PV system with and without sun tracking

date, time, and geographical position of panel (which means latitude, longitude, and zenith. The sun-tracking system (shown in Fig. 7) is set to start working at sunrise by a 24-hour timer. According to different reasons in defined and predetermined time, the command box starts working and based on the specified time, enables the motor and other components of the system to move. Thus, by rotating the main shaft which is done by an electromotor force, ratchet shaft will also revolve from east to west. In this status, two contrived gears on top of the ratchet shaft are unitized separately that will cause the panel moving from up to down and vice versa. Therefore the panel moves in two axes from east to west and up to down and vice versa. According to the calculated time and order of the command box, the panel moves towards the sun during the day. At the end of the day, using the contrived buttons while finishing the task of the system and pushing keys, the device will automatically turn to its initial dawn position and make it ready to work for the next day. Note that this system is able to rotate the panel 180° from sunrise to sunset (from east to west) during the day and 90° from morning to noon (from down to up) and 90° from noon to evening (from up to down) to move it completely towards the sun so that can make the panel benefit maximum form the incidence time of sunlight.

System components

This system is composed of four parts:

1. Base
2. Mechanical or rotator part
3. Propellant force or motor
4. Command motor.

As shown in Fig. 8, the semi-Fresnel lens is placed above the solar cell which the concentrator area is efficiently reduced. The optical segments are implemented on top of the two-axis open-loop sun-tracking system, which automatically controls the rotation of the module in order to have the orthogonal incidence angle.

Connection system

There are three kinds of connection systems for PV module. These systems are chosen for different applications based on their energy production. Here we will discuss them and for this paper we use off-grid system but PV modules with Fresnel-based refracting structure and sun-tracking method can be applied with grid-connected systems on order to compensate loses in power lines and networks.

A stand-alone system does not have a connection to the electricity mains (grid). Stand-alone systems vary widely in size and. If the load is to be supplied independently of solar insulation, the generated power is stored and buffered with a battery. In non-portable applications where weight is not an issue, lead acid batteries are most commonly used for their low cost and tolerance for abuse. A charge controller may be incorporated in the system to: (a) avoid battery damage by excessive charging or discharging and, (b) optimizing the production of the cells or modules by maximum power point tracking (MPPT). However, in simple

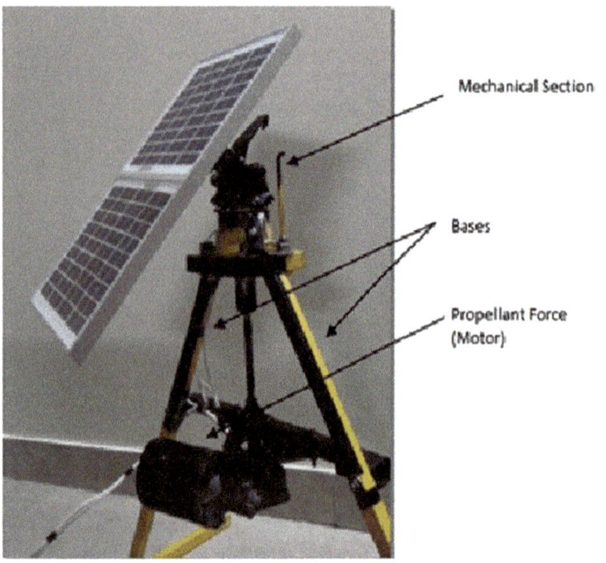

Fig. 7 Schematic of sun-tracking system and solar PV module

Fig. 8 Implementation of the system

PV systems where the PV module voltage is matched to the battery voltage, the use of MPPT electronics is generally considered unnecessary, since the battery voltage is stable enough to provide near-maximum power collection from the PV module (Fig. 9).

A grid-connected system is connected to a large independent grid (typically the public electricity grid) and feeds power into the grid. Grid-connected systems vary in size from residential (2–10 kWp) to solar power stations (up to 10 s of MWp). This is a form of decentralized electricity generation. In the case of residential or building mounted grid-connected PV systems, the electricity demand of the building is met by the PV system. Only the excess is fed into the grid when there is an excess. The feeding of electricity into the grid requires the transformation of DC into AC by a special, grid-controlled solar inverter [20]. In kW sized installations the DC side system voltage is as high as permitted (typically 1000 V) to limit ohmic losses. Most modules (72 crystalline silicon cells) generate about 160 W at 36 V. It is sometimes necessary or desirable to connect the modules partially in parallel rather than all in series [20].

A hybrid system combines PV with other forms of generation, usually a diesel generator. The other form of generation may be a type able to modulate power output as a function of demand. However more than one renewable form of energy may be used e.g. wind. The photovoltaic power generation serves to reduce the consumption of non renewable fuel. Hybrid systems are most often found on islands.

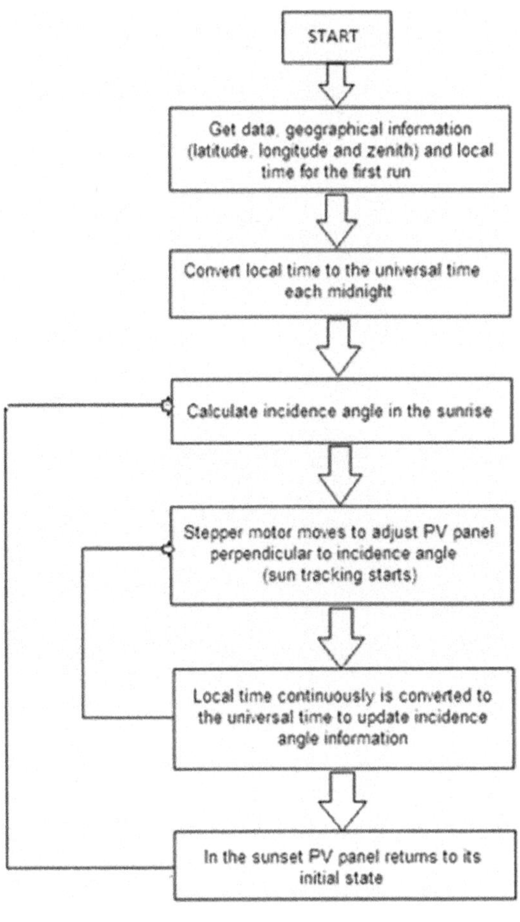

Fig. 9 Flowchart of sun-tracking system

Conclusion

In this paper we proposed a new concentrating structure for improving the performance of photovoltaic cells. To design the proposed semi-Fresnel lens, we used the simple trigonometry to determine the refraction and convergent angles of orthogonal incident ray. Then using Snell's law we determine the groove angle for each groove. As in practice the most efficiency is obtained for the orthogonal incident ray; we use a two-axis open-loop sun-tracking system to keep the module in a position at which the incoming ray is always orthogonal to the lens surface. This way the effects of TIR and spherical aberration in scattering the rays are removed. The outstanding feature of this concentrator is its lower loss and aberration compared to conventional structure with Fresnel lens. It also reduces the amount of solar cell needed to receive the sunlight. The refractive structure was tested experimentally with a narrow PV cell and the output power was calculated for different derived voltages. The resultant values were compared with the similar values for the PV cell with Fresnel concentrator. The results confirmed the superiority of the proposed method.

In future works and studies we can extend the system with proper connection method such as hybrid system. This will demonstrate higher performance for power systems and facing losses in networks while using modules connected together and working with another source of energy with respect to the pre-defined application.

We would also add automatic cleaning device –as there is space between the panel and Fresnel lens, the dust accumulated on the surface of the panel will affect the efficiency- in order to clean the panels regularly in order to enhance the efficiency by 15 % and shows higher performance in long-term usage and the automatic system is needed in cases where there is less physical access to panels.

References

1. Zhai, H., Dai, Y.J., Wu, J.Y., Wang, R.Z., Zhang, L.Y.: "Experimental investigation and analysis on a concentrating solar collector using linear Fresnel lens". Energy Convers. Manag. **51**, 48–55 (2010)
2. Mo, S., Chen, Z., Hu, P.: "Performance of a passively cooled

fresnel lens concentrating photovoltaic module. 3rd World Conference on Photovoltaic Energv Conversion", Osaka, Japan, pp. 2379–2382, May 2003
3. O'Neill, M.J., Piszczor, M.F., Eskenazi, M.I., McDanal, A.J., George, P.J., Botke, M.M., Brandhorst, H.W., Edwards, D.L., Hoppe, D.T.: Ultra-light stretched fresnel lens solar concentrator for space power applications. International symposium on optical science and technology, SPIE, No. 5179-17 (2003)
4. Piszczor, M.F., O'Neill, M.J.: Development of a dome fresnel fens/GaAs photovoltaic Cconcentrator for space applications. 19th IEEE-PVSC (1987)
5. Leutz, R., Suzuki, A., Akisawa, A., Kashiwagi, T.: Flux densities in optimum non imaging Fresnel lens solar concentrator for space. 0-7803-5772-8/00/$10.00 IEEE (2000)
6. Shvartz, M.Z., Chosta, O.I., Grilikhes, V.A., Rumyantsev, V.D., Andreev, V.M.: Space Fresnel lens concentrator modules with triple junction solar cells. 0-7803-8707-4/05/$20.00, IEEE (2005)
7. Wagner, D., He, L.: An innovative solar system with high efficiency and low cost. 978-1-4244-5892-9/10/$26.00, IEEE (2010)
8. Shvarts, M.Z., Soluyanov, A.A.: High quality fresnel lenses in a variety of sizes and focal lengths. In: Fresnel Technologies Advances in Science and Technology, Vol. 74, pp 188–195 (2010)
9. Rai, A.K., Kaushika, N.D., Singh, B., Agarwal, N.: Simulation model of ANN based maximum power point tracking controller for solar PV system. Sol. Energy Mater. Sol. Cells **95**, 773–778 (2011). doi:10.1016/j.solmat.2010.10.022
10. The Fresnel equations and Brewster law. www.Google.com
11. Omar, A.M., Shaari, S., Omar, A.R., Yahir, M.R.: An automated solar photovoltaic biaxial tracking system. In: First international power and energy conference PECon 2006, Malaysia, pp. 44–47 (2006)
12. Abu-Malouh, R., Abdallah, S., Muslih, I.M.: Design, construction and operation of spherical solar cooker with automatic sun tracking system. Energy Convers. Manag. **52**, 615–620 (2011)
13. Poullikkas, A., Kourtis, G., Hadjipaschalis, I.: Parametric comparative study for the installation of solar dish technologies in Cyprus. 7th Mediterranean Conference and Exhibition on Power Generation, Transmission, Distribution and Energy Conversion, paper No. MED10/110, Agia Napa, Cyprus, November 2010
14. Seme, S., Stumberger, G., Vorsic, J.: Maximum efficiency trajectories of a two axes sun tracking systems determined considering tracking system consumption. IEEE Transactions on power electronics, vol. 26, no. 4, April 2011, 0885-8993/$26.00 IEEE
15. Al-Mohamad, Ali: Efficiency improvements of photovoltaic panels using a sun tracking system. Appl. Energy **79**(0306–2619), 345–354 (2004)
16. Tejwany, R., Solanki, C.S.: 360° sun tracking with automated cleaning system for solar PV modules. In: Photovoltaic specialists conference (PVSC), 2010 35th IEEE. IEEE (2010). doi:10.1109/PVSC.2010.5614475
17. Hiramatsu, M., Miyazaki, Y., Egarni, T., Akisawa, A., Mizuta, Y.: Development of non-imaging Fresnel lens and sun tracking device. 3rd World Conference on Photovoltaic Energy Conversion, Osaka, Japan, pp. 2383–2385, May 2003
18. Sungur, C.: Multi-axes sun-tracking system with PLC control for photovoltaic panels in Turkey. Renew Energy **34**, 1119–1125 (2009)
19. Chong, K.-K., Wong, C.-W.: Open-loop azimuth-elevation sun tracking system using on-axis general sun tracking formula for achieving tracking accuracy of below 1 mrad. 978-1-4244-5892-9/10/$26.00 ©2010 IEEE
20. How Solar Power Works. http://www.solarcity.com/residential/how-solar-works.aspx

8

Instability due to trapped electrons in magnetized multi-ion dusty plasmas

M. M. Haider[1,2] · T. Ferdous[2] · S. S. Duha[2]

Abstract An attempt has been made to find out the effects of trapped electrons in dust-ion-acoustic solitary waves in magnetized multi-ion plasmas, as in most space plasmas, the hot electrons follow the trapped/vortex-like distribution. To do so, we have derived modified Zakharov–Kuznetsov equation using reductive perturbation method and its solution. A small-k perturbation technique was employed to find out the instability criterion and growth rate of such a wave.

Keywords Trapped electrons · Multi-ion dusty plasmas · Modified Zakharov-Kuznetsov equation

Introduction

Dust particles are found not only in laboratory plasmas (plasma processing and plasma crystal where low temperature plasmas are used) but also in astrophysical plasma systems (planetary rings, interstellar molecular clouds, protostellar disks, interstellar and circumstellar clouds, asteroid zones, planetary atmospheres, interstellar media, cometary tails, nebulae, Earth's ionosphere, etc) [1–9]. Therefore the electrostatic modes in dusty plasma have become a field of great interest. These dust grains in plasmas are very small (micron or sub-micron sized) and can have the opposite polarity due to the size effect on secondary emission; the smaller one is positively charged, whereas the larger one is negative charged [10]. The important elementary dust grain charging processes are (1) interaction of dust grains with gaseous plasma particles, (2) interaction of dust grains with energetic particles (electrons and ions) and (3) interaction of dust grains with photons. When dust grains are immersed in a gaseous plasma, the plasma particles are collected by the dust grains which act as probes. The dust grains are, therefore, charged by the collection of the plasma particles flowing onto their surfaces. When energetic plasma particles are incident onto a dust grain surface, they are either backscattered/reflected by the dust grain or they pass through the dust grain material. During their passage they may lose their energy partially or fully. A portion of the lost energy can go into exciting other electrons that in turn may escape from the material. The emitted electrons are known as secondary electrons. These secondary electrons make the grain surface positive. The interaction of photons incident onto the dust grain surface causes photoemission of electrons from the dust grain surface. The dust grains, which emit photoelectrons, may become positively charged. The emitted electrons collide with other dust grains and are captured by some of these grains which may become negatively charged. There are, of course, a number of other dust grain charging mechanisms [11], namely thermionic emission, field emission, radioactivity, impact ionization, etc. These are significant only in some different special circumstances. But both the ion temperature and ion-neutral collisions play important roles in the dust grain charging. With the decrease of ion temperature due to ion-neutral collisions, the magnitude of dust grain charge decreased accordingly. The grain charge in a collisionless regime is estimated to be much higher than that in a collisional regime [12] (Some dusty plasma parameters are shown in

✉ M. M. Haider
masum.phy@gmail.com; masum@mbstu.ac.bd

[1] Department of Physics, Mawlana Bhashani Science and Technology University, Santosh, Tangail 1902, Bangladesh

[2] Department of Physics, Jahangirnagar University, Savar, Dhaka 1342, Bangladesh

Table 1 Dusty plasma parameters [11, 16]

Characteristics	n_e (cm^{-3})	T_e (K)	n_d (cm^{-3})	r_d (μm)
Saturn's E-rings	10 to 20	10^5–10^6	10^{-7} to 10^{-6}	0.5–1.5
Interstellar clouds	10^{-4} to 10^{-3}	10–20	10^{-7} to 10^{-6}	0.1–0.5
Zodiacal dust disc	1 to 10	10^4–10^5	10^{-12} to 10^{-11}	1–10
Haley's comet	10^2 to 10^4	10^3–10^4	10^{-8} to 10^{-3}	0.1–10
Rocket exhaust	10^{12} to 10^{13}	10^3–10^4	10^7 to 10^8	0.1–1
Q-machine	10^6 to 10^7	10^3–10^4	10^3 to 10^4	10–20

n_e electron density, T_e electron temperature, n_d dust density, r_d dust radius

Table 1). The existence of positive and negative ion plasma are shown in the experimental investigation of Cooney et al. [13]. The latter may also support the formation of ion acoustic shocks when the ratio of the negative ion to positive ion number density exceeds about 0.9 [14]. It should be noted that collective interactions in positive and negative ion plasmas have potential applications in natural and technological environments like the D-region of the Earth's ionosphere, the Earth's mesosphere, the solar photosphere, and the microelectronics plasma processing reactors [15].

In 2008 Sayed et al. [17] considered non-magnetized dusty plasma mode having positive and negative ions with positive and negative dust and Maxwellian distributed electrons. But the most real plasmas are magnetized, and it can change its characteristics according to the wave direction. Considering this magnetic properties Haider et al. [18] have studied the instability of solitary structure containing positive and negative ions, Mxwellian electrons and positively and negatively charges stationary dust. These modes are only valid if a complete depletion of the background electrons and ions is possible, and both positive and negative dust fluids are cold. In practice, the hot electrons may not follow a Maxwellian distribution due to the formation of phase space holes caused by the trapping of hot electrons in a wave potential. Accordingly, in most space plasmas, the hot electrons are trapped following the vortex-like distribution [19–21]. Rahaman and Manun [22] have explained the effect of trapped electrons in dust-ion-acoustic (DIA) solitary waves (SWs) with arbitrarily charged dust. But they did not consider magnetic field. Later Haider et al. [23] have also studied the propagation of SWs in the presence of magnetized plasma with both positive and negative ions fluid, vortex-like distributed electrons and charge fluctuating stationary dusts. They have derived modified Korteweg-de Vries (mK-dV) equation and its solution. K-dV or mK-dV equation has studied for one dimensional cases. So, a necessary is felt to study the SWs in three dimensionally and its instability criterion as well as its growth rate.

In the present work, the propagation of DIA Solitary structures have been studied in magnetized dusty plasma consisting negatively and positively charged ion fluid, trapped electrons following vortex-like distribution, and arbitrary charged stationary dust where restoring force provided by the plasma thermal pressure of electrons and the inertia is due to the ion mass for instability analysis. Using reductive perturbation method [24] we have derived the modified Zakharov–Kuznetsov (mZK) equation which is also known as mK-dV equation in three dimension and its solution. We have also studied instability criterion and growth rate for such a SWs.

Basic equations

In the present work, collisionless magnetized dusty plasmas have been considered. We assume that (1) the ions (negatively and positively charged) are mobile, (2) the electrons follow the vortex-like distribution, and (3) charge fluctuating stationary dust. We have also consider that there is an external static magnetic field $\mathbf{B}_0$ acting along the z-direction ($\mathbf{B}_0 = \hat{k}B_0$), where $\hat{k}$ is the unit vector along the z-direction which is very strong that the electrons and dusts are moving along the magnetic field direction very fast, i.e. the response of electrons and dusts look like as that in the unmagnetized plasma. The nonlinear dynamics of the DIA SWs propagating in such a multi-component dusty plasma is governed by

$$\frac{\partial n_s}{\partial t} + \nabla \cdot (n_s \mathbf{u}_s) = 0, \quad (1)$$

$$\frac{\partial \mathbf{u}_n}{\partial t} + (\mathbf{u}_n \cdot \nabla)\mathbf{u}_n = \nabla\psi - \omega_{cn}\mathbf{u}_n \times \hat{k}, \quad (2)$$

$$\frac{\partial \mathbf{u}_p}{\partial t} + (\mathbf{u}_p \cdot \nabla)\mathbf{u}_p = -\beta\nabla\psi + \beta\omega_{cn}\mathbf{u}_p \times \hat{k}, \quad (3)$$

$$\nabla^2 \psi = \mu_0 n_e + n_n - \mu_p n_p - j\mu_d, \quad (4)$$

where n_s (n_n/n_p) is the ion number density (negative/positive) normalized by its equilibrium value n_{s0}, u_n (u_p) is the negative (positive) ion fluid speed normalized by $C_n = (k_B T_e/m_n)^{1/2}$, with k_B is the Boltzmann constant, T_e is the temperature of electrons and m_n being the rest mass of negative ions. ψ is the DIA wave potential normalized by $k_B T_e/e$, with e being the magnitude of the charge of an electron. ω_{cn} is the negative ion cyclotron frequency (eB_0/m_{nc}) normalized by plasma frequency $\omega_{pn} = (4\pi n_{n0}e^2/m_n)^{1/2}$ with c being the speed of light. The time variable (t) is normalized by $\omega_{pn}{}^{-1}$, the space variables are normalized by Debye radius $\lambda_D = (k_B T_e/4\pi n_{n0}e^2)^{1/2}$.

At equilibrium we have

$$n_{p0} + jn_{d0} = n_{e0} + n_{n0},$$

where, $jn_{d0} = n_{d+} - n_{d-}$ with n_{d+} being the positive dust number density and n_{d-} being the number density of negative dust. $j = 1$ for the condition $n_{d+} > n_{d-}$ and $j = -1$ for the condition $n_{d+} < n_{d-}$, i.e. the value of j dependents on net charge of dust grain and β is the mass ratio of negative ion to positive ion (m_n/m_p). We can also write

$$\mu_0 = \mu_p + j\mu_d - 1, \tag{5}$$

where, $\mu_0 = n_{e0}/n_{n0}$, $\mu_p = n_{p0}/n_{n0}$ and $\mu_d = n_{d0}/n_{n0}$.

To model the electron distribution in presence of trapped particles, we employ a vortex-like electron distribution of Schamel [19], which solves the Vlasov equation and using the similar procedure of Haider et al. [23] one can have the distribution of electron number density as

$$n_e = \left[1 + \psi - \frac{4(1-\sigma)}{3\sqrt{\pi}}\psi^{3/2} + \frac{1}{2}\psi^2\right]. \tag{6}$$

Derivation of mZK equation

To derive a dynamical equation for the nonlinear propagation of the electrostatic waves in a magnetized dusty plasma, under consideration, we use (1)–(4) and (6), and employ the reductive perturbation technique [24]. To do so, introducing the stretched coordinates [25–28]

$$\left.\begin{aligned} X &= \epsilon^{1/4}x,\\ Y &= \epsilon^{1/4}y,\\ Z &= \epsilon^{1/4}(z - V_0 t),\\ \tau &= \epsilon^{3/4}t, \end{aligned}\right\} \tag{7}$$

where ϵ is a smallness parameter $(0<\epsilon<1)$ measuring the weakness of the dispersion and V_0 is the Mach number (the phase speed normalized by C_i). n_i, u_i and ψ can be expand about their equilibrium values in power series of ϵ, viz.,

$$\left.\begin{aligned} n_s &= 1 + \epsilon n_s^{(1)} + \epsilon^{3/2} n_s^{(2)} + \cdots,\\ \psi &= \epsilon\psi^{(1)} + \epsilon^{3/2}\psi^{(2)} + \cdots,\\ u_{sx} &= \epsilon^{5/2}u_{sx}^{(1)} + \epsilon^{3/2}u_{sx}^{(2)} + \cdots,\\ u_{sy} &= \epsilon^{5/2}u_{sy}^{(1)} + \epsilon^{3/2}u_{sy}^{(2)} + \cdots,\\ u_{sz} &= \epsilon u_{sz}^{(1)} + \epsilon^{3/2}u_{sz}^{(2)} + \cdots, \end{aligned}\right\} \tag{8}$$

here, s represents the species (n for negative ions and p for positive ions).

Using the stretched coordinates and (8) in (1)–(4) and equating the coefficients of $\epsilon^{5/4}$ from the continuity and momentum equation, one can obtain the x-, y- and z-components of momentum equations, and first-order continuity equations as

$$\left.\begin{aligned} u_{nx}^{(1)} &= -\frac{1}{\omega_{cn}}\frac{\partial\psi^{(1)}}{\partial Y}, u_{ny}^{(1)} = \frac{1}{\omega_{cn}}\frac{\partial\psi^{(1)}}{\partial X},\\ u_{nz}^{(1)} &= -\frac{1}{V_0}\psi^{(1)}, n_n^{(1)} = -\frac{1}{V_0^2}\psi^{(1)},\\ u_{px}^{(1)} &= -\frac{1}{\omega_{cn}}\frac{\partial\psi^{(1)}}{\partial Y}, u_{py}^{(1)} = \frac{1}{\omega_{cn}}\frac{\partial\psi^{(1)}}{\partial X},\\ u_{pz}^{(1)} &= \frac{\beta}{V_0}\psi^{(1)}, n_p^{(1)} = \frac{\beta}{V_0^2}\psi^{(1)}. \end{aligned}\right\} \tag{9}$$

Equating the coefficients of ϵ from Poisson's equation, we get

$$0 = \mu_0\psi^{(1)} + n_n^{(1)} - n_p^{(1)}. \tag{10}$$

Using the value of $n_n^{(1)}$ and $n_p^{(1)}$ from (9) and into (10), we get the linear dispersion relation

$$V_0 = \sqrt{\frac{1 + \beta\mu_p}{\mu_0}}. \tag{11}$$

Again, following the same procedure, one can obtain the next higher order continuity equations as

$$\frac{\partial n_s^{(1)}}{\partial\tau} - V_0\frac{\partial n_s^{(2)}}{\partial Z} + \frac{\partial u_{sx}^{(2)}}{\partial X} + \frac{\partial u_{sy}^{(2)}}{\partial Y} + \frac{\partial u_{sz}^{(2)}}{\partial Z} = 0, \tag{12}$$

here, s represents the species (n for negative ions and p for positive ions).

The z-component of momentum equations are

$$\left.\begin{aligned} &\frac{\partial u_{nz}^{(1)}}{\partial\tau} - V_0\frac{\partial u_{nz}^{(2)}}{\partial Z} - \frac{\partial\psi^{(2)}}{\partial Z} = 0,\\ &\frac{\partial u_{pz}^{(1)}}{\partial\tau} - V_0\frac{\partial u_{pz}^{(2)}}{\partial Z} + \beta\frac{\partial\psi^{(2)}}{\partial Z} = 0. \end{aligned}\right\} \tag{13}$$

To the next higher order of ϵ, i.e. equating the coefficients of $\epsilon^{3/2}$, we can express Poisson's equation, and x- and y-components of the momentum equations for both negative and positive ions as

$$\left(\frac{\partial^2}{\partial X^2}+\frac{\partial^2}{\partial Y^2}+\frac{\partial^2}{\partial Z^2}\right)\psi^{(1)} = \mu_0\psi^{(2)} - \mu_0\frac{4(1-\sigma)}{3\sqrt{\pi}}\left[\psi^{(1)}\right]^{3/2} + n_n^{(2)} - \mu_p n_p^{(2)}, \quad (14)$$

$$\left.\begin{aligned} u_{nx}^{(2)} &= -\frac{V_0}{\omega_{cn}^2}\frac{\partial^2\psi^{(1)}}{\partial Z\partial X}, \quad u_{ny}^{(2)} = -\frac{V_0}{\omega_{cn}^2}\frac{\partial^2\psi^{(1)}}{\partial Y\partial Z},\\ u_{px}^{(2)} &= \frac{V_0}{\beta\omega_{cn}^2}\frac{\partial^2\psi^{(1)}}{\partial Z\partial X}, \quad u_{py}^{(2)} = \frac{V_0}{\beta\omega_{cn}^2}\frac{\partial^2\psi^{(1)}}{\partial Y\partial Z}.\end{aligned}\right\} \quad (15)$$

Now, using (9)–(15), we can readily obtain

$$\frac{\partial\psi^{(1)}}{\partial\tau} + AB\sqrt{\psi^{(1)}}\frac{\partial\psi^{(1)}}{\partial Z} + \frac{1}{2}A\frac{\partial}{\partial Z}\left[\frac{\partial^2}{\partial Z^2} + D\left(\frac{\partial^2}{\partial X^2}+\frac{\partial^2}{\partial Y^2}\right)\right]\psi^{(1)} = 0, \quad (16)$$

where

$$\left.\begin{aligned} A &= \left(\frac{V_0}{\mu_0}\right),\\ B &= \frac{\mu_0(1-\sigma)}{\sqrt{\pi}},\\ D &= 1 + \frac{1}{\omega_{cn}^2} + \frac{\mu_p}{\beta\omega_{cn}^2}.\end{aligned}\right\} \quad (17)$$

The Eq. (16) is known as the modified Zakharov–Kuznetsov (mZK) equation.

Solution of mZK equation

The solution of the mZK equation [29] is given by

$$\psi_0(\mathcal{Z}) = \psi_m \mathrm{sech}^4(\mathcal{Z}/\Delta), \quad (18)$$

where $\psi_m = (15U/8\delta_1)^2$ is the amplitude and $\Delta = \sqrt{16\delta_2/U}$ is the width of the solitary waves with U is soliton speed normalized by the positive ion-acoustic speed (C_i) and

$$\delta_1 = AB\cos\delta,$$
$$\delta_2 = \frac{1}{2}A(\cos^3\delta + D\sin^2\delta\cos\delta).$$

It has been found that the amplitude of the solitary waves is proportional to the soliton speed U and the width is inversely proportional to this soliton speed. That is, the profile of the faster soliton will be taller and narrower then slower one. From (17) we can say that V_0, μ_0 and $(1-\sigma)$ are always positive, i.e. A and B are positive. It means that the solitary waves associate with positive potential always. Similar result have found in the work of Rahman and Mamun [22] and Haider et al. [23]. In both the work only positive solitary waves are found. The soliton amplitude increases with increasing temperature ratio of free to trapped electrons for both positive and negative dust grains; but Fig. 1 indicates that amplitude goes higher for increasing the number density of positively charged dust grains whereas the amplitude decreases with increasing the negatively charged dust grain number density. The number density of positive or negative dust grains effect the net charge of the dust grains. As $jn_d = n_{d+} - n_{d-}$, the number density of positive dust grains increases in the system it causes the increase the positiveness or decreasing the negativeness of the net charge. Similarly number density of negative dust grains causes the richness of negativity of the net charge. That is the net charge of the dust grains effect the amplitude of the SWs but can't make it negative. Figure 2 indicates the variation of amplitude of SWs with propagation angle (δ) and mass ratio of negative to positive ions (β). For both the cases of charge density of dust grains ($j = 1$ or $j = -1$) the amplitude increases with propagation angle. It is found that the higher the value of β, the lower the amplitude. But the variation of width with propagation angle not similar to amplitude. For lower limit of the angle $(0^\circ - 50^\circ)$ the width increases with it and decreases for higher limits of the angle $(50^\circ - 90^\circ)$ as shown in Fig. 3 for both positively and negatively charged dust grains. It is also clear that an increase of the external magnetic field leads to a decrease in the potential width, i.e., a stronger magnetic field leads to steeper and thus narrower soliton profiles. This can be related to the effects of transverse perturbation. The Larmor radius for the ion motions are smaller at larger gyration frequency and then they contribute less to the nonlinearity of the plasma. Lower nonlinearity leads to lower dispersion which cause the reduction in the soliton width and soliton become more spiky. The width of the SWs is lower for positive dust

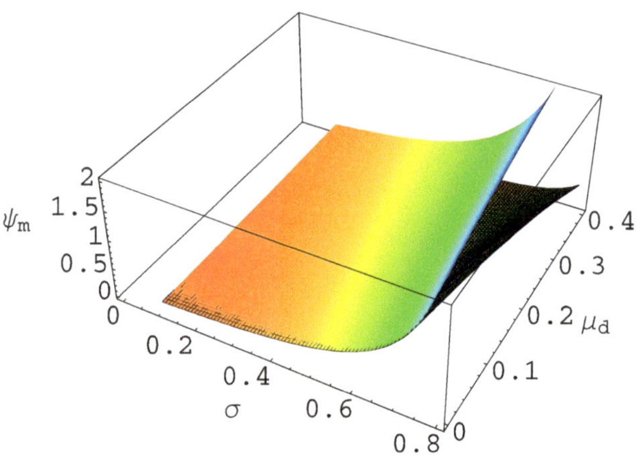

Fig. 1 (Color online) Variation of the amplitude of solitary wave σ and μ_d for $U = 0.1$, $\beta = 0.1$, $\mu_p = 1.5$ and $\delta = 45^\circ$ where *upper surface* represents the value of $j = 1$ and *lower surface* represent the value of $j = -1$

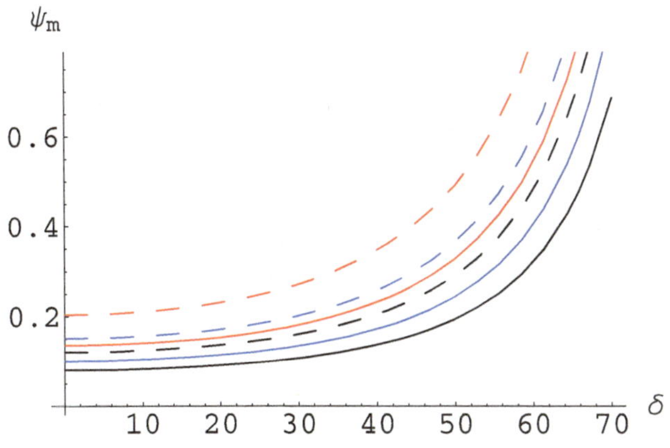

Fig. 2 (Color online) Variation of the amplitude of solitary wave (*dashed curve* for $j=1$ and *solid curve* for $j=-1$) with δ for $U=0.1$, $\sigma=0.5$, $\mu_p=1.5$ and $\mu_d=0.1$ having the values of $\beta=0.2$ (*red*), $\beta=0.5$ (*blue*) and $\beta=0.8$ (*black*)

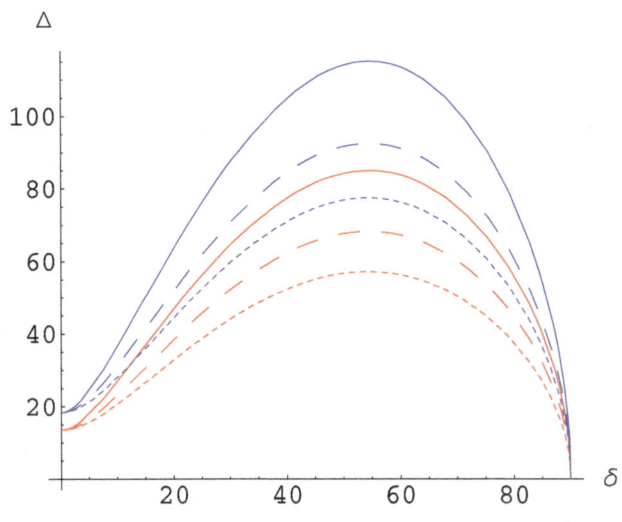

Fig. 3 (Color online) Variation of the width of solitary wave (*red curve* for $j=1$ and *blue curve* for $j=-1$) with δ for $U=0.1$, $\beta=0.1$, $\mu_p=1.5$ and $\mu_d=0.1$ having the values of $\omega_{ci}=0.2$ (*solid line*), 0.4 (*dashed line*) and 0.6 (*dotted line*)

grains ($j=1$) then negative dust grains ($j=-1$), thus the positive dust grains makes the soliton profile more steeper. From the Fig. 4 we have found that the width of the solitary waves decreases with both μ_p and β but it is higher for $j=-1$ then $j=1$.

Instability

Using the method of small-k perturbation expansion the instability criterion [29–36] of the obliquely propagating solitary waves can be express as $S_i>0$ where

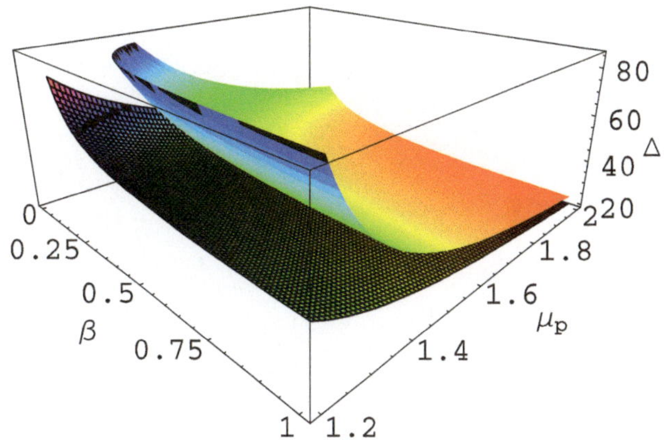

Fig. 4 (Color online) Variation of the width of solitary wave with β and μ_p for $U=0.1$, $\omega_{ci}=0.5$, $\delta=30^\circ$ and $\mu_d=0.1$ where *upper surface* represents the value of $j=-1$ and *lower surface* represent the value of $j=1$

$$S_i = l_\eta^2\left[\beta\omega_{cn}^2+(1+\mu_p)\sin^2\delta\right] + l_\zeta^2\left[\beta\omega_{cn}^2-\frac{9}{7}\left(\beta\omega_{cn}^2+\beta+\mu_p\right)\tan^2\delta\right]. \tag{19}$$

If this instability criterion $S_i>0$ is satisfied, the growth rate Γ of the unstable perturbation of these solitary waves is given by [29–36]

$$\Gamma=\frac{2U\sqrt{(\beta\omega_{ci}^2+\beta+\mu_p)S_i}}{\sqrt{63}(\beta\omega_{ci}^2+\beta+\mu_p\sin^2\delta)}. \tag{20}$$

The Eq. (20) represents that the growth rate Γ of the unstable perturbation is a linear function of DIA wave speed U, but a nonlinear function of propagating angle δ, ion-cyclotron frequency ω_{cn}, negative to positive ion mass ratio β, ratio of positive ion to negative ion number density μ_p and direction cosine (l_ζ, and l_η). The variation of growth rate (Γ) with nonlinear functions δ, ω_{ci}, β, μ_p, l_ζ and l_η are shown in Figs. 6, 7 and 8. Earlier, $S_i=0$ plot has been shown in Fig. 5, where the surface indicates that the critical condition of the SWs to be stable or unstable.

Figure 6 indicates the variation of the growth rate with δ and ω_{cn} for the value of $U=0.1$, $\mu_p=1.5$, $l_\eta=0.6$, $l_\zeta=0.5$ and $\beta=0.1$. The growth rate varies inversely with both propagating angle and negative ion ciclotron frequency. The growth rate (Γ) increases with increasing μ_p but decreases with increasing β (Fig. 7). The direction cosines l_ζ and l_η nonlinearly effect the growth rate which is shown in Fig. 8 for $U=0.1$, $l_\eta=0.6$, $l_\zeta=0.5$, $\omega_{cn}=0.5$ and $\delta=10^\circ$. l_η enhances the growth rate where as l_ζ does this inversely.

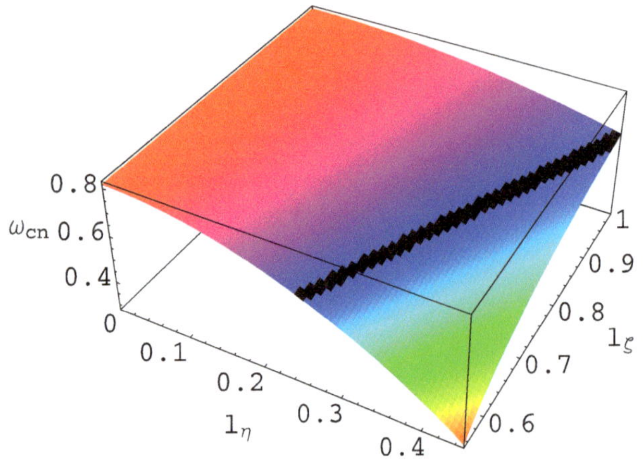

Fig. 5 (Color online) $S_i = 0$ *curve*, showing the variation of ω_{cn} with l_η and $l_\zeta = 0.5$ for $\mu_p = 1.5$, $\delta = 10°$ and $\beta = 0.1$

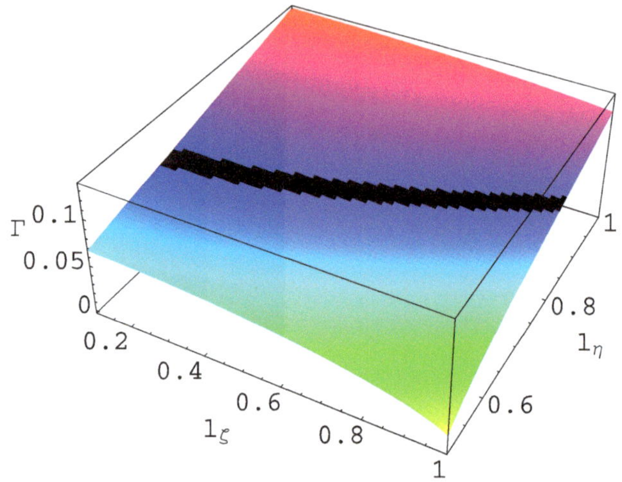

Fig. 8 (Color online) Variation of the growth rate (Γ) with l_ζ and l_η for $U = 0.1$, $\beta = 0.1$, $\mu_p = 1$, $\omega_{cn} = 0.5$ and $\delta = 10°$

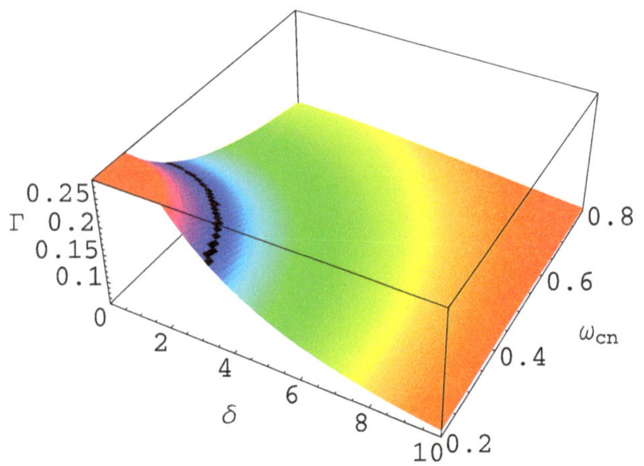

Fig. 6 (Color online) Variation of the growth rate (Γ) with δ and ω_{cn} for $U = 0.1$, $\mu_p = 1.5$, $l_\eta = 0.6$, $l_\zeta = 0.5$ and $\beta = 0.1$

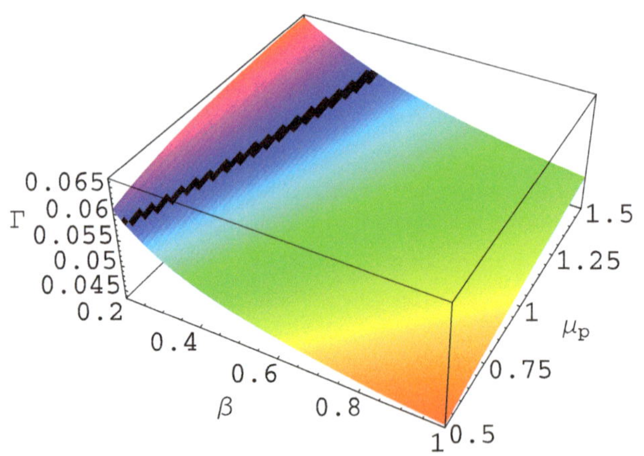

Fig. 7 (Color online) Variation of the growth rate (Γ) with β and μ_p for $U = 0.1$, $l_\eta = 0.6$, $l_\zeta = 0.5$, $\omega_{cn} = 0.5$ and $\delta = 10°$

Discussion

The nonlinear propagation of DIA solitary waves in multi component dusty plasma has analyzed where inertia provided by the positive and negative ions and restoring forces are provided by the hot trapped electrons in the presence of external magnetic field using reductive perturbation method. To analyzed this modified ZK equation has been derived as well as its solution. After that, the instability criterion and growth rate has also been studied using small-k perturbation technique. The results can be summarized as follows:

1. The trapped electrons are responsible for DA solitary waves which have smaller width, larger amplitude, and higher propagation velocity than that involving Maxwellian electrons, and that they can be represented in the form $\mathrm{sech}^4(\mathcal{Z}/\Delta)$, instead of $\mathrm{sech}^2(\mathcal{Z}/\Delta)$ which is the stationary solution of the standard ZK equation.
2. The dynamics of weakly dispersive non-linear DIA waves in the presence of the vortex-like distributed electrons is governed by the mZK equation instead of ZK equation, the stationary solution of which is represented in the form of an inverted secant hyperbolic fourth profile. Thus, the potential polarity of the DIA solitary waves in this system is different from the usual IA, DA or DIA solitary waves.
3. The solitary waves may associate with only positive potential.
4. The amplitude of SWs for dust grain having positively charged is higher then that for negatively charged but opposite picture is found in the case of width. It means that SWs associated with negative charged dust grain is narrower. It has also been

shown that the basic features (height and thickness) of such DIA solitary structures are completely different from those of the usual IA solitary structures.
5. The soliton speed U effects the amplitude linearly and the width inversely, the profile of the faster solitary wave will be taller and narrower then slower one.
6. The amplitude and width of the solitary wave are significantly modified by the parameter temperature ratio of free and trapped elections.
7. The width of the solitary waves decreases with increasing ratio of the positive and negative ion mass, positive ion number density and ion-cyclotron frequency. This indicates that higher the ratio of positive and negative ion mass, positive ion number density and ion-cyclotron frequency narrower the soliton profile. The width of SWs also increases with propagating angle for its lower range, but decreases for its upper range.
8. The SWs are more narrower for positively charged dust grain then for negative charged dust.
9. The magnitude of the external magnetic field $\mathbf{B}_0$ has no direct effect on the SW amplitude. However, it does have a direct effect on the width of the SWs and we have found that, as the magnitude of $\mathbf{B}_0$ increases, the width of the waves decreases, i.e. the magnetic field makes the solitary structures more spiky.
10. The parametric regimes for which the solitary waves become stable and unstable are identified. These are ion cyclotron frequency, direction of propagation and direction cosine.
11. Direction of propagation, ion cyclotron frequency, direction cosine (l_ζ and l_η), ratio of positive and negative mass and positive ion number density are the depending factors which can significantly modify the growth rate (Γ) of the unstable solitary structures.

It should be noted that the width of the SWs goes zero and amplitude goes to infinity and $\delta \rightarrow 90°$. It means that, for large angles, the assumption that the waves are electrostatic is no longer valid, and we should look for fully electromagnetic structures. Finally, The present work can provide a guideline to explain solitary structure of D-region of the Earth's ionosphere and mesosphere, solar photosphere and the microelectronics plasma processing reactors, which will be able to detect the DIA solitary structures, and to identify their basic features predicted in this theoretical investigation.

Acknowledgments One of the authors M. M. Haider acknowledges the financial support of the Dutch Bangla Bank Ltd. The authors would like to thank the 'Professor A. A. Mamun, Department of Physics, Jahangirnagar University' for his valuable suggestions during revising the manuscript.

References

1. Morfill, G.E., Ivlev, A.V.: Complex plasmas: an interdisciplinary research field. Rev. Mod. Phys. **81**, 1353 (2009)
2. Shukla, P.K., Eliasson, B.: Fundamentals of dust-plasma interactions. Rev. Mod. Phys. **81**, 25 (2009)
3. Ishihara, O.: Complex plasma: dusts in plasma. J. Phys. D **40**, R121 (2007)
4. Shukla, P.K.: Dust ion-acoustic shocks and holes. Phys. Plasmas **7**, 1044 (2000)
5. Pieper, J.B., Goree, J.: Dispersion of plasma dust acoustic waves in the strong-coupling regime. Phys. Rev. Lett. **77**, 3137 (1996)
6. Rosenberg, M., Merlino, R.L.: Ion-acoustic instability in a dusty negative ion plasma. Planet. Space Sci. **55**, 1464 (2007)
7. Goertz, C.K.: Dusty plasmas in the solar system. Rev. Geophys. **27**, 271 (1989)
8. Mendis, D.A., Horanyi, M.: Cometary plasma processes. AGU Monogr. **61**, 17 (1991)
9. Mendis, D.A., Rosenberg, M.: Cosmic Dusty Plasma. Anu. Rev. Astron. Astrophys. **32**, 418 (1994)
10. Chow, V.W., Mendis, D.A., Rosenberg, M.J.: Role of grain size and particle velocity distribution in secondary electron emission in space plasmas. Geophys. Res. (Space Phys.) **98**, 19065 (1993)
11. Shukla, P.K., Mamun, A.A.: Introduction to dusty plasma physics. IoP Publishing Ltd., Bristol (2002)
12. Sekine, W., Ishihara, O.: Cryogenic effect on dust grain charging in a plasma. J. Plasma Fusion Res. **9**, 416–421 (2010)
13. Cooney, I.L., Gavin, M.T., Tao, I., Lonngren, K.E., Trans, I.E.E.E.: A two-dimensional soliton in a positive ion-negative ion plasma. Plasma Sci. **19**, 1259 (1991)
14. Luo, Q.Z., D'Angelo, N., Merlino, R.L.: Shock formation in a negative ion plasma. Phys. Plasmas **5**, 2868 (1998)
15. Kim, S.H., Merlino, R.L.: Electron attachment to C-7 F-1-4 and SF-6 in a thermally ionized potassium plasma. Phys. Rev. E **76**, 035401 (2007)
16. Verheest, F.: Waves in dusty plasmas. Kluwer Academic, Dordrecht (2000)
17. Sayed, F., Haider, M.M., Mamun, A.A., Shukla, P.K., Eliasson, B., Adhikary, N.: Dust ion-acoustic solitary waves in a dusty plasma with positive and negative ions. Phys. Plasmas **15**, 063701 (2008)
18. Haider, M.M., Ferdous, T., Duha, S.S., Mamun, A.A.: Dust-ion-acoustic solitary saves in multi-component magnetized plasmas. Open J. Modern Phys. **1**(2), 13–24 (2014)
19. Schamel, H.: Stationary solitary, snoidal and sinusoidal ion acoustic waves. Plasma Phys. **14**, 905–924 (1972)
20. Schamel, H.: A modified Korteweg-de Vries equation for ion acoustic wavess due to resonant electrons. J. Plasma Phys. **9**, 377–387 (1973)
21. Schamel, H.: Theory of electron holes. Phys. Scr. **20**, 336 (1979)
22. Rahman, O., Mamun, A.A.: Dust-ion-acoustic solitary waves in dusty plasma with arbitrarily charged dust and vortex-like electron distribution. Phys. Plasmas **18**, 083703 (2011)
23. Haider, M.M., Ferdous, T., Duha, S.S.: The effects of vortex like distributed electron in magnetized multi-ion dusty plasmas. Cent. Eur. J. Phys. **1**(2), 13–24 (2014)
24. Washimi, H., Taniuti, T.: Propagation of ion-acoustic solitary waves of small amplitude. Phys. Rev. Lett. **17**, 996 (1966)
25. Mamun, A.A., Shukla, P.K.: Lower hybrid drift wave turbulence and associated electron transport coefficients and coherent

structures at the magnetopause boundary layer. J. Geophys. Res. **107**(A7), 1135 (2002)
26. Das, G.C., Dwivedi, C.B., Talukdar, M., Sarma, J.: A new mathematical approach for shock-wave solution in a dusty plasma. Phys. Plasmas **4**, 4236 (1997)
27. Anowar, M.G.M., Mamun, A.A.: Multidimensional instability of electron-acoustic solitary waves in a magnetized plasma with vortexlike electron distribution. Phys. Plasmas **15**, 102111 (2008)
28. Duha, S.S., Anowar, M.G.M., Mamun, A.A.: Dust ion-acoustic solitary and shock waves due to dust charge fluctuation with vortexlike electrons. Phys. Plasmas **17**, 103711 (2010)
29. Anowar, M.G.M., Mamun, A.A.: Effects of two-temperature electrons and trapped ions on multidimensional instability of dust-acoustic solitary waves. IEEE Trans. Plasma Sci. **37**, 8 (2009)
30. Witt, E., Lotko, W.: Ion-acoustic solitary waves in a magnetized plasma with arbitrary electron equationof state. Phys. Fluids **26**, 2176 (1983)
31. Rowlands, G.: Stability of non-linear plasma waves. J. Plasma Phys. **3**, 567 (1969)
32. Infeld, E.: On the stability of nonlinear cold plasma waves. J. Plasma Phys. **8**, 105 (1972)
33. Laedke, E.W., Spatschek, K.H.: Growth rates of bending KdV solitons. J. Plasma Phys. **28**, 469 (1982)
34. Mamun, A.A., Cairns, R.A.: Stability of solitary waves in a magnetized non-thermal plasma. J. Plasma Phys. **56**, 175 (1996)
35. Haider, M.M., Akter, S., Duha, S.S., Mamun, A.A.: Multi-dimensional instability of electrostatic solitary waves in ultra-relativistic degenerate electron-positron-ion plasmas. Cent. Eur. J. Phys. **10**(5), 1168–1176 (2012)
36. Haider, M.M., Mamun, A.A.: Ion-acoustic solitary waves and their multi-dimensional instability in a magnetized degenerate plasma. Phys. Plasmas **19**, 102105 (2012)

9

Temperature dependence of resistivity of RFeAsO compounds

S. Mukherjee[1] · Papri Dasgupta[2] · Asok Poddar[2] · Chandan Mazumdar[2]

Abstract The resistivity (ρ) data for RFeAsO compounds (R = Ce, Pr, Nd, Sm), in the temperature (T) range 35–315 K have been analyzed to identify the dominant scattering mechanisms. Close to the room temperature, the system appears to be a metal with low electron density, and the electron–phonon scattering is the dominant one. At lower temperatures, electron–electron scattering plays an important role. In an intermediate temperature region, unlike metallic system, $d\rho/dT$ is negative; and ρ^{-1} varies as $\ln T$ as in a state of weak localization. We look into the origin of negative $d\rho/dT$. The analysis of $\rho(T)$ data below the SDW transition temperature shows the presence of electron–electron interaction in addition to a SDW energy gap, and also gives an estimate of the SDW energy gap.

Keywords Oxypnictides · Resistivity · Transport · Weak localization

Introduction

Extensive research work has been done on the oxypnictides after the discovery of superconductivity in fluorine (F)-doped LaFeAsO by Kamihara et al. [1]. Interestingly, the superconducting compound contains the most familiar ferromagnetic atom "iron", and the observed superconducting transition temperature (T_c) for $LaFeAs(O_{0.89}F_{0.11})$ is relatively high (26 K) [2]. Takahashi et al. reported that T_c of this compound rises to 43 K under a high pressure of $\approx$ 4GPa [3]. Subsequently, superconductivity has been reported in other $RFeAs(O_{1-x}F_x)$ compounds with R = Ce, Pr, Nd, Sm, etc., and the reported values of T_c are 41 K in $CeFeAsO_{0.84}F_{0.16}$, 52 K in $PrFeAsO_{0.89}F_{0.11}$, 54.3 K in $NdFeAsO_{1-y}$, 55 K in $SmFeAsO_{0.9}F_{0.1}$, and 54 K in $GdFeAsO_{1-y}$ [4]. By now the iron-based superconductors have been extended to a much larger group containing variety of materials, mainly, four prototypical families 1111, 122, 111, and 11 types, and further variations such as 42622-type iron pnictides and 122-type iron chalcogenides [4].

RFeAsO compounds, with R=La or rare earth elements belong to the 1111-type family. At room temperature, they are of ZrCuSiAs-type structure belonging to the tetragonal (*T*) *P4/nmm* space group. The atomic structure consists of negatively charged Fe–As layers, and positively charged R–O layers. The structure may also be thought as a stacking of flat square nets of single atom types in the sequence 2O–R–As–2Fe–As–R. The layers are stacked along the c-axis, and such stacking of these layers gives tetrahedral coordination to Fe and O, while R and As are in square antiprismatic coordination [5, 6].

With the lowering of temperature, RFeAsO undergoes a structural phase transition (SPT) resulting into a stretching of the square Fe nets into rectangular nets, and the symmetry is reduced to orthorhombic (O) (*Cmma* space group) in the low-temperature phase. Polycrystalline RFeAsO samples show a broad continuous SPT [7]. Subsequently, an antiferromagnetically ordered state is formed with stripe-like magnetic order of small moments on the Fe sites. Regarding the origin of this antiferromagnetic (AFM) order, there are two propositions. One group identifies this order with the spin-density wave (SDW) [8] driven by

✉ S. Mukherjee
sanseb68@yahoo.co.in

[1] Department of Physics, The University of Burdwan, Golapbag, Burdwan, West Bengal 713 104, India

[2] Saha Institute of Nuclear Physics, 1/AF Bidhannagar, Kolkata, West Bengal 700 064, India

Fermi surface nesting between the electron pockets at M point and hole pockets at Γ point based on band structural calculations, while the other group suggests that the AFM order originates from the short-range magnetic exchange coupling between local moments [9].

Neutron diffraction studies for polycrystalline LaFeAsO suggest that the compound undergoes SPT at $T_S = 155$ K, and a transition to an AFM state at $T_N = 137$ K with an ordered magnetic moment of 0.36(5) μ_B [10]. Inelastic neutron scattering studies on polycrystalline LaFeAsO confirmed the presence of two-dimensional magnetic fluctuations even up to room temperature. Such fluctuations are expected to introduce dynamic disorder of the O/T mixed phase in an otherwise ordered T phase at temperatures above the SPT [11]. Elastic neutron scattering studies on single-crystalline LaFeAsO also concluded that the finite local O precursors appear to form at $T_S = 156$ K, although the O–T SPT occurs at 148 K. The SDW state is observed below $T_N = 140$ K with an ordered moment larger than the earlier mentioned value [11].

From neutron scattering studies of polycrystalline CeFeAsO samples, the SPT and SDW transition have been reported to occur near $T_S = 155$ K and $T_N = 140$ K, respectively [12], although the difference between T_S and T_N depends on the sample quality [13]. Specific heat data for CeFeAsO suggest that the onsets of SPT appear to be at the same temperature for both the single crystalline as well as poly crystalline sample; although the transition appears to be broader for the latter [13, 14]. For polycrystalline PrFeAsO (T_S, T_N) have been reported to be (153K, 127 K) [15], whereas the same is (142 K, 137 K) for single crystal of NdFeAsO [16]. For SmFeAsO (T_S, T_N) is (175 K, 140 K) [17]. The reported values of (T_S, T_N) for the measured samples can also be seen in Table 1.

The room temperature value of the resistivity and its behavior near room temperature for LaFeAsO suggest the material to be a metal with low carrier concentration [18]. In the temperature range (200–165 K), ρ increases with the lowering of temperature reaching a maximum around $T_\rho = 165$ K [18]. Similar behavior of increasing $\rho(T)$ with the lowering of temperature down to almost 150 K has been observed for the polycrystalline samples of CeFeAsO, PrFeAsO, NdFeAsO [5] and SmFeAsO [9] although McGuire et al. [5] consider the temperature variation of $\rho(T)$ to be negligible in the tetragonal phase. Upon further cooling, $\rho(T)$ drops sharply. In essence, the earlier reported data in the absence of magnetic field suggest the following characteristics:

a. ρ increases with the lowering of temperature down to a temperature T_ρ, irrespective of whether the sample is single crystalline or polycrystalline, indicating the property to be an intrinsic one for RFeAsO system [13].
b. ρ versus T curve shows a cusp at T_ρ.
c. ρ decreases rapidly below T_ρ.
d. The peak in dρ/dT has been pointed out to be an excellent mark of T_N, whereas the anomaly in dρ/dT at T_S is difficult to identify [13].

In earlier studies of the T-dependence of ρ for RFeAsO system, focus was mainly on the SPT and the subsequent AFM transition, and the emphasis was more on to find a signature of the said transitions from the $\rho(T)$ curve. However, we feel, certain important points like the underlying cause behind the negative dρ/dT observed in this system just above T_ρ, and the role of different scattering mechanisms [the idea of which can be obtained from the functional dependence of $\rho(T)$] in different phases have not been given due consideration. This paper is basically to get a comprehensive idea about the different scattering mechanisms and their relative strengths in RFeAsO system from the functional dependence of ρ on T in different temperature regions. We also wish to have a fresh look into the origin of negative dρ/dT, another important feature of $\rho(T)$ of such system, by comparing with a superlattice (SL) made of a ($LaVO_3$)[6 unit cells]/($SrVO_3$)[1 unit cell] [19].

Table 1 T_S (temperature corresponding to the structural phase transition), T_N (temperature corresponding to the SDW transition), T_ρ (temperature corresponding to the cusp of ρ(T) curve) and the temperature corresponding to the peak of dρ/dT versus T curve [(dρ/dT)$_{peak}$]

Sample	T_S (K)	T_N (K)	T_ρ (K)	$(d\rho/dT)_{peak}$ (K)
CeFeAsO	155	140	152	137
PrFeAsO	153	127	165	144
NdFeAsO	142	137	164	138
SmFeAsO	175	140	161	133
GdFeAsO	135	133	157	125

Experimental

The samples used in the measurement of resistance are the same as those used for the measurement of thermoelectric power in Ref. [20]. The samples were prepared using solid-state reaction technique from high purity (99.99 %) rare earth elements (R=Ce, Pr, Nd, Sm), As (chips), and Fe and Fe_2O_3 (powder). To obtain RFeAsO, we first prepared RAs by mixing stoichiometric amount of R and As chips inside glove box filled with inert Argon gas. The mixture was then sealed in an evacuated quartz tube, and given subsequent heat treatment at 900 °C with intermediate grindings inside the glove box. Prepared RAs were mixed with stoichiometric amounts of Fe and Fe_2O_3 and the mixture was pressed into pellets. The pellet obtained for each of the

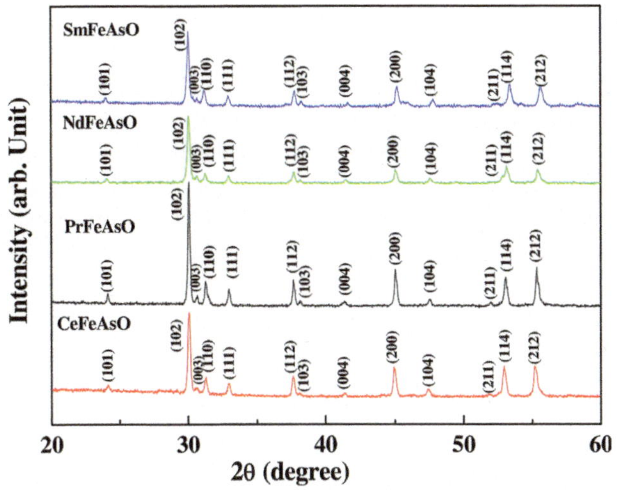

Fig. 1 Room temperature X-ray diffraction pattern for CeFeAsO, PrFeAso, NdFeAsO and SmFeAsO

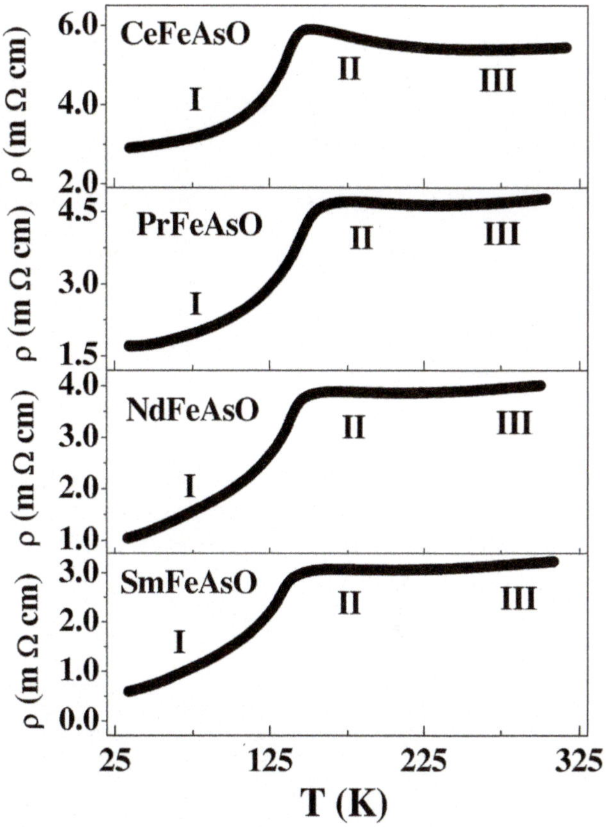

Fig. 2 Resistivity ρ as a function of temperature T for CeFeAsO, PrFeAsO, NdFeAsO, SmFeAsO

RFeAsO compounds was wrapped with Ta foil, and sealed in evacuated quartz tube. Finally, they were annealed at 1150–1200 °C for 48 h.

All the samples were characterized using powder X-Ray diffraction technique (PW/830, Philips, The Netherlands) with copper Cu K_α radiation at room temperature. Single phase natures of all the samples have been ascertained by indexing most of the peaks obtained in powder X-ray diffraction pattern (Fig. 1), assuming ZrCuSiAs-type (tetragonal) structure at room temperature (space group: *P4/nmm*). The standard four probe dc electrical resistivity measurements were performed using a close cycle refrigerator (25 K < T < 300).

Results and discussion

Figure 2 shows $\rho(T)$ for a number of members of the RFeAsO series, namely, CeFeAsO, PrFeAsO, NdFeAsO and SmFeAsO, all polycrystalline in nature, in the temperature range (35–315 K). The overall behavior of $\rho(T)$ for each sample is similar to the earlier reported data for RFeAsO [5, 9, 18], i.e., $\rho(T)$ increases with the lowering of temperature down to a temperature T_ρ (temperature corresponding to the cusp of $\rho(T)$ curve), and it decreases rapidly below the cusp at T_ρ. T_ρ for each sample has been given in Table 1.

For all of our samples belonging to the RFeAsO series, $\rho(T)$ shows metallic behavior with positive $d\rho/dT$ in a temperature region (Figs. 3, 4) close to room temperature. Figure 4 shows the behavior of $d\rho/dT$ for all of our samples in two temperature regions separately: a) 77 K—close to T_ρ (left panel), and b) temperature above T_ρ—room temperature (right panel). The range of temperature covering the linear region III (Fig. 3), the value of resistivity at 300 K ($\rho_{300\ K}$) and the value of $d\rho/dT$ in this region III are shown in Table 2 for each of our sample. In region III, $d\rho/dT$ is of the order of μΩ cm/K.

The negative and temperature-independent value of the Hall coefficient as well as the negative value of thermoelectric power (S) near room temperature for RFeAsO system identify electron as the charge carrier in this region [18]. The estimated Fermi energy E_F is 0.19 eV [20]. The linear dependence of $\rho(T)$ on T can be obtained from the Bloch–Grüneisen formula for metals,

$$\rho(T) = \rho(0) + A(T/\theta_R)^5 \int_0^{\frac{\theta_R}{T}} \frac{x^5}{(e^x - 1)(1 - e^x)} dx$$

as a high-temperature limit, i.e., at temperatures above the Debye temperature θ_R considering the resistance is due to scattering of electrons by phonons. The Debye temperatures for members of RFeAsO series are close to room temperature [21]. Hence, in the region III, the dominant scattering mechanism is the electron–phonon one.

From Fig. 4, we observe for each of our sample that in an intermediate temperature range (region II of Fig. 2, the range depends upon the choice of the element R) unlike

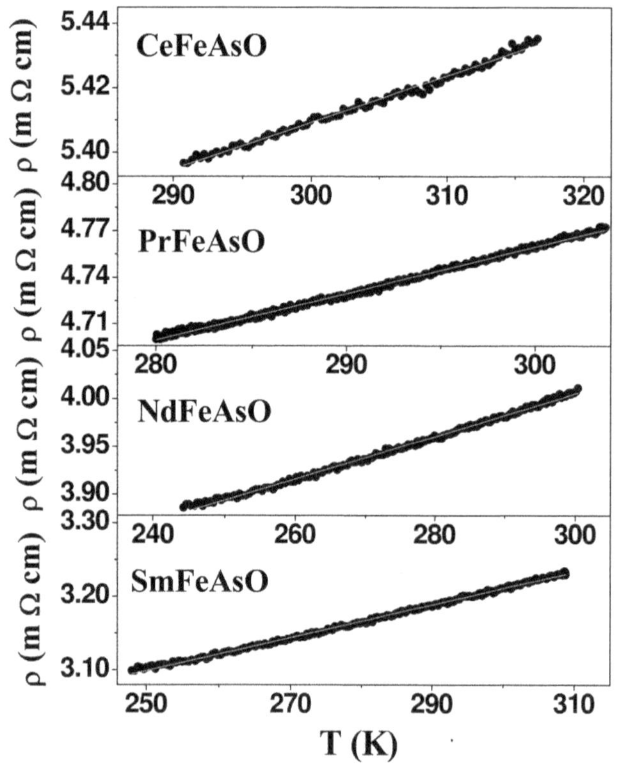

Fig. 3 $\rho(T)$ for all the samples in the region III, along with the fitted curves (*solid line*); ρ varies linearly with T

metallic system, $d\rho/dT$ is negative, although small. Opening of a gap with the lowering of temperature might cause such negative $d\rho/dT$. However, the weak temperature dependence in this region discards the possibility of opening a gap, and rather it suggests the system to be still in metallic phase with a loss of charge carrier coherence [19]. For polycrystalline LaFeAsO, such a negative $d\rho/dT$ below 200 K has been conjectured to be due to the increased charge carrier scattering by lattice fluctuations related to the onset of SPT [18]. Similar observation of negative $d\rho/dT$ above the SPT temperature in case of a superlattice (SL) made of a ($LaVO_3$)[6 unit cells]/($SrVO_3$)[1 unit cell] [19], and their analysis encourages us to think in a different way. Moreover, it may also be noted that RFeAsO and the SL have following interesting points of similarities:

1. In both the systems, the conducting zones are of two-dimensional (2D) characters. In the SL, a single unit cell of $SrVO_3$ is introduced between the insulating $LaVO_3$ layers, whereas in RFeAsO the conducting layers of $(FeAs)^-$ are interlaced by the insulating layers of $(RO)^+$.
2. In RFeAsO, we find SPT around $T_S = 150$ K, and likewise the SL undergoes a SPT at $T_S = 130$ K from

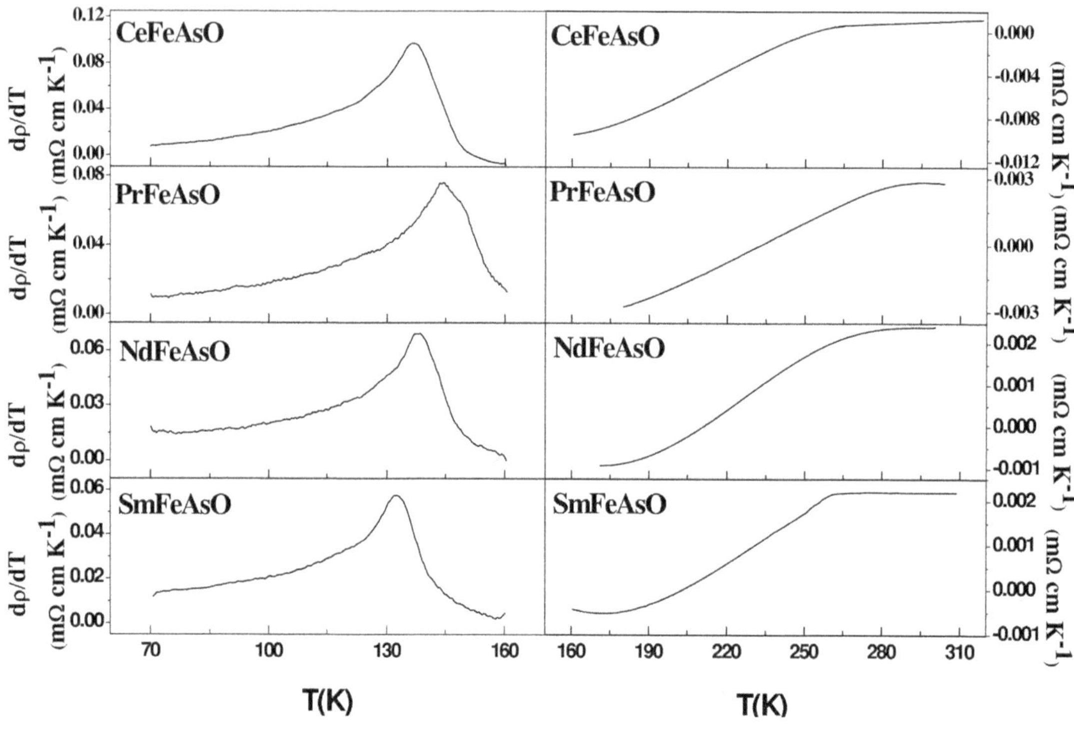

Fig. 4 $d\rho/dT$ as a function of temperature T: *left panel* covers the region (60–160 K), while the *right panel* describes the region (160–315 K) for each sample

Table 2 Temperature range corresponding to regions I, II & III; measured ρ at 300 K (ρ_{300K}), earlier reported values of ρ at 300 K (ρ_{300K} (reported)), value of $d\rho/dT$ in region III and the SDW energy gap Δ

Sample	Region III[a] (K)	ρ_{300K} (mΩ cm)	ρ_{300K} (reported) (mΩ cm)	$d\rho/dT$ (mΩ cm/K)	Region II (K)	Region I (K)	Δ (K)
CeFeAsO	250–318	5.4	4.9 [11]	8.4×10^{-4}	160–235	34–141	964
PrFeAsO	270–307	4.8	4.3 [5]	2.8×10^{-3}	175–235	34–148	968
NdFeAsO	244–301	4	3.3 [5]	2.2×10^{-3}	164–212	34–140	965
SmFeAsO	248–311	3.2	3.0 [28]	2.2×10^{-3}	164–211	34–138	968
GdFeAsO	201–312	6.0	2.4 [16]	4.2×10^{-3}	157–183	34–128	975

[a] Temperature range is taken from the linearly fitted region of Fig. 1

a high-temperature tetragonal phase to a low-temperature monoclinic one.

3. The value of the room temperature resistivity for both the systems corresponds to poor metals.
4. In both the cases, $d\rho/dT$ is negative and low indicating weak temperature dependence above T_S, and the same is positive and high below T_S.

In the SL, the negative $d\rho/dT$ above T_S has been explained by the weak localization in 2D. Physically, weak localization is the interference of pairs of waves into which conduction electrons split in the backscattering direction. Resistance is a measure of the intensity of interference. In the state of "weak localization" ρ^{-1} varies as ln T. For a number of two-dimensional systems such as thin films of Pd, Pt or Au alloys, such logarithmic increase of resistance has been interpreted as weak localization [22].

Figure 5 shows that the region II for each of our sample is well described by a linear variation of $1/\rho$ with ln T. The temperature range of this region has been noted in Table 2. This motivates us to conclude that the weak localization in 2D is responsible for such behavior like the mentioned SL [19]. However, such temperature dependence alone can not conclude decisively the state of weak localization. On the other hand, a similar behavior of resistivity ($1/\rho$–ln T) at low temperature in disordered 2D electron systems can be observed due to modified electron–electron (Coulomb) interaction also [22]. These two cases can be differentiated by performing the measurements in the presence of a magnetic field. In the state of weak localization, in which the conduction electrons split into pairs of interfering waves, application of a magnetic field (H) introduces a phase shift in the electronic wave function and suppresses the interference after a time interval proportional to 1/H [22]. Therefore, the magneto resistive (MR) effect should be dominant in this state. On the other hand, the MR of the second case which owes its origin to the Coulomb interaction is either negligible or only present in large fields. However, unlike the SL, for RFeAsO, the MR is negligible in this temperature range, although it is significant at temperatures below the SPT [23]. The MR is less

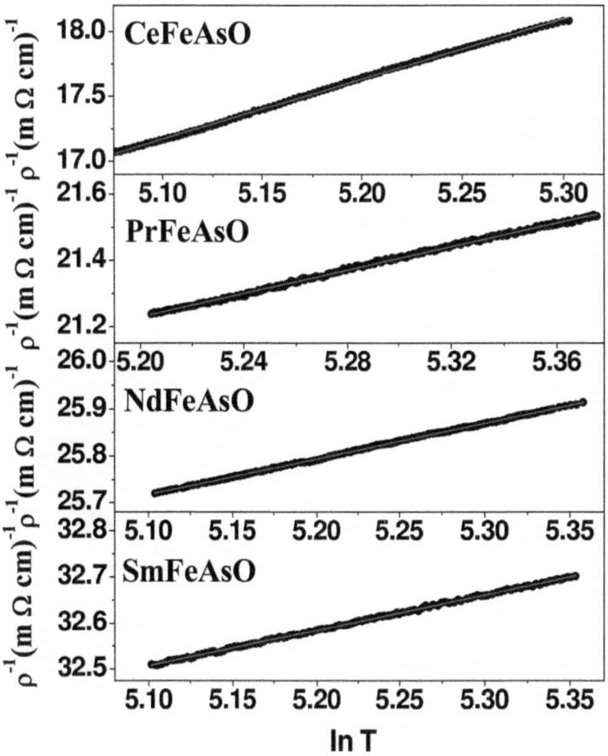

Fig. 5 ρ^{-1} ln T in the region II for all the samples; *curves* a linear variation of $1/\rho$ with ln T in this region

than 2 % for both SmFeAsO and PrFeAsO even in a field of 14 T at a temperature higher than 150 K [24] although a negative longitudinal in-plane MR has been observed in different $Ba(Fe_{1-x}Co_x)_2As_2$ single crystals in the paramagnetic phase [25]. Moreover, the localization theory deals with the noninteracting electrons [22], whereas the presence of strong electron–electron scattering in this intermediate temperature range has been predicted in RFeAsO systems [20]. Our study of temperature dependence of resistivity also confirms such observation, and we feel that electron–electron scattering is responsible for linear variation of $1/\rho$ with ln T in region II.

For the range of temperatures below the SDW transition, Cheng et al. [26] have analyzed their resistivity data with the functional dependence $\rho(T) = \rho_0 + AT^2 + BT(1 +$

$2T/\Delta)\exp(-\Delta/T)$; ρ_0 is the residual resistivity, A and B are the fitting parameters. The second term containing the T^2 term describes the Fermi liquid behavior, while the last term signifies the metallic ferromagnetic or antiferromagnetic state with an energy gap Δ. The associated theory has been given by Anderson and Smith [27]. The region I of each curve of Fig. 1 has been shown separately in Fig. 6, and fitted with the mentioned expression. The range of temperature, corresponding to region I for each sample has been noted in Table 2. In each case, the fitted parameter Δ which is a measure of the SDW energy gap has been given in Table 2. The SDW energy gap has been found to be of the order of 1000 K for any member of the RFeAsO series and comparable to the reported value of 1082 K for NdFeAsO single crystal [26].

$d\rho/dT$ for LaFeAsO shows a peak at 143 K, closer to the SDW transition temperature than T_S [18]. On the other hand, in the Ref. 5, $d\rho/dT$ has been claimed to show double-peaked structure for the samples CeFeAsO, PrFeAsO, NdFeAsO in addition to LaFeAsO. In explaining such double-peaked structure, one of the overlapping peaks has been associated with the crystallographic distortion and the other with the magnetic ordering [5] On the other hand, Luo et al. [9] report single peak in each of the $d\rho/dT$ versus T curves for polycrystalline LaFeAsO, SmFeAsO and GdFeAsO. The respective peak temperatures 132, 133, 128 K have been pointed out to be closer to AFM ordering temperatures rather than the SPT temperatures. For each of our samples, $d\rho/dT$ shows peak at a single temperature (Fig. 4). The peak temperatures $(d\rho/dT)_{peak}$ are noted in Table 1. $(d\rho/dT)_{peak}$ appears to be closer to T_{SDW} than T_S. Reported value of ρ_{300K} for SmFeAsO mentioned in Table 2 has been taken from Ref. [28].

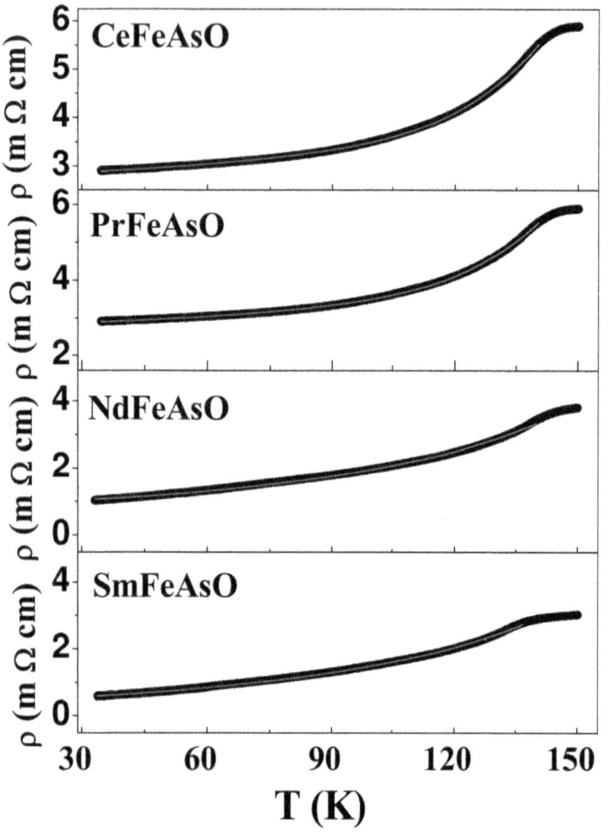

Fig. 6 $\rho(T)$ for all the samples in the region I, along with fitted curves, which show $\rho(T)$ varies as $\rho_0 + AT^2 + BT(1 + 2T/\Delta)\text{-}\exp(-\Delta/T)$

Conclusion

The overall behavior of $\rho(T)$ for any of our sample is identical to the earlier reported data. The whole temperature range of measurement can be divided into three regions. In the region III, close to room temperature, the system behaves as a metal with low carrier (electron) density, and the electron–phonon scattering is dominant. In the intermediate region II, $(d\rho/dT)$ is negative, and ρ^{-1} varies as ln T. Such a behavior cannot be explained by weak localization in absence of MR effect, rather electron–electron interaction appears to be responsible. Below the SDW transition temperature, the contribution to resistivity comes from electron–electron interaction, signified by a term proportional to T^2 in presence of SDW energy gap. The energy gap for the system has been identified to be of the order of 1000 K. $(d\rho/dT)$ for any of our samples shows a single peak at a temperature closer to T_{SDW} than T_S.

Acknowledgments Dr. Papri Dasgupta working as women scientist under Department of Science and Technology (DST), India for the project SR/WOS-A/PS-38/2013 acknowledges DST for financial support.

References

1. Kamihara, Y., Watanabe, T., Hirano, M., Hosono, H.: Iron-based layered superconductor La[$O_{1-x}F_x$]FeAs ($x = 0.05 - 0.12$) with $T_c = 26$ K. J. Am. Chem. Soc. **130**, 3296 (2008)
2. Ishida, K., Nakai, Y., Hosono, H.: To what extent iron-pnictide new superconductors have been clarified: a progress report. J. Phys. Soc. Jpn. **78**, 062001 (2009)
3. Takahashi, H., Igawa, K., Arii, K., Kamihara, Y., Hirano, M., Hosono, H.: Superconductivity at 43 K in an iron-based layered compound $LaO_{1-x}F_xFeAs$. Nature **453**, 376–378 (2008)
4. Oh, H., Moon, J., Shin, D., Moon, C.-Y., Choi, H.J.: Prog. Supercond. **13**, 65 (2011)
5. McGuire, M.A., Hermann, R.P., Sefat, A.S., Sales, B.C., Jin, R., Mandrus, D., Grandjean, F., Long, G.J.: Influence of the rare-earth element on the effects of the structural and magnetic phase

transitions in CeFeAsO, PrFeAsO and NdFeAsO. New J. Phys. **11**, 025011 (2009)
6. Lee, C.H., Kihou, K., Iyo, A., Kito, H., Shirage, P.M., Eisaki, H.: Relationship between crystal structure and superconductivity in iron-based superconductors. Solid State Commun. **152**, 644–648 (2012)
7. Ricci, A., Poccia, N., Joseph, B., Barba, L., Arrighetti, G., Ciasca, G., Yan, J.-Q., McCallum, R.W., Lograsso, T.A., Zhigadlo, N.D., Karpinski, J., Bianconi, A.: Structural phase transition and superlattice misfit strain of RFeAsO (R=La, Pr, Nd, Sm). Phys. Rev. B **82**, 144507 (2010)
8. Pan, M., Huang, Z., Cheng, C.H., Zhao, Y.: Possible coexistence of SDW state with superconductivity in Ir doped SmOFeAs. Solid State Commun. **152**, 458–461 (2012)
9. Luo, Y., Tao, Q., Li, Y., Lin, X., Li, L., Cao, G., Xu, Z., Xue, Y., Kaneko, H., Savinkov, A.V., Suzuki, H., Fang, C., Hu, J.: Evidence of magnetically driven structural phase transition in *R*FeAsO (R=La, Sm, Gd, and Tb): a low-temperature x-ray diffraction study. Phys. Rev. B **80**, 224511 (2009)
10. dela Cruz, C., Huang, Q., Lynn, J.W., Li, J.Y., Ratcliff, W., Zarestky, J.L., Mook, H.A., Chen, G.F., Luo, J.L., Wang, N.L., Dai, P.C.: Magnetic order close to superconductivity in the iron-based layered $LaO_{1-x}FxFeAs$ systems. Nat. (Lond.) **453**, 899–902 (2008)
11. Li, H.-F., Tian, W., Yan, J.-Q., Zarestky, J.L., McCallum, R.W., Lograsso, T.A., Vaknin, D.: Phase transitions and iron-ordered moment form factor in LaFeAsO. Phys. Rev. B **82**, 064409 (2010)
12. Zhao, J., Huang, Q., de la Cruz, C., Li, S., Lynn, J.W., Chen, Y., Green, M.A., Chen, G.F., Li, G., Li, Z., Luo, J.L., Wang, N.L., Dai, P.: Structural and magnetic phase diagram of $CeFeAsO_{1-x}F_x$ and its relation to high-temperature superconductivity. Nat. Mater. **7**, 953–959 (2008)
13. Jesche, A., Krellner, C., de Souza, M., Lang, M., Geibel, C.: Coupling between the structural and magnetic transition in CeFeAsO. Phys. Rev. B **81**, 134525 (2010)
14. Zocco, D.A., Baumbach, R.E., Hamlin, J.J., Janoschek, M., Lum, I.K., McGuire, M.A., Sefat, A.S., Sales, B.C., Jin, R., Mandrus, D., Jeffries, J.R., Weir, S.T., Vohra, Y.K., Maple, M.B.: Search for pressure-induced superconductivity in CeFeAsO and CeFePO iron pnictides. Phys. Rev. B **83**, 094528 (2011)
15. Zhao, J., Huang, Q., de la Cruz, C., Lynn, J.W., Lumsden, M.D., Ren, Z.A., Yang, J., Shen, X., Dong, X., Zhao, Z., Dai, P.: Lattice and magnetic structures of PrFeAsO, PrFeAsO0.85F0.15, and PrFeAsO0.85. Phys. Rev. B **78**, 132504 (2008)
16. Tian, W., Ratcliff, W., Kim, M.G., Yan, J.-Q., Kienzle, P.A., Huang, Q., Jensen, B., Dennis, K.W., McCallum, R.W., Lograsso, T.A., McQueeney, R.J., Goldman, A.I., Lynn, J.W., Kreyssig, A.: Interplay of Fe and Nd magnetism in NdFeAsO single crystals. Phys. Rev. B **82**, 060514(R) (2010)
17. Martinelli, A., Palenzona, A., Tropeano, M., Putti, M., Ferdeghini, C., Profeta, G., Emerich, E.: Retention of the tetragonal to orthorhombic structural transition in F-substituted SmFeAsO: a new phase diagram for $SmFeAs(O_{1-x}Fx)$. Phys. Rev. Lett. **106**, 227001 (2011)
18. McGuire, M.A., Christianson, A.D., Sefat, A.S., Sales, B.C., Lumsden, M.D., Jin, R., Payzant, E.A., Mandrus, D., Luan, Y., Keppens, V., Varadarajan, V., Brill, J.W., Hermann, R.P., Sougrati, M.T., Grandjean, F., Long, G.J.: Phase transitions in LaFeAsO: Structural, magnetic, elastic, and transport properties, heat capacity and Mössbauer spectra. Phys. Rev. B **78**, 094517 (2008)
19. David, A., Frésard, R., Boullay, Ph, Prellier, W., Lüders, U., Janolin, P.E.: Structural transition in $LaVO_3/SrVO_3$ superlattices and its influence on transport properties. Appl. Phys. Lett. **98**, 212106 (2011)
20. Poddar, A., Mukherjee, S., Samanta, T., Saha, R.S., Mukherjee, R., Dasgupta, P., Mazumdar, C., Ranganathan, R.: Thermoelectric power of RFeAsO (R=Ce, Pr, Nd, Sm and Gd). Phys. C **469**, 789–794 (2009)
21. Kaurav, N., Chung, Y.T., Kuo, Y.K., Liu, R.S., Chan, T.S., Chen, J.M., Lee, J.-F., Sheu, H.-S., Wang, X.L., Dou, S.X., Lee, S.I., Shi, Y.G., Belik, A.A., Yamaura, K., Takayama-Muromachi, E.: Crystal structure and electronic and thermal properties of $TbFeAsO_{0.85}$. Appl. Phys. Lett. **94**, 192507 (2009)
22. Bergmann, G.: Consistent temperature and field dependence in weak localization. Phys. Rev. B **28**, 515 (1983)
23. Yuan, H.Q., Jiao, L., Singleton, J., Balakirev, F.F., Chen, G.F., Luo, J.L., Wang, N.L.: The magnetoresistance and Hall effect in CeFeAsO: a high magnetic field study. J. Phys. Conf. Ser. **273**, 012110 (2011)
24. Meena, R.S., Singh, S.K., Pal, A., Kumar, A., Jha, R., Rao, K.V.R., Du, Y., Wang, X.L., Awana, V.P.S.: High field (14 T) magneto transport of Sm/PrFeAsO. J. Appl. Phys. **111**, 07E232–07E233 (2012)
25. Rullier-Albenque, F., Colson, D., Forget, A.: Longitudinal magnetoresistance in Co-doped BaFe2As2 and LiFeAs single crystals: interplay between spin fluctuations and charge transport in iron pnictides. Phys. Rev. B **88**, 045105 (2013)
26. Cheng, P., Yang, H., Jia, Y., Fang, L., Zhu, X., Mu, G., Wen, H.-H.: Hall effect and magnetoresistance in single crystals of $NdFeAsO_{1-x}Fx$ (x = 0 and 0.18). Phys. Rev. B **78**, 134508 (2008)
27. Andersen, N.H., Smith, H.: Electron-magnon interaction and the electrical resistivity of Tb. Phys. Rev. B **19**, 384–387 (1979)
28. Tropeano, M., Fanciulli, C., Ferdeghini, C., Marrè, D., Siri, A.S., Putti, M., Martinelli, A., Ferretti, M., Palenzona, A., Cimberle, M.R., Mirri, C., Lupi, S., Sopracase, R., Calvani, P., Perucchi, A.: Transport and infrared properties of $SmFeAs(O_{1-x}F_x)$: from SDW to superconducting ordering. Supercond. Sci. Technol. **22**, 034004 (2009)

10

The static properties and form factors of the deuteron using the different forms of the Wood–Saxon potential

B. Rezaei · A. Dashtimoghadam

Abstract The basic properties and form factors of the deuteron system are investigated for the different forms of the Wood–Saxon potential. We have used the Nikiforov-Uvarov (NU) method for analytical solution of the radial Schrodinger equation. A comparison of the calculated values with experimental results are given. It is shown that the obtained results for the modified form of the Wood–Saxon potential are very close to the experimental results in comparison with other forms of the potential.

Keywords Deuteron form factors · Wood–Saxon potential · Nikiforov-Uvarov (NU) method

Introduction

The internal structure of the nucleon has been the subject of intensive experimental and theoretical studies for several decads. The internal structure is conveniently described in terms of electromagnetic form factors (FFs). The form factors are a measurable and physical manifestation of the nature of the nucleons' constituents and the dynamics that binds them together. The elastic electron–deuteron scattering is one of the processes that probe mentioned structure. The deuteron electromagnetic form factors allow us to describe interaction between the deuteron and electromagnetic field [1]. Different aspects of the underlying nuclear structure of the deuteron can be obtained through analysis of scattering of various probes at different energy scales. As a relevant example, consider elastic electron–deuteron scattering. The form factor describes the structure and spatial extension of a nucleus, because in real life nuclei are not point like. The deuteron form factor $F_D(q^2)$ provides an ideal illustration of the continuality between nuclear and particle physics at the microscopic level. At low momentum transfer, where the nucleons can be treated as point-like objects and are the essential degrees of freedom, the usual effective potential Schrodinger theory is appropriate, and meson exchange effects provide the framework for the nuclear interaction. The deuteron electromagnetic form factors allow us to describe quantifiable an interaction between the deuteron and electromagnetic field. Form factors can be calculated in terms of information about electron elastic scattering by deuteron [2, 3]. Our goal in this project is to study the properties and form factors of deuteron system. So, this paper is organized as follows: "Deuteron form factors" introduces deuteron form factors. In "NU (Nikiforov-Uvarov) method", the theoretical framework of the NU method is briefly discussed. The solution of the Schrodinger equation for Standard, Generalized, and modified Wood–Saxon potentials via mentioned method are presented in "Calculations". In "Numerical results and applications", we will apply our results to calculate the statics properties and form factors of deuteron, and finally we summarize the paper and discuss future perspectives of the study in "Conclusion".

B. Rezaei (✉) · A. Dashtimoghadam
Physics Department, Razi University, Kermanshah 67149, Iran
e-mail: brezaei@razi.ac.ir

Deuteron form factors

The elastic electron-deuteron cross-section in the one-photon exchange approximation is given by [4–6]

$$\frac{d\sigma}{d\Omega} = \sigma_M \left[A(Q^2) + B(Q^2) \tan^2 \frac{\theta}{2} \right] \quad (1)$$

with

$$A(Q^2) = G_c^2(Q^2) + \frac{8}{9}\eta^2 G_Q^2(Q^2) + \frac{2}{3}\eta G_M^2(Q^2) \quad (2)$$

$$B(Q^2) = \frac{4}{3}\eta(1+\eta)G_M^2(Q^2) \quad (3)$$

$$\eta = \frac{Q^2}{4M_d^2}, \quad (4)$$

where $Q^2 = 4EE'\sin^2\frac{\theta}{2}$ is the four-momentum transfer squared, E is the incoming electron energy, θ is the electron scattering angle, E' is the scattered electron energy in the lab rest frame, $\eta = \frac{Q^2}{4M_d^2}$, and M_d is the deuteron mass.

The expression $\sigma_M = (\frac{d\sigma}{d\Omega})_{Mott}$ is a scattering cross-section by non-structure spin-free particle derived by Mott. Three terms (G_c, G_Q, G_M) are the form factors making contribution to the full cross-section. They are due to charge, quadrupole moment, and magnetic moment of the deuteron.

Calculation of deuteron form factors and static properties require a deuteron wave function [7, 8]. The best non-relativistic wave function are calculated from Schrodinger equation. An analytical solution of radial Schrodinger equation is of high importance in non-relativistic quantum mechanics, because the wave function contains all necessary information for full description of a quantum system. There are only few potentials for which the radial Schrodinger equation can be solved explicitly for n and l. So far, many methods were developed, such as supersymmetry (SUSY) [9, 10], Nikiforov-Uvarov (NU) method [11] the Pekeris approximation [12].

NU (Nikiforov-Uvarov) method

The NU method provides an exact solution of the non-relativistic Schrodinger equation for certain kinds of potentials [13, 14]. This method is based on the solution of generalized second-order differential linear equation with special orthogonal functions and for any given real or complex potential, the one-dimensional Schrodinger equation is reduced to a generalized equation of the hypergeometric with an appropriate $S = S(r)$ coordinate transformation. Thus it can be written in the following form [11]:

$$F''(s) + \frac{\tau(s)}{\sigma(s)}F'(s) + \frac{\tilde{\sigma}(s)}{\sigma^2(s)}F(s) = 0, \quad (5)$$

where $\sigma(s)$ and $\tilde{\sigma}(s)$ are polynomials at most second degree, and $\tilde{\tau}(s)$ is a first degree polynomials. We use the following transformation:

$$F(s) = \phi(s)y(s) \quad (6)$$

This selection reduces the Schrodinger equation (5), to an equation of the hypergeometric type:

$$\sigma(s)y''(s) + \tau(s)y'(s) + \lambda y(s) = 0 \quad (7)$$

$\phi(s)$ satisfies $\frac{\phi'(s)}{\phi(s)} = \frac{\pi(s)}{\sigma(s)}$ and $y(s)$ is the hypergeometric type function whose polynomial solutions are given by Rodrigues relation:

$$y_n(s) = \frac{B_n}{\rho(s)}\frac{d^n}{ds^n}[\sigma^n(s)\rho(s)] \quad (8)$$

B_n is the normalization constant and the weight function $\rho(s)$ must satisfy the condition:

$$\frac{d}{ds}(\sigma(s)\rho(s)) = \tau(s)\rho(s) \quad (9)$$

The function $\pi(s)$ and the parameter λ are defined as

$$\pi(s) = \frac{\acute{\sigma}(s) - \tilde{\tau}(s)}{2} \pm \sqrt{\left(\frac{\acute{\sigma}(s) - \tilde{\tau}(s)}{2}\right)^2 - \tilde{\sigma}(s) + k\sigma(s)} \quad (10)$$

and

$$\lambda = k + \pi'(s) \quad (11)$$

Here, $\pi(s)$ obviously is a polynomial depended on the transformation function $s(r)$. On the other hand, to find the value of k, the expression under the square root must be square of a polynomial. This is possible only if its discriminant is zero. Hence, a new eigenvalue equation for the Schrodinger equation becomes

$$\lambda_n + n\tau'(s) + \frac{n(n+1)}{2}\sigma''(s) = 0 \quad (n = 0, 1, 2, \ldots), \quad (12)$$

where

$$\tau(s) = \tilde{\tau}(s) + 2\pi(s). \quad (13)$$

Calculations

Standard Wood–Saxon potential

Woods and Saxon [15] introduced a potential to study elastic scattering of 20 MeV protons by medium and heavy nuclei. The Wood–Saxon potential is a reasonable potential for nuclear shell model and hence attracts lots of attention in nuclear physics. The spherical Wood–Saxon potential that was used as a major part of nuclear shell model, has received a lot of attention in nuclear mean field model, and plays an essential role in microscopic physics [16]. The standard Wood–Saxon potential is given by [17]:

$$V(r) = \frac{V_0}{1+\exp\left(\frac{r-R}{a}\right)}, \tag{14}$$

where R, V_0 and a are mean radius, wall depth potential, and skin thickness, respectively.

Let us consider the radial part of Schrodinger equation with mentioned potential:

$$\frac{d^2R}{dr^2} + \frac{2\mu}{\hbar^2}[E + V_{\text{total}}(r)]R(r) = 0 \tag{15}$$

$$V_{\text{total}}(r) = \frac{V_0}{1+\exp\left(\frac{r-R}{a}\right)} - \frac{\hbar^2 l(l+1)}{2\mu r^2}, \tag{16}$$

where l is the angular momentum number, μ is the reduced mass, $E_{n,l}$ is the appropriate energy eigenvalue, and $\frac{1}{r^2}$ is the centrifugal term.

Now, we want to solve the Eq. (15) for the $l \neq 0$ cases using NU method and Pekeris approximation to the centrifugal for l states [15–18]. The approximation is based on the expansion of the centrifugal barrier in the series of exponentials depending on the internuclear distance, keeping terms up to second order. We now replace the centrifugal term according to the Pekeris approximation with expression:

$$\tilde{V}_l(x) = \gamma\left(D_0 + D_1(1+\exp(\alpha x))^{-1} + D_2(1+\exp(\alpha x))^{-2}\right), \tag{17}$$

where $x = \frac{(r-R)}{R}$, $\alpha = \frac{R}{a}$, and D_i is the coefficients $(i = 0, 1, 2)$. In the $x = 0$, the expression of (16) can be expanded up to the terms x^3,

where

$$\tilde{V}_l(x) = \gamma\left(D_0 + \frac{D_1}{2} + \frac{D_2}{4} - x\left(\frac{D_1\alpha}{4} + \frac{D_2\alpha}{4}\right)\right) + \gamma\left(x^2\left(\frac{D_2\alpha^2}{16}\right) + x^3\left(\frac{D_1\alpha^3}{48} + \frac{D_2\alpha^3}{48}\right) + \ldots\right) \tag{18}$$

$$D_0 = 1 - \frac{4}{\alpha} + \frac{12}{\alpha^2}, \quad D_1 = \frac{8}{\alpha} - \frac{48}{\alpha^2}, \quad D_2 = \frac{48}{\alpha^2} \tag{19}$$

We now can take the potential $\tilde{V}_l$ instead of the centrifugal term in the V_{total} and solve the Schrodinger equation. We substitute (17) into (15) and using a new variable of the form

$$S(r) = \exp\left(\frac{r-R}{a}\right) \tag{20}$$

we obtain the following:

$$\frac{d^2R(s)}{ds^2} + \frac{(s+1)}{s(s+1)}\frac{dR(s)}{ds} + \frac{1}{s^2(s+1)^2} + \left[-\varepsilon^2 s^2(\beta^2 - 2\varepsilon^2)s + (\beta^2 - \chi^2 - \varepsilon^2)\right]R(s) = 0, \tag{21}$$

where

$$\varepsilon^2 = \frac{-2\mu a^2(E-\gamma D_0)}{\hbar^2}, \quad \beta^2 = \frac{2\mu a^2(V_0 - \gamma D_1)}{\hbar^2}$$
$$\chi^2 = \frac{2\mu a^2\gamma D_2}{\hbar^2}, \quad \gamma = \frac{\hbar^2 l(l+1)}{2\mu R^2} \tag{22}$$

are dimensionless parameters.

According to the NU method, we can obtain $\pi(s)$, λ and λ_n:

$$\pi(s) = \pm\left\{\frac{s}{2} \pm \frac{1}{2}\left[\sqrt{\alpha} + \left(\sqrt{1+4\chi^2} + 2\sqrt{\alpha}\right)s\right]\right\}$$
$$k_+ = 2\chi^2 - \beta^2 + \sqrt{\alpha(1+4\chi^2)}$$
$$\pi(s) = \pm\left\{\frac{s}{2} \pm \frac{1}{2}\left[\sqrt{\alpha} - \left(\sqrt{1+4\chi^2} - 2\sqrt{\alpha}\right)s\right]\right\}$$
$$k_- = 2\chi^2 - \beta^2 - \sqrt{\alpha(1+4\chi^2)}, \tag{23}$$

where $\alpha = \left(\varepsilon^2 + \chi^2 - \beta^2\right)$.

$$\lambda = 2\chi^2 - \beta^2 + \left(1 + \sqrt{1+4\chi^2}\right)\left(\frac{1}{2} + \sqrt{\chi^2 + \varepsilon^2 - \beta^2}\right) \tag{24}$$

$$\lambda_n = -n\left(1 + n + 2\sqrt{\chi^2 + \varepsilon^2 - \beta^2} + \sqrt{1+4\chi^2}\right) \tag{25}$$

Comparing (24) and (25), the exact energy eigenvalues of the radial Schrodinger equation with WS potential are obtained as follows:

$$E_{n,l} = \gamma D_0 + \beta^2 - \chi^2 - \frac{\hbar^2}{2\mu a^2}\left[\frac{\beta^2 - 2\chi^2 + n(n+1)}{\left(1 + 2n + \sqrt{1+4\chi^2}\right)} - \frac{(2n+1)}{2}\right]^2, \tag{26}$$

where substitute the values of $\gamma, D_0, D_1, D_2, \beta^2$ and χ^2 from (19) and (22). The corresponding eigenfunction as follows:

$$R_{n,l}(s) = C_n \frac{(1+s)^{\frac{1}{2}\left(1+\sqrt{1+4\chi^2}\right)} s^{\sqrt{\chi^2-\beta^2+\varepsilon^2}}}{n!(1+s)^{\sqrt{1+4\chi^2}} s^{2\sqrt{\chi^2-\beta^2+\varepsilon^2}}} \times \frac{d^n}{ds^n}\left[(1+s)^{n+\sqrt{1+4\chi^2}} s^{n+2\sqrt{\chi^2+\varepsilon^2-\beta^2}}\right], \tag{27}$$

where C_n is the new normalization constant and it is determined by the condition $\int_0^\infty R_n(s)ds = 1$.

Generalized Wood–Saxon potential

We consider the following potential which is a generalization of the Wood–Saxon potential and it is given by [19]:

$$V(r) = -V_1 \frac{\exp\left(-\left(\frac{r-R}{a}\right)\right)}{1+\exp\left(-\left(\frac{r-R}{a}\right)\right)} + V_2 \frac{\exp\left(-2\left(\frac{r-R}{a}\right)\right)}{1+\exp\left(-2\left(\frac{r-R}{a}\right)\right)}, \tag{28}$$

where r stands for the center-of-mass distance between the projectile nucleus and the target nucleus. The other parameters in the potential $R = r_0 A^{\frac{1}{3}}$ are the radius of the corresponding spherical nucleus or the width of the potential, A is the target mass number, r_0 is the radius parameter, V_1 is the potential depth of the Coulombic part [i.e., the first term on the right-hand side of (28)], a is the diffuseness of the nuclear surface, and finally V_2 is an introduced parameter for the second part of (28) (transforms like potential barrier).

Now, rewriting (15) by employing the convenient transformation

$$S(r) = \frac{1}{1+e^{-\left(\frac{r-R}{a}\right)}} \tag{29}$$

The radial Schrodinger equation for the generalized Wood–Saxon potential is as follows:

$$\frac{d^2R(s)}{ds^2} + \frac{(1-2s)}{s(1-s)}\frac{dR(s)}{ds} + \frac{1}{s^2(1-s)^2} \left[-\chi^2 s^2 + (2\chi^2 - \beta^2)s + (\beta^2 - \chi^2 - \varepsilon^2)\right] = 0, \tag{30}$$

where χ^2, β^2, ε^2 and γ are dimensionless parameters. According to the previous section, we calculate the $\pi(s)$, $E_{n,l}$, $R_{n,l}$ and write briefly

$$\begin{aligned} \pi(s) &= \pm\left[\sqrt{\varepsilon^2+\chi^2-\beta^2} + \left(\varepsilon - \sqrt{\varepsilon^2+\chi^2-\beta^2}\right)s\right] \\ k_+ &= \beta^2 - 2\varepsilon^2 + 2\varepsilon\sqrt{\varepsilon^2+\chi^2-\beta^2} \\ \pi(s) &= \pm\left[\sqrt{\varepsilon^2+\chi^2-\beta^2} + \left(\varepsilon - \sqrt{\varepsilon^2+\chi^2-\beta^2}\right)s\right] \\ k_- &= \beta^2 - 2\varepsilon^2 - 2\varepsilon\sqrt{\varepsilon^2+\chi^2-\beta^2} \end{aligned} \tag{31}$$

$$E_{n,l} = \gamma D_0 - \left(\frac{V_2 - \gamma D_2}{(n^2+\chi^2+(n+\frac{1}{2})(1\pm\sqrt{1+4\chi^2}))}\right) \times \left(\frac{(n^2+\chi^2+(n+\frac{1}{2})(1\pm\sqrt{1+4\chi^2}))+\beta^2-\chi^2}{2\chi}\right)^2 \tag{32}$$

$$R_{n,l}(s) = C_n \frac{(1-s)^{\varepsilon} s^{\sqrt{\chi^2-\beta^2+\varepsilon^2}}}{n!(1-s)^{2\varepsilon} s^{2\sqrt{\chi^2-\beta^2+\varepsilon^2}}} \times \frac{d^n}{ds^n}\left[s^{n+2\sqrt{\varepsilon^2+\chi^2-\beta^2}}(1-s)^{n+2\varepsilon}\right] \tag{33}$$

Modified form of Wood–Saxon potential

In order to modify this potential for $l \neq 0$, it is necessary to add two terms to generalized potential, so we have a modified potential as follows [20]:

$$V_{\text{mod}}(r) = \frac{-V_1}{1+\exp\left(\frac{r-R}{a}\right)} + \frac{V_2}{\left(1+\exp\left(\frac{r-R}{a}\right)\right)^2} + \xi\coth\left(\frac{r-R}{a}\right) + \eta\coth^2\left(\frac{r-R}{a}\right) \tag{34}$$

the parameters V_1, V_2, ξ and η are real constant values. It is required to remind the third and fourth terms in Eq. (34), in limit $r - R \ll a$ reformed as $\frac{1}{r}$ and $\frac{1}{r^2}$, respectively. These forms are corresponding to the coulombian repulsive potential and its square. In this section, using the NU method, we solve analytically the radial part of time-independent Schrodinger equation with angular momentum $l \neq 0$, for this modified shape of generalized Wood–Saxon potential.

By considering a new transformation as

$$S(r) = \frac{1}{1+e^{\left(\frac{r-R}{a}\right)}} \tag{35}$$

and according to the previous section the $E_{n,l}$, $R_{n,l}$ are calculated and write briefly:

$$E_{n,l} = \gamma D_0 - \left(\frac{V_1 - \gamma D_1}{\beta^2}\right) \left(\left(\frac{\alpha^2+\beta^2-\chi^2}{\sqrt{1+4\lambda^2+4\chi^2}+2n-1}\right)^2 - (\alpha^2+\lambda^2-\beta^2)\right) \tag{36}$$

where

$$\varepsilon^2 = \frac{-2\mu a^2(E-\gamma D_0)}{\hbar^2}, \quad \beta^2 = \frac{2\mu a^2(V_1-\gamma D_1)}{\hbar^2}$$
$$\chi^2 = \frac{2\mu a^2(V_2+\gamma D_2)}{\hbar^2}, \quad \gamma = \frac{\hbar^2 l(l+1)}{2\mu R^2}$$
$$\alpha^2 = \frac{2\mu a^2\xi}{\hbar^2}, \quad \lambda^2 = \frac{2\mu a^2\eta}{\hbar^2} \tag{37}$$

$$R_{n,l}(s) = C_n \frac{(s(1-s))^{\sqrt{\alpha^2+\varepsilon^2+\lambda^2-\beta^2}}(1-2s)^{\frac{1}{2}(1+\sqrt{1+4\chi^2+4\lambda^2})}}{n!(s(1-s))^{2\sqrt{\alpha^2+\varepsilon^2+\lambda^2-\beta^2}}(1-2s)^{\sqrt{1+4\chi^2+4\lambda^2}}} \times \frac{d^n}{ds^n}\left[(s(1-s))^{n+2\sqrt{\alpha^2+\varepsilon^2+\lambda^2-\beta^2}}(1-2s)^{n+\sqrt{1+4\chi^2+4\lambda^2}}\right] \tag{38}$$

The S and D state wave functions determined from these of three potentials are shown in Figs. 1, 2 and 3, standard,generalized, and modified forms of the potential, respectively. Also S and D states of the deuteron for the wave function obtained from experimental data are shown in Fig. 4. In the next section, we will apply these results to calculate the statis properties and form factors of deuteron.

Numerical results and applications

In the non-relativistic impulse approximation, the charge monopole, $G_C(q)$, charge quadrupole, $G_Q(q)$, and magnetic dipole, $G_M(q)$, form factors of the deuteron that may be expressed in terms of the deuteron wave function and nucleon isoscalar form factor as follows [21, 26, 27]:

$$G_C(q) = 2G_E \int [u^2(r) + w^2(r)] j_0\left(\frac{qr}{2}\right) dr \tag{39}$$

$$G_Q(q) = 4G_E \int \left[u(r)w(r) - \frac{w^2(r)}{\sqrt{8}}\right] j_2\left(\frac{qr}{2}\right) dr \tag{40}$$

$$G_M(q) = \frac{M_d}{M_p}(2G_{M_0}C_S + G_E C_L) \tag{41}$$

where

$$C_L = \frac{3}{2}\int w^2(r)\left[j_0\left(\frac{qr}{2}\right) + j_2\left(\frac{qr}{2}\right)\right] dr \tag{42}$$

$$\begin{aligned} C_S = & \int \left[u^2(r) - \frac{w^2(r)}{2}\right] j_0\left(\frac{qr}{2}\right) dr \\ & + \frac{1}{\sqrt{2}}\int w(r)\left[u(r) + \frac{w(r)}{\sqrt{2}}\right] j_2\left(\frac{qr}{2}\right) dr \end{aligned} \tag{43}$$

$$2G_{M_0} = G_{M_p} + G_{M_n}, \quad 2G_E = G_{E_p} + G_{E_n}, \tag{44}$$

where $u(r)$ and $w(r)$ are the radial wave functions of the bound state, while j_0, j_2 are spherical Bessel functions.

The charge form factor for the neutron was assumed to be zero, while the charge form factor for the proton was parameterized as

$$G_{E_p} = \frac{1}{(1 + 0.054844q^2)^2}, \tag{45}$$

where q is the momentum transfer in fm^{-1} units and the magnetic form factors for the nucleon were determined on the basic of the scaling law, $G_{M_p} = \mu_p G_{E_p}$ and $G_{M_n} = \mu_n G_{E_p}$. Now, substituting the obtained wave functions (Eqs. 27, 33, and 38) into Eqs. (39), (40), and (41) leads to the form factors of the deuteron. The mentioned form factors of the deuteron as a function of q for various of the Wood–Saxon potential are calculated and the obtained results are shown in Figs. 5, 6 and 7.

One can evaluate the electric quadropole moment using the obtained wave functions which is expressed as

$$Q_d = \frac{1}{\sqrt{50}}\int \left(u(r)w(r) - \frac{w^2(r)}{\sqrt{8}}\right) r^2 dr \tag{46}$$

The obtained results are given in Table 1.

The deuteron magnetic moment can be expressed in terms of the nucleon moments and the D-state probability P_D [22], as follows:

$$\begin{aligned} \mu_d &= \mu_n + \mu_p - \frac{3}{2}\left(\mu_n + \mu_p - \frac{1}{2}\right) P_D \\ P_D &= \int w^2(r) dr \end{aligned} \tag{47}$$

With use of the obtained wave functions, the deuteron magnetic moment is determined (see Table 1).

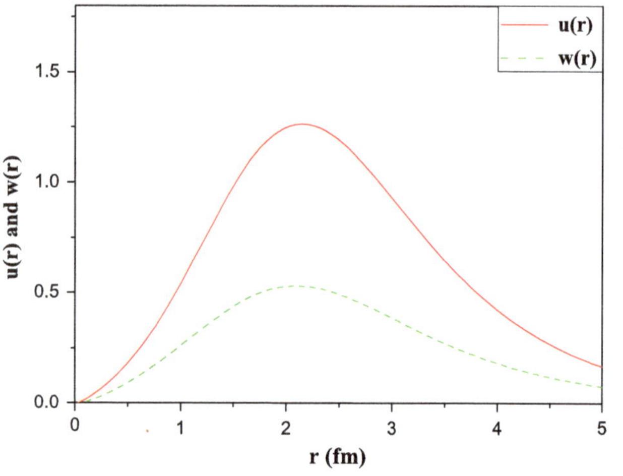

Fig. 1 S and D states of the deuteron wave function for the standard Wood–Saxon potential

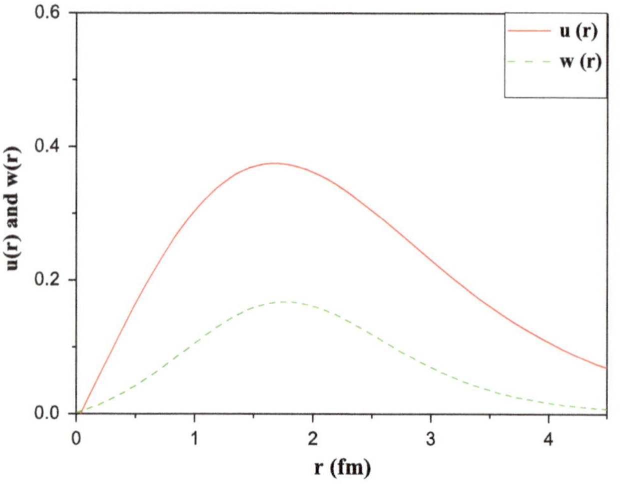

Fig. 2 S and D states of the deuteron wave function for the generalized Wood–Saxon potential

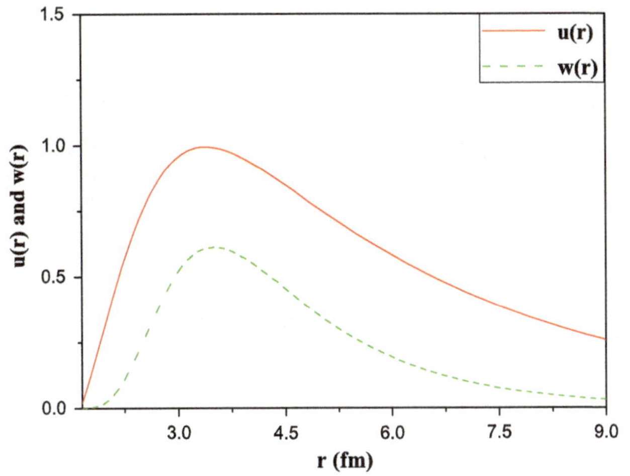

Fig. 3 S and D states of the deuteron wave function for the modified Wood–Saxon potential

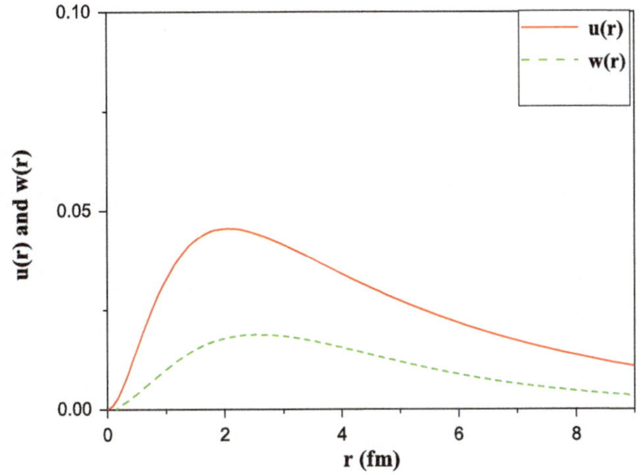

Fig. 4 S and D states of the deuteron wave function from experimental data [23]

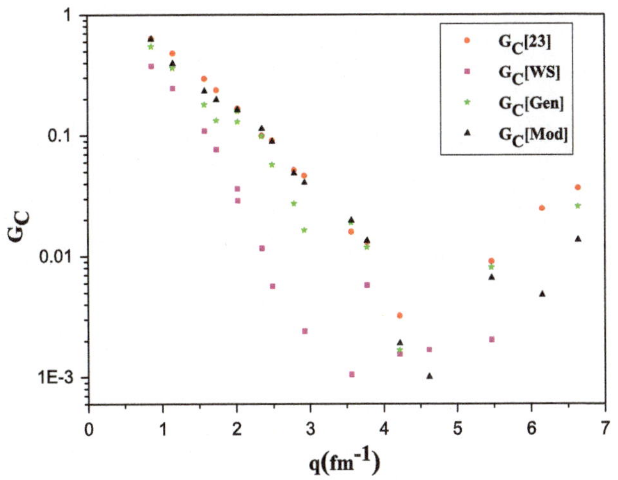

Fig. 5 Charge monopole form factor of the deuteron, G_C, as a function of q for the various versions of the WS potential

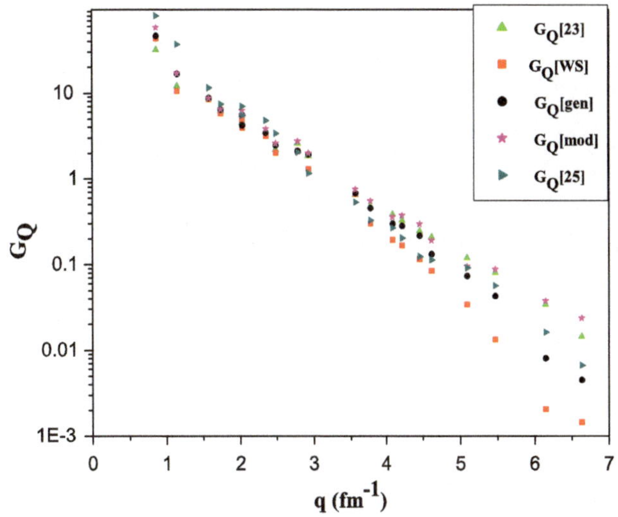

Fig. 6 Charge quadrupole form factor of the deuteron, G_Q, as a function of q for the various versions of the WS potential

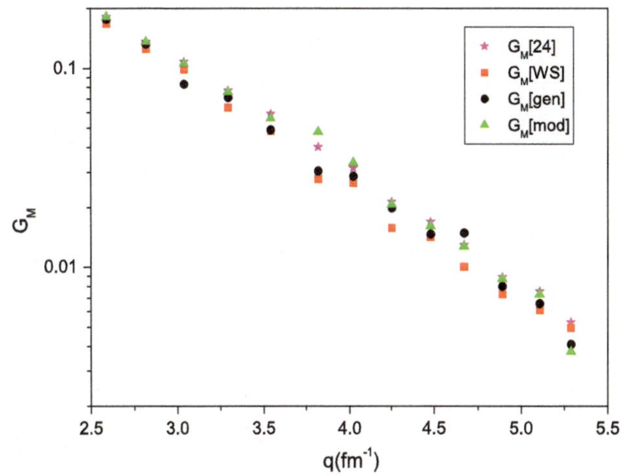

Fig. 7 Magnetic dipole form factor of the deuteron, G_M, as a function of q for the various versions of the WS potential

Finally, in the non-relativistic form the deuteron size maybe characterized by a charge reduced r_{ch} and by a matter radius r_m. The rms charge radius r_{ch} is related to the matter rms-radius r_m by [22]

$$r_{ch}^2 = r_m^2 + \Delta r_m^2 + r_p^2 + r_n^2 + \frac{3}{4}\left(\frac{\hbar}{m_p}\right)^2 \quad (48)$$

$$r_m^2 = \frac{1}{4}\int \left[u^2(r) + w^2(r)\right] r^2 dr, \quad (49)$$

where $u(r)$ and $w(r)$ are the S and D components of the NR deuteron wavefunction, respectively. r_m is the matter radius, Δr_m^2 is the effect of the meson exchange current plus other non-classical effects, and r_p and r_n are the root-mean-square radii of the proton and neutron, respectively. The obtained results are shown in Table 1.

Table 1 The calculated values of electric quadrupole moment, magnetic dipole moment, D state percentage, matter and charge radius for deuteron system comparison with experimental values

Quantity	$V_{WS}(r)$	$V_{Gen}(r)$	$V_{Mod}(r)$	Exp. values [22]
Electric quadrupole moment $Q_d(\mathrm{fm}^2)$	$0.2769 \pm (0.00622)$	$0.2818 \pm (0.00281)$	$0.2836 \pm$	0.2859(3) fm
Magnetic dipole moment μ_d	$0.8266 \pm (0.00957)\mu_N$	$0.8424 \pm (0.00621)\mu_N$	$0.8519 \pm (0.0072)\mu_N$	$0.857438(94)\mu_N$
Matter radius $r_m(\mathrm{fm})$	$1.9532 \pm (0.00475)$	$1.7269 \pm (0.00358)$	$1.9532 \pm (0.00475)$	1.975(3) fm
Charge radius $r_{ch}(\mathrm{fm})$	2.1161(4)	1.9092(3)	2.1354(9)	2.130(10) fm
D state percentage $P_d\%$	6.659	5.056	4.86	[2–6]

Conclusion

We have used the Nikiforov-Uvarov (NU) method for analytical solution of the radial Schrodinger equation. The NU method provides an exact solution of the non-relativistic Schrodinger equation for certain kinds of potentials. With the use of the mentioned method, the wave function are obtained for Standard, Generalized, and modified Wood–Saxon potentials. A comparison of the calculated values with experimental values is given. Then we applied our results to calculate the static properties and form factors of deuteron system. It is shown that the obtained results for modified form of the Wood–Saxon potential are very close to the experimental results with other forms of the potential. Our estimations indicate that the obtained wave function for modified potential provides relatively accurate values for form factors and static properties of deuteron system, emphasizing on the importance of the wave function and the nucleon-nucleon interaction model.

References

1. Wong, S.S.M.: Introductory nuclear physics. Wiley (2008)
2. Arrington, J., Roberts, C.D., Zanotti, J.M.: Nucleon electromagnetic form factors. J. Phys.G Nucl. Part. Phys. **34**, S23 (2007)
3. Elias, J.E., Friedman, J.I., Hartmann, G.C., Kendall, H.W., Nirk, P.N., Sogard, M.R., Van Speyerbroeck, L.P., de Pageter, J.K.: Measurements of elastic electron-deuteron scattering at high momentum transfers. Phys. Rev. **177**, 2075 (1969)
4. Garon, M., Arvieux, J., Beck, D.H., Beise, E.J., Boudard, A., Cairns, E.B., Cameron, J.M., Dodson, G.W., Dow, K.A., Farkhondeh, M., Fielding, H.W., Flanz, J.B., Goloskie, R., Hoibraten, S., Jourdan, J., Kowalski, S., Lapointe, C., McDonald, W.J., Nia, B., Pham, L.D., Redwine, R.P., Rodning, N.L., Roy, G., Schulze, M.E., Souder, P.A., Soukup, J., The, I., Turchinetz, W.E., Williamson, C.F., Wilson, K.E., Wood, S.A., Ziegler, W.: Tensor polarization in elastic electron-scattering in the momentum transfer range $3.8 \leq Q \leq 4.6$ fm-1. Phys. Rev. C **49**, 2516 (1994)
5. Bimbot, L.: Measurement of the deuteron electric and magnetic form factors at Jefferson Laboratory. Nucl. Phys. **A684**, 513c (2001)
6. Arnold, R.G., Chertok, B.T., Dally, E.B., Grigorian, A., Jordan, C.L., Schtz, W.P., Zdarko, R., Martin, F., Mecking, B.A.: Measurement of the electron-deuteron elastic-scattering cross section in the range $0.8 \leq q_2 \leq 6$ GeV_2. Phys. Rev. Lett. **35**, 776 (1975)
7. Gross, F., Gilman, R.: The deuteron: a mini-review. Conf. A.I.P. Proc. **603**, 55 (2001)
8. Alexa, L.C., Anderson, B.D., Aniol, K.A., Arundell, K., Auerbach, L., Baker, F.T., Berthot, J., Bertin, P.Y., Bertozzi, W., Bimbot, L., Boeglin, W.U., Brash, E.J., Breton, V., Breuer, H., Burtin, E., Calarco, J.R., Cardman, L.S., Cavata, C., Chang, C.C., Chen, J.P., Chudakov, E., Cisbani, E., Dale, D.S., Degrande, N., De Leo, R., Deur, A., dHose, N., Diederich, B., Domingo, J.J., Epstein, M.B., Ewell, L.A., Finn, J.M., Fissum, K.G., Fonvieille, H., Frois, B., Frullani, S., Gao, H., Gao, J., Garibaldi, F., Gasparian, A., Gilad, S., Gilman, R., Glamazdin, A., Glashausser, C., Gomez, J., Gorbenko, V., Hansen, J.O., Holmes, R., Holtrop, M., Howell, C., Huber, G.M., Hyde-Wright, C., Iodice, M., de Jager, C.W., Jaminion, S., Jardillier, J., Jones, M.K., Jutier, C., Kahl, W., Kato, S., Katramatou, A.T., Kelly, J.J., Kerhoas, S., Ketikyan, A., Khayat, M., Kino, K., Kramer, L.H., Kumar, K.S., Kumbartzki, G., Kuss, M., Lavessiere, G., Leone, A., LeRose, J.J., Liang, M., Lindgren, R.A., Liyanage, N., Lolos, G.J., Lourie, R.W., Madey, R., Maeda, K., Malov, S., Manley, D.M., Margaziotis, D.J., Markowitz, P., Marroncle, J., Martino, J., Martoff, C.J., McCormick, K., McIntyre, J., Mehrabyan, S., Meziani, Z.E., Michaels, R., Miller, G.W., Mougey, J.Y., Nanda, S.K., Neyret, D., Offermann, E.A.J.M., Papandreou, Z., Perdrisat, C.F., Perrino, R., Petratos, G.G., Platchkov, S., Pomatsalyuk, R., Prout, D.L., Punjabi, V.A., Pussieux, T., Quemener, G., Ransome, R.D., Ravel, O., Roblin, Y., Rowntree, D., Rutledge, G., Rutt, P.M., Saha, A., Saito, T., Sarty, A.J., Serdarevic, A., Smith, T., Soldi, K., Sorokin, P., Souder, P.A., Suleiman, R., Templon, J.A., Terasawa, T., Todor, L., Tsubota, H., Ueno, H., Ulmer, P.E., Urciuoli, G.M., Van Hoorebeke, L., Vernin, P., Vlahovic, B., Voskanyan, H., Watson, J.W., Weinstein, L.B., Wijesooriya, K., Wilson, R., Wojtsekhowski, B.B., Zainea, D.G., Zhang, W.M., Zhao, J., Zhou, Z.L.: Measurements of the deuteron elastic structure function $A(Q_2)$ for $0.7 \leq Q_2 \leq 6.0$ $(\mathrm{GeV/c})^2$ at Jefferson laboratory. Phys. Rev. Lett. **82**, 1372 (1999)
9. Cooper, F., Khare, A., Sukhatme, U.: Supersymmetry and quantum mechanics. Phys. Rep. **251**, 267 (1995)
10. Morales, D.A.: Supersymmetric improvement of the Pekeris approximation for the rotating Morse potential. Chem. Phys. Lett. **68** (2004)
11. Nikiforov, A.F., Uvarov, V.B.: Special Function of Mathematical Physics. Birkhauser, Basel (1988)
12. Pekeris, C.L.: The rotation-vibration coupling in diatomic molecules. Phys. Rev. **45**, 98 (1934)
13. Antia, A.D., Ikot, A.N., Hassanabadi, H., Maghsoodi, E.: Bound state Indian solutions of Klein–Gordon equation with Mobius square plus Yukawa potentials. Indian J. Phys. **87**, 1133 (2013)
14. Ikot, A.N., Yazarloo, B.H., Antia, A.D., Hassanabadi, H.: Relativistic treatment of spinless particle subject to generalized Tiez-Wei oscillator. Indian J. Phys. **87**, 913 (2013)
15. Woods, R.D., Saxon, D.S.: Diffuse surface optical model for nucleon-nuclei scattering. Phys. Rev. **95**, 577 (1954)
16. Berkdemir, C., Berkdemir, A., Sever, R.: Shape-invariance approach and Hamiltonian hierarchy method on the Woods–

Saxon potential for L $\neq$ 0 states. J. Math. Chem. **43**, 944 (2008)
17. Krane, K.S.: Introductory nuclear physics. Wiley, New York (1988)
18. Berkdemir, C., Han, J.: Any *l*-state solutions of the Morse potential through the Pekeris approximation and Nikiforov–Uvarov method. Chem. Phys. Lett. **203** (2005)
19. Berkdemir, C., Berkdemir, A., Sever, R.: Deformed woods-saxon potential in the frame of supersymmetric quantum mechanics for any *l*-state. In: Internal Report, United Nations Educational Scientific and Cultural Organization and International Atomic Energy Agency, ICTP, Trieste, Italy, 019 (2005)
20. Pahlavani, M.R., Sadeghi, J., Ghezelbash, M.: Solutions of the central Woods-Saxon potential in l $\neq$ 0 case using mathematical modification method. Appl. Sci. **11**, 106 (2009)
21. Dubovichenko, S.B.: Deuteron form factors for the nijmegen potentials. Phys. Atomic Nucl. **63**, 734 (2000)
22. Garcon, M., Van Orden, J.W.: The deuteron: structure and form factors. Adv. Nucl. Phys. **26**, 293 (2001)
23. Abbott, D., Ahmidouch, A., Anklin, H., Arvieux, J., Ball, J., Beedoe, S., Beise, E.J., Bimbot, L., Boeglin, W., Breuer, H., Carlini, R., Chant, N.S., Danagoulian, S., Dow, K., Ducret, J.E., Dunne, J., Ewell, L., Eyraud, L., Furget, C., Garcon, M., Gilman, R., Glashausser, C., Gueye, P., Gustafsson, K., Hafidi, K., Honegger, A., Jourdan, J., Kox, S., Kumbartzki, G., Lu, L., Lung, A., Markowitz, P., McIntyre, J., Meekins, D., Merchez, F., Mitchell, J., Mohring, R., Mtingwa, S., Mrktchyan, H., Pitz, D., Qin, L., Ransome, R., Real, J.S., Roos, P.G., Rutt, P., Sawafta, R., Stepanyan, S., Tieulent, R., Tomasi-Gustafsson, E., Turchinetz, W., Vansyoc, K., Volmer, J., Voutier, E., Williamson, C., Wood, S.A., Yan, C., Zhao, J., Zhao, W.: Phenomenology of the deuteron electromagnetic form factors. Eur. Phys. J. **A7**, 421 (2000)
24. Auffret, S., Cavedon, J.M., Clemens, J.C., Frois, B., Goutte, D., Huet, M., Leconte, P.H., Martino, J., Mizuno, Y., Phan, X.H., Platchkov, S., Sick, I.: Magnetic form factor of the deuteron. Phys. Rev. Lett. **54**, 649 (1985)
25. Yu, A., Berezhnoy, V.Y., Korda, U., Gakh, A.G.: Deuteron structure and diffractive deuteron–nucleus interaction. Phys. At. Nucl. **69**, 947 (2006)
26. McGurk, N.J.: The deuteron wave function at short range and the triton. Nucl. Phys. **A281**, 310 (1977)
27. Akhiezer, A.I., Sitenko, A.G., Tartakovskii, V.K.: Nuclear Electrodynamics. Springer, Berlin (1994)

11

Tensor coupling and relativistic spin and pseudospin symmetries of the Pöschl–Teller-like potential

H. Tokmehdashi · A. A. Rajabi · M. Hamzavi

Abstract In this research, we have been obtained the Dirac equation for second Pöschl–Teller-like potential including a Coulomb-like tensor interaction with arbitrary spin–orbit coupling quantum number κ. Under the condition of spin and pseudospin (p-spin) symmetries, we use the basic concept of the supersymmetric shape invariance formulism in quantum mechanics and the function analysis method to obtain energy eigenvalues and corresponding two-component spinors of the Dirac particle. We have also shown that tensor interaction removes degeneracies between spin and p-spin doublets. Some numerical results are also given.

Keywords Dirac equation · Spin symmetry · Pseudospin symmetry · Pöschl–Teller-like potential · Coulomb-like tensor potential · Supersymmetric quantum mechanics (SUSYQUM)

Introduction

The spin and pseudospin symmetry concepts introduced in nuclear theory [1, 2] have been used to explain the features of deformed nuclei [3] and superdeformation [4], and to establish an effective shell-model coupling scheme [5]. Within the framework of the relativistic mean field theory, Ginocchio [6, 7] has found that a Dirac Hamiltonian with scalar and vector harmonic oscillator potentials in the case of $(V(r) - S(r) = 0)$ possesses not only a spin symmetry but also a $U(3)$symmetry, but a Dirac Hamiltonian in the case of $(V(r) + S(r) = 0)$ possesses a pseudospin symmetry and a pseudo-$U(3)$symmetry. Meng et al. [8] have showed that the pseudospin symmetry is exact under the condition $(\mathrm{d}(V(r) + S(r))/\mathrm{d}r = 0)$. In addition, Alhaidari et al. [9] have investigated in detail physical interpretation on the three-dimensional Dirac equation in the case of spin symmetry limit $(V(r) - S(r) = 0)$ and pseudospin symmetry limit $(V(r) + S(r) = 0)$. In recent years, by considering the importance of spin and pseudospin symmetries, some authors have contributed many works in this field. For more review of this, one can read the recent works by Wei and Dong [10–13].

The p-spin symmetry refers to a quasidegeneracy of single nucleon doublets with non-relativistic quantum number $(n, l, j = l + 1/2)$ and $(n - 1, l + 2, j = l + 3/2)$, where n, l and j are single nucleon radial, orbital and total angular quantum numbers, respectively [1, 2]. The total angular momentum $j = \tilde{l} + \tilde{s}$ with $\tilde{l} = l + 1$ is a pseudo-angular momentum and $\tilde{s}$ is p-spin angular momentum [14–18].

In this paper, we attempt to study the spin and pseudospin symmetry solutions of the Dirac equation for arbitrary quantum number κ with the Pöschl–Teller-like potential. This is given by

$$V(r) = \frac{V_1 - V_2 \cos h\,\alpha r}{\sin h^2 \alpha r}. \tag{1}$$

The potential parameters V_1 and V_2 describe the property of the potential well, $V_1 > V_2$, while α is related to the range of the potential [19–24]. The behavior of this potential with respect to four different values of α is shown in Fig. 1.

H. Tokmehdashi (✉) · A. A. Rajabi
Department of Physics, Shahrood University of Technology, Shahrood, Iran
e-mail: h.tokmehdashi@gmail.com

M. Hamzavi
Department of Physics, University of Zanjan, Zanjan, Iran

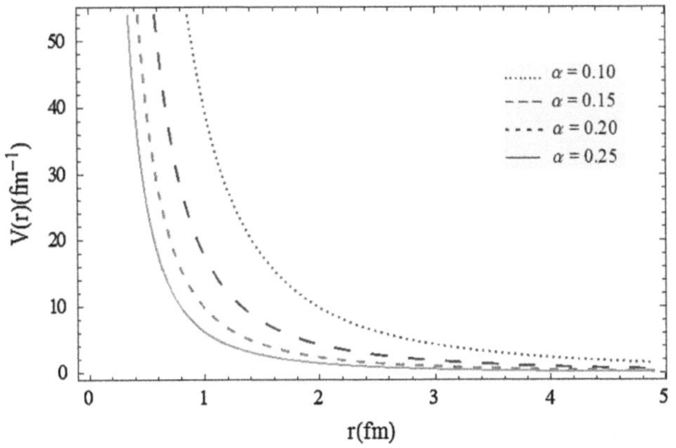

Fig. 1 The behavior of the Pöschl–Teller-like potential with respect to four different values of α, and for $V_1 = 0.5$, $V_2 = 0.1$

Also Tensor potentials have been introduced into the Dirac equation with the substitution $\vec{p} \to \vec{p} - im\omega\beta \cdot \hat{r}\,U(r)$ and a spin–orbit coupling is added to the Dirac Hamiltonian [25, 26]. For more review of tensor interaction, one can refer to the [27–34] that authors used different potential and different kinds of tensor potential. Here we study a tensor potential in the Coulomb-like form as follows:

$$U(r) = -\frac{U_{0C}}{r}, \quad U_{0C} = \frac{Z_a Z_b e^2}{4\pi\varepsilon_0}, \quad r \geq R_C, \tag{2}$$

where $R_C = 7.78\,fm$ is the coulomb radius, and Z_a and Z_b denote the charges of the projectile a and the target nuclei b, respectively.

The Potential in Eq. (1) is also one of the important examples for the special case of the multiparameter exponential-type potential model [35, 36]. By solving the Klein–Gordon equation and Dirac equation with equal scalar and vector Pöschl–Teller-like potentials, the exact relativistic energy equations have been obtained for the s-wave bound states ($l = 0$) [37, 38]. Using the conventional approximation scheme proposed by Greene and Aldrich [39] to deal with the centrifugal term, Dong et al. [40, 41] have investigated the arbitrary l-wave bound-state solutions of the Schrödinger equation and Klein–Gordon equation with the Pöschl–Teller-like potential in terms of the standard function analysis method. However, as far as we know, one has not reported the investigation of the spin and pseudospin symmetries solutions of the Dirac equation with the Pöschl–Teller-like potential including a Coulomb-like potential as a tensor interaction for the arbitrary spin–orbit quantum number κ. In this paper, we solve approximately the Dirac equation with the Pöschl–Teller-like potential for the spin–orbit quantum number κ. Under the condition of spin and pseudospin symmetries, we study the bound-state energy equation and corresponding spinor wave functions in terms of the basic concept of the supersymmetric shape invariance formalism [42, 43] and the function analysis method.

Dirac equation including tensor coupling

The Dirac equation for fermionic massive spin-1/2 particles moving in attractive scalar $S(r)$, repulsive vector $V(r)$ and tensor $U(r)$ potentials is (in units $\hbar = c = 1$)

$$\begin{aligned}&[\vec{\alpha}\cdot\vec{p} + \beta(M + S(r)) - i\beta\vec{\alpha}\cdot\hat{r}\,U(r)]\psi(\vec{r})\\&\quad = [E - V(r)]\psi(\vec{r}),\end{aligned} \tag{3}$$

where E is the relativistic energy of the system, $\vec{p} = -i\vec{\nabla}$ is the three-dimensional momentum operator and M is the mass of the fermionic particle. Further, $\vec{\alpha}$ and β are the 4×4 Dirac matrices given by

$$\vec{\alpha} = \begin{pmatrix} 0 & \vec{\sigma} \\ \vec{\sigma} & 0 \end{pmatrix}, \quad \beta = \begin{pmatrix} I & 0 \\ 0 & -I \end{pmatrix}, \tag{4}$$

where I is 2×2 unitary matrix and $\vec{\sigma}$ are three-vector spin matrices

$$\sigma_1 = \begin{pmatrix} 0 & 1 \\ 1 & 0 \end{pmatrix}, \quad \sigma_2 = \begin{pmatrix} 0 & -i \\ i & 0 \end{pmatrix}, \quad \sigma_3 = \begin{pmatrix} 1 & 0 \\ 0 & -1 \end{pmatrix}. \tag{5}$$

The total angular momentum operator $\vec{J}$ and spin–orbit $K = (\vec{\sigma}\cdot\vec{L} + 1)$, where $\vec{L}$ is orbital angular momentum of the spherical nucleons, commute with the Dirac Hamiltonian. The eigenvalues of spin–orbit coupling operator are $\kappa = (j + \frac{1}{2}) > 0$ and $\kappa = -(j + \frac{1}{2}) < 0$ for unaligned spin $j = l - \frac{1}{2}$ and the aligned spin $j = l + \frac{1}{2}$, respectively. (H^2, K, J^2, J_z) can be taken as the complete set of the conservative quantities. Thus, the spinor wave functions can be classified according to their angular momentum j; spin–orbit quantum number κ and the radial quantum number n can be written as follows:

$$\psi_{n\kappa}(\vec{r}) = \begin{pmatrix} f_{n\kappa}(\vec{r}) \\ g_{n\kappa}(\vec{r}) \end{pmatrix} = \begin{pmatrix} \frac{F_{n\kappa}(r)}{r} Y_{jm}^{l}(\theta,\varphi) \\ i\frac{G_{n\kappa}(r)}{r} Y_{jm}^{\tilde{l}}(\theta,\varphi) \end{pmatrix}, \tag{6}$$

where $f_{n\kappa}(\vec{r})$ is the upper (large) component and $g_{n\kappa}(\vec{r})$ is the lower (small) component of the Dirac spinors. $Y_{jm}^{l}(\theta,\phi)$ and $Y_{jm}^{\tilde{l}}(\theta,\phi)$ are spin and p-spin spherical harmonics, respectively, and m is the projection of the angular momentum on the z-axis. Substituting Eq. (6) into Eq. (3) and using the following relations:

$$\left(\vec{\sigma}\cdot\vec{A}\right)\left(\vec{\sigma}\cdot\vec{B}\right) = \vec{A}\cdot\vec{B} + i\vec{\sigma}\cdot\left(\vec{A}\times\vec{B}\right), \tag{7a}$$

$$\left(\vec{\sigma}\cdot\vec{P}\right) = \vec{\sigma}\cdot\hat{r}\left(\hat{r}\cdot\vec{P} + i\frac{\vec{\sigma}\cdot\vec{L}}{r}\right) \tag{7b}$$

together with the following properties:

$$
\begin{aligned}
&(\vec{\sigma}\cdot\vec{L})Y^{\tilde{l}}_{jm}(\theta,\phi) = (\kappa-1)Y^{\tilde{l}}_{jm}(\theta,\phi),\\
&(\vec{\sigma}\cdot\vec{L})Y^{l}_{jm}(\theta,\phi) = -(\kappa-1)Y^{l}_{jm}(\theta,\phi),\\
&(\vec{\sigma}\cdot\hat{r})Y^{\tilde{l}}_{jm}(\theta,\phi) = -Y^{l}_{jm}(\theta,\phi),\\
&(\vec{\sigma}\cdot\hat{r})Y^{l}_{jm}(\theta,\phi) = -Y^{\tilde{l}}_{jm}(\theta,\phi),
\end{aligned}
\tag{8}
$$

one obtains two coupled differential equations for upper and lower radial wave functions $F_{n\kappa}(r)$ and $G_{n\kappa}(r)$ as:

$$\left(\frac{\mathrm{d}}{\mathrm{d}r}+\frac{\kappa}{r}-U(r)\right)F_{n\kappa}(r) = (M+E_{n\kappa}-\Delta(r))G_{n\kappa}(r), \tag{9a}$$

$$\left(\frac{\mathrm{d}}{\mathrm{d}r}-\frac{\kappa}{r}+U(r)\right)G_{n\kappa}(r) = (M-E_{n\kappa}+\Sigma(r))F_{n\kappa}(r). \tag{9b}$$

where

$$\Delta(r) = V(r)-S(r), \tag{10a}$$

$$\Sigma(r) = V(r)+S(r), \tag{10b}$$

are the difference and the sum potentials, respectively. Eliminating $F_{n\kappa}(r)$ and $G_{n\kappa}(r)$ from Eqs. (9), we finally obtain the following two Schrödinger-like differential equations for the upper and lower radial spinor components, respectively:

$$
\begin{aligned}
&\left[\frac{\mathrm{d}^2}{\mathrm{d}r^2}-\frac{\kappa(\kappa+1)}{r^2}+\frac{2\kappa}{r}U(r)-\frac{\mathrm{d}U(r)}{\mathrm{d}r}-U^2(r)\right]F_{n\kappa}(r)\\
&\quad+\frac{\frac{\mathrm{d}\Delta(r)}{\mathrm{d}r}}{M+E_{n\kappa}-\Delta(r)}\left(\frac{\mathrm{d}}{\mathrm{d}r}+\frac{\kappa}{r}-U(r)\right)F_{n\kappa}(r)\\
&\quad=[(M+E_{n\kappa}-\Delta(r))(M-E_{n\kappa}+\Sigma(r))]F_{n\kappa}(r),
\end{aligned}
\tag{11}
$$

$$
\begin{aligned}
&\left[\frac{\mathrm{d}^2}{\mathrm{d}r^2}-\frac{\kappa(\kappa-1)}{r^2}+\frac{2\kappa}{r}U(r)+\frac{\mathrm{d}U(r)}{\mathrm{d}r}-U^2(r)\right]G_{n\kappa}(r)\\
&\quad+\frac{\frac{\mathrm{d}\Sigma(r)}{\mathrm{d}r}}{M-E_{n\kappa}+\Sigma(r)}\left(\frac{\mathrm{d}}{\mathrm{d}r}-\frac{\kappa}{r}+U(r)\right)G_{n\kappa}(r)\\
&\quad=[(M+E_{n\kappa}-\Delta(r))(M-E_{n\kappa}+\Sigma(r))]G_{n\kappa}(r)
\end{aligned}
\tag{12}
$$

where $\kappa(\kappa-1)=\tilde{l}(\tilde{l}+1)$ and $\kappa(\kappa+1)=l(l+1)$.

The quantum number κ is related to the quantum numbers for spin symmetry l and p-spin symmetry $\tilde{l}$ as:

$$
\kappa=\begin{cases}
-(l+1)=-\left(j+\frac{1}{2}\right) & (s_{1/2},p_{3/2},\text{etc.})\quad j=l+\frac{1}{2}, \quad \text{aligned spin }(\kappa<0)\\
+l=+\left(j+\frac{1}{2}\right) & (p_{1/2},d_{3/2},\text{etc.})\quad j=l-\frac{1}{2}, \quad \text{unaligned spin }(\kappa>0),
\end{cases}
$$

and the quasidegenerate doublet structure can be expressed in terms of a p-spin angular momentum $\tilde{s}=1/2$ and pseudoorbital angular momentum $\tilde{l}$, which can be defined as:

$$
\kappa=\begin{cases}
-\tilde{l}=-\left(j+\frac{1}{2}\right) & (s_{1/2},p_{3/2},\text{etc.})\quad j=\tilde{l}-\frac{1}{2}, \quad \text{alinged p}-\text{spin }(\kappa<0),\\
+(\tilde{l}+1)=+\left(j+\frac{1}{2}\right) & (d_{3/2},f_{5/2},\text{etc.})\quad j=\tilde{l}+\frac{1}{2}, \quad \text{unaligned p}-\text{spin }(\kappa>0),
\end{cases}
$$

where $\kappa=\pm 1, \pm 2, \ldots$. For example, $(1s_{1/2}, 0d_{3/2})$ and $(1p_{3/2}, 0f_{5/2})$ can be considered as p-spin doublets.

Spin symmetry limit

In this section, we will solve Dirac equation under spin symmetry limit with Pöschl–Teller-like potential and Coulomb-like potential as a tensor interaction. The exact spin symmetry occurs in Dirac equation when ($\mathrm{d}[V(r)-S(r)]/\mathrm{d}r=\mathrm{d}\Delta(r)/\mathrm{d}r=0$) or $\Delta(r)=C_s=$ constant [44–47]. Substituting Eqs. (1), (2) into Eq. (11) and taking $\Sigma(r)$ as the Pöschl–Teller-like potential, the equation obtained for the upper component of the Dirac spinor, $F_{n\kappa}(r)$, becomes

$$\left[\frac{d^2}{dr^2}-\frac{\lambda_\kappa(\lambda_\kappa+1)}{r^2}-\gamma\left(\frac{V_1-V_2\cos h\,\alpha r}{\sin h^2\alpha r}\right)-\beta^2\right]F_{n\kappa}(r)=0, \tag{13}$$

where $\lambda_\kappa=\kappa+U_{0C}$, $\gamma=M+E_{n\kappa}-C_s$ and $\beta^2=(M-E_{n\kappa})(M+E_{n\kappa}-C_s)$. Also, $\kappa=l$ and $\kappa=-l-1$ for $\kappa<0$ and $\kappa>0$, respectively.

P-spin symmetry limit

Within the pseudospin symmetry case, $(d[V(r)+S(r)]/dr=d\Sigma(r)/dr=0)$ or $\Sigma(r)=C_{ps}=$ constant and p-spin symmetry is exact in the Dirac equation [8, 48–50]. In this part, we consider $\Delta(r)$ as the Pöschl–Teller-like potential, the equation obtained for the lower component of the Dirac spinor, $G_{n\kappa}(r)$, becomes

$$\left[\frac{d^2}{dr^2}-\frac{\lambda_\kappa(\lambda_\kappa-1)}{r^2}-\tilde{\gamma}\left(\frac{V_1-V_2\cos h\alpha r}{\sin h^2\alpha r}\right)-\tilde{\beta}^2\right]G_{n\kappa}(r)=0, \tag{14}$$

where $\lambda_\kappa=\kappa+U_{0C}$, $\tilde{\gamma}=M-E_{n\kappa}+C_{ps}$, $\tilde{\beta}^2=(M+E_{n\kappa})(M-E_{n\kappa}+C_{ps})$. Also, $\kappa=-\tilde{l}$ and $\kappa=\tilde{l}+1$ for $\kappa<0$ and $\kappa>0$, respectively.

Equations (13) and (14) can be solved analytically only for the case of $\lambda_\kappa=-1$ and $\lambda_\kappa=1$ due to the pseudo-centrifugal terms, $\lambda_\kappa(\lambda_\kappa+1)/r^2$ and $\lambda_\kappa(\lambda_\kappa-1)/r^2$, respectively. Using the approximation scheme suggested by Greene and Aldrich [39], we can express approximately the pseudocentrifugal term in the following form [51–54]:

$$\frac{1}{r^2}\approx\alpha^2\frac{e^{-\alpha r}}{(1-e^{-\alpha r})^2} \tag{15}$$

This is a good approximation for small values of the parameter α and it breaks down for large values of α. For the case of $\alpha r\ll 1$, one can show that (see Fig. 2)

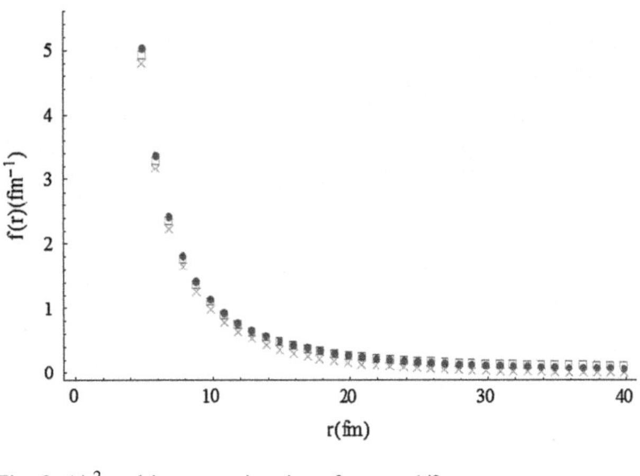

Fig. 2 $1/r^2$ and its approximations for $\alpha=1/2$

$$\alpha^2\left(c_0+\frac{e^{-\alpha r}}{(1-e^{-\alpha r})^2}\right)=\alpha^2\left(c_0+\frac{1}{4\sin h^2\frac{\alpha r}{2}}\right)\rightarrow\frac{1}{r^2} \tag{16}$$

where $c_0=1/12$ is a dimensionless constant.

Bound States of the Pöschl–Teller-like potential with Coulomb tensor interaction

Spin symmetry solution

Substituting Eq. (16) into Eq. (13) leads us to obtain the following Schrödinger-like equation for the upper spinor component:

$$\left[-\frac{d^2}{dr^2}+\frac{\lambda_\kappa(\lambda_\kappa+1)\alpha^2+\gamma(V_1-V_2)}{4\sin h^2\frac{\alpha r}{2}}-\frac{\gamma(V_1+V_2)}{4\cos h^2\frac{\alpha r}{2}}\right]F_{n\kappa}(r)=\tilde{E}_{n\kappa}F_{n\kappa}(r) \tag{17}$$

where $\tilde{E}_{n\kappa}$ is defined as $\tilde{E}_{n\kappa}=(E_{n\kappa}-M)(M+E_{n\kappa}-C_s)-\lambda_\kappa(\lambda_\kappa+1)\alpha^2c_0$. Using the basic concept of the super-symmetric shape invariance formulism [42, 43], we solve Eq. (17). The ground-state upper component $F_{0,\kappa}(r)$ can be written as:

$$F_{0,\kappa}(r)=\exp\left(-\int W(r)dr\right), \tag{18}$$

where $W(r)$ is called a superpotential in supersymmetric quantum mechanics [43]. Substituting Eq. (18) into Eq. (17), we have the following equation for $W(r)$

$$W^2(r)-\frac{dW(r)}{dr}=\frac{\lambda_\kappa(\lambda_\kappa+1)\alpha^2+\gamma(V_1-V_2)}{4\sin h^2\frac{\alpha r}{2}}-\frac{\gamma(V_1+V_2)}{4\cos h^2\frac{\alpha r}{2}}-\tilde{E}_{0,\kappa}, \tag{19}$$

where $\tilde{E}_{0,\kappa}$ is the ground-state energy. Considering the compatibility between the superpotential function $W(r)$ and the right-hand side of Eq. (19), we write the superpotential $W(r)$ in the following form:

$$W(r)=A\tan h\frac{\alpha r}{2}+B\cot h\frac{\alpha r}{2} \tag{20}$$

Substituting Eq. (20) into (18) leads us to obtain the ground-state upper component

$$F_{0,\kappa}(r),\quad F_{0,\kappa}(r)=\left(\cos h\frac{\alpha r}{2}\right)^{-2A/\alpha}\left(\sin h\frac{\alpha r}{2}\right)^{-2B/\alpha} \tag{21}$$

For the bound-state solutions, the upper component $F_{n\kappa}(r)$ must satisfy the boundary conditions that $F_{n\kappa}(r)/r$ becomes zero when $r\rightarrow\infty$, and $F_{n\kappa}(r)/r$ is finite if $r=0$. In view of these regularity conditions, we have the restriction conditions: $A>0$, $B<0$ and $A>-B$.

Substituting Eq. (20) into Eq. (19) and comparing equal powers of two sides in Eq. (19), we obtain the following relationships:

$$(A+B)^2 = -\tilde{E}_{0,\kappa} \tag{22}$$

$$A^2 + \frac{\alpha}{2}A = \frac{\gamma(V_1+V_2)}{4} \tag{23}$$

$$B^2 + \frac{\alpha}{2}B = \frac{\gamma(V_1-V_2)+\lambda_\kappa(\lambda_\kappa+1)\alpha^2}{4} \tag{24}$$

Considering the regularity conditions, $A > 0$ and $B < 0$, we obtain the coefficients A and B by solving Eqs. (23) and (24),

$$A = \frac{\alpha}{4}\left[-1+\sqrt{1+\frac{4\gamma(V_1+V_2)}{\alpha^2}}\right] \tag{25}$$

$$B = \frac{\alpha}{4}\left[-1-\sqrt{(1+2\lambda_\kappa)^2+\frac{4\gamma(V_1-V_2)}{\alpha^2}}\right] \tag{26}$$

Using the expression given in Eq. (20), we construct the following two supersymmetric partner potentials:

$$\begin{aligned} V_\pm(r) &= W^2(r) \pm \frac{dW(r)}{dr} \\ &= (A+B)^2 + \frac{B^2 \mp \frac{1}{2}\alpha B}{\sin h^2\frac{\alpha r}{2}} - \frac{A^2 \mp \frac{1}{2}\alpha A}{\cos h^2\frac{\alpha r}{2}} \end{aligned} \tag{27}$$

Setting $(a_0, b_0) = (A, B)$, one can get the following shape-invariant relationship,

$$V_+(r, a_0, b_0) = V_-(r, a_1, b_1) + R(a_1, b_1), \tag{28}$$

where a_1 and b_1 are the functions of a_0 and b_0, respectively, i.e., $a_1 = f(a_0) = a_0 - \alpha/2$, $b_1 = f(b_0) = b_0 - \alpha/2$, and the remainder $R(a_1, b_1)$ is independent of r, $R(a_1, b_1) = (a_0 + b_0)^2 - (a_1 + b_1)^2 = (A + B)^2 - (A + B - \alpha)^2$.

The energy eigenvalues of the potential $V_-(r)$ can be determined using the shape invariance approach [42]. The energy eigenvalues of the potential $V_-(r)$ are given by

$$\tilde{E}^{(-)}_{0,\kappa} = 0, \tag{29}$$

$$\begin{aligned} \tilde{E}^{(-)}_{n\kappa} &= \sum_{k=1}^{n} R(a_k, b_k) = R(a_1,b_1) + R(a_2,b_2) + \cdots + R(a_n,b_n) \\ &= (a_0+b_0)^2 - (a_1+b_1)^2 + (a_1+b_1)^2 - (a_2+b_2)^2 \\ &\quad + \cdots + (a_{n-1}+b_{n-1})^2 - (a_n+b_n)^2 \\ &= (a_0+b_0)^2 - (a_n+b_n)^2 = (A+B)^2 - (A+B-n\alpha)^2, \end{aligned} \tag{30}$$

where the quantum number $n = 0, 1, 2, \ldots$. From Eqs. (19) and (27), we have the following relation:

$$\frac{\lambda_\kappa(\lambda_\kappa+1)\alpha^2+\gamma(V_1-V_2)}{4\sin h^2\frac{\alpha r}{2}} - \frac{\gamma(V_1+V_2)}{4\cos h^2\frac{\alpha r}{2}} = V_-(r) + \tilde{E}_{0,\kappa}. \tag{31}$$

From Eqs. (17) and (31), we can find the solution for $\tilde{E}_{n\kappa}$ in Eq. (17),

$$\tilde{E}_{n\kappa} = \tilde{E}_{0,\kappa} + \tilde{E}^{(-)}_{n\kappa} = -(A+B-n\alpha)^2, \tag{32}$$

where we have employed the relation $\tilde{E}_{0,\kappa} = -(A+B)^2$. Substituting Eqs. (25) and (26) into Eq. (32) and using $\tilde{E}_{n\kappa} = (E_{n\kappa} - M)(M + E_{n\kappa} - C_s) - \lambda_\kappa(\lambda_\kappa + 1)\alpha^2 c_0$, we can find the energy eigenvalue equation of the relativistic Pöschl–Teller-like potential under the condition of spin symmetry,

$$\begin{aligned} &E^2_{n\kappa} - M^2 - C_s(E_{n\kappa} - M) - (\kappa + U_{0C})(\kappa + U_{0C} + 1)\alpha^2 c_0 \\ &= -\alpha^2\left(-n - \frac{1}{2} + \frac{1}{4}\sqrt{1+\frac{4(M+E_{n\kappa}-C_s)(V_1+V_2)}{\alpha^2}}\right. \\ &\left.\quad -\frac{1}{4}\sqrt{(1+2(\kappa+U_{0C}))^2+\frac{4(M+E_{n\kappa}-C_s)(V_1-V_2)}{\alpha^2}}\right)^2 \end{aligned} \tag{33}$$

where the quantum number $n = 0, 1, 2, \ldots, <\frac{1}{\alpha}(A + B)$. Using the recursion operator approach [55, 56], we can determine the excited state upper components from the superpotential $W(r)$ given in Eq. (20) and the ground-state upper component, $F_{0,\kappa}(r)$ given in Eq. (21).

To find the corresponding wave functions, we take the function analysis method to calculate the unnormalized excited state upper components. Substituting Eq. (32) into Eq. (17), we obtain

$$\begin{aligned} &\left(-\frac{d^2}{dr^2} + \frac{\lambda_\kappa(\lambda_\kappa+1)\alpha^2+\gamma(V_1-V_2)}{4\sin h^2\frac{\alpha r}{2}} - \frac{\gamma(V_1+V_2)}{4\cos h^2\frac{\alpha r}{2}}\right)F_{n\kappa}(r) \\ &= -(A+B-n\alpha)^2 F_{n\kappa}(r). \end{aligned} \tag{34}$$

Defining a new variable of the form $s = -\sin h^2\frac{\alpha r}{2}$ and making a transformation of the upper spinor component of the form $F_{n\kappa}(r) = (1-s)^{\frac{-A}{\alpha}} s^{\frac{-B}{\alpha}} f_{n\kappa}(s)$, Eq. (34) can be written as follows:

$$\begin{aligned} &(1-s)s\frac{d^2 f_{n\kappa}(s)}{ds^2} + \left[\frac{1}{2} - \frac{2B}{\alpha} - \left(1 - \frac{2A}{\alpha} - \frac{2B}{\alpha}\right)s\right]\frac{df_{n\kappa}(s)}{ds} \\ &\quad + n\left(n - \frac{2A}{\alpha} - \frac{2B}{\alpha}\right)f_{n\kappa}(s) = 0. \end{aligned} \tag{35}$$

This equation is the well-known differential equation satisfied by the hypergeometric function ${}_2F_1(a, b; c; s)$, i.e.,

$$\begin{aligned} f_{n\kappa}(s) &= {}_2F_1(a, b; c; s) \\ &= \frac{\Gamma(c)}{\Gamma(a)\Gamma(b)} \sum_{k=0}^{\infty} \frac{\Gamma(a+k)\Gamma(b+k)}{\Gamma(c+k)} \frac{s^k}{k!}, \end{aligned} \tag{36}$$

where $a = -n$, $b = n - \frac{2A}{\alpha} - \frac{2B}{\alpha}$, and $c = \frac{1}{2} - \frac{2B}{\alpha}$. Using the original variable r, the upper component $F_{n\kappa}(r)$ corresponding to energy level $E_{n\kappa}$ can be expressed as follows:

$$F_{n\kappa}(r) = \left(1 + \sinh^2\frac{\alpha r}{2}\right)^{-\frac{A}{\alpha}}\left(-\sinh^2\frac{\alpha r}{2}\right)^{-\frac{B}{\alpha}} \times {}_2F_1\left(-n, n - \frac{2A}{\alpha} - \frac{2B}{\alpha}; \frac{1}{2} - \frac{2B}{\alpha}; -\sinh^2\frac{\alpha r}{2}\right), \tag{37}$$

where A and B are given in Eqs. (25) and (26), respectively. Using Eq. (9a) and the expression of $F_{n\kappa}(r)$ given in Eq. (37), we obtain the lower spinor component $G_{n\kappa}(r)$ corresponding to the upper component $F_{n\kappa}(r)$ and energy level $E_{n\kappa}$,

$$\begin{aligned} G_{n\kappa}(r) &= \frac{1}{(M+E_{n\kappa}-C_s)}\left(\frac{d}{dr} + \frac{\kappa}{r} - U(r)\right)F_{n\kappa}(r) \\ &= \frac{1}{(M+E_{n\kappa}-C_s)}\left(1+\sinh^2\frac{\alpha r}{2}\right)^{-\frac{A}{\alpha}}\left(-\sinh^2\frac{\alpha r}{2}\right)^{-\frac{B}{\alpha}} \\ &\times\left[-\left(\frac{A\sinh\alpha r}{2\left(1+\sinh^2\frac{\alpha r}{2}\right)} + \frac{B\cosh\frac{\alpha r}{2}}{\sinh\frac{\alpha r}{2}} - \frac{\kappa+U_{0C}}{r}\right)\right. \\ &\times {}_2F_1\left(-n, n-\frac{2A}{\alpha}-\frac{2B}{\alpha}; \frac{1}{2}-\frac{2B}{\alpha}; -\sinh^2\frac{\alpha r}{2}\right) \\ &+ \frac{n\alpha\sinh\alpha r\left(n-\frac{2A}{\alpha}-\frac{2B}{\alpha}\right)}{2\left(\frac{1}{2}-\frac{2B}{\alpha}\right)} \\ &\left.\times {}_2F_1\left(-n+1, n+1-\frac{2A}{\alpha}-\frac{2B}{\alpha}; \frac{3}{2}-\frac{2B}{\alpha}; -\sinh^2\frac{\alpha r}{2}\right)\right], \end{aligned} \tag{38}$$

where $E_{n\kappa} \neq -M + C_s$. From Eqs. (37) and (38), we can observe that the upper component $F_{n\kappa}(r)$ and lower component $G_{n\kappa}(r)$ can satisfy the regularity conditions for the bound states when $A > 0$ and $B < 0$ and $A > -B$.

The energy level $E_{n\kappa}$ is given implicitly by energy eigenvalue Eq. (33) which is a rather complicated transcendental equation. To show the procedure of determining the bound-state energy eigenvalues from Eq. (33), we take a set of physical parameter values, $C_s = 2$, $\alpha = 1.2$, $M = 10$, $V_1 = 5$, $V_2 = 3$, and $c_0 = 1/12$, to give a numerical example. When $n = 0$ and $\kappa = 1$, Eq. (33) yields the following values of $E_{0,1}$: 9.727533, −7.948904. We choose $E_{0,1} = 9.727533$ as the solution of Eq. (33), and find that the values of A and B are $A = 5.66197$ and $B = -3.41027$, respectively. These values satisfy the regularity conditions: $A > 0$, $B < 0$ and $A > -B$. If we take $E_{0,1} = -7.948904$ as the solution of Eq. (33), the values of A and B are $A = 0.138395$ and $B = -1.21408$, which do not satisfy the regularity condition: $A > -B$. Therefore, we can only take the positive energy value $E_{0,1} = 9.727533$ as the solution of Eq. (33). Using the same parameter values of α, M, V_1, V_2 and C_s, the numerical solutions of Eq. (33) for the other values of n and κ are presented in Table 1. In Fig. 3, we have investigated the effect of the tensor potential on the bound states.

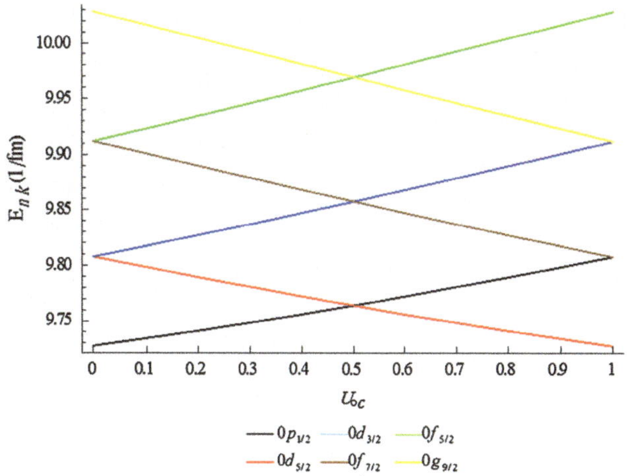

Fig. 3 (color online) Contribution of the tensor Coulomb potential parameter to the energy levels in the case of spin symmetry

P-spin symmetry solution

In this subsection, we will obtain the energy eigenvalues and the corresponding wave functions for the p-spin symmetric limit by substituting Eq. (16) into Eq. (14) that

Table 1 The spin symmetric bound-state energy levels (in unit of fm^{-1}) of the Coulomb potential taking several values of n and κ

l	$n, \kappa < 0$	$(l, j = l + 1/12.2)$	$E_{n,\kappa<0}$, $U_{0C} = 0$	$E_{n,\kappa<0}$, $U_{0C} = 1$	$n, \kappa > 0$	$(l, j = l - 1/2)$	$E_{n,\kappa>0}$, $U_{0C} = 0$	$E_{n,\kappa>0}$, $U_{0C} = 1$
1	0, −2	$0p_{3/2}$	9.727533734	9.683443227	0, 1	$0p_{1/2}$	9.727533734	9.807727942
2	0, −3	$0d_{5/2}$	9.807727942	9.727533734	0, 2	$0d_{3/2}$	9.807727942	9.911973645
3	0, −4	$0f_{7/2}$	9.911973645	9.807727942	0, 3	$0f_{5/2}$	9.911973645	10.02835022
4	0, −5	$0g_{9/2}$	10.02835022	9.911973645	0, 4	$0g_{7/2}$	10.02835022	10.14718163
1	1, −2	$1p_{3/2}$	9.949464374	9.921517656	1, 1	$1p_{1/2}$	9.949464374	9.999543951
2	1, −3	$1d_{5/2}$	9.999543951	9.949464374	1, 2	$1d_{3/2}$	9.999543951	10.06287108
3	1, −4	$1f_{7/2}$	10.06287108	9.999543951	1, 3	$1f_{5/2}$	10.06287108	10.13048571
4	1, −5	$1g_{9/2}$	10.13048571	10.06287108	1, 4	$1g_{7/2}$	10.13048571	10.19492768

leads us to obtain the following Schrödinger-like equation for the lower spinor component:

$$\left[-\frac{d^2}{dr^2}+\frac{\lambda_\kappa(\lambda_\kappa-1)\alpha^2+\tilde{\gamma}(V_1-V_2)}{4\sinh^2\frac{\alpha r}{2}}-\frac{\tilde{\gamma}(V_1+V_2)}{4\cosh^2\frac{\alpha r}{2}}\right]G_{n\kappa}(r)$$
$$=\tilde{E}_{n\kappa}G_{n\kappa}(r), \tag{39}$$

where $\tilde{E}_{n\kappa}$ is defined as $\tilde{E}_{n\kappa}=(E_{n\kappa}+M)(E_{n\kappa}-M-C_{ps})-\lambda_\kappa(\lambda_\kappa-1)\alpha^2c_0$. We write the super-potential $\tilde{W}(r)$ in the following form:

$$W(r)=\tilde{A}\tanh\frac{\alpha r}{2}+\tilde{B}\coth\frac{\alpha r}{2}, \tag{40}$$

that leads us to obtain the ground-state upper component

$$G_{0,\kappa}(r),\quad G_{0,\kappa}(r)=\left(\cosh\frac{\alpha r}{2}\right)^{-2\tilde{A}/\alpha}\left(\sinh\frac{\alpha r}{2}\right)^{-2\tilde{B}/\alpha}, \tag{41}$$

where

$$\tilde{A}=\frac{\alpha}{4}\left[-1+\sqrt{1+\frac{4\tilde{\gamma}(V_1+V_2)}{\alpha^2}}\right], \tag{42}$$

$$\tilde{B}=\frac{\alpha}{4}\left[-1-\sqrt{(1-2\lambda_\kappa)^2+\frac{4\tilde{\gamma}(V_1-V_2)}{\alpha^2}}\right], \tag{43}$$

To avoid repetition in our solution to Eq. (14), We follow the same procedures explained in the previous section to obtain the energy eigenvalue equation,

$$E_{n\kappa}^2-M^2-C_{ps}(E_{n\kappa}+M)-(\kappa+U_{OC})(\kappa+U_{OC}-1)\alpha^2$$
$$c_0=-\alpha^2\left(-n-\frac{1}{2}+\frac{1}{4}\sqrt{1+\frac{4(M-E_{n\kappa}+C_{ps})(V_1+V_2)}{\alpha^2}}-\frac{1}{4}\sqrt{(1-2(\kappa+U_{OC}))^2+\frac{4(M-E_{n\kappa}+C_{ps})(V_1-V_2)}{\alpha^2}}\right)^2 \tag{44}$$

and the corresponding wave functions for the lower Dirac spinor as:

$$G_{n\kappa}(r)=\left(1+\sinh^2\frac{\alpha r}{2}\right)^{-\frac{\tilde{A}}{\alpha}}\left(-\sinh^2\frac{\alpha r}{2}\right)^{-\frac{\tilde{B}}{\alpha}}$$
$$\times{}_2F_1\left(-n,n-\frac{2\tilde{A}}{\alpha}-\frac{2\tilde{B}}{\alpha};\frac{1}{2}-\frac{2\tilde{B}}{\alpha};-\sinh^2\frac{\alpha r}{2}\right). \tag{45}$$

Finally, the upper-spinor component of the Dirac equation can also be obtained via Eq. (9b) as:

$$F_{n\kappa}(r)=\frac{1}{(M-E_{n\kappa}+C_{ps})}\left(\frac{d}{dr}-\frac{\kappa}{r}+U(r)\right)G_{n\kappa}(r)$$
$$=\frac{1}{(M-E_{n\kappa}+C_{ps})}\left(1+\sinh^2\frac{\alpha r}{2}\right)^{-\frac{\tilde{A}}{\alpha}}\left(-\sinh^2\frac{\alpha r}{2}\right)^{-\frac{\tilde{B}}{\alpha}}$$
$$\left[-\left(\frac{\tilde{A}\sinh\alpha r}{2\left(1+\sinh^2\frac{\alpha r}{2}\right)}+\frac{\tilde{B}\cosh\frac{\alpha r}{2}}{\sinh\frac{\alpha r}{2}}+\frac{\kappa+U_{OC}}{r}\right)\right.$$
$$\times{}_2F_1\left(-n,n-\frac{2\tilde{A}}{\alpha}-\frac{2\tilde{B}}{\alpha};\frac{1}{2}-\frac{2\tilde{B}}{\alpha};-\sinh^2\frac{\alpha r}{2}\right)$$
$$+\frac{n\alpha\sinh\alpha r\left(n-\frac{2\tilde{A}}{\alpha}-\frac{2\tilde{B}}{\alpha}\right)}{2\left(\frac{1}{2}-\frac{2\tilde{B}}{\alpha}\right)}$$
$$\left.\times{}_2F_1\left(-n+1,n+1-\frac{2\tilde{A}}{\alpha}-\frac{2\tilde{B}}{\alpha};\frac{3}{2}-\frac{2\tilde{B}}{\alpha};-\sinh^2\frac{\alpha r}{2}\right)\right], \tag{46}$$

where $E_{n\kappa}\neq M+C_{ps}$.

By taking a set of physical parameter values, $C_{ps}=2$, $\alpha=1.2$, $M=10$, $V_1=5$, $V_2=3$, $c_0=1/12$, when $n=1$ and $\kappa=2$, Eq. (44) yields the following values of $E_{1,2}$: −9.92187466, 11.814382. We choose $E_{1,2}=-9.92187466$ as the solution of Eq. (44), and find that the values of $\tilde{A}$ and $\tilde{B}$ are $\tilde{A}=6.32825$ and $\tilde{B}=-3.73088$, respectively. These values satisfy the regularity conditions: $\tilde{A}>0$, $\tilde{B}<0$ and $\tilde{A}>-\tilde{B}$. If we take $E_{1,2}=11.814382$ as the solution of Eq. (44), the values of $\tilde{A}$ and $\tilde{B}$ are $\tilde{A}=0.379142$ and $\tilde{B}=-1.25016$, which do not satisfy the regularity condition: $\tilde{A}>-\tilde{B}$. Therefore, we can only take the negative energy value $E_{1,2}=-9.92187466$ as the solution of Eq. (44). Using the same parameter values of α, M, V_1, V_2 and C_{ps}, the numerical solutions of Eq. (44) for the other values of n and κ are presented in Table 2.

Table 2 The p-spin symmetric bound-state energy levels (in unit of fm^{-1}) of the Coulomb potential taking several values of n and κ

$\tilde{l}$	$n, \kappa<0$	(l, j)	$E_{n,\kappa<0}$, $U_{OC}=0$	$E_{n,\kappa<0}$, $U_{OC}=1$	$n-1, \kappa>0$	$(l+2, j+1)$	$E_{n,\kappa>0}$, $U_{OC}=0$	$E_{n,\kappa>0}$, $U_{OC}=1$
1	1, −1	$1s_{1/2}$	−9.921874665	−9.896961773	0, 2	$0d_{3/2}$	−9.921874665	−9.967517561
2	1, −2	$1p_{3/2}$	−9.967517561	−9.921874664	0, 3	$0f_{5/2}$	−9.967517561	−10.02721223
3	1, −3	$1d_{5/2}$	−10.02721223	−9.967517561	0, 4	$0g_{7/2}$	−10.02721223	−10.09381971
4	1, −4	$1f_{7/2}$	−10.09381971	−10.02721223	0, 5	$0h_{9/2}$	−10.09381971	−10.16103413
1	2, −1	$2s_{1/2}$	−10.00901073	−9.995644024	1, 2	$1d_{3/2}$	−10.00901073	−10.03267841
2	2, −2	$2p_{3/2}$	−10.03267841	−10.00901073	1, 3	$1f_{5/2}$	−10.03267841	−10.06171352
3	2, −3	$2d_{5/2}$	−10.06171352	−10.03267841	1, 4	$1g_{7/2}$	−10.06171352	−10.09077285
4	2, −4	$2f_{7/2}$	−10.09077285	−10.06171352	1, 5	$1h_{9/2}$	−10.09077285	−10.11504725

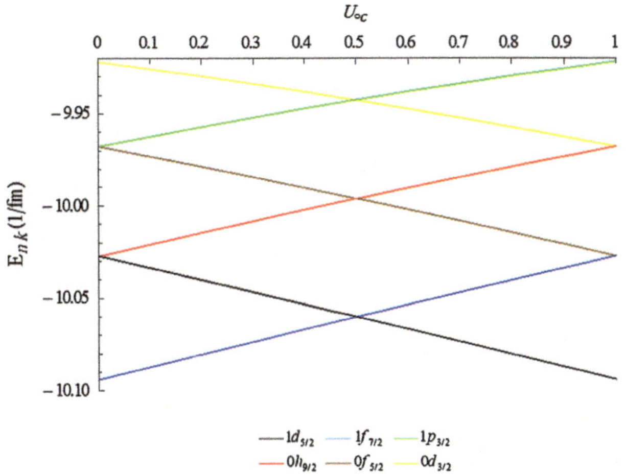

Fig. 4 (color online) Contribution of the tensor Coulomb potential parameter to the energy levels in the case of p-spin symmetry

Also in Fig. 4, we have investigated the effect of the tensor potential on the p-spin doublet splitting by considering some pairs of orbitals.

Conclusion

In this paper, we have approximately studied the bound-state solutions of the Dirac equation for the Pöschl–Teller-like potential with a Coulomb-like tensor interaction within the framework of spin and pseudospin symmetry limits. By employing an improved approximation scheme to deal with the pseudocentrifugal term $1/r^2$ and the SUSYQUM technique, We have obtained the energy levels in a closed form and the corresponding wave functions in terms of the hypergeometric function ${}_2F_1(a, b; c; s)$. Some numerical values of the energy levels are reported in Tables 1 and 2 under the condition of the spin and p-spin symmetries, respectively. Obviously, the degeneracy between the members of doublet states in spin and p-spin symmetries is removed by tensor interaction. The p-spin spectra of the present potential are identical to those ones obtained in Ref. [57] as the potential parameters $U_{0C} = 0$, $\alpha = 0.15$, $M = 1.0$, $C_{ps} = -5$.

Acknowledgments We would like to thank the kind referees for positive suggestions which have improved the present manuscript.

References

1. Arima, A., Harvey, M., Shimizu, K.: Pseudo LS coupling and pseudo-SU(3) coupling schemes. Phys. Lett. B **30**, 517 (1969)
2. Hecht, K.T., Adeler, A.: Generalized seniority for favored J-0 pairs in mixed configurations. Nucl. Phys. A **137**, 129 (1969)
3. Bohr, A., Hamamoto, I., Mottelson, B.R.: Pseudospin in rotating nuclear potentials. Phys. Scr. **26**, 267 (1982)
4. Dudek, J., Nazarewicz, W., Szymanski, Z., Leander, G.A.: Abundance and systematics of nuclear superdeformed states: relation to the pseudospin and pseudo-SU(3) symmetries. Phys. Rev. Lett. **59**, 1405 (1987)
5. Troltenier, D., Bahri, C., Draayer, J.P.: Generalized pseudo-SU(3): model and pairing. Nucl. Phys. A **586**, 53 (1995)
6. Ginocchio, J.N.: U(3) and pseudo-U(3) symmetry of the relativistic harmonic oscillator. Phys. Rev. Lett. **95**, 252501 (2005)
7. Ginocchio, J.N.: Pseudospin as a relativistic symmetry. Phys. Rev. Lett. **78**, 436 (1997)
8. Meng, J., Sugawara-Tanabe, K., Yamaji, S., Ring, P., Arima, A.: Pseudospin symmetry in relativistic mean field theory. Phys. Rev. C **58**, R628 (1998)
9. Alhaidari, A.D., Bahlouli, H., Al-Hasan, A.: Dirac and Klein–Gordon equations with equal scalar and vector potentials. Phys. Lett. A **349**, 87 (2006)
10. Wei, G.F., Dong, S.H.: Approximately analytical solutions of the Manning–Rosen potential with the spin–orbit coupling term and spin symmetry. Phys. Lett. A **373**, 49 (2008)
11. Wei, G.F., Dong, S.H.: The spin symmetry for deformed generalized Pöschl–Teller potential. Phys. Lett. A **373**, 2428 (2009)
12. Wei, G.F., Dong, S.H.: A novel algebraic approach to spin symmetry for Dirac equation with scalar and vector second Pöschl–Teller potentials. Euro. Phys. J. A **43**, 185 (2010)
13. Wei, G.F., Dong, S.H.: Spin symmetry in the relativistic symmetrical well potential including a proper approximation to the spin–orbit coupling term. Phys. Scr. **81**, 035009 (2010)
14. Ikhdair, S.M., Sever, R.: Approximate bound state solutions of Dirac equation with Hulthén potential including Coulomb-like tensor potential. Appl. Math. Comput. **216**, 911 (2010)
15. Berkdemir, C.: Pseudospin symmetry in the relativistic Morse potential including the spin–orbit coupling term. Nucl. Phys. A **770**, 32 (2006)
16. Dong, S.H., Wei, G.F.: Algebraic approach to pseudospin symmetry for the Dirac equation with scalar and vector modified Pöschl–Teller potentials. Europhys. Lett. **87**, 40004 (2009)
17. Dong, S.H., Wei, G.F.: Pseudospin symmetry in the relativistic Manning–Rosen potential including a Pekeris-type approximation to the pseudocentrifugal term. Phys. Lett. B **686**, 288 (2010)
18. Dong, S.H., Wei, G.F.: Pseudospin symmetry for modified Rosen–Morse potential including a Pekeris-type approximation to the pseudo-centrifugal term. Eur. Phys. J. A **46**, 207 (2010)
19. Dong, S.H., Qiang, W.C.: Analytical approximations to the *l*-wave solutions of the Klein Gordon equation for a second Pöschl–Teller like potential. Phys. Lett. A **372**, 4789 (2008)
20. Bagrov, V.G., Gitman, D.M.: Exact Solution of Relativistic Wave Equations. Kluwer Academic, Dordrecht (1990)
21. Miranda, M.G., Sun, G.H., Dong, S.H.: The solution of the second Pöschl–Teller like potential by Nikiforov–Uvarov method. Int. J. Mod. Phys. E **19**, 123 (2010)
22. Dong, S.H., Qiang, W.C., Gracía-Ravelo, J.: Analytical approximations to the Schrödinger equation for a second Pöschl–Teller-like potential with centrifugal term. J. Int. Mod. Phys. A **23**, 1537 (2008)
23. Sun, G.H., Aoki, M.A., Dong, S.H.: Quantum information entropies of the eigenstates for the Pöschl–Teller-like potential. Chin. Phys. B **22**, 050302 (2013)
24. Dong, S.H., Cisneros, A.G.: Energy spectra of the hyperbolic and second Pöschl–Teller like potentials solved by new exact quantization rule. Annal. Phys. **323**, 1136 (2008)
25. Moshinsky, M., Szczepaniak, A.: The Dirac oscillator. J. Phys. A: Math. Gen. **22**, L817–L820 (1989)
26. Mao, G.: Effect of tensor couplings in a relativistic Hartree approach for finite nuclei. Phys. Rev. C **67**, 044318-12 (2003)
27. Lisboa, R., Malheiro, M., de Castro, A.S., Alberto, P., Fiolhais, M.: Pseudospin symmetry and the relativistic harmonic oscillator. Phys. Rev. C **69**, 024319 (2004)

28. Alberto, P., Lisboa, R., Malheiro, M., de Castro, A.S.: Tensor coupling and pseudospin symmetry in nuclei. Phys. Rev. C **71**, 034313 (2005)
29. Akcay, H.: Dirac equation with scalar and vector quadratic potentials and Coulomb-like tensor potential. Phys. Lett. A **373**, 616 (2009)
30. Akcay, H.: The Dirac oscillator with a Coulomb-like tensor potential. J. Phys. A: Math. Theor. **40**, 6427 (2007)
31. Aydoğdu, O., Sever, R.: Exact pseudospin symmetric solution of the Dirac equation for pseudoharmonic potential in the presence of tensor potential. Few-Body Syst. **47**, 193 (2010)
32. Aydoğdu, O., Sever, R.: Pseudospin and spin symmetry in the Dirac equation with Woods–Saxon potential and tensor potential. Eur. Phys. J. A **43**, 73 (2010)
33. Hamzavi, M., Rajabi, A.A., Hassanabadi, H.: Exact pseudospin symmetry solution of the Dirac equation for spatially-dependent mass Coulomb potential including a Coulomb-like tensor interaction via asymptotic iteration method. Phys. Lett. A **374**, 4303 (2010)
34. Hamzavi, M., Rajabi, A.A., Hassanabadi, H.: Exactly complete solutions of the Dirac equation with pseudoharmonic potential including Linear plus Coulomb-like tensor potential. Int. J. Mod. Phys. A **26**, 1363 (2011)
35. Jia, C.S., Zeng, X.L., Sun, L.T.: PT symmetry and shape invariance for a potential well with a barrier. Phys. Lett. A **294**, 185 (2002)
36. Jia, C.S., Li, Y., Sun, Y., Liu, J.Y., Sun, L.T.: Bound states of the five-parameter exponential-type potential model. Phys. Lett. A **311**, 115 (2003)
37. Diao, Y.F., Yi, L.Z., Jia, C.S.: Bound states of the Klein–Gordon equation with vector and scalar five-parameter exponential-type potentials. Phys. Lett. A **332**, 157 (2004)
38. Zhang, X.C., Liu, Q.W., Jia, C.S., Wang, L.Z.: Bound states of the Dirac equation with vector and scalar Scarf-type potentials. Phys. Lett. A **340**, 59 (2005)
39. Greene, R.L., Aldrich, C.: Variational wave functions for a screened Coulomb potential. Phys. Rev. A **14**, 2363 (1976)
40. Dong, S.S., Gracía-Ravelo, J., Dong, S.H.: Analytical approximations to the *l*-wave solutions of the Schrödinger equation with an exponential-type potential. Phys. Scr. **76**, 393 (2007)
41. Qiang, W.C., Dong, S.H.: Analytical approximations to the solutions of the Manning–Rosen potential with centrifugal term. Phys. Lett. A **368**, 13 (2007)
42. Gendenshtein, L.E.: Derivation of exact spectra of the Schrodinger equation by means of supersymmetry. Sov. Phys. JETP Lett. **38**, 356 (1983)
43. Cooper, F., Khare, A., Sukhatme, U.: Supersymmetry and quantum mechanics. Phys. Rep. **251**, 267 (1995)
44. Zhou, S.G., Meng, J., Ring, P.: Spin symmetry in the antinucleon apectrum. Phys. Rev. Lett. **91**, 262501 (2003)
45. He, X.T., Zhou, S.G., Meng, J., Zhao, E.G., Scheid, W.: Test of spin symmetry in anti nucleon spectra. Eur. Phys. J. A **28**, 265 (2006)
46. Song, C.Y., Yao, J.M., Meng, J.: Spin symmetry for antilambda spectrum in atomic nucleus. Chin. Phys. Lett. **26**, 122102 (2009)
47. C.Y., Song, J.M., Yao: Polarization effect on the spin symmetry for anti-lambda spectrum in $^{16}O + \bar{\lambda}$-system. Chin. Phys. C **34**, 1425 (2010)
48. Ginocchio, J.N.: The relativistic foundations of pseudospin symmetry in nuclei. Nucl. Phys. A **654**, 663c (1999)
49. Ginocchio, J.N.: A relativistic symmetry in nuclei. Nucl. Phys. Rep. **315**, 231 (1999)
50. Meng, J., Sugawara-Tanabe, K., Yamaji, S., Arima, A.: Pseudospin symmetry in Zr and Sn isotopes from the proton drip line to the neutron drip line. Phys. Rev. C **59**, 154 (1999)
51. Zhang, L.H., Li, X.P., Jia, C.S.: Analytical approximation to the solution of the Dirac equation with the Eckart potential including the spin–orbit coupling term. Phys. Lett. A **372**, 2201 (2008)
52. Soylu, A., Bayrak, O., Boztosun, I.: k state solutions of the Dirac equation for the Eckart potential with pseudospin and spin symmetry. J. Phys. A: Math. Theor. **41**, 065308 (2008)
53. Xu, Y., He, S., Jia, C.S.: Approximate analytical solutions of the Dirac equation with the Pöschl–Teller potential including the spin-orbit coupling term. J. Phys. A: Math. Theor. **41**, 255302 (2008)
54. Dong, S.H., Qiang, W.C., Sun, G.H., Bezerra, V.B.: Analytical approximations to the *l* wave solutions of the Schrödinger equation with the Eckart potential. J. Phys. A: Math. Theor. **40**, 10535 (2007)
55. Dabrowska, J.W., Khare, A., Sukhatme, U.P.: Explicit wavefunctions for shape-invariant potentials by operator techniques. J. Phys. A: Math. Gen. **21**, L195 (1988)
56. Jia, C.S., Wang, X.G., Yao, X.K., Chen, P.C., Xiao, W.: A unified recurrence operator method for obtaining normalized explicit wavefunctions for shape-invariant potentials. J. Phys. A: Math. Gen. **31**, 4763 (1998)
57. Jia, C.S., Chen, T., Cui, L.G.: Approximate analytical solutions of the Dirac equation with the generalized Pöschl–Teller potential including the pseudo-centrifugal term. Phys. Lett. A **373**, 1621 (2009)

12

Evolution of shapes in even–even nuclei using the standard interacting boson model

Ali M. Khalaf · Mahmoud M. Taha

Abstract The sd-version of the interacting boson model (IBM) is used to establish the shape phase transitional structure. A simplified Hamiltonian is used which is intermediate between the three dynamical symmetries of U(6), namely the spherical U(5), the prolate and oblate deformed SU(3) and the γ-unstable O(6) limits. The potential energy surfaces (PESs) to the IBM Hamiltonian have been obtained using the intrinsic state formalism which introduces the shape variables β and γ. The Gadolinium (Gd) and Ruthenium (Ru) isotopic chains have been taken as examples in illustrating the U(5)–SU(3) and U(5)–O(6) shape phase transitions, respectively. We used the standard χ^2 test to get the IBM Hamiltonian parameters. The fit is performed by minimizing the χ^2 function for some selected experimental low-lying energy levels, the two neutron separation energies and B(E2) transition rates.

Keywords Nuclear structure · Potential energy surface · Interacting boson model

Introduction

The interacting boson model (IBM) [1] was designed to describe the collective quadrupole degrees of freedom in nuclei. The IBM Hamiltonian was written from the beginning in second quantization form in terms of the generators of the unitary Lie algebra U(6) subtended by s and d bosons which carry angular momentum 0 and 2, respectively. The model present three special limits that can be solved easily. These three limits are U(5), SU(3) and O(6) dynamical symmetries appropriate for an harmonic vibrator, axial deformed rotor and γ-unstable deformed rotor, respectively. Each of these limits is assigned to spherical, axially deformed and deformed with γ-instability shapes, respectively. Among the transitional Hamiltonian a specially interesting case occurs when it describes a critical point in the transition from a given shape to another. A shape phase transition may be of first or second order, depending on the discontinuity of the first- or second-order derivatives of the order parameter as a function of the control parameters [2–10].

The classical limit of the IBM is defined as its expectation value in a coherent state [10–12] and yields a function of shape variables β and γ which can be interpreted as a potential energy surface (PES) depending on these parameters. The symmetry E(5) [13] is designed for the critical point of the transition from spherical to deformed γ-unstable shapes. Later [14, 15] X(5) and Y(5) describe the critical points between spherical and axially deformed shapes and between axial and triaxial deformed shapes, respectively.

In the present work, we use the two parameters sd-IBM-1 Hamiltonian with the intrinsic coherent state method to produce the potential energy surfaces (PESs) for chains of isotopes $^{150-162}$Gd and $^{96-114}$Ru involving nuclei suggested [16–18] to be good candidates for X(5) and E(5) critical point symmetries.

A. M. Khalaf
Physics Department, Faculty of Science, Al-Azhar University, Cairo, Egypt

M. M. Taha (✉)
Mathematics and Theoretical Physics Department, Nuclear Research Center, Atomic Energy Authority, P.No. 13759, Cairo, Egypt
e-mail: mahmoudmt@hotmail.com

A simplified Hamiltonian of the sd-IBM in the intrinsic condensate state

In this work, we adopt a simplified two-parameter IBM-1 Hamiltonian in the form

$$H = \epsilon \hat{n}_d + a_2 \hat{Q}(\chi) \cdot \hat{Q}(\chi) \quad (1)$$

where the d-boson number operator $\hat{n}_d$ and the quadrupole operator $Q(\chi)$ are defined as

$$\hat{n}_d = d^\dagger \cdot \tilde{d} \quad (2)$$

$$\hat{Q}(\chi) = s^\dagger \tilde{d} + d^\dagger \tilde{s} + \chi [d^\dagger \times \tilde{d}]^{(2)} \quad (3)$$

with χ the structure parameter of the quadrupole operator of the IBM, and $[d^\dagger \times \tilde{d}]^{(2)}$ stands for the $l = 2$ tensor coupling of the d-boson creation and annihilation operators ($\tilde{d}_\mu = (-1)^\mu d_{-\mu}$), where $\mu = -2, -1, 0, +1, +2$ is the angular momentum projection, and $(\cdot)$ denoting the scalar product.

The geometric interpretation of the Hamiltonian (1) can be derived using the intrinsic condensate state $|N\beta\gamma\rangle$ defined by [10, 11]

$$|N\beta\gamma\rangle = \frac{1}{\sqrt{N!}} (\Gamma_c^\dagger)^N |0\rangle \quad (4)$$

with

$$\Gamma_c^\dagger = \frac{1}{\sqrt{1+\beta^2}} [s^\dagger + \beta \cos \gamma d_0^\dagger + \frac{1}{\sqrt{2}} \sin \gamma (d_2^\dagger + d_{-2}^\dagger)] \quad (5)$$

where β and γ are the geometric parameters. We get the ground state potential energy surface (PES) by calculating the expectation value of the Hamiltonian on the intrinsic boson condensate state:

$$\begin{aligned} E(N,\beta,\gamma,\chi) &= \langle N\beta\gamma | H | N\beta\gamma \rangle \\ &= \epsilon \frac{N\beta^2}{1+\beta^2} + a_2 \Big\{ \frac{N}{1+\beta^2} [5 + (1+\chi^2)\beta^2] \\ &\quad + \frac{N(N-1)}{(1+\beta^2)^2} \Big[4\beta^2 - 4\sqrt{\frac{2}{7}} \chi \beta^3 \cos 3\gamma + \frac{2}{7}\chi^2 \beta^4 \Big] \Big\} \end{aligned} \quad (6)$$

If we eliminate the contributions of the one-body terms of the quadrupole–quadrupole interaction, then

$$E(N,\beta,\gamma,\chi) = \frac{A_2\beta^2 + A_3\beta^3 \cos 3\gamma + A_4\beta^4}{(1+\beta^2)^2} \quad (7)$$

with

$$\begin{aligned} A_2 &= [\epsilon + 4a_2(N-1)]N \\ A_3 &= -4\sqrt{\frac{2}{7}} \chi a_2 (N-1) N \\ A_4 &= \Big[\epsilon + \frac{2}{7}\chi^2 (N-1) a_2\Big] N \end{aligned} \quad (8)$$

Minimization of the PES with respect to β for a given value of χ, gives the equilibrium value β_m defining phase of the system; $\beta_m = 0$ corresponding to the symmetric phase while $\beta_m \neq 0$ to the broken symmetry phase. Since γ enters the potential (7) only through the $\cos 3\gamma$ dependence in the cubic term, the minimization in this variable can be performed separately. The global minimum is either at $\gamma_m = 0\,(2\pi/3, 4\pi/3)$ for $A_3 < 0$ or at $\gamma_m = \pi/3\ (\pi, 5\pi/3)$ for $A_3 > 0$. The second possibility can be expected via changing the sign of the corresponding β_m and simultaneously setting $\gamma_m = 0$.

The IBM phase can be described as follows:

1. For $A_3^2 < 4A_2|A_4|$, phase with $\beta_m = 0$ interpreted as spherical U(5) dynamical symmetry.
2. For $A_3^2 > 4A_2|A_4|$; $A_3 < 0$, phase with $\beta_m > 0$, $\gamma_m = 0$ interpreted as prolate deformed SU(3) dynamical symmetry.
3. For $A_3^2 > 4A_2|A_4|$; $A_3 > 0$, phase with $\beta_m > 0$, $\gamma_m = \pi/3$ interpreted as oblate deformed $S\bar{U}(3)$ dynamical symmetry.
4. The dynamical symmetries O(6) and $\bar{O}(6)$ do not represent separate phases.

For β non-zero, the first derivative of Eq. (7) must be zero and the second derivative positive. This gives

$$\begin{aligned} &\beta^3(4A_4\beta^2 + 3A_3\beta\cos 3\gamma + 2A_2) = 0 \\ &12A_4\beta^2 + 6A_3\beta\cos 3\gamma + 2A_2 > 0 \end{aligned} \quad (9)$$

The solution for $\beta \neq 0$ are obtained if we set $A_3 = 0$ in equation (9). Then equation (9) gives

$$4A_4\beta^2 + 2A_2 = 0 \quad (10)$$

or

$$\beta = \pm\sqrt{\frac{-A_2}{2A_4}} \quad (11)$$

This requires A_4 and A_2 to have opposite signs. Since A_4 must be positive for bound solutions, A_2 must be negative in deformed phase.

The finite β solutions corresponds to prolate ($\beta > 0$) and oblate ($\beta < 0$) equilibrium shapes, separated by a curve corresponding to $A_3 = 0$. Thus, there are in total, three phases. The spherical-deformed phase transition is generated by a change in sign of A_2, while the prolate–oblate correspond to changing the sign of A_3.

For $\gamma = 0$ (to study the β dependence) and in the large N-limit (i.e., $N - 1 = N$), yield the following coefficients in the PESs

$$\begin{aligned} A_2 &= [\epsilon + 4a_2N]N \\ A_3 &= -4\sqrt{\frac{2}{7}}\chi a_2 N^2 \\ A_4 &= \Big[\epsilon + \frac{2}{7}\chi^2 a_2 N\Big] N \end{aligned} \quad (12)$$

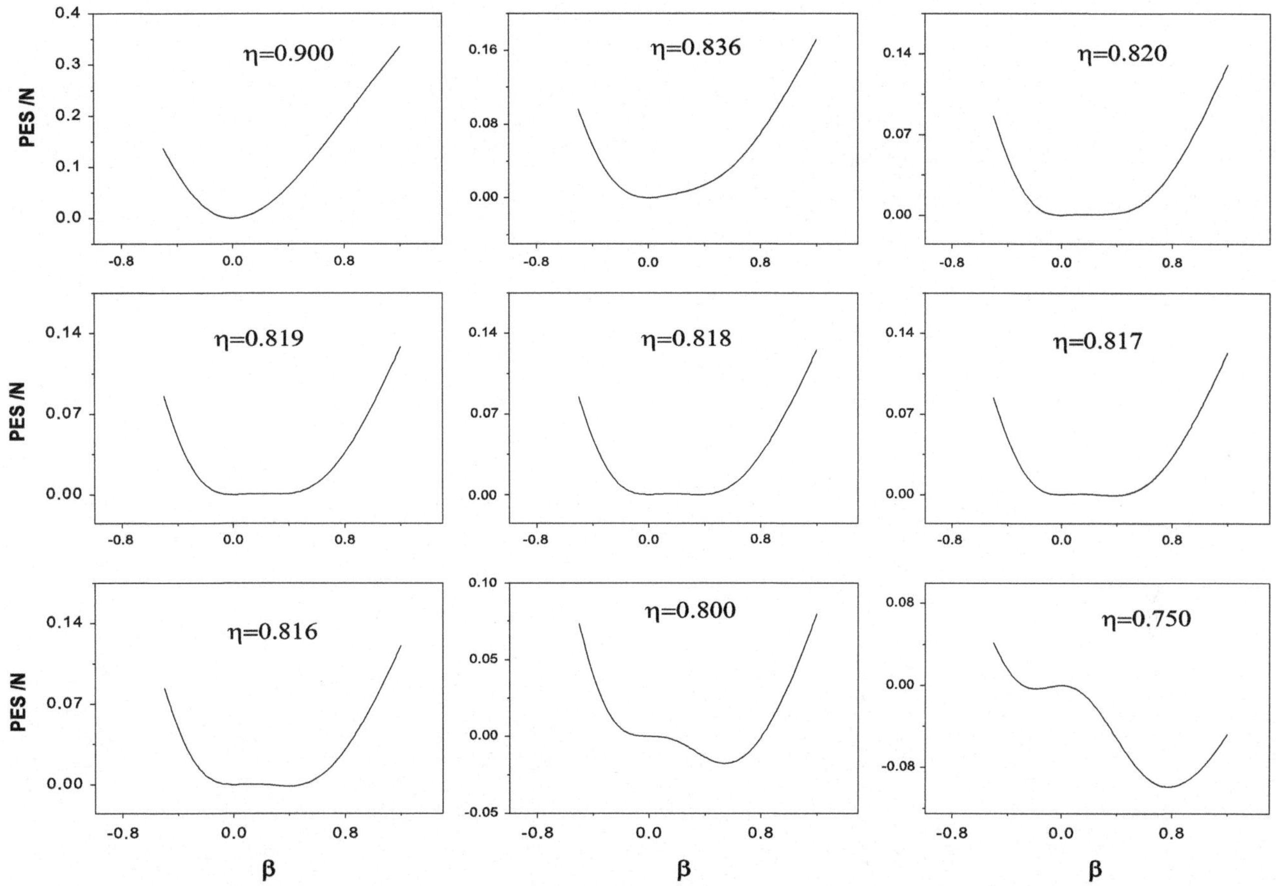

Fig. 1 Potential energy surface (PES) as a function of deformation parameters β corresponding to U(5)–SU(3) transition for large N limit for different values of control parameter η

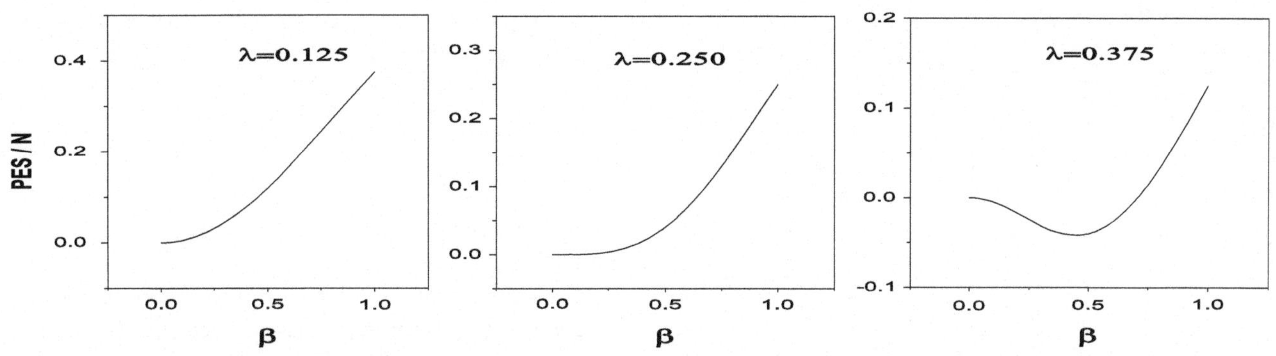

Fig. 2 Potential energy surface (PES) as a function of deformation parameters β corresponding to U(5)–O(6) transition for large N limit for three values of control parameter λ

providing that $A_2 > 0$ and $A_3 > 0$, then the critical point is located at $A_3^2 = 4A_2A_4$, which leads to

$$1 = -\frac{a_2N}{\epsilon}\left(4 + \frac{2}{7}\chi^2\right) \tag{13}$$

If we introduce the control parameter η such that

$$\frac{1-\eta}{\eta} = -\frac{a_2N}{\epsilon} \tag{14}$$

Then the critical point is located at

$$\eta_c = \frac{4 + \frac{2}{7}\chi^2}{5 + \frac{2}{7}\chi^2} \tag{15}$$

At η_c the depth of the $\beta = 0$ and $\beta \neq 0$ minima is equal.

The antispinodal point for the transition U(5)–SU(3) appears when $A_2 = 0$, *i.e.*,

$$\frac{-a_2N}{\epsilon} = \frac{1}{4} \tag{16}$$

or

$$\eta = 4/5 \tag{17}$$

For $\chi = -\sqrt{7}/2$, the critical point is located at $\eta_c = 9/11$ and the equilibrium value of the deformation parameter at the critical point is $\beta_0 = 1/2\sqrt{2}$. At $\eta = 1$ the system is in the symmetric phase since the PES has a unique minimum at $\beta = 0$. When η decreases, one reaches the spinodal point at $\eta < \eta_c$ ($\eta = 0.820$) where a second local minimum arises.

The two minima have the same depth at the critical point $\eta_c = 9/11$ (0.818). Beyond this value for $\eta > \eta_c$ the symmetric minimum at $\beta = 0$ becomes a local minimum till $\eta = 4/5$ (0.800) where it becomes antispinodal point. A sketch of this evolution is illustrated in Fig. 1.

For $\chi = 0$ ($A_3 = 0$), and for large N limit the PES reduces to

$$E(N, \beta) = \frac{A_2\beta^2 + A_4\beta^4}{(1+\beta^2)^2} \tag{18}$$

with

$$A_2 = (\epsilon + 4a_2N)N \tag{19}$$

$$A_4 = \epsilon N \tag{20}$$

At the critical point $A_2 = 0$, which leads to

Table 1 The optimized fitted parameters for the Gd isotopic chain

Nucleus	N_B	A_2	A_3	A_4
^{150}Gd	9	4.0693	−1.6933	6.1707
^{152}Gd	10	2.4414	−2.6035	5.6722
^{154}Gd	11	−0.9785	−3.460	4.4119
^{156}Gd	12	−3.9732	−3.6407	3.5284
^{158}Gd	13	−5.1200	−3.8485	3.4606
^{160}Gd	14	−7.8990	−3.0936	2.8020
^{162}Gd	15	−12.4621	−2.2000	1.1697

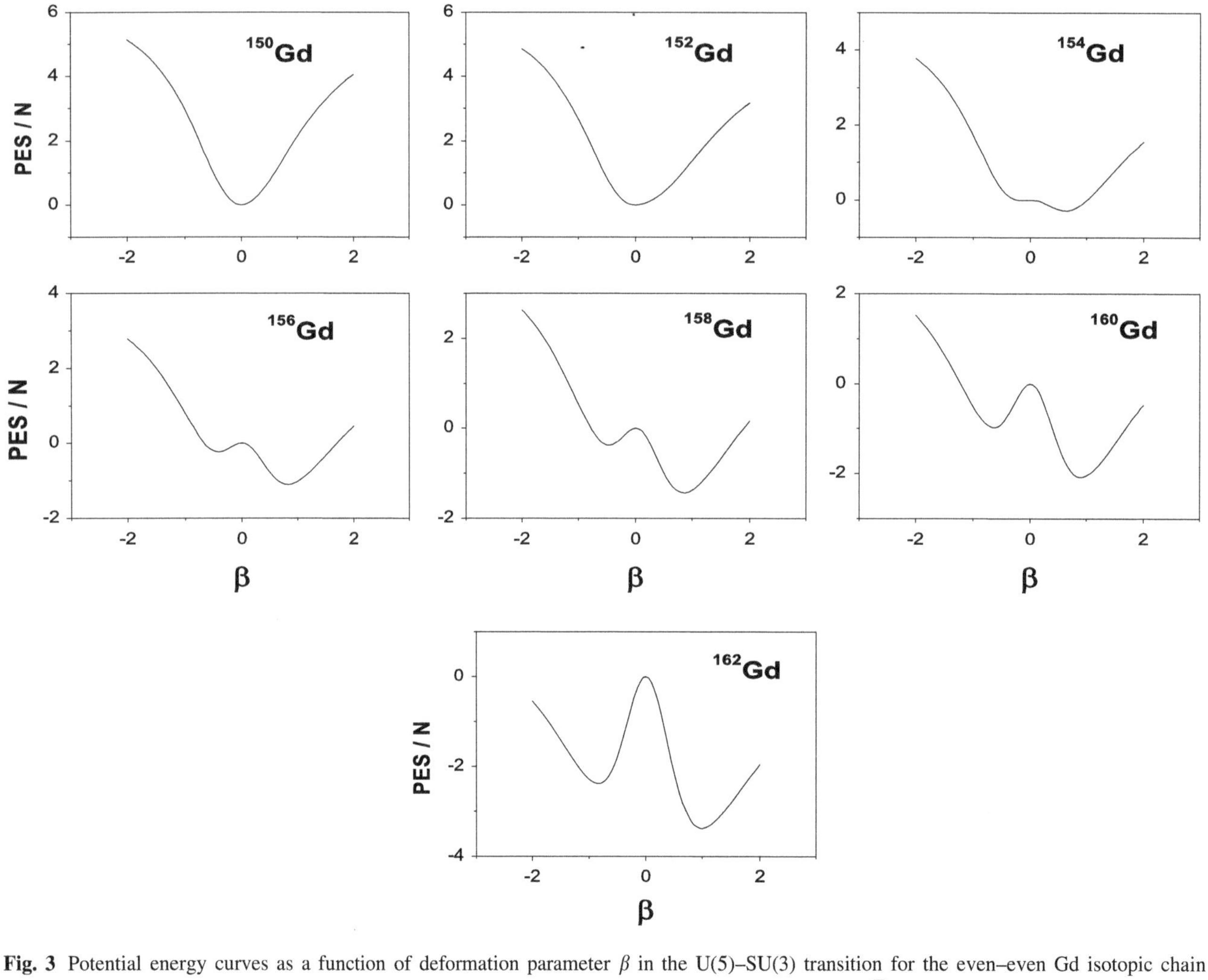

Fig. 3 Potential energy curves as a function of deformation parameter β in the U(5)–SU(3) transition for the even–even Gd isotopic chain obtained from IBM with intrinsic state formalism. The total number of bosons $N = 9$–15 and $\chi = -\sqrt{7}/2$

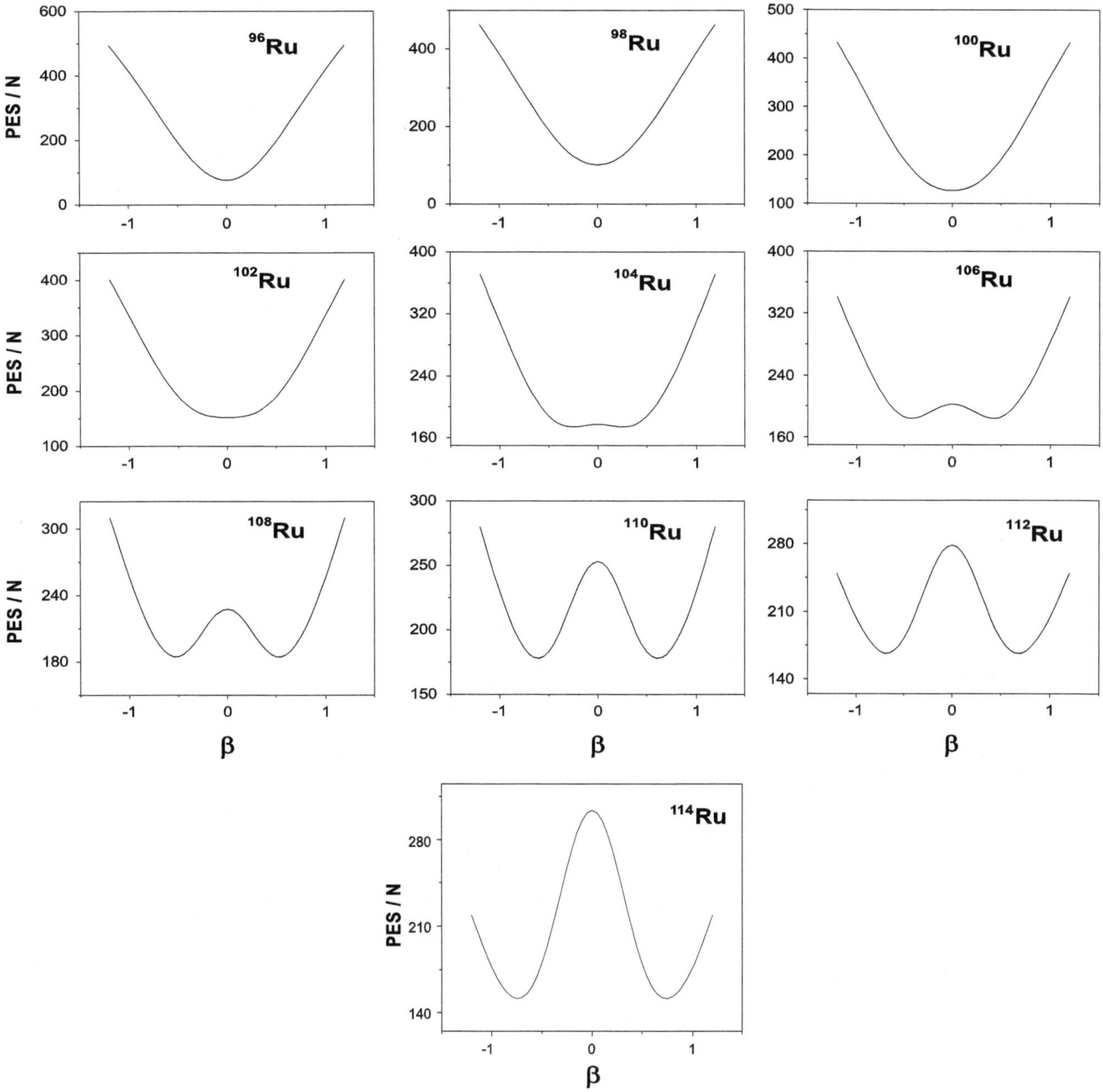

Fig. 4 The same as in Fig. 3 but for U(5)–O(6) transition for the even-even Ru isotopics chain. The total number of bosons $N = 4$–13 and $\chi = 0$

$$-\frac{a_2 N}{\epsilon} = \frac{1}{4} \tag{21}$$

or

$$\frac{1-\eta}{\eta} = \frac{1}{4} \tag{22}$$

Then the critical point is located at $\eta_c = 4/5$. All the three points, spinodal, critical and antispinodal are located at $\eta = 4/5$ and the quantum phase transition between spherical and deformed phase is of the second order. A sketch of this case is illustrated in Fig. 2.

Application of the model to the Gd and Ru isotopic chains

We used the χ^2 test to get the coefficients A_1, A_2 and A_3 of the PESs. The χ^2 function is defined in the standard way,

$$\chi^2 = \left[\frac{1}{N_{\text{data}} - N_{\text{par}}} \sum_{i=1}^{N_{\text{data}}} \left(\frac{\chi_i^{\text{exp}}(I) - \chi_i^{\text{IBM}}}{\sigma_i}\right)\right]^2 \tag{23}$$

where N_{data} is the number of data to be fitted, N_{par} is the number of parameters used in the IBM fit, χ_i^{exp} is an experimental energy level [19] or the two-neutron

Table 2 The optimized fitted parameters for the Ru isotopic chain

Nucleus	N_B	A_2	A_4	a_0
^{96}Ru	4	527.6	831.2	75.9
^{98}Ru	5	273.4	778.2	101.2
^{100}Ru	6	219.2	725.2	126.5
^{102}Ru	7	65.0	672.2	151.8
^{104}Ru	8	−89.2	619.2	177.1
^{106}Ru	9	−243.4	566.2	202.4
^{108}Ru	10	−397.6	513.2	227.7
^{110}Ru	11	−551.8	460.2	253.0
^{112}Ru	12	−706.0	407.2	278.3
^{114}Ru	13	−860.2	354.2	303.6

separation energies or the $B(E2)$ values and $\chi_i^{\rm IBM}$ is the corresponding calculated IBM value and σ_i is the experimental error assigned to each χ_i^{exp}. To perform the fit, we minimized χ^2 function for the selected energy levels ($2_1^+, 4_1^+, 6_1^+, 8_1^+, 0_2^+, 2_3^+, 4_3^+, 2_2^+, 3_1^+$ and 4_2^+) using A_1, A_2 and A_3 as free parameters.

We apply now the formalism outlined in the previous section to Gd and Ru isotopic chains. In case of Gadolinium chain $^{150-162}_{\ \ \ \ \ 64}$Gd we assumed the structure parameters χ of the quadrupole operator to be equal to $-\sqrt{7}/2$, because some Gd isotopes clearly exhibit the character of the SU(3) dynamical symmetry. Using the optimized fitted parameters shown in Table 1, we calculated the PESs which are illustrated in Fig. 3. The phase diagram exhibits first-order shape phase transition from spherical U(5) to deformed axial symmetric prolate SU(3) when moving from light isotopes to heavy ones. The ^{150}Gd nucleus still shows a vibrational structure while $^{156-162}$Gd are considered as rather good SU(3) examples.

As an example of isotopic chain in which the structure changes from spherical U(5) to γ-unstable O(6) symmetry, we have studied the Ruthenium chain $^{96-114}_{\ \ \ \ \ 44}$Ru. The corresponding PESs are plotted in Fig. 4 using the coefficients in Table 2. We considered $\chi = 0$ and performed the same process of fitting as applied in Gd chain. The lighter isotopes exhibit a vibrational structure, while the heavier ones present a γ-unstable behavior with the critical point located of ^{104}Ru isotope.

Conclusion

In the present work, we have analyzed the shape phase transitions in the framework of a two-parameter simplified truncated sd-IBM1 as a very tractable model for theoretical studies. We used the boson intrinsic coherent state formalism which provides a connection between the IBM and the geometric collective model (GCM) to calculate the PESs. We have analyzed the critical points of the shape phase transitional regions U(5)–SU(3) and U(5)–O(6) in the space of the model parameters. We have investigated the first-order quantum phase transition from spherical U(5) to deformed axial symmetric prolate SU(3) along the chain of $^{150-162}$Gd isotopes and the second-order quantum phase transition from spherical U(5) to γ-unstable rotor O(6) along the chain of $^{96-114}$Ru isotopes. We find that the critical nuclei characterizing the phase transitions are ^{154}Gd and ^{104}Ru. To get the optimized parameter of the PES for each nucleus, a computer-simulated search program is performed to obtain a minimum root mean squared (rms) deviation between calculated and some experimentally selected low-lying energy levels, B(E2) transition rates and two-neutron separation energies.

References

1. Iachello, F., Arima, A.: The Interacting Boson Model. Cambridge University Press, Cambridge (1987)
2. Khalaf, A.M., Taha, M.M., Kotb, M.: Identical bands and ΔI = 2 staggering in superdeformed nuclei in A ~ 150 mass region using three parameters rotational model. Prog. Phys. **8**, 39–44 (2012)
3. Khalaf, A.M., Ismail, A.M.: The nuclear shape phase transitions studied within the geometric collective model. Prog. Phys. **9**, 51–55 (2013)
4. Khalaf, A.M., Ismail, A.M.: Structure shape evolution in lanthanide and actinide nuclei. Prog. Phys. **9**, 98–104 (2013)
5. Khalaf, A.M., Hamdy, H.S., El Sawy, M.M.: Nuclear shape transition using interacting boson model with the intrinsic coherent state. Prog. Phys. **9**, 44–51 (2013)
6. Turner, P.S., Rowe, D.J.: Phase transitions and quasidynamical symmetry in nuclear collective models. II. The spherical vibrator to gamma-soft rotor transition in an SO(5)-invariant Bohr model. Nucl. Phys. A **756**, 333–355 (2005)
7. Rowe, D.J., Turner, P.S.: The many relationships between the IBM and the Bohr model. Nucl. Phys. A **760**, 59–81 (2005)
8. Cejnar, P., Heinze, S., Dobes, J.: Thermodynamic analogy for quantum phase transitions at zero temperature. Phys. Rev. C **71**, 011304(R) (2005)
9. Heinze, S., et al.: Evolution of spectral properties along the O(6)-U(5) transition in the interacting boson model. I. Level dynamics. Phys. Rev. C **73**, 014306–014316 (2006)
10. Dieperink, A.E.L., Scholten, O., Iachello, F.: Classical limit of the interacting-boson model. Phys. Rev. Lett. **44**, 1747–1750 (1980)
11. Ginocchio, J.N.: An exactly solvable anharmonic Bohr hamiltonian and its equivalent boson hamiltonian. Nucl. Phys. A **376**, 438–450 (1982)
12. Ginocchio, J.N., Kirson, M.W.: Relationship between the Bohr collective Hamiltonian and the interacting-boson model. Phys. Rev. Lett. **44**, 1744 (1980)
13. Iachello, F.: Dynamic symmetries at the critical point. Phys. Rev. Lett. **85**, 3580–3583 (2000)

14. Iachello, F.: Analytic description of critical point nuclei in a spherical-axially deformed shape phase transition. Phys. Rev. Lett. **87**, 052502–052506 (2001)
15. Iachello, F.: Phase transitions in angle variables. Phys. Rev. Lett. **91**, 132502–132507 (2003)
16. Raduta, A.A., Faessler, A.: A coherent state description of the shape phase transition in even–even Gd isotopes. J. Phys. G Nucl. Part Phys. **31**, 873–901 (2005)
17. Garcia-Ramos, J.E., et al.: Phase transitions and critical points in the rare-earth region. Phys. Rev. C **68**, 024307 (2003)
18. Frank, A., Alonso, C.E., Aria, J.M.: Search for E(5) symmetry in nuclei: the Ru isotopes. Phys. Rev. C **65**, 014301 (2001)
19. Singh, B., Zywina, R., Firestone, R.B.: Table of superdeformed nuclear bands and fission isomers. Nucl. Data Sheets **97**, 241–592 (2002)

13

An experimental study of thermal diffusivity of Au nanoparticles: effects of concentration particle size

Esmaeil Shahriari[1,2] · **Mohammad Moradi**[1,2] · **Morteza Raeisi**[1]

Abstract In this article, Au nanoparticles in polyvinylpyrrolidone (PVP) solution were prepared by gamma radiation at different concentrations. The solutions were irradiated at doses of 50 kGy for making different sizes. The average sizes of particle in the prepared samples were measured using the nanophox machine. A dual-beam mode-mismatched thermal lens (TL) method was used to investigate the effect of thermal diffusivity of samples. The TL measurement was carried out using a green diode laser (wavelength 532 nm, 60 mW) and a He–Ne laser (wavelength 632.8 nm, 0.5 mW) for excitation source and probe beam, respectively. The results showed that the thermal diffusivity of samples enhances with the growth of particle size and density of solutions. This increment can be attributed to phonon scattering at interface of particles–liquid and contact between the nanoparticles and surrounded liquid.

Keywords Au nanoparticles · Thermal lens method · Thermal diffusivity

Introduction

Nanoparticles, such as metals, are of great interest for a wide variety of applications in electronics [1] and optical components [2], heat transfer fluids in automotive [3], electronic cooling, medical therapies, and drug delivery [4–7] due to their unique or improved properties determined primarily by size, composition, and structure along with their self-organized film structures. In fluids with metal nanoparticles, laser excitation leads to generate hot electrons which thermalized by electron phonon scattering. This phenomenon can help to increase thermal diffusivity of nanoparticles. The suspension of solid particles in liquid media enhances its influences thermal diffusivity [8]. The changes in the thermal diffusivity of the suspension mainly depend on shape and size of the suspended particles [9, 10]. In addition, the nature of the surface, such as charge, its distribution, and the presence of surfactant plays an important role [11, 12]. The subject of heat transport through dilute suspensions of solid particles has received a renewed boost owing to the reports of enhancement in thermal conductivity of suspensions carrying nanoparticles [13–16].

The TL [17–19] is a type of photothermal effect, which a laser with Gaussian intensity profile is used as an excitation laser that induces the beam temperature distribution in a sample. The heat generated is strongest at the center of the beam, because the beam intensity is strongest at that point. This temperature gradient produces a refractive index gradient which behaves like a converging or diverging lens depending on whether the change rate of refractive index with respect to temperature, dn/dT, is positive or negative [20, 21]. The first measurement of the TL effect was performed by Gordon et al. in 1965 using a simple single-beam apparatus [22].

The enhancement in the thermal diffusivity values of solvents, such as ethanol, water, ethylene glycol, and culture medium, by the presence of Au nanoparticles has been studied [23, 24]. For example, with 0.033-mmol Ag in 50 ml of ethanol, water and ethylene glycol made the

✉ Esmaeil Shahriari
esmaeil.phy@gmail.com

1 Department of Physics, Faculty of Science, Shahrekord University, Shahrekord, Iran

2 Photonic Research Group, Shahrekord University, Shahrekord, Iran

thermal diffusivity increased by 0.7, 7.0, and 8.4 %, respectively. It was also reported that with 4 vol. % of nanoparticles of CuO (23.6 nm diameter) in ethylene glycol made the thermal diffusivity increased by 22 % [25]. This phenomenon could be due to the electrostatic interaction between cationic solutions and negatively charged metal nanoparticles.

We have previously investigated the influence of particle size on thermal diffusivity of Au nanoparticles prepared by γ irradiation in PVA solution [26], but in present work, we studied the effect of concentration and particle size of Au/PVP nanoparticles which prepared using γ-radiation on thermal diffusivity by using the TL spectrometric measurements.

Material and method

For preparing the Au/PVP nanoparticles samples, hydrogen tetrachloroaurate (III) hydrate, premion (metals basis), (HAuCl4.3H2O with 99.999 % purity), 4-g PVP (MW 29,000, Aldrich), and 1-ml isopropanol were used for preparing Au nanoparticles samples. The PVP and isopropanol were used as a colloidal stabilizer and hydroxyl radical scavenger, respectively. The PVP solutions were made by dissolving PVP powder in 50-ml deionized water at room temperature. The solution was magnetically stirred for 2 h and was bubbled with nitrogen gas (99.5 %) to remove oxygen. γ-radiation (^{60}Co-rays) was used as an effective tool for polymerization process and reducing agent same as procedure reported in our previous paper [27]. Hydrogen tetrachloroaurate (III) hydrate, $HAuCl_4.3H_2O$, at different weights (i.e. 2, 5, 10, 15, 20 and 30 mg) was added into PVP solution to produce Au nanoparticles samples. The Au ions with different concentrations in PVP solution were then irradiated with γ-radiation at dose of 50 kGy to produce various concentrations and nanoparticle sizes, and were labeled as $S1 = 1.17 \times 10^{-4}$ M; $S2 = 2.94 \times 10^{-4}$ M; $S3 = 5.88 \times 10^{-4}$ M; $S4 = 8.83 \times 10^{-4}$ M; $S5 = 11.76 \times 10^{-4}$ M; and $S6 = 17.64 \times 10^{-4}$ M. In this process, γ-irradiation produces hydrated electrons that reduce the Au ions to Au atoms, and then, the aggregated Au atoms are distributed in the solution. The characteristic of Plasmon peak of Au nanoparticle inside the fluid was detected using UV visible spectrometer (Lambda 25-Perkin Elmer), and the size of silver nanoparticles was determined using nanophox machine (Sympatec GmbH, D-38678).

The average diameter of Au nanoparticles was measured using the nanophox machine (Sympatec GmbH, D-38678) and listed in Table 1. The shape of the particles and their polydispersity of a few samples were determined via transmission electron microscopy (TEM) (Hitachi H-7100). We found that the particles distribution in the solution are in uniform shape.

Schematic diagram of the TL experimental setup is shown in Fig. 1. The measurements were carried out using a diode laser (532 nm at 60 mW, DPSS) as an excitation source and a He–Ne laser (632.8 nm, 0.5 mW, HRP050) as a probe beam. The excitation light beam was modulated by a variable frequency chopper (SR540), and then, it was focused by a lens of 21-cm focal length onto a liquid sample contained in a 5-mm length quartz cuvette positioned at its focal plane. The Rayleigh parameter was found to be 16 mm which is larger than thickness of samples (5 mm), and hereby, the Rayleigh condition is satisfied. The probe beam was focused by a lens of 5 cm focal length away from the cuvette and aligned at an angle smaller than 1.5° with respect to the excitation beam. The transmitted beam was detected by a photodiode detector (PD). An optical bandpass filter is placed in front of an aperture to remove the stray light from excitation beam entering PD. The PD output was coupled to a digital storage oscilloscope (Tektronix TDS 210) to record the time evolution of TL signal. The LabView software was used to capture the TL data from oscilloscope and to normalize it with respect to TL signal at time $t = 0$.

A mechanical chopper was used for limiting the exposure of the sample to the light; when the hole is opened, the sample is illuminated, and light sent a signal to initialize the data acquisition. The maximum intensity of the probe beam was centered on the photodiode detector, and then signals, as a function of time, were registered and recorded by a detector, oscilloscope, and sent to a PC, through a GPIB interface bus. The analysis data must be extracted throughout the one period of whole signal and then normalized. The experimental data are fitted using theoretical Eq. (1) with θ and characteristic thermal time constant (t_c) as adjustable parameters. Now, thermal diffusivity of the sample can be easily calculated from Eq. (1). This process was applied for all experimental data to obtain the thermal diffusivity of samples.

Theory

The TL signal is defined as the normalized change of the transmission of the probe beam, centered at the beam axis and located at a distance much larger than the pump Rayleigh parameter. Shen et al. in 1995 have derived as expression for the signal using a diffraction approximation for Gaussian beams, which is given by [28]:

Table 1 Au nanoparticles samples in PVP polymer prepared at different concentrations and irradiated for dose of 50 kGy

Nanofluid samples	Concentration ($\times 10^{-4}$ M)	Average particle size (nm)	t_c ($\times 10^{-3}$ s)	θ	D ($\times 10^{-3}$ cm^2/s)
S1	1.17	10	9.43 ± 0.24	1.254 ± 0.06	1.23 ± 0.04
S2	2.94	15	6.91 ± 0.21	1.121± 0.02	1.68 ± 0.14
S3	5.88	19	5.95 ± 0.22	1.232 ± 0.04	1.95 ± 0.07
S4	8.83	32	4.58 ± 0.22	1.152± 0.05	2.53 ± 0.11
S5	11.76	45	4.01 ± 0.19	2.245 ± 0.02	2.89 ± 0.13
S6	17.64	60	3.72 ± 0.19	2.321 ± 0.03	3.12 ± 0.13

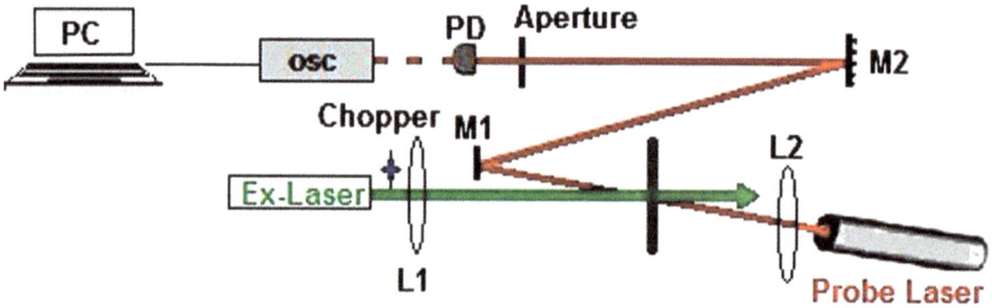

Fig. 1 Schematic diagram of TL experimental setup

$$I(z,t) = I_0\left(1 - \theta \tan^{-1}\left(\frac{2\,mV}{[(1+2m)^2 + V^2]\frac{t_c}{2t} + 1 + 2m + V^2}\right)\right)^2 \qquad (1)$$

where

$$V = \frac{z_1}{z_c};\ m = \left(\frac{w_p}{w_e}\right)^2;\ t_c = \frac{w_e^2}{4D},\ \theta = \frac{P_e.\alpha.l}{\kappa\lambda_P}\frac{\mathrm{d}s}{\mathrm{d}T} \qquad (2)$$

in which λ_p is the wavelength of the probe beam, D and κ are the thermal diffusivity and thermal conductivity, P_e is the excitation beam power, α is the absorption coefficient, and l is the sample thickness. $z_c = w_0^2/\lambda$ is the confocal distance (cm), w_o is the probe beam waist radius, w_p and w_e are the probe and the excitation beam radii at sample, respectively, z_1 is the distance of laser beam waist to sample, and t_c is the characteristic thermal time constant. In Eq. (1), $I(t)$ is the temporal dependence of the probe beam intensity at the detector, I_o is the initial value of, $I(t)$. $\mathrm{d}s/\mathrm{d}T$ is the temperature coefficient of the optical path length change, at the probe beam wavelength, which is written as [29]:

$$\frac{\mathrm{d}s}{\mathrm{d}T} = (n-1)(1+\nu)\alpha_T + \frac{\mathrm{d}n}{\mathrm{d}T} \qquad (3)$$

where α_T is the sample linear thermal expansion coefficient, ν is the Poisson ratio, and $\mathrm{d}n/\mathrm{d}T$ is the temperature coefficient of its refractive index.

Results and discussion

The absorption spectra of the Au nanoparticle prepared using γ-radiation is shown in Fig. 2. The measurements of absorption spectra were carried out at room temperature for visible wavelength, 380–800 nm. As it can be seen, absorbance values increase with increasing the concentration of Au nanoparticles which indicated the total number of particles in the solution. Furthermore, the absorption peak shifted to lower energy or higher wavelengths with growth of nanoparticle sizes [30].

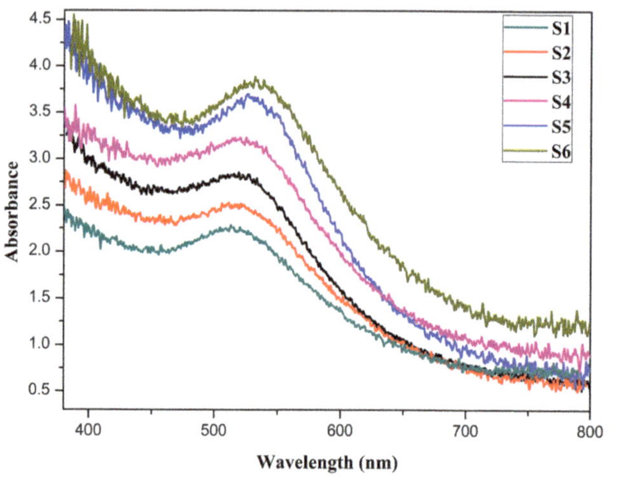

Fig. 2 Absorption spectra of the different concentration Au nanoparticles

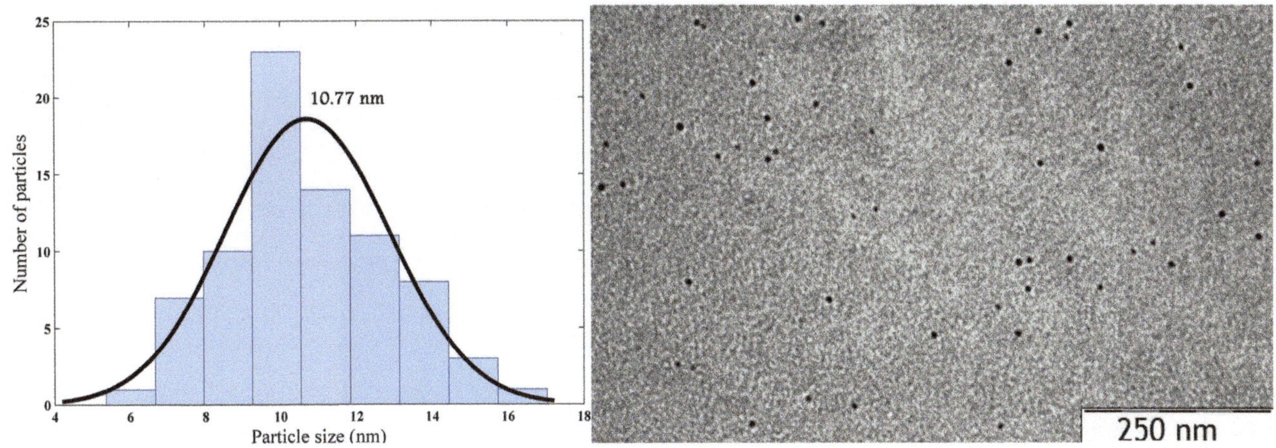

Fig. 3 TEM image and particles size histogram of Au particles with a nominal size of 10 nm

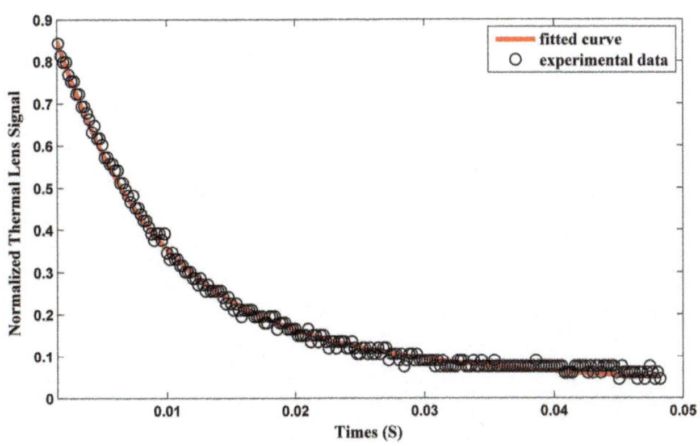

Fig. 4 Typical time evolution of the TL signal for Au nanoparticles (S1). Symbol (O) represents the experimental data and solid line is the best fitting line to Eq. (1)

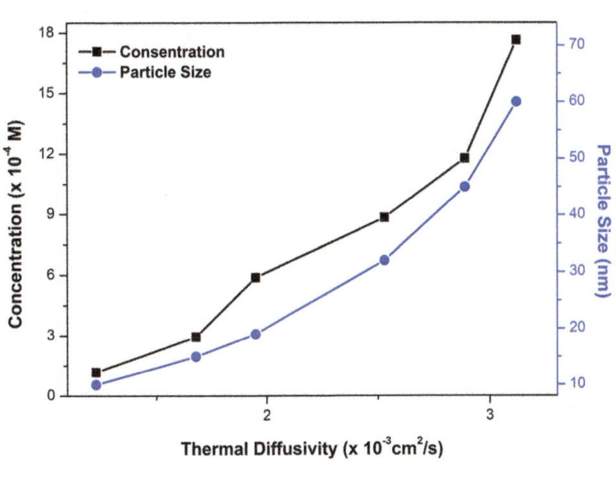

Fig. 5 Variations of thermal diffusivity of Au nanoparticles as a function of concentration and particle size

The particle size can be controlled by controlling the doses of γ-radiation, however, in the present work, the specimens were prepared at different concentration levels and radiated at 50 kGy. A typical TEM image of Au nanoparticles distributions with a nominal size of 10 nm and shape of these particles displayed in Fig. 3. Other samples of various concentrations or sizes of nanoparticles after irradiated at 50 kGy can be seen in Table 1.

Figure 4 shows a typical TL signal as a function of time for Au nanoparticles dispersed in PVP at concentration of 1.17×10^{-4} M and irradiated with dose of 50 kGy. The theoretical curves in Fig. 4 calculated using Eq. (1) agree very well with experimental data, which give the thermal diffusivity value for the sample as 1.23×10^{-3} cm^2/s.

Thermal diffusivity value as a function of concentrations and nanoparticles sizes plotted in Fig. 5 which shows a non-linear increasing with the increasing of concentration of nanoparticles and particles sizes. This enhancement can be due to the phonon scattering phenomena between the particles-liquid interfaces. Since, as the particle size becomes small, the phonon scattering transfers less energy to the surrounding liquid [31]. Moreover, the energy deposited into the phonon modes is subsequently transferred to the surrounding medium. In addition, when the particle concentration is increased, the optical absorption intensity is increased, as well as the thermal diffusivity of the surrounding medium. This interaction implies a higher nanoparticle concentration per volume unit in the solution [32].

Conclusion

We investigated thermal diffusivity treatment of Au nanoparticle at different concentrations and particle sizes. According to the experimental results with the increment of concentration and particle size of nanoparticle, the thermal diffusivity of samples increased. This increment in thermal diffusivity can be attributed to phonon scattering at interface of particles–liquid and contact between the nanoparticles and surrounded liquid.

References

1. Gittins, D.I., Bethell, D., Schiffrin, D.J., Nichols, R.J.: A nanometer scale electronic switch consisting of a metal cluster and redox-addressable groups. Nat. (London) **408**, 67–69 (2000)
2. Ashkarran, A.A.: Synthesis and characterization of gold nanoparticles via submerged arc discharge based on a seed-mediated approach. J. Theo. Appl. Phys. **6**, 14–16 (2012)
3. Neal, D.P.O., Hirsch, L.R., Halas, N.J., Payne, J.D., West, J.L.: Photo-thermal tumor ablation in mice using near infrared-absorbing nanoparticles. Cancer Lett. **209**, 171–176 (2004)
4. Wen, D.S., Ding, W.: Natural convective heat transfer of suspensions of titanium dioxide nanoparticles (Nanofluids). IEEE Trans. Nanotechnol. **5**, 220–227 (2006)
5. Jang, S.P., Choi, S.U.S.: Cooling performance of a microchannel heat sink with nanofluids. Appl. Therm. Eng. **26**, 2457–2463 (2006)
6. Hamad-Schifferli, K., Schwartz, J.J., Santos, A.T., Zhang, S.G., Jacobson, J.M.: Remote electronic control of DNA hybridization through inductive coupling to an attached metal nanocrystal antenna. Nat. (London) **415**, 152–155 (2002)
7. Loo, C., Lin, A., Hirsch, L., Lee, M.H., Barton, J., Halas, N., West, J., Drezek, R.: Nanoshell-enabled photonics-based imaging and therapy of cancer. Tech. Cancer Res. Treat. **3**, 33–40 (2004)
8. Deen, W.M.: Analysis of transport phenomena. Oxford University Press, New York (1998)
9. Abdi, F., Siabi-Gerjan, A., Savaloni, H.: On the discrete dipole approximation investigation of the extinction spectra of Ag/glass nano-flower thin film with threefold symmetry. J. Theo. Appl. Phys. **6**(4), 2–10 (2012)
10. Chopkar, M., Sudarshan, S., Das, P., Manna, I.: Effect of particle size on thermal conductivity of nanofluids. Metall. Mater. Trans. A **39**, 1535 (2008)
11. Pilkington, G.A., Briscoe, W.H.: Nanofluids mediating surface forces. Adv. Coll. Interface Sci. **68**, 179–182 (2012)
12. Gao, J.W., Zheng, R.T., Ohtani, H., Zhu, D.S., Chen, G.: Experimental investigation of heat conduction mechanisms in nanofluids clue on clustering. Nano Lett. **9**, 4128 (2009)
13. Eastman, J.A., Phillpot, S.R., Choi, S.U.S., Keblinski, P.: Thermal transport in nanofluids. Annu. Rev. Mater. Res. **34**, 219 (2004)
14. Fan, J., Wang, L.: Review of heat conduction in nanofluids. J. Heat Transfer **133**, 040801 (2011)
15. Keblinski, P., Prasher, R., Eapen, J.: Thermal conductance of nanofluids: is the controversy over. J. Nanopart. Res. **10**, 1089 (2008)
16. Kleinstreuer, C., Feng, Y.: Experimental and theoretical studies of nanofluids thermal conductivity enhancement: a review. Nanoscale Res. Lett. **6**, 229 (2011)
17. Franko, M.: Thermal lens spectrometric detection in flow injection analysis and separation techniques. Appl. Spectrosc. Rev. **43**, 358–388 (2008)
18. Kitamori, T., Tokeshi, M., Hibara, A., Sato, K.: Thermal lens microscopy and microchip chemistry. Anal. Chem. **76**, 52A–60A (2004)
19. Cognet, A., Berciaud, S., Lasne, D., Lounis, B.: Photothermal methods for single nonlumimescent nano objects. Anal. Chem. **80**, 2289–2294 (2008)
20. Bialkowski, S.: Photothermal spectroscopy methods for chemical analysis. Wiley, New York (1996)
21. Shen, J., Lowe, R.D., Snook, R.D.: Two-beam thermal lens spectrometer for. Ultra-trace Analysis. Chem. Phys. **18**, 403–408 (1998)
22. Gordon, J.P., Leite, R.C.C., Moore, R.S., Porto, S.P.S., Whinnery, J.R.: Long-transient effects in lasers with inserted liquid samples. J. Appl. Phys. **36**, 3–8 (1965)
23. Perez, J.L.J., Fuentes, R.G., Ramirez, J.F.S., Orea, A.C.: Study of gold nanoparticles effect on thermal diffusivity of nanofluids based on various solvents by using thermal lens spectroscopy. Eur. Phys. J. Spec. Top. **153**, 159–161 (2008)
24. Perez, J.L.J., Fuentes, R.G., Alvarad, E.M., Gallegos, E.R., Orea, A.C., Cordova, J.T., Alvarez, J.G.M.: Enhancement of the thermal transport in a culture medium with Au nanoparticles. Appl. Surf. Sci. **255**, 701–702 (2008)
25. Xue, Q., Xu, W.M.: A model of thermal conductivity of nanofluids with interfacial shells. Mater. Chem. Phys. **90**, 298–301 (2005)
26. Shahriari, E., Yunus, W.M., Zamiri, R.: The effect of nanoparticle size on thermal diffusivity of gold nano-fluid measured using thermal lens technique. J. Eur. Opt. Soc. Rap. Public. **8**, 13026 (2013)
27. Shahriari, E., Yunus, W.M., Saion, E.: Effect of particle size on nonlinear refractive index of au nano particle in PVA solution. Braz J. Phys. **40**, 256–260 (2010)
28. Shen, J., Lowe, R.D., Snook, R.D.: Two-beam thermal lens spectrometer for ultra-trace analysis. Chem. Phys. **18**, 403–408 (1998)
29. Andrade, A.A., Catunda, T.: Thermal lens determination of the temperature coefficient of optical path length in optical materials. Rev. Sci. Inst. **74**, 877–880 (2003)
30. Naghavi, K., Saion, E., Yunus, W.M.: Influence of dose on particle size of colloidal silver nanoparticles synthesized by gamma radiation. Radiat. Phys. Chem. **79**, 1203–1208 (2010)
31. Ge, Z.B., Cahill, D.G., Braun, P.V.: Thermal conductance of hydrophilic and hydrophobic interfaces. Phys. Rev. Lett. **96**, 186101–186105 (2006)
32. Rotkin, S.V., Perebeinos, V., Petrov, A.G., Avouris, P.: An essential mechanism of heat dissipation in carbon nanotube electronics. Nano Lett. **9**, 1850–1855 (2009)

14

Simultaneous estimation of plasma parameters from spectroscopic data of neutral helium using least square fitting of CR-model

Jalaj Jain · Ram Prakash · Gheesa Lal Vyas · Udit Narayan Pal · Malay Bikas Chowdhuri · Ranjana Manchanda · Nilanjan Halder · Yaduvendra Choyal

Abstract In the present work an effort has been made to estimate the plasma parameters simultaneously like—electron density, electron temperature, ground state atom density, ground state ion density and metastable state density from the observed visible spectra of penning plasma discharge (PPD) source using least square fitting. The analysis is performed for the prominently observed neutral helium lines. The atomic data and analysis structure (ADAS) database is used to provide the required collisional-radiative (CR) photon emissivity coefficients (PECs) values under the optical thin plasma condition in the analysis. With this condition the estimated plasma temperature from the PPD is found rather high. It is seen that the inclusion of opacity in the observed spectral lines through PECs and addition of diffusion of neutrals and metastable state species in the CR-model code analysis improves the electron temperature estimation in the simultaneous measurement.

J. Jain
Universidad de Talca, Talca, Chile

J. Jain
Comisión Chilena de Energía Nuclear, Santiago, Chile

R. Prakash · U. N. Pal
Microwave Tubes Division, CSIR-Central Electronics and Engineering Research Institute, Pilani 333031, India

G. L. Vyas (✉) · N. Halder
Manipal University Jaipur, Jaipur 303007, India
e-mail: glvyas27@gmail.com

M. B. Chowdhuri · R. Manchanda
Institute for Plasma Research, Bhat, Gandhinagar 382428, India

Y. Choyal
School of Physics, Devi Ahilya Vishwavidyalaya, Indore 452017, India

Keywords Penning discharge · Collisional-radiative model · Least square fitting · Opacity

Introduction

Penning plasma discharge (PPD) devices have been studied significantly during last few decades. This device was originally proposed by Penning as a low-pressure manometer (i.e., cold cathode or penning gauge) [1]. This has later been adopted to function as an ion source [2–4]. Penning ion sources have been used for variety of applications, such as sputtering and evaporation of surfaces, electromagnetic separation of isotopes and fusion applications [5, 6]. The PPD devices are also attractive due to their extensive use as charged particles trapping [7, 8], light emitting sources [9], etc. In the applications of light emitting sources it has recently been proposed that the PPD device can be used to calibrate a VUV spectrometer-detector system [10]. In fact, the PPD device can generate spectral radiations in the visible and VUV region simultaneously and visible radiations analysis can help in VUV intensity calibration once the desired basic plasma parameters are obtained from the visible spectra.

In a penning discharge, the plasma is produced between two end cathodes and a shell anode, with power derived from DC high voltage power supply. The electron confinement is maintained very high in this geometry by end cathodes (electrostatic confinement) and a pair of permanent magnets, which produces an axial magnetic field (magnetic confinement). The high electron energies in the source gas results in highly ionized gas and secondary electron release from the cathode. The source produces a large number of bright emission lines even at low pressures and the spectral distribution depends on the combination of discharge gas, applied magnetic field, applied power, cathode material, etc. [11–13]. Although penning discharge sources have been used for many applications and different sources have been researched, there exist no direct diagnostic which can provide the basic plasma parameters measurement simultaneously like—electron density, electron temperature, ground state atom density, ground state ion density, metastable state density, etc.

In the case of magnetic field environment, and small geometries of the penning discharge devices where sheath–sheath intersection could be prominent, the metallic probe diagnostics can lead erroneous results and are not much useful [14–16]. The optical emission spectroscopy (OES) [17–19], on the other hand, is an advanced technique which plays an important role to complement the metallic probe diagnostic techniques because this is a passive diagnostic and does not perturb the plasma environment locally. Nevertheless, the interpretation of the spectra of OES is not straight-forward. The delicacy lies in the complex nature of the plasma and subsequently the manifold atomic processes. In fact, it requires an estimation of the population density of the specific excited energy state wherefrom specific wavelength radiation is emitted. This is possible by solving a set of coupled rate equitation's by including all the atomic processes at a particular condition.

Based on the assumptions and requirements there exist three models which are quite popular in the plasma physics community for the plasma parameter estimations, namely, Corona, local thermodynamic equilibrium (LTE) and collisional-radiative (CR) models [20, 21]. The corresponding Corona and LTE models are best used for low ($<10^{10}$ cm^{-3}) and high (10^{15} cm^{-3}>) electron density regimes of the plasmas. The CR-model is a general model which works in the intermediate density regime and it is also applicable in the Corona and the LTE model regimes self-consistently even if the density is of the order of above-mentioned low and high density limits [21]. This makes it more useful in the most spectroscopic studies.

The CR-model based line-ratio technique is often used to estimate the electron plasma density and temperature [22, 23]. The line ratio method is used by the earlier workers to estimate the electron plasma density and temperature in the PPD source, which are typically $\sim 10^{11}$ cm^{-3} and $\sim$2–5 eV [24], respectively. However, the line-ratio method is constrained by the fact that identified line-ratios must be sensitive either to electron density or to electron temperature for estimation of one plasma parameter at a time. Also, we have to identify suitable spectral line pairs to be used, which should follow either ionizing or recombining plasma conditions [25]. In this work, an effort has been made to develop a spectroscopic method which works on least square fitting of the absolute intensities of the observed visible neutral helium spectral lines from a PPD source and the CR-model-based estimated intensities using atomic data and analysis structure (ADAS) code [26] to find out the basic plasma parameters simultaneously from a large number of visible spectral lines. This method is applicable for all the spectral lines observed at a time.

The analysis is performed with thin plasma condition and electron density, electron temperature, ground state atom density, ground state ion density and 2 ^{3}S metastable state density are the parameters estimated. The estimated plasma temperature from the PPD is found high in the simultaneous measurement at which the neutral helium may not even survive and seems to be unrealistic. The analysis is improved by the inclusion of opacity in the observed spectral lines through photon emissivity coefficients (PECs) and also addition of diffusion of neutrals and metastable state species in the CR-model code. This helps in adequate electron temperature estimation in the simultaneous measurement.

Experimental setup

The PPD source (VUV Source SD-01 from Jobin-Yvon, France) consists of two cathodes with an anode ring between them. The two cathodes face each other and the emitted electrons that oscillate between the cathodes fall through the central ring. Because of the field of the permanent magnets located behind the cathodes the charged particles are confined within the discharge volume. Electrons that leave the cathode are confined by the magnetic field. The experimental arrangement of 1 cm cylindrical plasma column length is shown in the Fig. 1.

Helium gas discharge is produced in the penning plasma source. The gas pressure is controlled by the needle valve and discharge current is controlled with the control knob in the PPD source power supply. The PPD source is cooled during operation by flowing demineralised water at 1 lit/min. During the experiment the working pressure is kept $\sim 4 \times 10^{-4}$–1×10^{-2} mbar and the discharge current $\sim$20–150 mA. The spectroscopic data are recorded in the visible region using visible spectrometer at various filled pressures for fixed discharge current and also at different discharge currents and at fixed fill pressure. The analysis of the spectra which is recorded at different discharge currents 20, 64, 80, 90 and 100 mA and at constant filled pressure 4×10^{-3} mbar is discussed.

The light emitted during the discharge is transmitted by an optical fiber and fed to the visible monochromator (wavelength resolution $\sim$2.5 Å) fitted with a charge coupled device (CCD) camera-based acquisition system to record the visible emission lines. The recorded helium lines are He I 7281.3 Å ($2p^1P^0$–$3s^1S$), He I 7065.1 Å

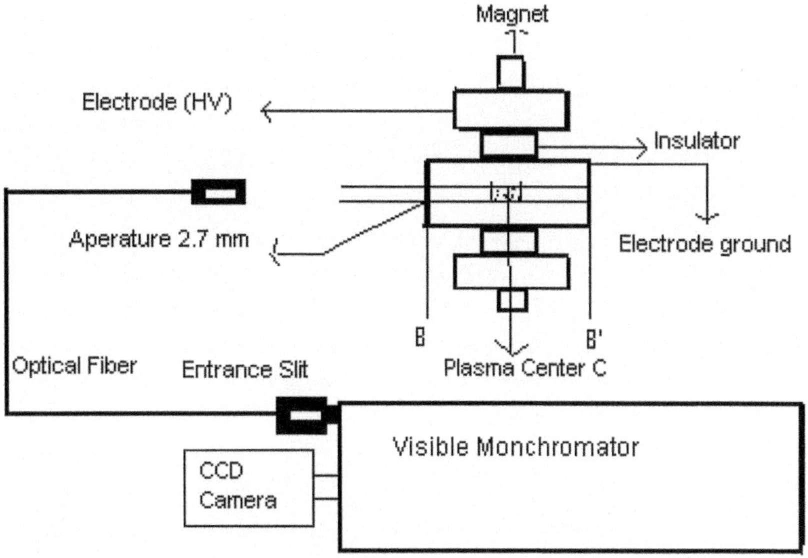

Fig. 1 Schematic diagram of the penning discharge source

($2p^3P^0$–$3s^3S$), He I 6678.1 Å ($2p^1P^0$–$3d^1D^0$), He I 5875.6 Å ($2p^3P^0$–$3d^3D$), He I 5047.7 Å ($2p^1P^0$–$4s^1S$), He I 5015.7 Å ($2s^1S$–$3p^1P^0$), He I 4921.9 Å ($2p^1P^0$–$4d^1D$), He I 4713.1 Å ($2p^3P^0$–$4s^3S$), He I 3964.7 Å ($2s^1S$–$4p^1P^0$), and He I 3888.6 Å ($2s^3S$–$3p^3P^0$). Since He I 5015.7 Å ($2s^1S$–$3p^1P^0$) line is affected by resonance scattering [27], we have worked with 9 lines out of these 10 observed lines. In fact, this effect has not been considered in the CR-model calculations of the ADAS code [26]. Using remaining nine lines a simple procedure using least square fitting is developed to determine the plasma parameters.

Model analysis

With an assumption of average electron density and temperature in 1 cm plasma column length, the photon intensity of a spectral line for such an averaged measurement can be written from the CR-model as [22, 28],

$$\tilde{I}(\lambda_{ul}) = \{\tilde{N}_e\tilde{N}_i PE\tilde{C}_{reco} + \tilde{N}_e\tilde{N}_g PE\tilde{C}_{exc} + \tilde{N}_e\tilde{N}_{m1} PE\tilde{C}_{meta,1} + \tilde{N}_e\tilde{N}_{m2} PE\tilde{C}_{meta,2}\}. \quad (1)$$

Here $PE\tilde{C}_{reco}$, $PE\tilde{C}_{exc}$, $PE\tilde{C}_{meta,1}$ and $PE\tilde{C}_{meta,2}$ are the average values of PECs, having units cm^3/s for recombination, excitation, singlet metastable state ($2\ ^1S$) and triplet metastable state ($2\ ^3S$), respectively. The values $\tilde{N}_e$, $\tilde{N}_i$, $\tilde{N}_g$, $\tilde{N}_{m1}$ and $\tilde{N}_{m2}$ represent the average densities of electron, ground state ion, ground state atom, singlet and triplet metastable states of helium atoms, respectively. Moreover, PECs are the complicated functions of the electron density and temperature and can be obtained from the ADAS database for a given value of electron density and temperature. Since the life time of singlet metastable state is of the order of characteristic time scale of change of plasma, ADAS does not consider $2\ ^1S$ level as a metastable state and hence Eq. (1) can be written as,

$$\tilde{I}(\lambda_{ul}) = \{\tilde{N}_e\tilde{N}_i PE\tilde{C}_{reco} + \tilde{N}_e\tilde{N}_g PE\tilde{C}_{exc} + \tilde{N}_e\tilde{N}_{m2} PE\tilde{C}_{meta,2}\}. \quad (2)$$

Assuming quasi-neutral and purely ionizing condition for plasma, a merit function S is defined to use the least square fitting technique for all nine observed spectral lines as,

$$S = \sum_{j=1}^{9}\left[I_{\exp_j} - \left(\tilde{N}_e\tilde{N}_g PE\tilde{C}_{excj} + \tilde{N}_e\tilde{N}_{m2} PE\tilde{C}_{meta,2j}\right)\right]^2, \quad (3)$$

where, $I_{\exp_j}$ represents experimentally observed intensities and the terms in the small bracket represent the theoretical intensities derived from the CR-model. In Eq. 3 there are two fitting parameters $\tilde{N}_g$ and $\tilde{N}_{m2}$ for a given set of electron density and temperature. The most suitable values of these fitting parameters can be obtained by minimizing the merit function S with respect to the parameters $\tilde{N}_g$ and $\tilde{N}_{m2}$, i.e.,

$$\frac{\partial S}{\partial N_g} = 0 \text{ and } \frac{\partial S}{\partial N_{m2}} = 0 \Rightarrow \tilde{N}_g = \frac{I_2 - (C/B)I_1}{B - AC/B}$$

$$\text{and } \tilde{N}_{m2} = \frac{I_1 - \tilde{N}_g A}{B}, \quad (4)$$

$$A = \sum_j \left(\tilde{N}_e PE\tilde{C}_{excj}\right)^2, B = \sum_j \tilde{N}_e PE\tilde{C}_{excj} \times \tilde{N}_e PE\tilde{C}_{meta,2_j},$$

$$C = \sum_j \left(\tilde{N}_e PE\tilde{C}_{meta,2_j}\right)^2,$$

$$I_1 = \sum_j \tilde{N}_e PE\tilde{C}_{excj} I_{\exp_j}, I_2 = \sum_j \tilde{N}_e PE\tilde{C}_{meta,2_j} I_{\exp_j}.$$

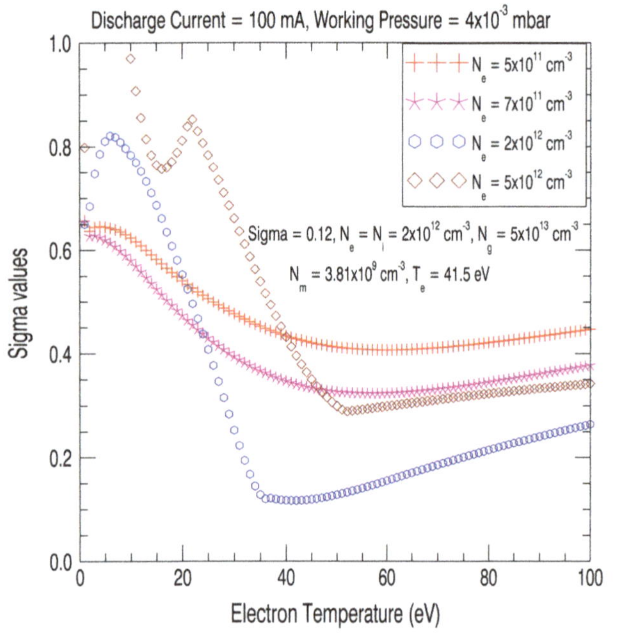

Fig. 2 Mismatch parameter σ for the discharge current 100 mA at constant pressure of 4×10^{-3} mbar under thin plasma condition

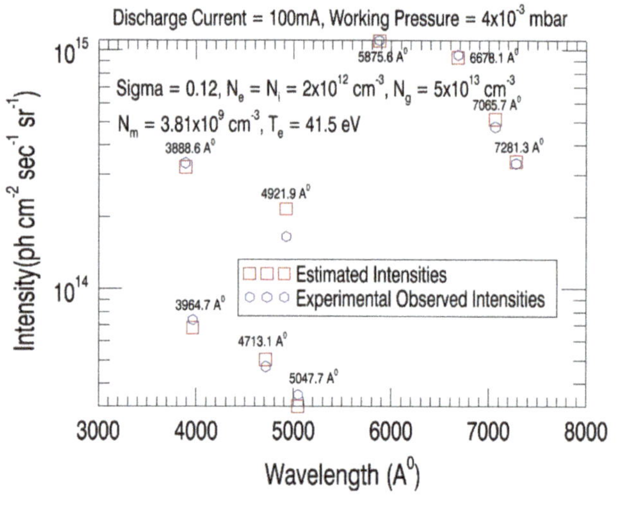

Fig. 3 Comparison of theoretical and experimental intensities at 100 mA and 4×10^{-3} mbar of pressure under thin plasma condition

From Eqs. (4) and (5) the values of $\tilde{N}_g$ and $\tilde{N}_{m2}$ are estimated. Theoretical intensities $\tilde{I}(\lambda_{\text{ul}})$ are calculated for all the nine lines, which are termed as $\tilde{I}_{\text{cal}j}$, using the obtained values of $\tilde{N}_g$ and $\tilde{N}_{m2}$ for larger grid points of $\tilde{N}_e$ and $\tilde{T}_e$. To predict the best fit basic plasma parameters, these calculated intensities $\tilde{I}_{\text{cal}j}$ are compared with the experimental values ($I_{\text{exp}j}$) for all the nine lines ($j = 1,9$) through a mismatch parameter,

$$\sigma = \sqrt{\frac{1}{9}\sum_{j=1}^{9}\left(\frac{I_{\text{exp}j} - I_{\text{cal}j}}{I_{\text{exp}j}}\right)^2}, \tag{6}$$

The minimum in the mismatch parameter (σ) gives the best-fit values of $\tilde{N}_e$, $\tilde{T}_e$, $\tilde{N}_i$, $\tilde{N}_g$ and $\tilde{N}_{m2}$.

Results and discussion

The mismatch parameters obtained at fixed working pressure 4×10^{-3} mbar and different discharge currents 20, 64, 80, 90 and 100 mA are 0.35, 0.33, 0.14, 0.15 and 0.12, respectively. Figure 2 shows the plot of mismatch parameter at different temperatures and densities for fixed discharge current 100 mA and working pressure 4×10^{-3} mbar. For this operating plasma condition, the predicted basic plasma parameters are $\tilde{N}_g = 5 \times 10^{13}$ cm^{-3}, $\tilde{N}_{m,2} = 3.8 \times 10^9$ cm^{-3}, $N_e \approx \tilde{N}_i = 2 \times 10^{12}$ cm^{-3}, $T_e = 41.5$ eV with a minimum $\sigma \sim 0.12$. At this condition a comparison between estimated intensities and

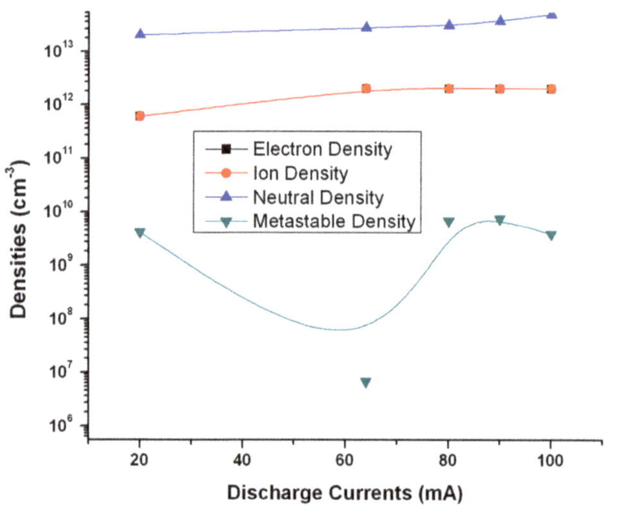

Fig. 4 The variation of neutral helium, triplet metastable state density, ion and electron densities at fixed pressure 4×10^{-3} mbar and with different discharge currents under thin plasma condition

experimentally observed intensities is also shown in Fig. 3. Though the fit with intensities is good but the estimated electron temperature is quite high. At such temperature neutral helium may not survive that shows a discrepancy in temperature estimation.

To further elaborate these results the variation of the metastable state density and ground state atom density for different discharge currents and fixed pressure is shown in Fig. 4. It is observed that the metastable state density decreases up to 64 mA discharge current and then increases. This behavior of metastable state does not seem obvious. Indeed, at fixed pressure, an increment in the discharge current suggest the consumption of more neutrals in creation of larger electrons to draw the higher discharge currents, which may enhance the effective recombination process over the excitation/ionization process. Moreover, the transitions happening due to recombination process may lead in production of larger metastable populations. It is also seen from the figure that the neutral density increases with discharge current. This seems ambiguous because for larger discharge current the neutrals have to be ionized more and hence the neutral density should decrease. Therefore, this analysis needs one or more considerations to predict the correct behavior of the plasma discharge.

It is to be emphasized that the above calculations are carried out assuming optically thin and diffusion less plasma conditions which may not necessarily be correct for all the applications. In fact, most spectroscopic diagnostic techniques require optically thin plasma for model simplifications, which is very near homogeneous along the line-of-sight and remains in steady-state for the duration of the observation. However, if the plasma is optically thick for the radiation coming from the plasma, the re-absorption (opacity) of the photons may cause the wrong information about the estimated plasma parameters [29, 30]. So, the aforesaid discrepancies in temperature and neutral density obtained in the results may be due to the effect of the opacity and also the diffusion of neutral and metastable states. The ADAS code calculates opacity affected PECs on the basis of escape factor methodology [26, 29–33]. Basically, inclusion of escape factors causes a net reduction in Einstein's A coefficients, i.e., it modifies the atomic transition probability values. The effect of self-absorption not only changes the net emergent flux but also changes the population of excited states and hence the effective rate coefficients.

The changes in excited state population and ionization balance due to re-absorption of radiation are then derived using the ADAS 208 code [26]. According to the used experimental setup the following parameters are used to calculate the opacity affected photo emissivity coefficients in the ADAS code: density distribution (parabolic), plasma geometry (cylindrical), line profile (Doppler), neutral temperature (600 K), plasma length (1 cm) and neutral density (5×10^{13} cm^{-3}). The detailed description of the model used for opacity calculations is given in the ADAS user manual [26]. The opacity effect has been considered for the following transitions and spectral lines: He I 7281.3 Å ($2p^1P^0$–$3s^1S$), He I 7065.1 Å ($2p^3P^0$–$3s^3S$), He I 6678.1 Å ($2p^1P^0$–$3d^1D^0$), He I 5875.6 Å ($2p^3P^0$–$3d^3D$), He I 5047.7 Å ($2p^1P^0$–$4s^1S$), He I 5015.7 Å ($2s^1S$–$3p^1P^0$), He I 4921.9 Å ($2p^1P^0$–$4d^1D$), He I 4713.1 Å ($2p^3P^0$–$4s^3S$), He I 3964.7 Å ($2s^1S$–$4p^1P^0$), and He I 3888.6 Å ($2s^3S$–$3p^3P^0$).

A diffusion time scale $\sim 10^{-4}$ s is also taken into account in the ADAS code to introduce the diffusion of neutral and metastable states to the wall. The diffusion time scale was estimated for our geometry using radius of the cylinder ($a = 0.5$ cm), zero order Bessel's function J_0 and the diffusion coefficient D for helium discussed earlier [34]. To illustrate it clearly, we have assumed purely diffusive transport in cylindrical symmetry. In fact, if the impurity species' are no longer fed by external source in the plasma, their density decays and the radial impurity profile first rearrange their shape and eventually decrease self-similarly with a time (i.e., diffusion time scale) given by the following expression [34],

$$\tau_0 = \frac{a^2}{J_0^2(0)D} \tag{7}$$

In this expression, the particle diffusion coefficients are still not very well-known from theory because they are often determined by instabilities in plasma (i.e., anomalous nature in general) and are usually described from the experiments. One basically requires diagnostic line excited data and atomic coefficients for total radiation to compare final results with bolometer kind of diagnostics. All such calculations are quite complicated and are done through various plasma transport codes, e.g., STRAHL [35, 36].

In spectroscopy, the diffusion overall effects the population of the excited species, which is directly related with the ground state and metastable populations and indirectly related with the purely or partially ionized plasmas [37]. In our case, we have added diffusion for partially ionized plasma and only for metastable state, which can affect the populations of the particular energy level (i.e., excited state) during the measurements. This may be the region for wrong estimation of plasma parameter during model understanding without opacity and diffusion.

After this, the theoretical intensities and accordingly the mismatch parameters (σ) with the experimental values are recalculated by repeating the same procedure as explained above while including the effect of opacity in the PEC's and diffusion of neutrals and metastable states to the wall. The obtained minimum mismatch parameters (σ) at working pressure 4×10^{-3} mbar and different discharge currents 20, 64, 80, 90 and 100 mA are 0.62, 0.66, 0.65, 0.65 and 0.69, respectively. The variation of the σ with electron temperatures and different electron densities is shown in Fig. 5. At this condition the estimated values of $\tilde{N}_g$ and $\tilde{N}_{m,2}$ are 4.22×10^{13} and 1.02×10^{13} cm^{-3}, respectively, while the calculated values of electron temperature, electron density and ion density are $T_e = 2.0$ eV, $N_e \approx \tilde{N}_i = 7 \times 10^{10}$ cm^{-3}, respectively.

A comparison between experimental and estimated intensities is shown in Fig. 6. From this figure it can be seen that the miss-match between the two spectral lines 6678.1 and 7281.3 Å has increased in the simultaneous measurement but it improves the temperature estimation.

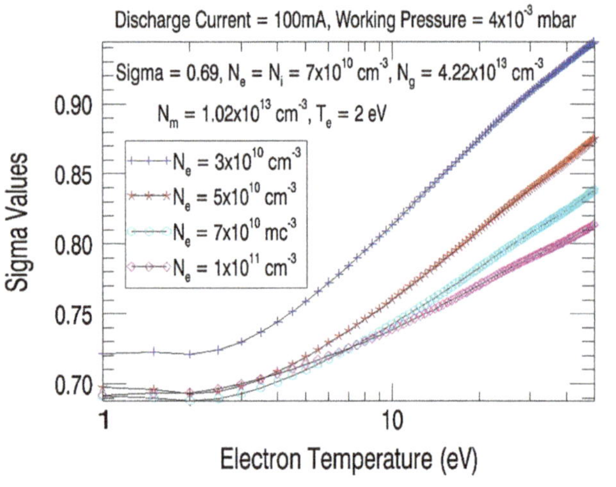

Fig. 5 Mismatch parameter σ for the discharge current 100 mA at constant pressure of 4×10^{-3} mbar with opacity and diffusion effects considerations

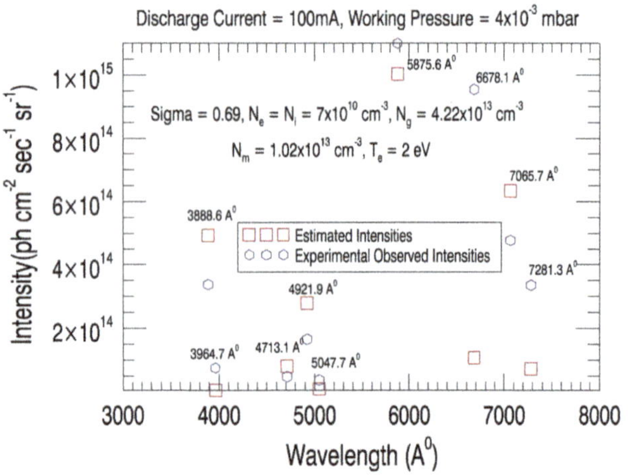

Fig. 6 Comparison of theoretical and experimental intensities at 100 mA discharge current and 4×10^{-3} mbar working pressure under opacity and diffusion considerations

More efforts are needed to understand the exact behavior of 6678.1 and 7281.3 Å spectral lines in penning type of plasma discharges. The non-Maxwellian electron consideration in the CR-model may provide an exact solution, however, such state-of-art researches are underway [38].

Based on the improved calculations the variation of metastable and neutral densities with the discharge currents at fixed working pressure is shown in the Fig. 7. It is seen that with the increasing discharge current the neutral density decreases and the metastable state density increases. This is the predictable behavior of neutrals and metastable states because the enhancement in the plasma discharge current requires more ionization which may increase the recombination processes and hence there might be an increment in metastable state densities and also decrement in the neutral density.

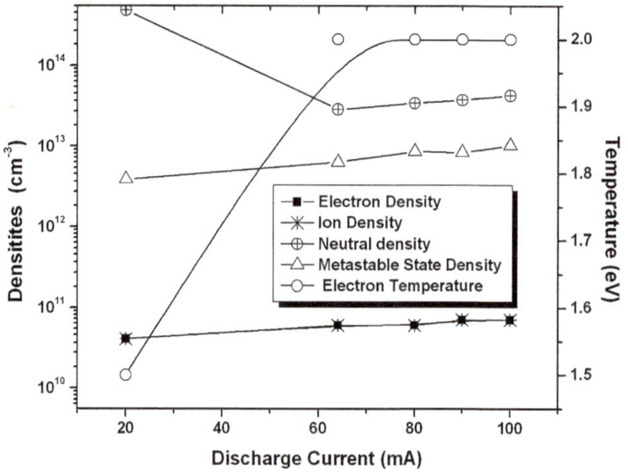

Fig. 7 The variation of neutral helium, triplet metastable state density, ion and electron densities and electron temperature at fixed pressure 4×10^{-3} mbar and with different discharge currents including opacity and diffusion effects

Conclusions

In this study, a procedure to calculate the basic plasma parameters like—N_e, T_e, $\tilde{N}_i$, $\tilde{N}_g$ and $\tilde{N}_m$ simultaneously from a penning discharge source using least square fitting of spectroscopic data and ADAS code is presented. In case of the optically thin and diffusion less plasma assumption, the obtained electron temperature is rather high. Hence the assumption of optically thin and diffusion less plasma is not applicable. The opacity calculations and diffusion of metastable and neutral are then added in the method using ADAS code to estimate the plasma parameters correctly. With this condition, the neutral density decreases and metastable population of helium 2 ^{3}S state increases with discharge currents, which is quite consistent.

Acknowledgments The authors thankfully acknowledge the Spectroscopy Diagnostics Division of Institute for Plasma Research, Bhat, Gandhinagar, India for providing the experimental facility and data. Thanks are also due the ADAS consortium, Prof. H. P. Summers, Prof. Kurt Behringer and Prof. Martin O' Mullane for helping us to use and learn the ADAS database. Authors also thankfully acknowledge Board of Research in Fusion Science and Technology (BRFST) for the financial support for this work under projects NFP/DIAG/2 and NFP/DAQ/A09/02.

References

1. Penning, F.M.: Ein neues manometer für niedrige gasdrucke, insbesondere zwischen 10^{-3} und 10^{-5} mm. Physica. (Amsterdam) **4**, 71 (1937)
2. Rikoski, R.A.: Compact penning discharge ion source for plasma diffusion studies. J. Phys. E: Sci. Instrum. **5**, 540 (1972)
3. Roth, J.R.: Industrial Plasma Engineering, IOP publishing, London, pp. 204 (1995)

4. Gow, J.D., Foster, J.S.: A high-intensity pulsed ion source. Rev. Sci. Instrum. **24**, 606 (1953)
5. Hillis, D.L., Morgan, P.D., Groth, J.K.M., Stamp, M.F., Hellermann, M., Ehrenberg, Von., Kumar, V.: Tritium concentration measurements in the Joint European torus by optical spectroscopy of a penning gauge. Rev. Sci. Instrum. **70**, 359 (1999)
6. Hillis, D.L., Klepper, C.C., Hellermann, M., Ehrenberg, Von J., Finken, K.H., Mank, G.: Deuterium-tritium concentration measurements in the divertor of a tokamak via a modified Penning gauge. Fusion Eng. Des. **347**, 34–35 (1997)
7. Brown, L.S., Gabrielse, G.: Geonium theory: physics of a single electron or ion in a penning trap. Rev. Mod. Phys. **58**, 233 (1986)
8. Major, F.G., Gheorghe, V.N., Werth, G.: Charged Particle Traps: Physics and Techniques of Charged Particle Field Confinement. Springer, Berlin (2005)
9. Heise, C., Hollandt, J., Kling, R., Kock, M., Kuehne, M.: Radiometric characterization of a penning discharge in the vacuum ultraviolet. Appl. Opt. **33**, 5111 (1994)
10. Prakash, R., Jain, J., Kumar, V., Manchanda, R., Agarwal, B., Banerjee, S., Chowdhuri, M.B., Vasu, P.: Calibration of a VUV spectrograph by collisional-radiative modelling of a discharge plasma. J. Phys. B At. Mol. Opt. Phys. **43**, 144012 (2010)
11. Hirsch, E.H.: On the mechanism of the Penning discharge. Br. J. Appl. Phys. **15**, 1535 (1964)
12. Feng, P.X., Weiner, B.: High abundance of metastable helium atoms for diagnostic applications. J. Phys. B At. Mol. Opt. Phys. **37**, 3265 (2004)
13. Finley, S.D., Bowyer, S., Paresce, F., Malina, R.F.: Continuous discharge penning source with emission lines between 50 a and 300 a. Appl. Opt. **18**, 649 (1979)
14. Hutchinson, I.H.: Principle of Plasma Diagnostics, Cambridge Univ. Press, Cambridge, p. 60 (2002)
15. Godyak, V., Demidov, V.I.: Probe measurements of electron-energy distributions in plasmas: what can we measure and how can we achieve reliable results? J. Phys. D Appl. Phys. **44**, 233001 (2011)
16. Chen, F.F.: Langmuir Probe Diagnostics, Mini-Course on Plasma Diagnostics, IEEE-ICOPS meeting, Jeju (2003)
17. Griem, H.R.: Principles of Plasma Spectroscopy, Cambridge Univ Press, Cambridge (1997)
18. Kunze, H.J.: Introduction to Plasma Spectroscopy. Springer, Heidelberg (2009)
19. Fujimoto, T.: Plasma Spectroscopy. Clarendon Press, Oxford (2004)
20. Bates, D.R., Kingston, A.E., McWhirter, R.W.P.: Recombination between electrons and atomic ions. I. optically thin plasmas. Proc. R. Soc. Lond. A **267**, 297 (1962)
21. Fantz, U.: Basics of plasma spectroscopy. Plasma Sources Sci. Technol. **15**, S137 (2006)
22. Prakash, R., Vasu, P., Kumar, V., Manchanda, R., Chowdhuri, M.B., Goto, M., Aditya Team: Characterization of helium discharge cleaning plasmas in ADITYA tokamak using collisional-radiative model code. J. Appl. Phys. **97**, 43301 (2005)
23. Goto, M.: Collisional-radiative model for neutral helium in plasma revisited. J. Quant. Spectrosc. Radiat. Transfer **76**, 331 (2003)
24. Feng, P.X., Andruczyk, D., James, B., Takiyama, W.K., Namba, S., Oda, T.: High-density metastable helium atoms produced by Penning-type discharges. Plasma Source Sci. Technol. **12** 142 (2003)
25. Sasaki, S., Goto, M., Kato T., Takamura, S.: Line Intensity Ratios of Helium Atom in an Ionizing Plasma. NIFS-DATA-**49** (2008)
26. Summers, H.P.: Atomic data and analysis structure user's manual. JET—IR **06**, JET Joint undertaking, Abingdon (1994)
27. Sasaki, S., et al.: Helium I Line Intensity Ratios in a Plasma for the Diagnostics of Fusion Edge Plasmas. NIFS-DATA-**346** (1995)
28. Prakash, R., Pütterich, T., Dux, R., Kallenbach, A., Fischer, R., Behringer, K., ASDEX Upgrade Team.: Spectroscopic Divertor Characterization and Impurity Influx Measurements in ASDEX Upgrade Tokamak, vol. **10/31**, IPP (2006)
29. Behringer, K.: Escape Factors for Line Emission and Population Calculations, vol. **10/11**, Report IPP, (1998)
30. Behringer, K., Fantz, U.: The influence of opacity on hydrogen excited-state population and applications to low-temperature plasmas. New J. Phys. **2**, 23 (2000)
31. Irons, F.E.: The escape factor in plasma spectroscopy—I. The escape factor defined and evaluated. JQSRT **22**, 1 (1979)
32. Irons, F.: E, The escape factor in plasma spectroscopy—II. The case of radiative decay JQSRT **22**, 21 (1979)
33. Boivin, R.F., Kline, J.L., Scime, E.E.: Electron temperature measurement by a helium line intensity ratio method in helicon plasmas. Phys. Plasmas **8**, 5303 (2001). (and references therein)
34. Behringer, K.: Spectroscopy of fusion plasmas: methods and atomic data requirements. In: Ram Prakash. (ed) IAEA International Workshop on Challenges in Plasma Spectroscopy for Future Fusion Research Machines, Jaipur, India (2008) vol. **1,** pp. 3. Excel India Publisher, Delhi (2008)
35. Behringer, K.: Description of the Impurity Transport Code "STRAHL", Report **JET-R(87) 08**, (1987)
36. Dux, R.: STRAHL User Manual, vol. **10/30**, Report IPP (2006)
37. Prakash, R.: Plasma and the Role of Atomic Physics in its Spectroscopic Diagnostic. In: Srivastava, R., Choubisa, R. (eds.) Atomic and Molecular Physics: Introduction to Advanced Topics, pp. 203–235. Narosa Publishing House, New Delhi (2012)
38. Bingham, R., Kellett, B.J., Bryans, P., Summers, H.P., Torney, M., Shapiro, V.D., Spicer, D.S., O'Brien, M.: A new mechanism explaining the asymmetric x-ray emission from Supernova Remnant PKS 1209–52. Astrophys. J. **601**, 896 (2004)

15

Effects of electric field on magnetic properties of Mn_xGe_{1-x} diluted magnetic semiconductors

Gezahegn Assefa[1,2] · P. Singh[1]

Abstract We report the effect of external electric field (EEF) on the magnetic properties of Mn_xGe_{1-x}, diluted magnetic semiconductor. We present a Kondo Lattice Model type Hamiltonian with exchange coupling between localized spins, itinerant holes and the EEF. The magnetization, the dispersion and critical temperature (T_c) are calculated for different values of EEF parameters (α) using double time temperature-dependent Green function formalism. The enhancement of the (T_c) with the EEF is shown to be very distinct and is in agreement with recent experimental observation and much required for spintronics applications and devices.

Keywords Diluted magnetic semiconductor · Spintronics · Electric field control of ferromagnetism · Green function

Introduction

The carrier-mediated ferromagnetism observed in dilute magnetic semiconductors (DMS) has gained interest over the past decade due to the possibility of controlling the ferromagnetism and their potential applications in spintronics which use both the charge and spin degrees of freedom for electronic applications [1]. DMS materials such as (Ga, Mn) As, (In, Mn) As, (Ga, Mn) N have combined the magnetic and semiconducting properties in a single material [2].

Even though the origin of ferromagnetism in diluted magnetic semiconductor is not fully understood and explained yet, there are different theoretical models proposed to explain the ferromagnetism in DMS materials. The magnetic interactions between magnetic moments in (III, Mn) V and (Mn, IV) DMS systems are hole mediated in which the substitutional Mn produces the holes and acts as acceptors [3].

Ge (Group IV)-based DMS system have attracted a great deal of attention due to its high hole mobility and compatibility with the present silicon technology [4–6]. Recently Mn-doped Ge DMS system showing magnetic ordering was successfully grown by molecular beam epitaxy (MBE) technique which are a good candidates of Ge-based DMSs [7, 8]. Recently Xiu et al. [9] successfully fabricated self-assembled Mn_xGe_{1-x} DMSQDs (with $x = 0.05$) without metallic precipitates such as Mn_5Ge_3 and $Mn_{11}Ge_5$ which showed electric field-controlled ferromagnetism, T_C above 400 K. The presence of such metallic clusters strongly affect the magnetic properties of the Mn-Ge DMS system [5]. In addition to the MBE technique, Mn-Ge DMS systems were prepared by ion implantation techniques, which shows carrier-mediated ferromagnetism due to high hole concentration [10, 11].

Control of the magnetic phase in DMS is one of the most important processes for magnetic recording and information storage. The use of electric field-controlled magnetization reduces power consumption for storage devices [12, 13]. Electric field control of ferromagnetism was so far demonstrated in a field effect transistor (FET) structure which have been used for non-volatile spin logic devices via carrier-mediated effect [9, 14].

✉ Gezahegn Assefa
gzhgnassefa2@gmail.com

1 Department of Physics, Addis Ababa University, P.O.Box 1176, Addis Ababa, Ethiopia

2 Department of Physics, Wollo University, P.O.Box 1145, Dessie, Ethiopia

In this work, we propose the Kondo Lattice model (KLM) type Hamiltonian to explain the ferromagnetism of the DMS Mn-Ge system, and we theoretically study the influence of electric field on the magnetic properties of Mn_xGe_{1-x} DMS, using the double time Green functions formalism. Magnetization and the critical temperature are studied in relation to the EEF. The paper is organized as follows. In section II, we present the model Hamiltonian which describes the DMS charier–charier interaction, carrier–Mn interaction and the Mn–Mn interaction. In the next section, the numerical results of various physical quantities showing the effect of the electric field are discussed. Finally the conclusion and summary are presented in the last section.

Model Hamiltonian

Our system consists of an Mn-Ge DMS system containing Mn as acceptor impurity inside. The Hamiltonian of the system will have the form

$$H = H_{ee} + H_I + H_{mag} \quad (1)$$

We assume a parabolic energy band for the carriers and the second quantized electron Hamiltonian, H_{ee} is given by

$$H_{ee} = \sum_{k\sigma} \epsilon_{k\sigma} c^{\dagger}_{k\sigma} c_{k\sigma}, \quad (2)$$

where $c^{\dagger}_{k\sigma}(c_{k\sigma})$ are Fermionic creation (annihilation) operators for carriers in the DMS. $\epsilon_{k\sigma}$ is the energy of carriers with momentum k, band σ (moments ↑ or ↓).

H_I is the interaction Hamiltonian between the carriers and the localized magnetic ions which is Kondo Lattice Model (KLM) type Hamiltonian with the interaction of electric field with carrier spin and localized spin included, and can be written as

$$H_I = -2J_{sd} \sum_m \mathbf{S}_m \cdot \sigma_m - \mu_e \mathbf{E} \sum_m \mathbf{S}^z_m \sigma^z_m \quad (3)$$

The first term in H_I is due to the interaction of the band carriers (holes) and the localized spin Mn^{2+} ions, expressed as Heisenberg type Hamiltonian, where J_{sd} is exchange coupling between the confined holes and the Mn^{2+} ions impurity spins S_m. σ_m represents the spin of the confined holes, which can be written as $\sigma_\alpha = \frac{1}{2}\sum_{\sigma\sigma'} c^{\dagger}_{m\sigma} \boldsymbol{\tau}_{\sigma\sigma'} c_{m\sigma'}$, $(\alpha = x, y, z)$, where $\boldsymbol{\tau}$ are the matrix elements of the Pauli spin matrices. The second term arises from interaction of the holes and the localized moments with the external electric field (EEF) **E**. It is Ising type Hamiltonian. The mean electric polarization is assumed to be proportional to the z-component of the carriers and the localized spins.

$\mathbf{H}_{mag}$ is given by

$$H_{mag} = -\frac{1}{2} \sum_{mn} I_{mn} \mathbf{S_m} \cdot \mathbf{S_n} - g\mu_\beta \mathbf{H} \sum_{\mathbf{m}} \mathbf{S^z_m}, \quad (4)$$

The first term in $\mathbf{H}_{mag}$ is the Hamiltonian of the localized Mn^{+2} spins, where I_{mn} is the exchange coupling between the localized spins $S_m(S_n)$ at different sites. The second term gives the Zeemann energy when the external magnetic field H has been applied in the z-direction, where g is Landé g factor and μ_β is the Bohr magneton.

The spin operators in Hamiltonian can be written using the Holstein–Primakoff (HP) transformation as

$$S^-_m = \left(\sqrt{2xS - a^{\dagger}_m a_m}\right) a^{\dagger}_m \quad (5)$$

$$S^+_m = \left(\sqrt{2xS - a^{\dagger}_m a_m}\right) a_m \quad (6)$$

$$S^z_m = xS - a^{\dagger}_m a_m, \quad (7)$$

where $a^{\dagger}_m(a_m)$ are bosonic creation (annihilation) operators, x is the mean number of magnetic ions in the DMS system.

It is reasonable to use the approximation $\sqrt{2xS - a^{\dagger}_m a_m} \approx \sqrt{2xS}$ in HP transformation at low temperature. Restricting the involved momentum in the first Brillouin zone and Fourier transforming the Fermionic and bosonic operators, the effective Hamiltonian can then be written in second quantized form as

$$\begin{aligned} H_{eff} = {} & \left(\frac{\mu_e \mathbf{E}}{2N} - \frac{J_{sd}}{N}\right) \sum_{\sigma,k,q,q'} \sigma a^{\dagger}_q a_{q'} c^{\dagger}_{k-q\sigma} c_{k-q\sigma} \\ & - \frac{J_{sd}\sqrt{2xS}}{\sqrt{N}} \sum_{kq} \left(a^{\dagger}_q c^{\dagger}_{k\uparrow} c_{k+q\downarrow} + a_q c^{\dagger}_{k+q\downarrow} c_{k\uparrow} \right) \\ & - g\mu_\beta H \sum_q a^{\dagger}_q a_q + xS \sum_q (I_0 - I_q) a^{\dagger}_q a_q \end{aligned} \quad (8)$$

For the quasiparticle spectrum of the system described by the Hamiltonian in Eq. (8), we consider the following Green function (GF).

$$G_q(t, t') = \langle\langle a_q(t); a^{\dagger}_q(t')\rangle\rangle \quad (9)$$

To evaluate the GF we use the general equation of motion method [15].

$$\omega\langle\langle a_q(t); a^{\dagger}_q(t')\rangle\rangle_\omega = \frac{1}{2\pi}\langle[a_q, a^{\dagger}_q]\rangle + \langle\langle [a_q(t), H]; a^{\dagger}_q(t')\rangle\rangle_\omega \quad (10)$$

$$[a_q(t), H] = \left(\frac{\mu_e E}{2N} - \frac{J_{\rm sd}}{N}\right)\sum_{kq'\sigma} a_{q'} c^{\dagger}_{k-q\sigma} c_{k-q'\sigma} - \frac{J_{\rm sd}\sqrt{2xS}}{\sqrt{N}}\sum_k c^{\dagger}_{k\uparrow} c_{k+q\downarrow} + \left[xS\sum_q (I_0 - I_q) + g\mu_\beta H\right] a_q \tag{11}$$

Using Eq. (11) into Eq. (10), the GF becomes

$$\omega\langle\langle a_q(t); a^{\dagger}_q(t')\rangle\rangle_\omega = \frac{1}{2\pi} + \Lambda_q \langle\langle a_q(t); a^{\dagger}_q(t')\rangle\rangle_\omega - \frac{J_{\rm sd}\sqrt{2xS}}{\sqrt{N}}\sum_k \langle\langle c^{\dagger}_{k\uparrow} c_{k+q\downarrow}; a^{\dagger}_q(t')\rangle\rangle_\omega + \left(\frac{\mu_e E}{2N} - \frac{J_{\rm sd}}{N}\right)\sum_{kq'\sigma}\langle\langle a_{q'} c^{\dagger}_{k-q\sigma} c_{k-q'\sigma}; a^{\dagger}_q(t')\rangle\rangle_\omega, \tag{12}$$

where $\Lambda_q = xS\sum_q (I_0 - I_q) + g\mu_\beta H$.

Employing the following decoupling procedure on the higher order GFs [15],

$$\langle\langle c^{\dagger}_{k\uparrow} c_{k+q\downarrow}; a^{\dagger}_q(t')\rangle\rangle_\omega = \langle\langle [c^{\dagger}_{k\uparrow} c_{k+q\downarrow}, H]; a^{\dagger}_q(t')\rangle\rangle_\omega = (\langle c^{\dagger}_{k\uparrow} c_{k\uparrow}\rangle - \langle c^{\dagger}_{k+q\downarrow} c_{k+q\downarrow}\rangle)\,\langle\langle a_q; a^{\dagger}_q(t')\rangle\rangle_\omega \tag{13}$$

$$\langle\langle a_{q'} c^{\dagger}_{k-q\sigma} c_{k-q'\sigma}; a^{\dagger}_q(t')\rangle\rangle_\omega = \delta_{qq'}\langle c^{\dagger}_{k-q\sigma} c_{k-q'\sigma}\rangle\langle\langle a_{q'}; a^{\dagger}_q(t')\rangle\rangle_\omega = \langle c^{\dagger}_{k-q\sigma} c_{k-q\sigma}\rangle\langle\langle a_q; a^{\dagger}_q(t')\rangle\rangle_\omega \tag{14}$$

Plugging Eqs. (13) and (14) into Eq. (12) the GF at $q \to 0$ can be written in the form

$$G_q(\omega) = \frac{1}{2\pi}\frac{1}{\omega - \Lambda_q + m(2xSJ^2_{\rm sd} - J_{\rm sd} - \frac{\mu_e E}{2})}, \tag{15}$$

where

$$m = \frac{1}{N}\sum_k (c^{\dagger}_{k\uparrow} c_{k\uparrow} - c^{\dagger}_{k-q\downarrow} c_{k-q\downarrow}) = \frac{1}{N}\sum_k (c^{\dagger}_{k-q\uparrow} c_{k-q\uparrow} - c^{\dagger}_{k-q\downarrow} c_{k-q\downarrow}), \tag{16}$$

is the carrier spin polarization at $q \to 0$.

From the poles of the GF we can obtain the excitation spectrum of the magnon as

$$\omega_q = xS\sum_q (I_0 - I_q) + g\mu_\beta H - m\left(2xSJ^2_{\rm sd} + J_{\rm sd} - \frac{\mu_e E}{2}\right) \tag{17}$$

The exchange integral I_q is given by

$$I_q = \frac{1}{N}\sum_{mn} I_{mn} e^{i\mathbf{q}\cdot(\mathbf{R_m} - \mathbf{R_n})} \tag{18}$$

The dispersion of the localized spin subsystem includes the simplest spin wave result when expanding I_q in the limit $q \to 0$ which leads to the dispersion law $\omega_q = Dq^2$, where D is spin wave stiffness and is given by $D = xS\sum_m (\hat{\mathbf{q}}\mathbf{R_m})^2 \mathbf{I}(|\mathbf{R_m}|))$ and $\hat{q} = \mathbf{q}/|\mathbf{q}|$ is the unit vector.

The dispersion of the localized spin subsystem can then be written as

$$\omega_q = Dq^2 + g\mu_\beta H + m\alpha - m\gamma, \tag{19}$$

where

$$\alpha = \frac{\mu_e E}{2} \tag{20}$$

is called the external electric field parameter (EEFP), and

$$\gamma = 2xSJ^2_{\rm sd} + J_{\rm sd} \tag{21}$$

is a parameter associated with the coupling of carrier spin and the localized magnetic spin.

The number of magnon excited at temperature T, can be calculated using the correlation function,

$$\langle a^{\dagger}_q(t') a_q(t)\rangle = \lim_{\epsilon \to 0} i\int_{\infty}^{\infty} \frac{e^{-i\omega(t-t')}}{e^{\beta\omega} - 1}(G_q(\omega + \epsilon) - G_q(\omega - \epsilon))\mathrm{d}\omega \tag{22}$$

The equal time correlation gives the number of excited magnon and is obtained as

$$\tilde{n}_q = \frac{1}{e^{\beta\omega_q} - 1} \tag{23}$$

At a temperature T, the magnetization per site is given by

$$M(T) - M(0) = -g\mu_\beta \sum_q \langle\tilde{n}_q\rangle, \tag{24}$$

where $M(0) = g\mu_\beta nS$ is the magnetization at absolute zero where all spins are parallel.

For small values of q, the sum in (24) can be replaced by an integral over the whole value of the q-space and can be written as

$$M(T) - M(0) = -g\mu_\beta \frac{V}{(2\pi)^3}\int_{-\infty}^{\infty} \frac{d^3q}{\exp\,[\beta(Dq^2 + g\mu_\beta H + m\alpha - m\gamma)]} \tag{25}$$

$$\frac{M(T)}{M(0)} = 1 - AT^{3/2}\exp[-\beta(g\mu_\beta H + m\alpha - m\gamma)], \tag{26}$$

where $\frac{M(T)}{M(0)}$ is the reduced magnetization and A is a constant given by

$$A = \frac{1}{16\pi^2 N_m S}\left(\frac{k}{IxS}\right)^{3/2}. \quad (27)$$

Based on the above frame work, we can calculate the critical temperature, the spin wave dispersion and the susceptibility of the localized spin.

Numerical result and discussion

Controlling the magnetization of magnetic devices using electric field will lead to new devices that will be energy efficient, fast and compact as compared to those devices tuned by electromagnets. The following material parameters are used for the analysis. $J_{sd} = 2$ meV, $I = 20$ meV, $a = 5.77$ Å, $m = 0.9$, $S = 5/2$.

In Fig. 1, we plot the reduced magnetization as a function of temperature in different values of the electric field parameters. The magnetization of the system tends to increase towards the temperature as the electric field parameter increases. The electric field increases the coupling between localized spin and the itinerant holes spins and produces spin-polarized carriers which in turn creates molecular field which is responsible to the higher magnetization of the Mn spins. The slow drop of the magnetization as compared to the Weiss mean field approximation is due to low temperature approximation and the ignorance of the strong spin fluctuation. This is in good agreement with experimental work of Ohno et al. [12] and Chiba et al. [16]. They reported that in the study of the anomalous Hall effect of the semiconductor field effect transistor (FET) structure, the hole-induced ferromagnetism can be turned on and off by EEF.

As seen from Fig. 2 the spontaneous magnetization is favored by negative electric field and the T_C will increase for higher magnitude of the (negative) electric field.

The other property that is affected by the application of EEF is spin wave dispersion gap. As can be seen from Fig. 3 the spin wave dispersion gap increases when the electric field parameters increases.

The dependence of the reduced magnetization on the electric field is observed on Fig. 4. One can notice that the increase of magnetization of the system as electric field increases, which subsequently enhances the critical temperature, T_C.

Figure 5 is the plot of the variation of the critical temperature with the EEF. An increase in critical temperature is observed as the EEF increases. As can be seen from the curve, it is not a monotonic increase rather it is a parabolic increase, which shows there is a saturation for critical

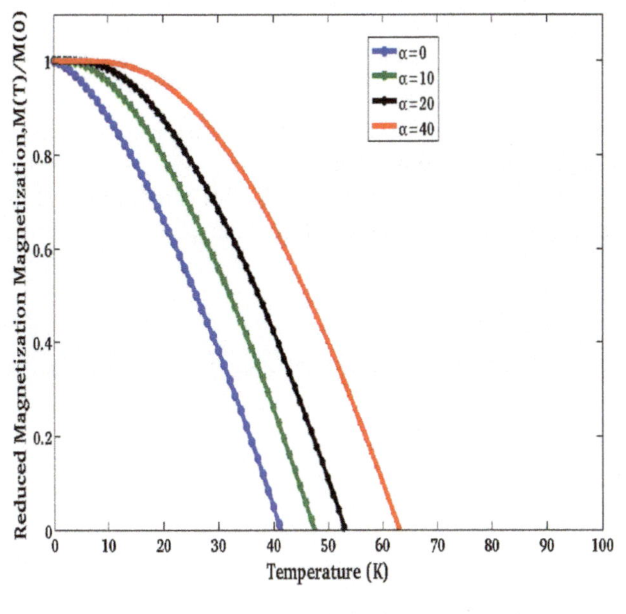

Fig. 1 Reduced magnetization (M(T)/M(0)) as a function of temperature for different values of EEF parameter, α

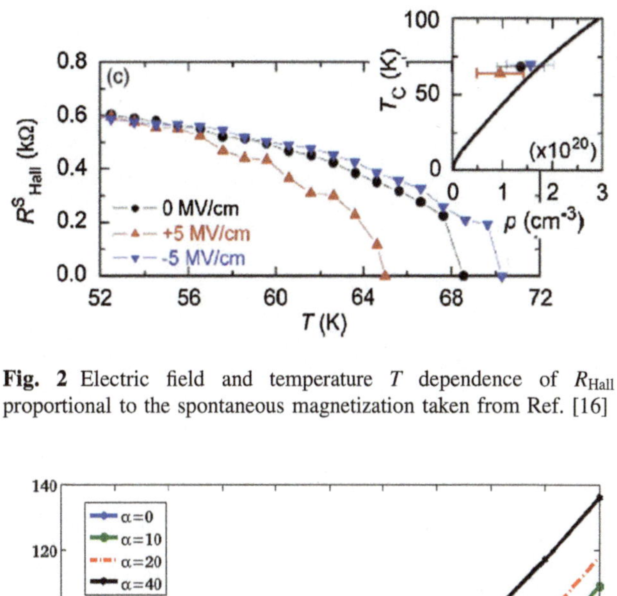

Fig. 2 Electric field and temperature T dependence of R_{Hall} proportional to the spontaneous magnetization taken from Ref. [16]

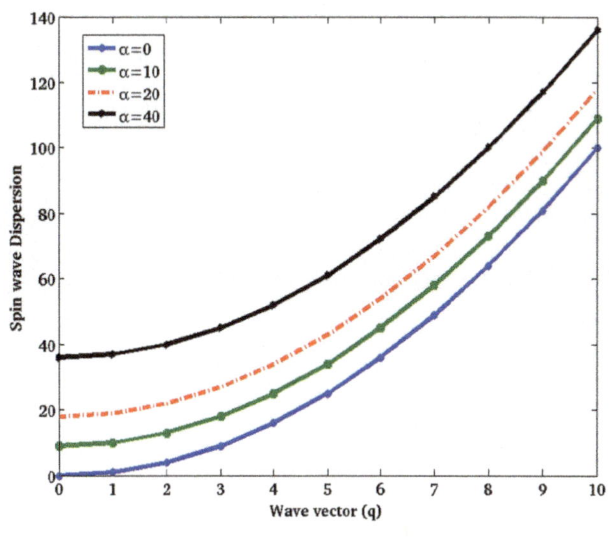

Fig. 3 Spin wave dispersion as a function of wave vector, q

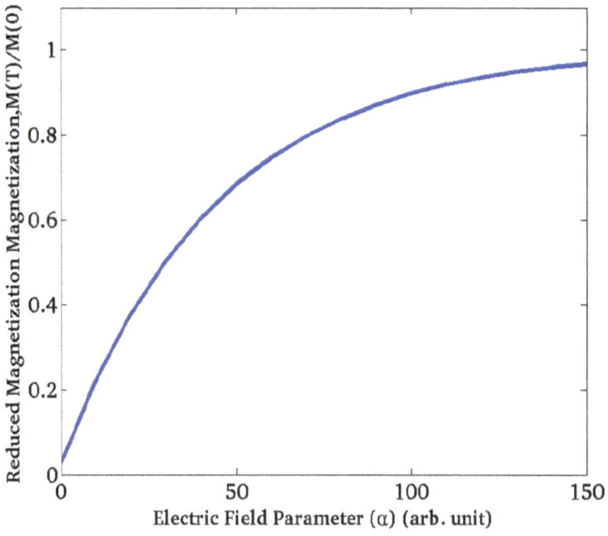

Fig. 4 Reduced magnetization (M(T)/M(0)) as a function of EEF parameter, α

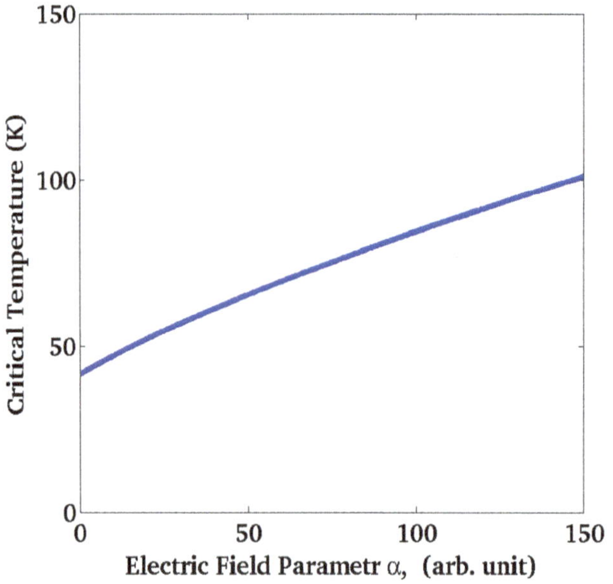

Fig. 5 Critical temperature as a function of EEF parameter, α

temperature at high values of the EEF, and this is also in agreement with [16] as shown in the inset of Fig. 2, which shows the variation of T_C with hole concentration p. (p is dependent on the electric field, i.e. the electric field increases the hole concentration).

Conclusions

We have shown that the magnetic properties of the DMS Mn-Ge system were greatly affected by the application of EEF. The magnetization shows an increase with increasing EEF. This confirms that the internal molecular field due to the carrier–Mn coupling leads to high curie temperature ferromagnetism. This could be explained by the fact that the EEF would accumulate large enough spin-polarized carriers which affect the orientation of the Mn ions and induce ferromagnetism. In addition, the critical temperature (T_C) and the spin wave dispersion gap increases with increasing EEF. These electric field-dependent magnetic properties will lead to the control of the magnetism by an electric field which is essential for the spintronics applications. In conclusion, we see that our system can be tuned by the external applied electric field.

Acknowledgments This work was supported by the school of Graduate studies of Addis Ababa university, Addis Ababa, Ethiopia and Wollo university, Dessie, Ethiopia.

References

1. Wolf, S.A., Awschalom, D.D., Buhrman, R.A., Daughton, J.M., Roukes, M.L., Chtchelkanova, A.Y., Treger, D.M.: Spintronics: a spin-based electronics vision for the future. Science **294**, 1488 (2001)
2. Ohno, H.: Making nonmagnetic semiconductors ferromagnetic. Science **281**, 951 (1998)
3. Chen, Hua, Zhu, Wenguang, Kaxiaras, Efthimois, Zhang, Zhenyu: Optimaization of Mn doping in group IV based diluted magnetic semiconductors by electronic co-dopants. Phys. Rev. B **79**, 235202 (2009)
4. Song, S.H., Lim, S.H., Jung, M.H., Santos, T.S., Moodera, J.S.: Magnetic and transport properties of amorphous Ge-Mn thin films. J. Korean Phys. Soc. **49**(6), 23–86 (2006)
5. Dung, D.D., Mua, N.T., Chung, H.V., Cho, S.: Novel Mn-doped Ge: from basis to application. J. Mater. Sci. Eng. A **2**(2), 248–271 (2012)
6. Li, A.P., Shen, J., Thompson, J.R., Weitering, H.H.: Ferromagnetic percolation in Mn_xGe_{1-x} dilute magnetic semiconductor. Appl. Phys. Lett. **86**, 152507 (2005)
7. Park, Y.D., Hanbicki, A.T., Erwin, S., C, Hellberg C.S., Sullivan J.M., Mattson J.E., Ambrose T.F., Wilson A, Spanos G, Jonker B.T.: A Group-IV Ferromagnetic Semiconductor: Mn_xGe_{1x}. Science **295**, 651–654 (2002)
8. Cho, S., Choi, S., Hong, S.C., Kim, Y., Ketterson, J.B., Kim, B.J., Kim, Y.C., Jung, J.H.: Ferromagnetism in Mn-doped Ge. Phys. Rev. B **66**, 033303 (2002)
9. Xiu, F., Wang, Y., Kim, J., Hong, A., Tang, J., Jacob, A.P., Zou, J., Wang, K.L.: Electric-field controlled ferromagnetism in high-Curie-temperature $Mn_{0.05}Ge_{0.95}$ quantum dots. Nat. Mater. **9**, 337–344 (2010)
10. Zhou, Shengqiang, Burger, Danilo, Skorupa, Wolfgang, Oesterlin, Peter, Helm, Manfred, Schmidt, Heidemarie: The importance of hole concentration in establishing carrier-mediated ferromagnetism in Mn doped Ge. Appl. Phys. Lett. **96**, 202105 (2010)
11. Zhou, S., Burger, D., Mucklich, A., Baumgart, C., Skorupa, W., Timm, C., Oesterlin, P., Helm, M., Schmidt, H.: Hysteresis in the magnetotransport of manganese-doped germanium: Evidence for carrier-mediated ferromagnetism. Phys. Rev. B **81**, 165204 (2010)

12. Ohno, H., Chiba, D., Matsukura, F., Omiya, T., Abe, E., Dietl, T., Ohno, Y., Ohtani, K.: Electric-Field control of ferromagnetism. Nature **408**, 21 (2000)
13. Climente, J.I., Korkusinski, M., Hawrylak, P., Planelles, J.: Voltage control of the magnetic properties of charged semiconductor quantum dots containing magnetic ions. Phys. Rev. B **71**, 125321 (2005)
14. Boukari, H., Kossacki, P., Bertolini, M., Ferrand, D., Cibert, J., Tatarenko, S., Wasiela, A., Gaj, J.A., Dietl, T.: Light and Electric Field Control of Ferromagnetism in Magnetic Quantum Structures. Phys. Rev. B **88**, 207204 (2002)
15. Zubarev, D.N.: Usp. Fiz. Nauk **71**, 71 (1960) [English Translation: Sov. Phs. Usp. **3**, 320 (1960)]
16. Chiba, D., Matsukura, F., Ohno, H.: Electric-field control of ferromagnetism in $(Ga,Mn)As$. Appl. Phys. Lett. **89**, 162505 (2006)

16

First principle investigations on structural, mechanical, spin polarized electronic and magnetic properties of TmZn and TmCd

Satish Chand[1] · R. P. Singh[2] · A. Govindan[3]

Abstract Structural, mechanical, spin polarized electronic and magnetic properties of TmZn and TmCd intermetallic compounds have been studied using the full-potential linearize augmented plane-wave plus local orbital method. The structural and mechanical properties have been studied in terms of lattice parameter (a_0), bulk modulus (B_0) and its first-order pressure derivative (B_0'), elastic constants (C_{ij}), Young's modulus (Y), shear modulus (G) and Poisson's ratio (υ) in equilibrium state which are found to be consistent with available experimental/theoretical values. Spin polarized electronic properties have been investigated in terms of band structure and density of state histograms for spin up and spin down channel. Electronic behavior of TmZn and TmCd shows that studied materials are metallic ferromagnets with high spin polarization in which Tm-f state electrons have dominant character.

Keywords Binary intermetallics · Mechanical properties · Electronic structure · Magnetic properties

Introduction

Binary rare earth intermetallic compounds viz. TmZn and TmCd exhibit regular and systematic changes in the physical properties. Binary rare earth intermetallics offers high ductility and high-temperature mechanical strength due to which they are widely used in automobile and aerospace applications, viz. commercial aircraft turbines, etc. [1–4]. Binary rare earth intermetallics are extensively studied also due to their partially filled 4f shells in rare earth atoms and availability of their single crystals. TmZn and TmCd crystallize in space group Pm3m (number 221). In Pm3m space group (221) cubic crystals, the more electronegative atoms are located at the corners of the unit cell, while the more electropositive atoms are located in the center of the unit cell [5]. The typical crystal unit cell for TmZn and TmCd is shown in Fig. 1. From this figure, it can be seen that Zn (or Cd) sits at the center of the cubic unit cell while the rare earth atoms are at the corners.

Neutron diffraction, temperature dependence of thermoelectric power and magneto-elastic studies has been made on several rare earth intermetallic compounds [6–8]. Recently, some more studies on structural, elastic and electronic behavior have been carried out on many rare earth intermetallics [9–13]. But mechanical, spin polarized electronic and magnetic studies have not been made on TmZn and TmCd. Thus, these compounds have been considered for mechanical, spin polarized electronic and magnetic studies. The knowledge of mechanical, spin polarized electronic and magnetic properties of these compounds will help in further understanding and controlling the material properties which are necessary for promoting the future applications of the compounds.

Computational approach

Density functional calculations have been made to investigate the structural, mechanical, spin polarized electronic and magnetic properties of TmZn and TmCd intermetallic

✉ R. P. Singh
rishisingh79@gmail.com

[1] Department of Physics, Mewar University, Chittorgarh, Rajasthan, India

[2] Department of Physics, S. S. V. Degree College (Affiliated to C. C. S. University, Meerut U. P.), Hapur, India

[3] Department of Physics, M. M. H. College (C. C. S. University-Meerut), Ghaziabad, India

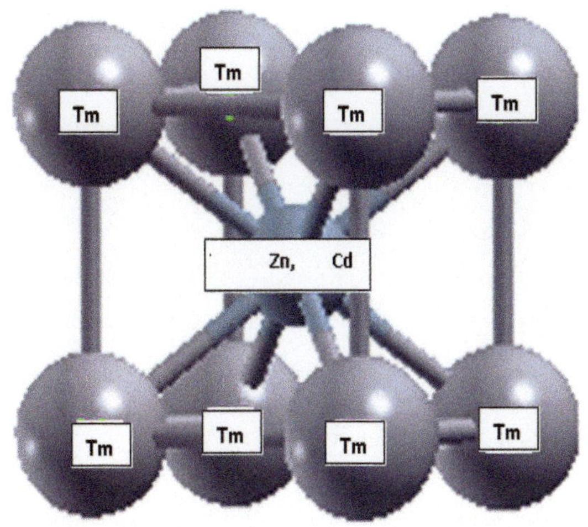

Fig. 1 Typical crystal unit cell for TmZn (or TmCd)

compounds. The calculations have been performed using an accurate full-potential linearize augmented plane-wave plus local orbital method (FP-LAPW + lo), implemented in WIEN2k package [14–18]. In this approach, the unit cell is divided into two distinct regions: (i) first region contain all points inside the muffin-tin spheres around the atom (ii) second region is the interstitial region outside the muffin-tin spheres. The potential and the density in the interstitial region are expanded in terms of plane waves while inside the muffin-tin sphere they are expanded in terms of spherical harmonics. Inside the muffin-tin spheres, the expansion of the non-spherical potential and charge density were carried out up to $l_{max} = 10$. The generalized gradient approximation (GGA) has been used to calculate the total energy, which is based on exchange–correlation energy optimization [15]. The k and E convergences were checked by increasing the number of k points and the energy convergences criteria. In the irreducible part of the Brillouin zone, 14 × 14 × 14 k points were used to calculate the total and partial density of states. The lattice parameters were taken to be 3.516 Å for TmZn and 3.665 Å for TmCd [19] for the calculations. The radius of muffin-tin spheres was chosen to be 2.5 a.u. for Tm and 2.4 a.u. for Zn and Cd.

Elastic constants provide the detailed information about mechanical behavior of the materials. The elastic constants and hence other mechanical parameters viz. bulk modulus, shear modulus, Young's modulus, and Poisson's ratio can be determined by calculating the total energy of the crystal system [20, 21]:

Total energy of a crystal in strained state is given by:

$$E_{\text{total}} = E^0_{\text{total}} + P(V - V_0) + \phi_{\text{elastic}}, \tag{1}$$

where E^0_{total} is the total energy of unstrained crystal, V_0 is the volume of the crystal in initial state, V is the volume in strained state, P and ϕ_{elastic} are the pressure and elastic energy, respectively, defined by:

$$P = -\left(\frac{\partial E^0_{\text{total}}}{\partial V}\right) \tag{2}$$

$$\phi_{\text{elastic}} = \frac{V}{2}.C_{ijkl}\varepsilon_{ik}\varepsilon_{jl} \tag{3}$$

Here, (i, j, k, l = 1, 2, 3)

In Voigt two suffix notations, ϕ_{elastic} is given by:

$$\phi_{\text{elastic}} = \frac{V}{2}.C_{ij}\varepsilon_i\varepsilon_j, \tag{4}$$

where C_{ij} are elastic moduli derived from second-order derivative of E_{total} w.r.t. finite strain is given by:

$$C_{ij} = \left(\frac{1}{V_0}\frac{\partial^2 E_{total}}{\partial\varepsilon_i\partial\varepsilon_j}\right)_{\varepsilon=0}. \tag{5}$$

A cubic crystal has only three independent elastic constants, namely, C_{11}, C_{12} and C_{44}. As a result, a set of three equations is needed to determine all the constants. Hence, three types of strain must be applied to the starting crystal.

The first type strain involves calculating the bulk modulus, given by the formula

$$B_0 = (C_{11} + 2C_{12})/3 \tag{6}$$

Second type strain involves performing volume conservative tetragonal strain given by the following tensor:

$$\begin{bmatrix} \varepsilon & 0 & 0 \\ 0 & \varepsilon & 0 \\ 0 & 0 & \frac{1}{(1+\varepsilon)^2} - 1 \end{bmatrix}. \tag{7}$$

This strain has an effect on the total energy from its unstrained state given by following equation:

$$E(\varepsilon) = E(0) + 3(C_{11} - C_{12})V_0\varepsilon^2 + o(\varepsilon^3) \tag{8}$$

Finally, for the third type of deformation, we use the volume-conserving rhombohedral strain tensor given by

$$\frac{\varepsilon}{3}\begin{bmatrix} 1 & 1 & 1 \\ 1 & 1 & 1 \\ 1 & 1 & 1 \end{bmatrix} \tag{9}$$

This volume-conserving rhombohedral strain transforms the total energy to

$$E(\varepsilon) = E(0) + \frac{1}{6}(C_{11} + 2C_{12} + 4C_{12})V_0\varepsilon^2 + o(\varepsilon^3). \tag{10}$$

Here, O (ε^3) indicates that the neglected terms in the polynomial expansion are cubic and higher power of the ε.

Using the three elastic constants C_{11}, C_{12} and C_{44}, other mechanical parameters viz. bulk modulus (B) Young's modulus (Y) isotropic shear modulus (G) and Poisson ratio,

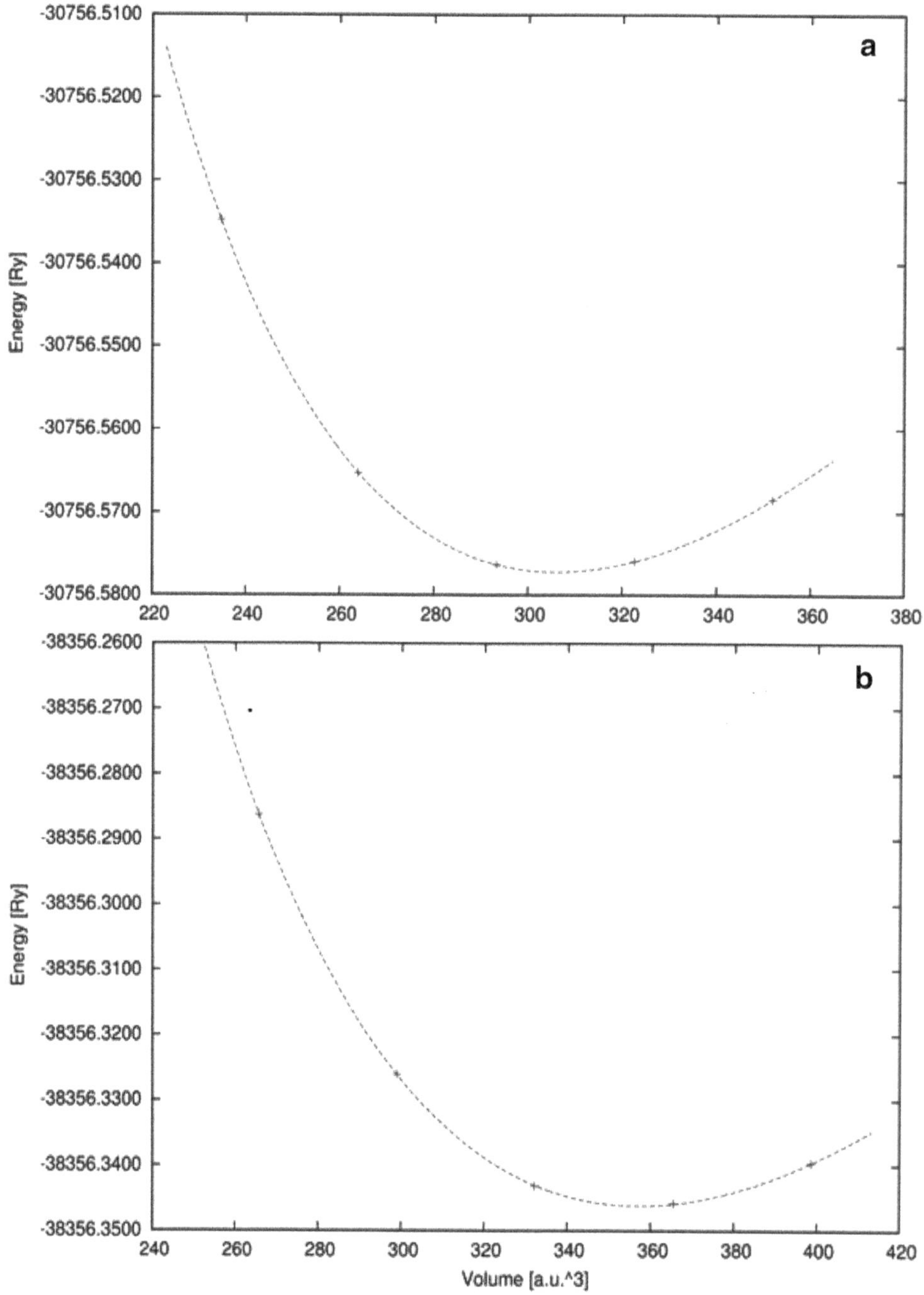

Fig. 2 Total energy as a function of unit cell volume for **a** TmZn **b** TmCd with GGA approximation

Table 1 Lattice constant, a_0 (Å), bulk modulus, B_0 (GPa), pressure derivative of bulk modulus, B'_0 (GPa) equilibrium condition (at 0 K) for TmZn and TmCd using GGA

Compounds	a_0	B_0	B'_0
TmZn	3.49	81.23	4.01
Expt.	3.51[a]	–	–
TmCd	3.63	54.33	3.54
Expt.	3.66[a]	–	–

[a] Buschow et al. [19]

(υ) can be calculated to show the mechanical behavior using the following explicit expressions:

$$G = (G_V + G_R)/2. \quad (11)$$

Here G_V is Voigt's shear modulus corresponding to the upper bound of G values, and G_R is Reuss's shear modulus for cubic crystals corresponding to the lower bound values, expressed as:

$$G_V = (C_{11} - C_{12} + 2C_{44})/4 \quad (12)$$

$$G_R = 5[(C_{11} - C_{12})C_{44})]/[4C_{44} + 3(C_{11} - C_{12})]. \quad (13)$$

Table 2 Calculated elastic moduli, C_{ij} (GPa), bulk modulus, B (GPa), Young's modulus, Y (GPa), shear modulus, G (GPa) and Poisson ratio, υ in equilibrium condition (at 0 K) for TmZn and TmCd using GGA

Compounds	C_{11}	C_{12}	C_{44}	C_{11}–C_{12}	B	Y	G	B/G	ν	A
TmZn (at 0 K)	117.08	63.37	56.10	53.44	81.45	106.71	41.63	1.95	0.28	2.10
Expt. (at 300 K)	103.03[a,c]	55.23[a,b]	52.20[a,b]	47.80[a,b]	70.70[a,b]	97.00	38.10	1.86	0.27	2.17
TmCd (at 0 K)	99.11	64.42	43.90	33.60	75.65	79.26	29.90	2.52	0.32	2.61
Expt. (at 300 K)	–	–	–	31.37[a]	–	–	–	–	–	–

[a] Wohlfarth et al. [25], [b] Morin et al. [8]

And the expressions for the bulk modulus, Young's modulus and Poisson's ratio are given by:

$$B = \frac{C_{11} + 2C_{12}}{3} \tag{14}$$

$$Y = \frac{9GB}{G + 3B} \tag{15}$$

$$\upsilon = (3B - 2G)/(6B + 2G). \tag{16}$$

Elastic anisotropic factor (A) is an important parameter which is the measure of anisotropy of elastic wave velocity in a crystal. For an isotropic material, anisotropic factor (A) is unity and expressed as [22]:

$$A = \frac{2C_{44}}{C_{11} - C_{12}}. \tag{17}$$

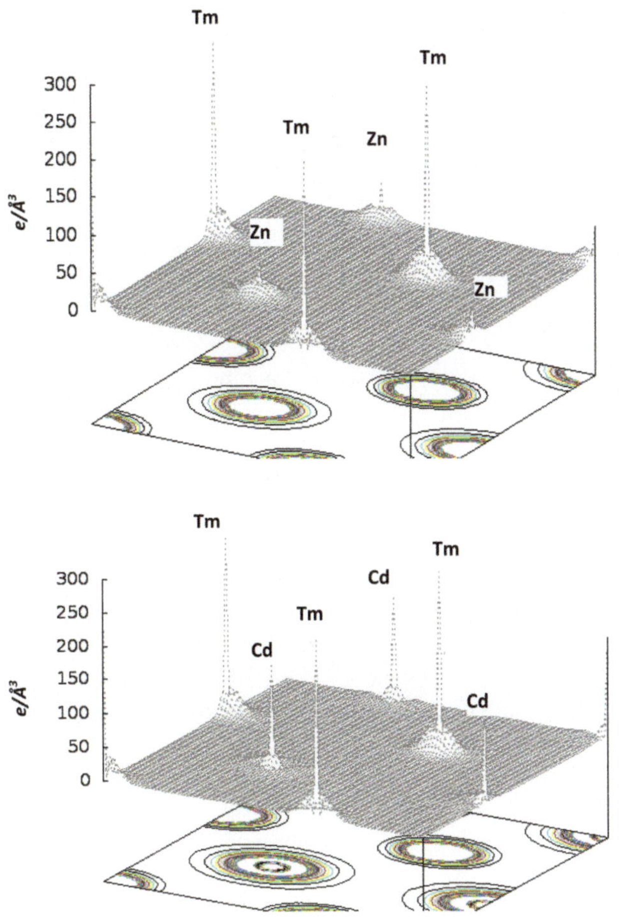

Fig. 3 3d-contour plots of the total valence charge density in (100) plane for **a** TmZn, **b** TmCd

Results and discussion

Structural and mechanical properties

The theoretical lattice parameters, bulk modulus and its first-order pressure derivative have been obtained through fitting the total energy data with the Murnaghan equation of state [23, 24] given by:

$$E_{\text{total}} = E_0(V) + \frac{B_0 V_0}{B_0'(B_0' - 1)}\left[B_0\left(1 - \frac{V_0}{V}\right) + \left(\frac{V_0}{V}\right)^{B_0'} - 1\right], \tag{18}$$

where $E_0(V)$ and V_0 are the energy and volume at equilibrium. B_0 and B_0' are the equilibrium bulk modulus and its first-order pressure derivative.

The energy versus volume curves (obtained using optimization method) for TmZn and TmCd are shown in Fig. 2a, b. It is clear from these figures that equilibrium volume increases while equilibrium energy decreases from TmZn → TmCd. Its reason is increasing the molecular weight from TmZn → TmCd. Because molecular stability increases with increasing the molecular weight due to which molecules vibrate in the least volume of the unit cell for occupying the equilibrium state. Hence, equilibrium unit cell volume increases and equilibrium energy decreases from TmZn → TmCd.

The calculated lattice parameter (a_0), bulk modulus (B_0) and its first-order pressure derivative (B_0') are shown in Table 1. Calculated values of lattice parameter show good agreement with experimental values [8, 25]. Calculated values of elastic constants (C_{11}, C_{12}, C_{44} and C_{11}–C_{12}) are shown in Table 2. To best our knowledge, experimental values of elastic constants for TmZn are given in the

Fig. 4 **a** Spin polarized electron dispersion curves along high symmetry directions in the Brillouin zone for **a** TmZn_up, **b** TmCd_dn

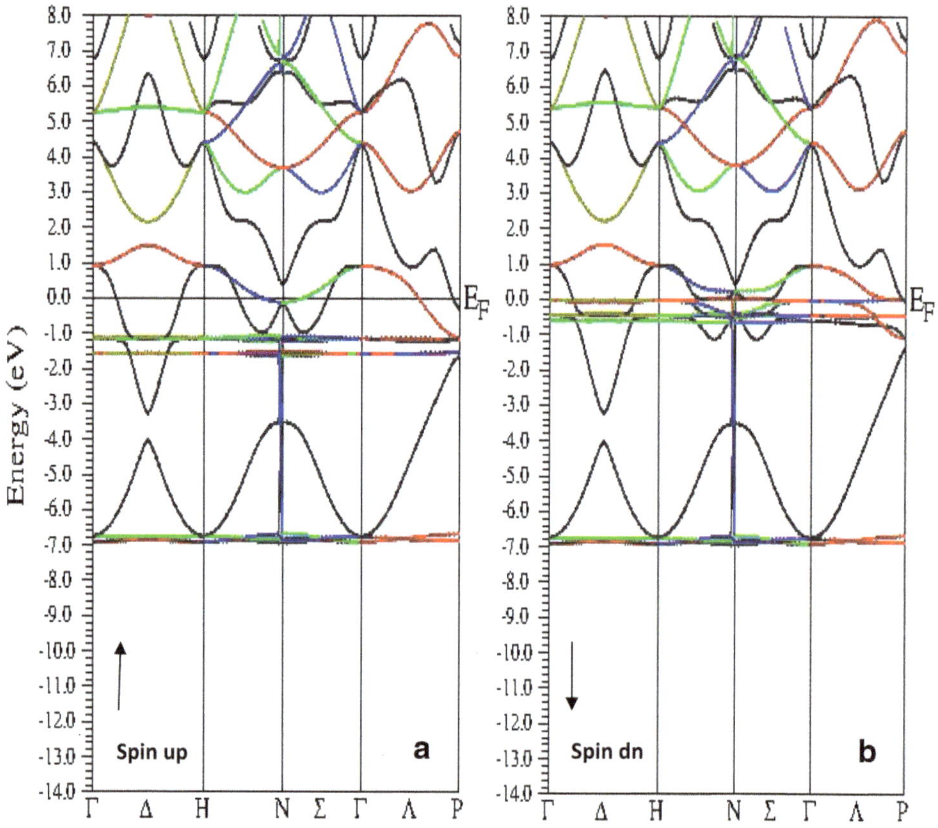

literatures which show good agreement with available experimental values. Our C_{11}, C_{12}, C_{44} and C_{11}–C_{12} values are found to be underestimated by about ≈12 %, ≈13 %, ≈07 % and ≈11 %, respectively, for TmZn compared to the experimental value of Wohalfarath et al. [25]. Furthermore, calculated values of elastic constants (C_{ij}) satisfy all mechanical stability conditions (C_{11}–$C_{12} > 0$, $C_{11} > 0$, $C_{44} > 0$, $C_{11} + 2C_{12} > 0$) and cubic stability condition ($C_{12} < B < C_{11}$) [26] that lead to the validity of calculated elastic constants.

The bulk modulus (B) is a material property indicating the degree of resistance of a material to compression. Larger the bulk modulus, greater is the degree of resistance. Shear modulus, G represents the resistance to plastic deformation. The calculated values of B and G are shown in Table 2. It can be observed from Table 2 that $B_{TmZn} > B_{TmCd}$ which indicates that degree of resistance of TmZn is larger than TmCd. The ratio B/G is a parameter which indicates the brittle or ductile behavior of the material proposed by Pugh [27].The critical value of B/G is 1.75. If $B/G > 1.75$, then material indicates ductile behavior. Otherwise, the material behavior is brittle in nature. In our case, B/G value for both the compounds TmZn and TmCd is greater than 1.75 (Table 2). Therefore, both the compounds are ductile and TmCd is found to be more ductile compared to TmZn. Poisson's ratio (υ) also show ductile/brittle behavior of a material. If $\upsilon > 0.27$, compounds are ductile otherwise brittle in nature. In our case, $\upsilon \approx 0.28$ for TmZn and $\upsilon \approx 0.32$ for TmCd. Therefore, values of υ for TmZn and TmCd also justify that both the compounds are ductile in nature. Furthermore, Poisson's ratio also indicates the compressibility of a material. As $\upsilon \rightarrow 1/2$, material tends to incompressibility [28] and at $\upsilon = 1/2$ the material is nearly incompressible. In our case, value of Poisson's ratio is $\upsilon \approx 0.28$ for TmZn and $\upsilon \approx 0.32$ for TmCd which show that both the compounds are compressible. The stiffness of a material is explained using Young's modulus (Y) [29]. Higher value of Young's modulus indicates larger stiffness. In our case, $Y_{TmZn} > Y_{TmCd}$, indicating TmZn is stiffer than TmCd. Another important parameter is elastic anisotropic factor (A) which is the measure of anisotropy of elastic wave in a crystal. For an isotropic material, $A = 1$. In our case, $A > 1$ indicating that both the compounds are anisotropic [22].

Spin polarized electronic and magnetic properties

Electronic charge density in terms of 3d-contour plots has been calculated to know the nature of chemical bonding in TmZn and TmCd which are shown in Fig. 3a, b. From Fig. 3a, b, it can be seen that there is no bonding charge to link the Tm and Zn (or Cd) and charge density distribution

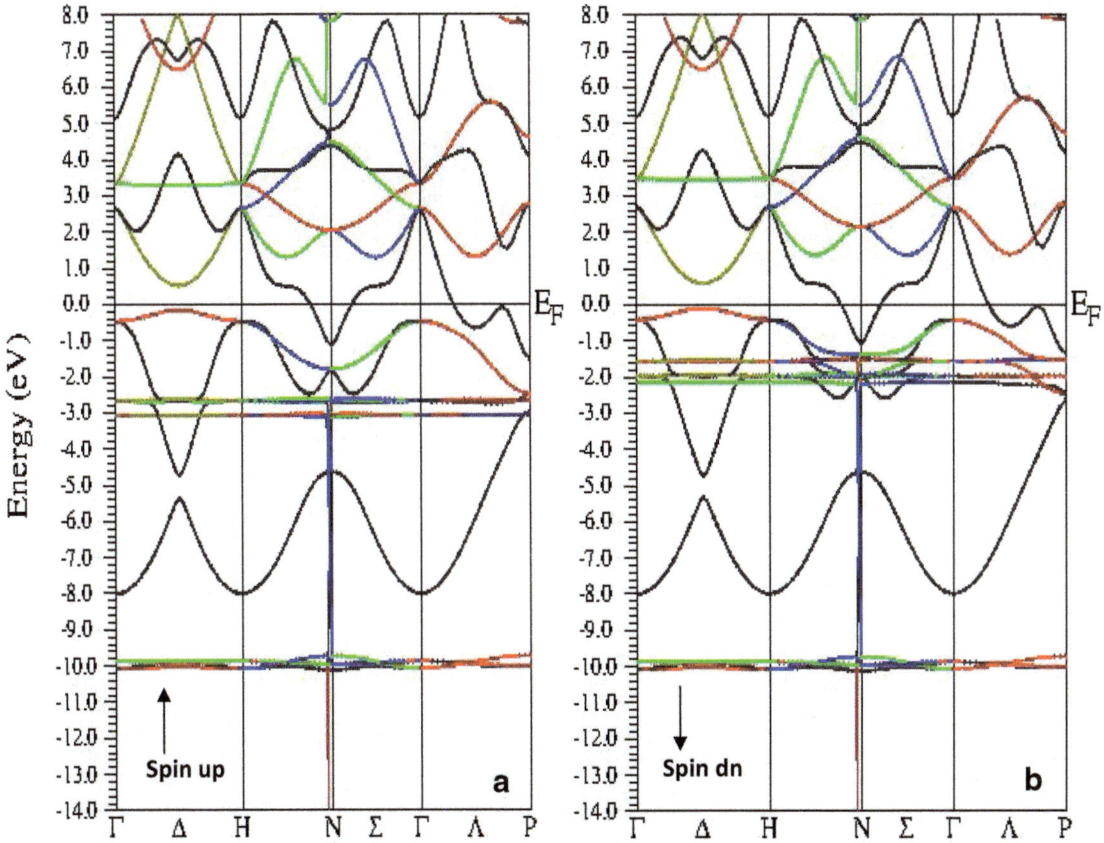

Fig. 5 **a** Spin polarized electron dispersion curves along high symmetry directions in the Brillouin zone for **a** TmZn_up, **b** TmCd_dn

is spherically symmetric around each atom which indicates that atoms have ionic character.

Electronic behavior of TmZn and TmCd has been shown in terms of energy bands and total, partial density of states. The calculated band structures along the high symmetry directions Γ, ΔH, N, $\sum$, Λ and P in the Brillouin zone for spin up and spin down channel have been shown in Figs. 4a, 3b for TmZn and Fig. 5a, b for TmCd. Zero value of energy gap in the both spin up and spin down band structures indicates the metallic character, whereas zero energy gap in one band structure (either spin up or spin down) is the consequence of half metallic character of the material [30]. It can be seen from these figures that most of the valence bands lie from −7.0 to 0.0 eV for TmZn and −9.0 to 0.0 eV for TmCd (where Fermi level, E_F is considered at the origin). In both the cases (spin up and spin down configuration) and for both the compounds TmZn and TmCd, the conduction bands are crossing the Fermi level (E_F) and overlapping significantly at the Fermi level (E_F) (i.e., zero energy gap), indicating typical conducting behavior. This behavior indicates that TmZn and TmCd exhibit the metallic character.

Figures 6a, f and 7a–f show the total density of states (TDOS) and partial density of states (PDOS) plots for TmZn and TmCd, respectively. In case of TmZn, it is observed that there are three peaks in the majority spin channel below the Fermi level at around −7.0, −1.6 and −1.1 eV. The peak at around −7.0 eV is due to mainly Zn-d states (see Fig. 6e). The sharp peaks at −1.6 and −1.1 eV are due mainly due to hybridization of Zn-p and Tm-f states of majority channel (see Fig. 6d, f). For spin down, three peaks were also observed in the minority spin channel below the Fermi level at around −7.0, −0.5 eV and third one peak is just near the Fermi level. The DOS at around −7.0 eV is due to Zn-d states. DOS at around −0.5 eV is due to hybridization of Zn-p and Tm-f (see Fig. 6d, f) states and DOS near the Fermi level are due to Tm-f states.

Similarly, in case of TmCd, three peaks were observed in the majority spin channel below the Fermi level at around −9.0, −1.5 and −1.0 eV. The peak at around −9.0 eV is due to Cd-d states (see Fig. 7e). The sharp peaks at around −1.5 and −1.0 eV are mainly due to hybridization of Cd-p and Tm-f states (see Fig. 7d, f). Three peaks were also observed in the minority spin channel below the Fermi level at around −9.0, 0.8 eV and third one peak is just near the Fermi level. The sharp peak at around −9.0 eV is the result of Cd-d states while the

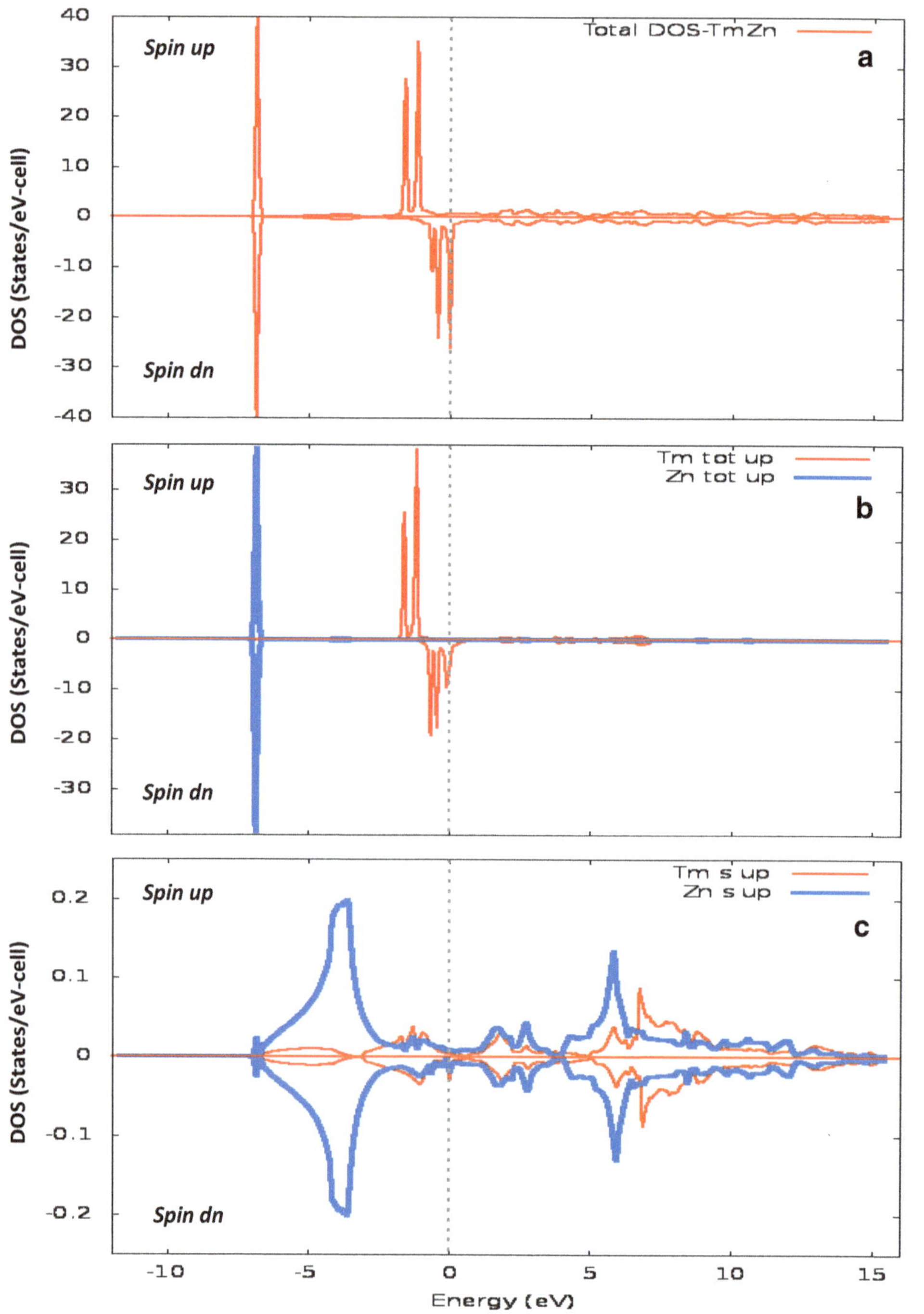

Fig. 6 Calculated total density of states for **a** TmZn; **b** Tm and Zn; **c** partial density of states for Tm-s, Zn-s orbital; **d** partial density of states for Tm-p, Zn-p orbital; **e** partial density of states for Tm-d, Zn-d orbital; **f** partial density of states for Tm-f orbital

peak at around −0.8 eV is due to hybridization of Cd-p and Tm-f states (see Fig. 7d, f). The sharp peak near the Fermi level is dominated by Tm-f states. It is evident from Figs. 6a–f) and 7a–f that minority spin channel of Tm-f states are dominant near the Fermi level which is responsible for the magnetic moment. A little DOS of Zn-p (or Cd-p) are found to be empty above the Fermi level for the conduction.

The electron spin polarization at the Fermi energy (E_F) for a material can be calculated by the equation [31, 32]:

$$P = \frac{D\uparrow(E_F) - D\downarrow(E_F)}{D\uparrow(E_F) + D\downarrow(E_F)}. \tag{19}$$

Here, $D\uparrow(E_F)$ and $D\downarrow(E_F)$ are the density of states for majority and minority spin channel at the Fermi level, respectively. Here, ↑ and ↓ denote the spin up and spin down. The value of P identifies the type of materials, viz. zero value of P shows paramagnetic and anti-ferromagnetic character of materials even below the magnetic transition temperature. The finite value of P indicates the

Fig. 6 continued

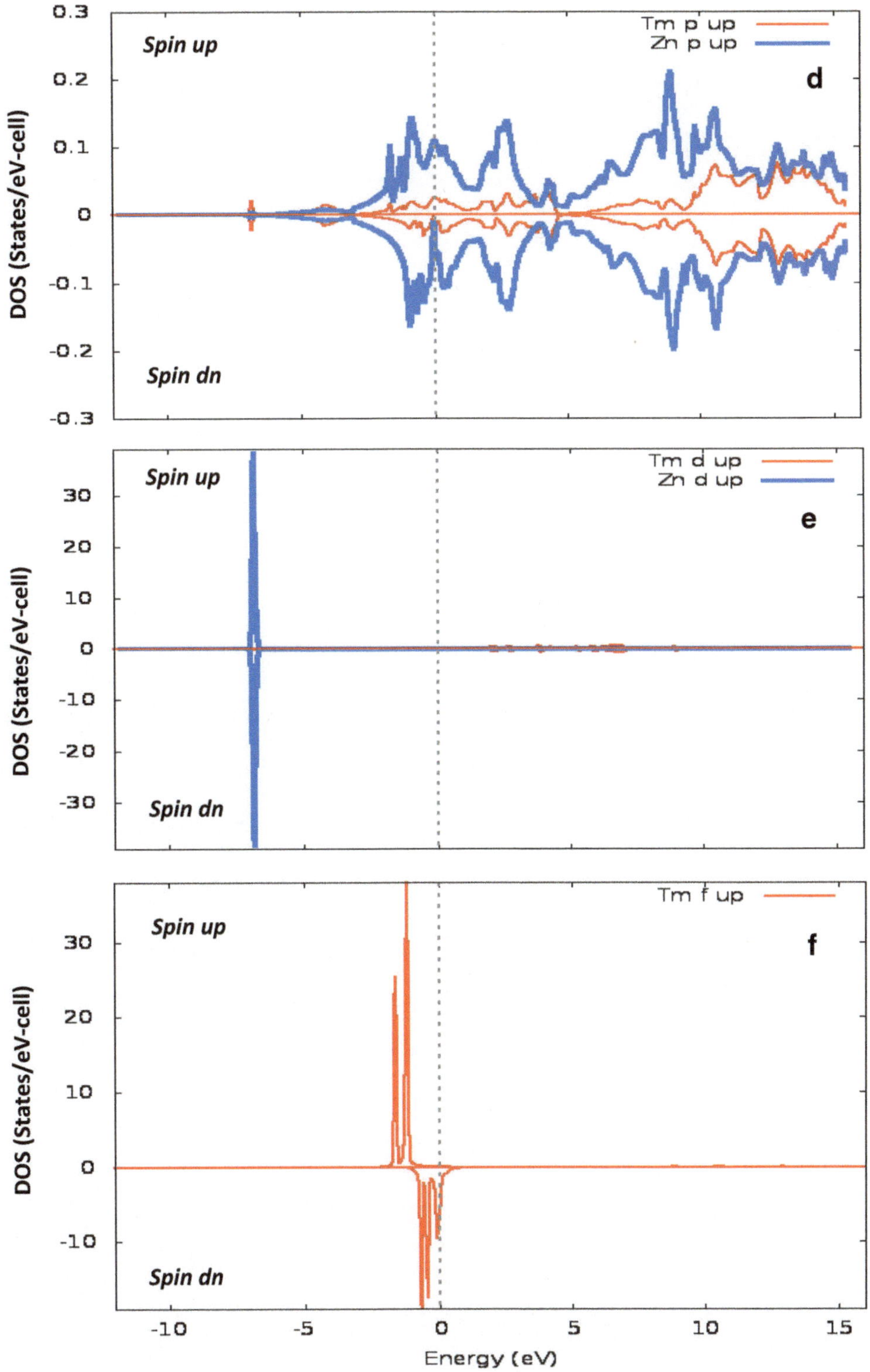

ferromagnetic material below Curie temperature [33]. Furthermore, 100 % spin polarization of electrons at Fermi level is obtained when the value of $D\uparrow(E_F)$ or $D\downarrow(E_F)$ is zero and the material is said to be fully spin polarized.

In our both the cases, neither $D\uparrow(E_F)$ nor $D\downarrow(E_F)$ vanish at the Fermi level (see Fig. 6a for TmZn and Fig. 7a for TmCd) which gives finite value of P, implying both the compounds are typically metallic ferromagnets. The spin polarization is found to be $P\ \approx 92\ \%$ for TmZn and $\approx 93\ \%$ for TmCd. High percentage values of P for both the compounds indicate that spin polarization is high in which mainly 4f-electrons of rare earth atom are

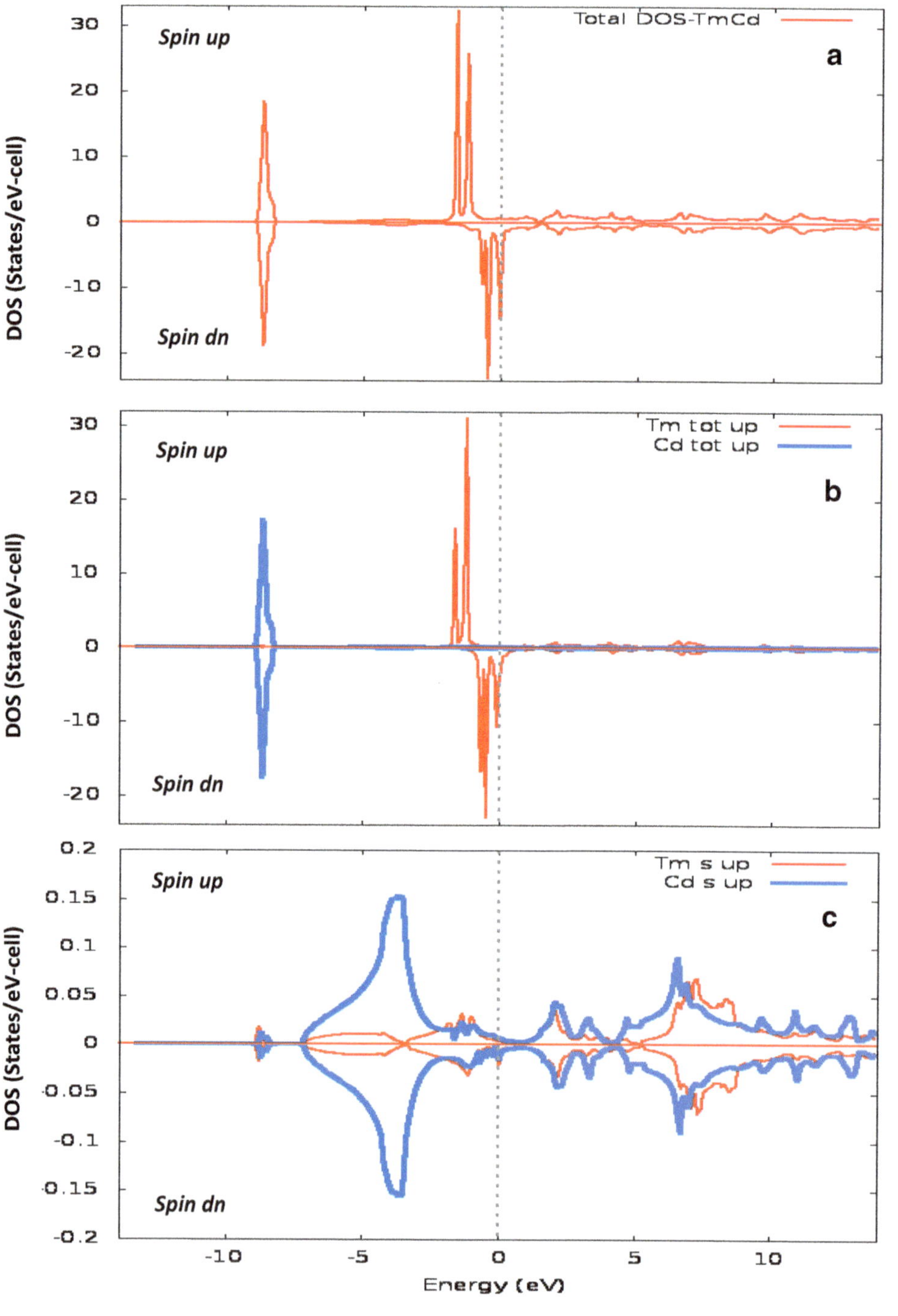

Fig. 7 Calculated total density of states for **a** TmCd; **b** Tm and Cd; **c** partial density of states for Tm-s, Cd-s orbital; **d** partial density of states for Tm-p, Cd-p orbital; **e** partial density of states for Tm-d, Cd-d orbital; **f** partial density of states for Tm-f orbital

contributed mostly (see Fig. 6a, f for TmZn and Fig. 7a, f for TmCd).

In the present research paper, we have performed spin polarized electronic calculations. Therefore, it is necessary to discuss about the magnetic character of these compounds. In magnetic properties, the integer value of magnetic moment is a characteristic feature of half metallic ferromagnets, whereas fractional value of magnetic moment indicates the characteristic feature of metallic ferromagnets [30]. Calculated magnetic moments on individual atoms and molecules are shown in Table 3. The total magnetic moment is found to be 1.2939 μ_B/formula unit for TmZn and 1.2961 μ_B/formula for TmCd. The calculated magnetic moments have been shown up to four digits after decimal. However, it does not mean that the magnetic moment is accurate up to four digits. The accuracy of magnetic moment calculated by WIEN2k package depends on the criteria of convergence and value of RMT which we are taking. These criteria have been used in the present study up to good accuracy. From Table 3, it can be

Fig. 7 continued

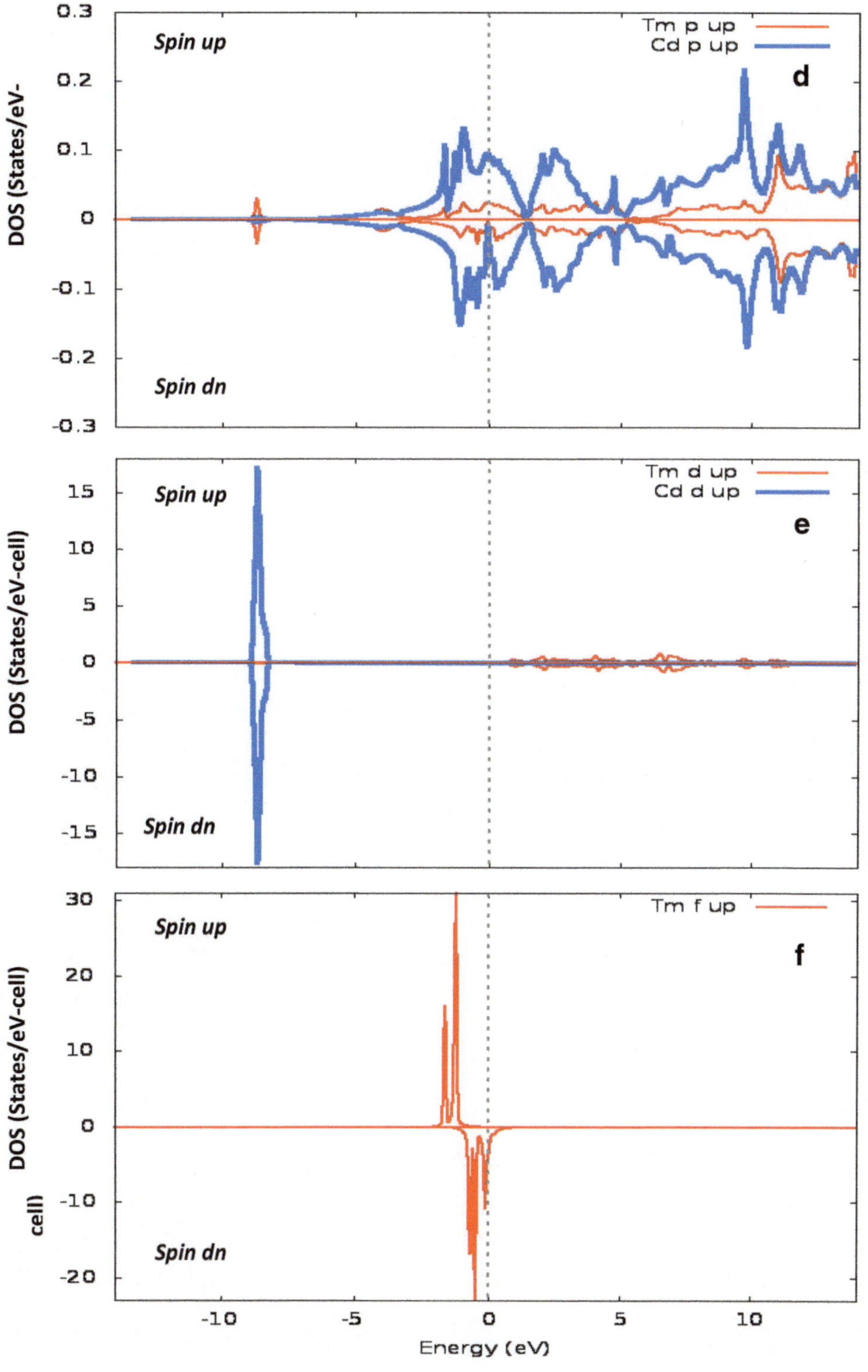

Table 3 Calculated spin magnetic moments (μ_B) of TmZn and TmCd

	Rare earth	Interstitial region	Zn	Cd	Total magnetic moment
TmZn	1.3133	0.0145	−0.00478	–	1.2939
TmCd	1.2996	−0.0014	–	−0.0021	1.2961
TmCu [30]	1.3626	−0.0080	−0.00815	–	1.3465

seen that total magnetic moments have fractional value which favors our results of metallic ferromagnets [30]. The total magnetic moments contain three contributions: one from the rare earth atom (Tm), Zn (or Cd) atoms and the interstitial region. Zn (or Cd) atoms and the interstitial region are found with negligible negative magnetic moments. Thus, it is concluded that net spin magnetic moment is dominated by the 4f-state electrons of the rare earth atom (Tm). The validation of calculated magnetic moment has been made by comparing present values with available theoretical values of magnetic moment for similar type of compounds [34]. Our calculated values match with good accuracy to similar type of compounds. Thus, our calculated values of magnetic moments are justified.

Conclusions

An accurate method named full-potential linearized augmented plane-wave plus local orbitals (FP-LAPW + lo) using generalized gradient approximation (GGA) has been used to study the structural, mechanical, spin polarized electronic and magnetic properties of TmZn and TmCd intermetallic compounds. The calculated lattice parameter and bulk modulus for TmZn and TmCd compounds are consistent with experimental/theoretical values. Calculated values of elastic constants and other mechanical parameters are found to be in good agreement with available theoretical/experimental values and indicate the ductile nature of TmZn and TmCd. 3d-contour plots and band structure histograms indicate the ionic character of TmZn and TmCd which is consequence of metallic character. Spin polarized electronic calculations show that studied compounds are metallic ferromagnets with high spin polarization in which mainly 4f-electrons of Tm are taking part. Magnetic behavior of TmZn and TmCd is found to be consistent with ferromagnetic nature and the total spin magnetic moment is also dominated by 4f-state electrons of the thulium.

References

1. Gschneidner Jr, K.A., Russell, A.M., Pecharsky, A.O., Zhang, Z., Lograsso, T.A., Morris, J.R., Hsu, D.K., Lo, C.H.C., Ye, Y.Y., Slager, A.J., Kesse, D.C.: New family of ductile intermetallic compounds. Nat Mater **2**, 5910 (2003)
2. Gschneidner Jr, K.A., Russell, A.M., Biner, S.B., Lo, C.H.C., Lograsso, T.A., Ye, Y., Barnard, D., Tsokol, A.O.: US Department of Energy, Basic Energy Science (2004)
3. Russell, A.M.: Ductility in intermetallic compounds. Adv. Eng. Mater. **5**, 639 (2003)
4. Russell, A.M., Zhang, Z., Lograsso, T.A., Lo, C.H.C., Pecharsky, A.O., Morris, J.R., Ye, Y.Y., Gschneidner Jr, K.A., Slager, A.J.: Mechanical properties of single crystal YAg. Acta Mater. **52**, 40333 (2004)
5. Gschneidner Jr, K.A., Russell, A.M., Biner, S.B., Harmon, B.N., Hsu, D.K., Lo, C.H.C., Lograsso, T.A., Pecharsky, A.O.: US Department of Energy, Basic Energy Science (2005)
6. Cable, J.W., Koehler, W.C., Wollan, E.O.: Magnetic order in rare-earth intermetallic compounds. Phys. Rev. **136**(1A), A240 (1964)
7. Sousa, J.B., Minto, P.P., Amado, M.M., Moreira, I.M., Braga, M.E., Morin, P.: Temperature dependence and critical behavior of the thermoelectric power in ferromagnetic TbZn compound. J. Phys. F: Met. Phys. **10**, 1809 (1980)
8. Morin, P., Williamson, S.J.: Isotropic magneto-elastic properties in thulium intermetallic compounds. Physical Review B **29**, 1425 (1984)
9. Pasha, S.K., Sundareswari, M., Rajagopalan, M.: Ab initio study of the electronic structure of some B2 intermetallic compounds. Phys. B **348**, 206 (2004)
10. Singh, R.P., Singh, R.K., Rajagopalan, M.: First-principle study on structural, elastic and electronic properties of rare-earth intermetallic compounds: TbCu and TbZn. Intermetallics **19**, 1359 (2011)
11. Shugani, M., Chouhan, S.S., Aynyas, M., Sanyal, S.P.: Ab-initio study of structural, electronic and elastic properties of ErCu. Adv. Phys. Theor. Appl. **19**, 83 (2013)
12. Singh, R.P., Singh, V.K., Singh, R.K., Rajagopalan, M.: Elastic, acoustical and electronic behaviour of the RM (R = Dy, Ho, Er; M = Cu, Zn) compounds. Am. J. Condens Matter Phys. **3**(5), 123 (2013)
13. Pagare, G., Devi, H., Chouhan, S.S.: First principles study of electronic, elastic and thermal properties of B2 -type RECd (RE = La, Ce and Pr) compounds. Int. J. Advance. Elect. Comp. Engg. **2**(8), 248 (2013)
14. Blaha, P., Schwarz, K., Madsen, G.K.H., Kvasnicka, D., Luitz, J.: WIEN2k, An Augmented Plane Wave Plus Local Orbitals Program for Calculating Crystal Properties. University of Technology, Vienna-Austria (2001)
15. Perdew, J.P., Burke, K.: Generalized gradient approximation made simple. Phys. Rev. Lett. **77**, 3865 (1996)
16. Sjostedt, E., Nordstrom, L., Singh, D.J.: An alternative way of linearizing the augmented plane wave method. Sol. Stat. Comm. **114**, 15 (2000)
17. Schwarz, K., Blaha, P., Madsen, G.K.H.: Electronic structure calculations of solids using WIEN2k package for material sciences. Comput. Phys. Comm. **147**, 71 (2002)
18. Monkhorst, H.J., Pack, J.D.: On special points for Brillouin zone integration. Phys. Rev. B **13**, 5188 (1976)
19. Buschow, K.: Rep. Prog. Phys. **42**(8), 1373 (1979). doi:10.1088/0034-4885/42/8/003
20. Charpin, T.: A Package for Calculating Elastic Tensors of Cubic Phase Using WIEN. Laboratory of geometrix F-75252 Paris, France (2001)
21. Shafaay, B.A., Hassan, F.E.H., Korek, M.: First principle investigation of mercury chalcogenides and their HgS_xSe_{1-x} and HgS_xSe_{1-x} ternary alloys. Comput. Mater. Sci. **83**, 107 (2014)
22. Mayer, B., Anton, H., Bott, E., Methfessel, M., Sticht, J., Schmidt, P.C.: Ab-initio calculation of the elastic constants and thermal expansion coefficients of Laves phases. Intermetallics **11**, 23 (2003)
23. Birch, F.: Finite elastic strain of cubic crystals. Phys. Rev. **71**, 809 (1947)
24. Murnaghan, F.D.: The compressibility of media under extreme pressures. Proc. Nat. Acad. Sci. USA **30**, 244 (1944)
25. Wohalfarth, E.P, Buschow, K.H.J.: Handbook of Magnetic Materials 5. Elsevier, North Holland (1990)
26. Yasui, M., Terai, T., Kakeshita, T., Hagiwara, M.: Investigation of a quadrupolar interaction in DyCu by elastic constants measurement. J. Phys. Conf. Ser. **165**, 012059 (2009)

27. Pugh, S.F.: Relations between elastic moduli and plastic properties of polycrystalline pure metals. Philos. Mag. **45**, 823 (1954)
28. Mott, P.H., Dorgan, J.R., Roland, C.M.: The bulk modulus and Poisson's ratio of incompressible materials. J. Sound and Vib. **312**, 572 (2008)
29. Yuan, P.F., Ding, Z.J.: Ab initio calculation of elastic properties of rock-salt and zinc-blend MgS under pressure. Phys. B **403**, 1996 (2008)
30. Gao, G.Y., Yao, K.L., Şaşıoğlu, E., Sandratskii, L.M., Liu, Z.L., Jiang, J.L.: Half-metallic ferromagnetism in zinc-blende CaC, SrC, and BaC from first principles. Phys. Rev. B **75**, 174442 (2007)
31. Soulen Jr, R.J., Byers, J.M., Osofsky, M.S., Nadgorny, B., Ambrose, T., Cheng, S.F., Broussard, P.R., Tanaka, C.T., Nowak, J., Moodera, J.S., Barry, A., Coey, J.M.D: Measuring the spin polarization of a metal with a superconducting point contact. Science **282**, 85 (1998)
32. Rai, D.P., Thapa, R.K.: A density functional theory study of half-metallic ferromagnets (HMFs) behavior in Co2YSb (Y = Sc, Ti). Chin. J. Phy. **51**, 812 (2013)
33. Kandpal, H.C., Fecher, G.H., Felser, C.: Calculated electronic and magnetic properties of the half-metallic transition metal based Heusler compounds. J. Phys. D Appl. Phys. **40**, 1507 (2006)
34. Chand, S., Singh, R.P., Govindan, A., Singh, S.K.: Mechanical, spin polarized electronic and magnetic properties of TmX (X = Cu, Ag): first principle study. Int. J. Mod. Phys. B **29**, 1550007 (2015)

17

Investigation of sensitivity and selectivity of ZnO thin film to volatile organic compounds

F. Teimoori[1] · K. Khojier[1] · N. Z. Dehnavi[1]

Abstract This research addresses a detailed study on the sensitivity and selectivity of ZnO thin film to volatile organic compound (VOC) vapors that can be used for the development of VOC sensors. The ZnO thin film of 100 nm thickness was prepared by post-annealing of e-beam evaporated Zn thin film. The sample was structurally, morphologically, and chemically characterized by X-ray diffraction and field emission scanning electron microscopy analyses. The sensitivity, selectivity, and detection limit of the sample were tested with respect to a wide range of common VOC vapors, including acetone, formaldehyde, acetic acid, formic acid, acetylene, toluene, benzene, ethanol, methanol, and isopropanol in the temperature range of 200–400 °C. The results show that the best sensitivity and detection limit of the sample are related to acetone vapor in the studied temperature range. The ZnO thin film-based acetone sensor also shows a good reproducibility and stability at the operating temperature of 280 °C.

Keywords ZnO · Thin film · Sensitivity · Selectivity · VOCs

✉ K. Khojier
khojier@iauc.ac.ir; k_khojier@yahoo.com

[1] Department of Physics, Faculty of Science, Central Tehran Branch, Islamic Azad University,
P.O. Box 13185-786, Tehran, Iran

Introduction

Volatile organic compound or VOC is the name given to a substance that contains carbon and evaporates at room temperature. VOCs are widely used in household and commercial products such as cleansers, disinfectants, waxes, glues, cosmetics, dry cleaning products, paints, varnishes and preservatives, gasoline, kerosene, and other fuels [1]. The ability of VOCs to cause harmful health effects varies greatly [1, 2]. As with other chemical vapors, the effects of VOC exposure depends on several factors including the type of VOC, the amount of VOC, and the length of time a person is exposed [1, 2]. There are also remarkable correlations between VOC emissions and different kinds of cancers [2, 3]. Although VOCs can be found in both outdoor and indoor settings, the levels of VOCs found indoors can be much higher than those found outdoors (up to 10 times higher), and it is estimated that 50–300 different VOCs may be detected in the air of homes, schools, offices, and commercial buildings at any given time [4]. Therefore, the development of a reliable and low-cost VOC sensor is highly necessary. Compared to conventional techniques, low-dimensional nanostructured ZnO due to the remarkable chemical, physical, and sensing properties [5–7] can be a good candidate to detect the VOCs [3, 8–12]. Many researchers have been reported the study of different volatile vapors sensing properties of nano-structured ZnO but it seems that there is a lack of detailed and more deeply study on VOCs sensing properties of nano-structured ZnO. For example, Al-Hardan et al. [3] produced ZnO thin films using the RF technique and heat treated them at 500 °C for 6 h. They studied the sensitivity of the ZnO thin films to ethanol, isopropanol, and acetone. Muthukrishnan et al. [9] used the sol–gel dip coating method to obtain ZnO thin films and investigated the sensitivity of the sample to acetone, ethanol, and acetaldehyde. Luo et al. [10] employed a hydrothermal method to produce mesoporous ZnO and Sn-doped ZnO thin films and studied the sensitivity of the samples to different volatile vapors including ethanol, methanol, acetone, formaldehyde, and ammonia. Peng et al. [8] also prepared 3D flower-like hierarchical ZnO nanostructure by the hydrothermal method and investigated the sensitivity of the sample to different gases and vapors such as acetone, hydrogen, acetic acid, ethanol, methanol, formaldehyde, and toluene. In contrast to the previous works and to obtain a deep understanding of the VOC sensing properties of nanostructured ZnO, we focused on sensitivity, selectivity and detection limit of the ZnO thin film with respect to a wide range of common VOCs including acetone, formaldehyde, acetic acid, formic acid, acetylene, toluene, benzene, ethanol, methanol, and isopropanol. In addition to these, we studied the reliability of the ZnO thin film-based VOC sensor.

Experimental details

In previous research, we studied the sensitivity of the ZnO thin films to hydrogen and methane gases and optimized the sensitivity of the ZnO thin films with respect to the deposition method [6], post-annealing conditions [7], as well as thickness [13]. Based on the knowledge obtained from our previous works, Zn thin film of 100 nm thickness was deposited on SiO_2/Si substrate by the e-beam evaporation technique and then post-annealed in a horizontal tube furnace at 500 °C with an oxygen flow of 200 standard cubic centimeter per minute (sccm) for 60 min. Details of the substrate preparation, deposition condition, and post-annealing processes are given in Ref. [7].

The sample was structurally, morphologically, and chemically characterized by means of a Philips X-ray diffractometer (X'pert multi-purpose diffractometer (MPD); CuKa radiation) with a step size of 0.02° and step time of 1 s, and field emission scanning electron microscopy (FESEM) (CamScan MV2300, Czech and England). To fabricate the gas-sensing element based on ZnO thin film, a pair of Au electrodes with dimensions of 3×3 mm^2 was deposited on the ZnO thin film by the e-beam evaporation method using a mask made to be used in this application. To measure the response of the sample, it was positioned in a domestic-made chamber and the electrical resistance of the sample was measured in dry air and in the presence of a wide range of VOC vapors including acetone (C_3H_6O), formaldehyde (CH_2O), acetic acid (CH_3COOH), formic acid (CH_2O_2), acetylene (C_2H_2), toluene (C_7H_8), benzene (C_6H_6), ethanol (C_2H_5OH), methanol (CH_3OH), and isopropanol (C_3H_7OH) in the temperature range of 200–400 °C. Schema and setup of the gas sensing measurement system and the method of calculating the concentration and injection volume of vapors are given in [7, 10], respectively. The main parameters which will be used to investigate the sensing properties are also listed below:

- Response (S) is defined as the ratio of R_a/R_g ($S = R_a/R_g$), where R_a and R_g are the resistance of the metal oxide semiconductors in the air and the presence of volatile vapors, respectively [8].
- Response or recovery time was estimated as the time taken for the sensor output to reach 90% of its saturation after applying or switching off the target gas in a step function [8].
- Selectivity of a sensor is defined as the ratio of its response to a certain object to that of other objects as $K = S_A/S_B$. S_A and S_B are the responses of a sensor to a target gas (gas A) and an interfering gas (gas B), respectively [14].
- Detection limit is the lowest amount of the object/vapor that the sensor can detect [6].
- Reproducibility of a sensor is evaluated by the change of sensing behavior after switching numerous times between the 'ON' state and 'OFF' state [14].
- Stability is a parameter to evaluate the feature of a sensor to maintain its properties when it is continuously employed in hostile environments for a long time [12].

Results and discussion

Figure 1 depicts the X-ray diffraction (XRD) pattern of the ZnO thin film prepared in this research. The diffraction lines are matched very well with those given by the JCPDS

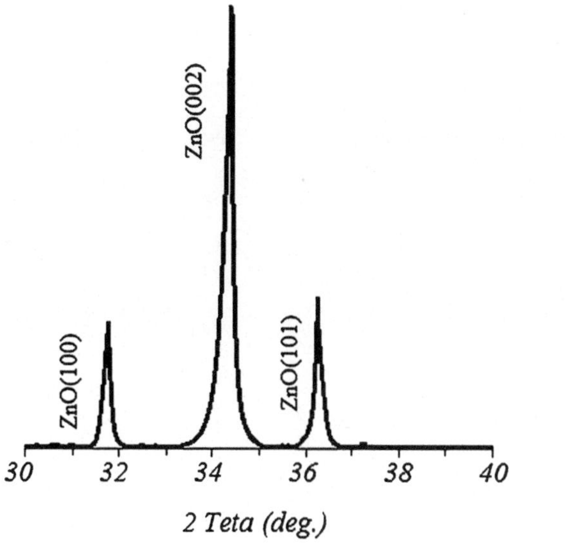

Fig. 1 X-ray diffraction (XRD) pattern of the ZnO thin film prepared in this work

Card no.: 36–1451 attributed to a hexagonal wurtzite structure. It can be distinguished easily that the (002) is a preferred orientation and the produced sample has a large fraction of (002) facets. The chemical composition of the ZnO thin film is obtained by energy dispersive X-ray (EDX) analysis and the EDX spectrum of the sample is shown in Fig. 2. The result shows that the ratio of O to Zn is 1.15 and confirms the ZnO structure. Figure 3 depicts the FESEM micrograph of the ZnO thin film prepared in this work. It can be seen that the sample shows a structure of nanosheets. A detailed study of the dependence of crystallographic structure, chemical composition, and surface morphology on the deposition method and the post-annealing process has been given in our previous research [5–7].

To obtain the sensitivity of the sample produced in this work for the detection of different VOC vapors mentioned in the previous sections, the electrical resistance was measured in air and in the presence of 100 ppm concentration of different VOC vapors in the temperature range of 200–400 °C. The response of the sample to various VOCs at different temperatures is given in Table 1, while a bar chart of this data is shown in Fig. 4. It is noteworthy that the highest response value for various vapors is obtained at different temperatures in the following order: formaldehyde (11) at 360 °C, acetic acid (11) at 320 °C, acetylene (8) at 360 °C, formic acid (7) at 360 °C, toluene (8) at 360 °C, benzene (7) at 360 °C, acetone (30) at 280 °C, ethanol (14) at 320 °C, methanol (10) at 360 °C, and isopropanol (23) at 280 °C. This behavior can be due to the fact that the energy of adsorption, reaction, and desorption of various vapors on the film's surface is different. From this data, it can also be deduced that the most sensitivity is attributed to acetone vapor at different temperatures; however, it becomes maximum at the operating temperature of 280 °C. Figure 5 depicts the selectivity of the ZnO thin film to acetone relative to the other vapors at the operating temperature of 280 °C. It can be seen that the ZnO thin film prepared in this work shows a remarkable selectivity to acetone vapor with respect to other VOCs. To explain these behaviors, we start with reviewing of the VOC vapors sensing mechanism of the ZnO thin film. When ZnO is exposed to air, oxygen molecules are chemisorbed on the surface of the film in various forms such as O_2^-, O^-, or O^{2-} ions by capturing electrons from the conduction band. Transfer of electrons from the conduction band reduces the free carrier density and causes the generation of the electron depletion region at the surface of the film, which in turn increases the resistance of the ZnO film [15]. After that, the ZnO thin film (an n-type metal oxide semiconductor) is exposed to VOC vapors (reducing vapors), the adsorbed vapor reacts with the chemisorbed oxygen anions on the film surface, and then oxygen ions release the trapped electrons back to the conduction band, which result in the decrease of sensor resistance. It is shown in the XRD analysis that the produced sample has a large fraction of (002) facets, and it is well known that all of the Zn atoms at (002) the surface of the hexagonal ZnO are unsaturated coordinated, exhibiting the most dangling bonds. These dangling bonds result in the predominant exposed polar (002) facets that are highly polarized and contain more of the above-mentioned oxygen vacancies [8, 16]. On the other hand, acetone compared with the other VOCs investigated in this work has the largest dipole moment (Table 2) which in turn facilitates the adsorption and the reaction processes with polar (002) facets at lower temperature [8].

To investigate the detection limit, the ZnO thin film was exposed to VOC vapors with different concentrations in the

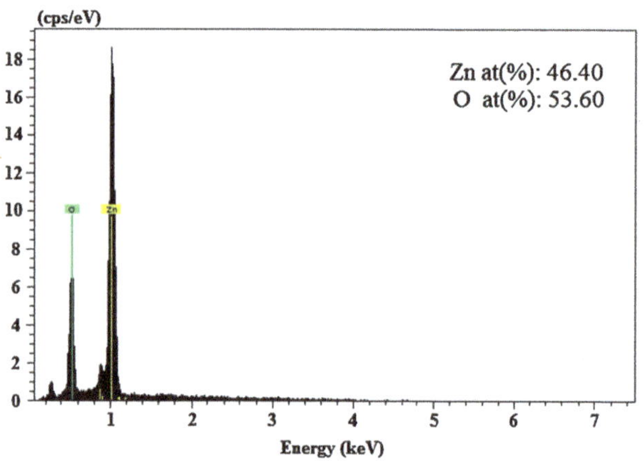

Fig. 2 Energy dispersive X-ray (EDX) spectrum of the ZnO thin film prepared in this work

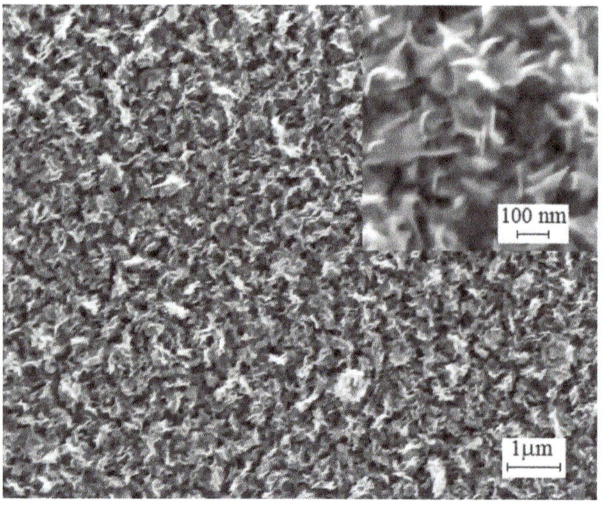

Fig. 3 FESEM micrograph of the ZnO thin film prepared in this work

Table 1 The response of the ZnO thin film prepared in this work to 100 ppm of various VOC vapors at different temperatures

Vapors	Formula	200 °C	240 °C	280 °C	320 °C	360 °C	400 °C
Formaldehyde	CH_2O	–	2	5	8	11	8
Acetic acid	$C_2H_4O_2$	1.5	3	6	11	8	7
Acetylene	C_2H_2	–	–	2	5	8	6
Formic acid	$C_2H_2O_2$	–	–	1.5	4	7	6
Toluene	C_7H_8	–	–	3	5	8	5.5
Benzene	C_6H_6	–	–	2	4	7	5
Acetone	C_3H_6O	11	20	30	22	16	13
Ethanol	C_2H_5OH	2	4	7	14	10	8
Methanol	CH_3OH	–	1.5	4	8	10	8
Isopropanol	C_3H_7OH	7	14	23	17	12	9

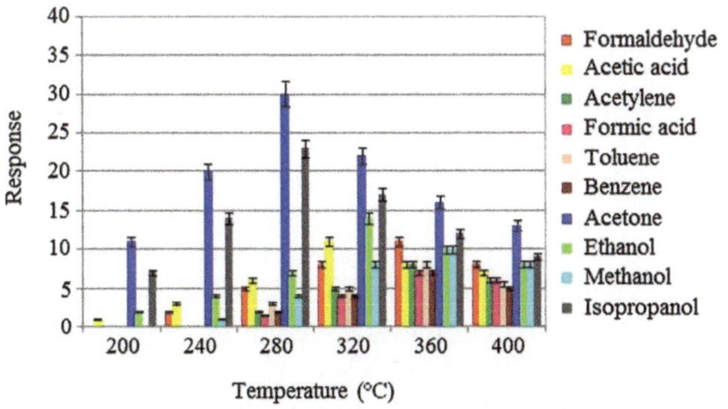

Fig. 4 The electrical response of the ZnO thin film prepared in this work to various VOC vapors at different temperatures

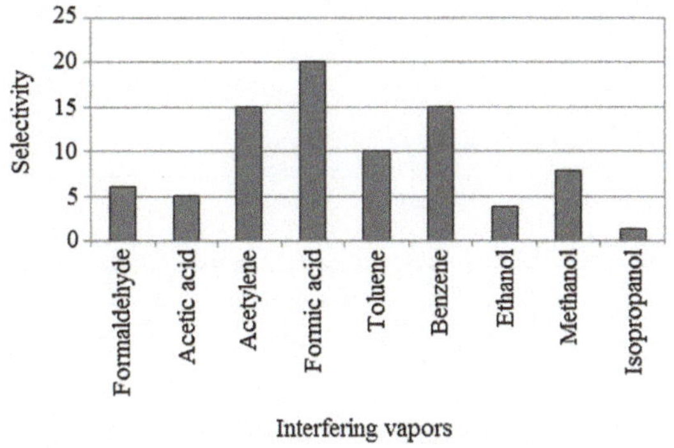

Fig. 5 Selectivity of the ZnO thin film prepared in this work. Target vapor: acetone, operating temperature: 280 °C

range of 5–400 ppm at the attributed operating temperature and the best responses were obtained in the following order: 2.5 for 20 ppm of formaldehyde at 360 °C, 2 for 20 ppm of acetic acid at 320 °C, 4 for 50 ppm of acetylene at 360 °C, 3 for 50 ppm of formic acid at 360 °C, 4 for 50 ppm of toluene at 360 °C, 3 for 50 ppm of benzene at 360 °C, 2 for 10 ppm of acetone at 280 °C, 3 for 20 ppm of ethanol at 320 °C, 2.5 for 20 ppm of methanol at 360 °C, and 5 for 20 ppm of isopropanol at 280 °C. From these investigations it can be concluded that the ZnO thin film-based sensor prepared in this research shows the best

Table 2 The dipole moment of various VOC vapors used in this work [17]

Vapors	Dipole moment
Formaldehyde	2.33
Acetic acid	1.7
Acetylene	0
Formic acid	1.41
Toluene	0.0375
Benzene	0
Acetone	2.88
Ethanol	1.69
Methanol	1.70
Isopropanol	1.58

sensitivity, selectivity, and detection limit to the acetone vapor compared with the other VOC vapors studied in this work. Hence, we focus on the acetone sensing performance of the ZnO thin film-based sensor.

Figure 6 depicts the dynamic variations of the response of the ZnO thin film exposed to acetone vapor with different concentrations (10–300 ppm) at the operating temperature of 280 °C. The response, response time, and recovery time as a function of acetone concentration obtained from the curves in Fig. 6 are shown in Fig. 7. It can be observed that an increase in acetone concentration causes the increase in the response and decrease in the response time. It is quite logical that the increase of acetone concentration provides more vapor atoms to react with the ZnO film surface per unit time and as a result changes in resistance are more and fast. However, it is expected that the sensor becomes saturated when all of the surface atoms react with vapor atoms.

Details of published works on acetone gas sensitivity of the ZnO thin films prepared by various methods [3, 8, 15, 18–21] and the results of this research are given in Table 3. A comparison of the results shows many

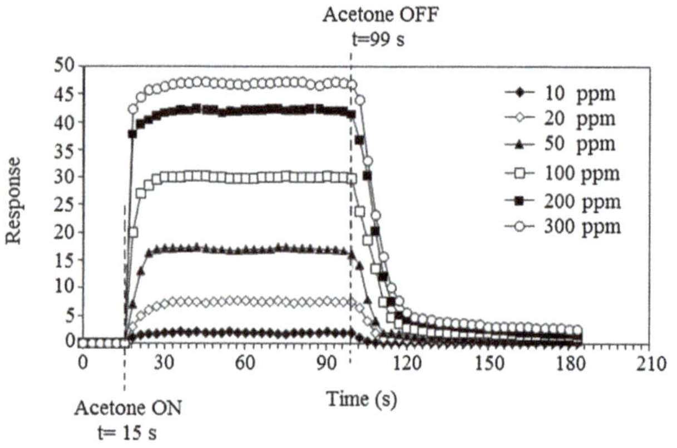

Fig. 6 Dynamic response of the ZnO thin film prepared in this work for different concentrations of acetone vapor at the operating temperature of 280 °C

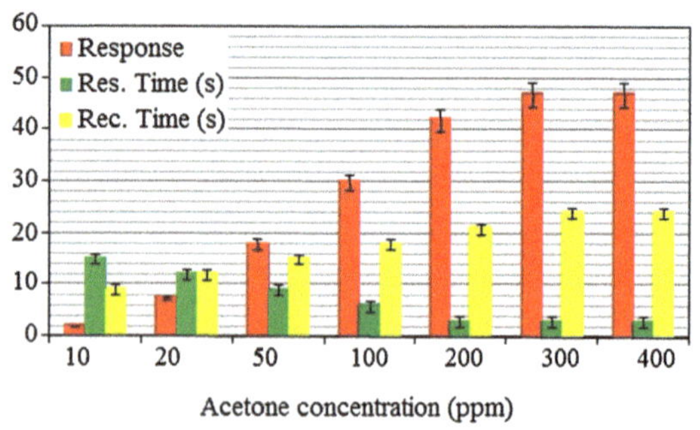

Fig. 7 The response, response time, and recovery time of the ZnO thin film prepared in this work as a function of acetone concentration at the operating temperature of 280 °C

differences between the produced samples by various methods; however, it seems that the sample prepared in this work may be a good candidate as an acetone sensor. The differences between the results listed in Table 3 can be due to the different factors such as the growth or preparation method and conditions, substrate, thickness, and electrode which in turn affect the crystallinity, porosity, surface morphology, as well as sensing mechanism.

In addition to pure ZnO, there are many reports on acetone vapor sensitivity of other metal oxide semiconductors such as WO_3 [22], α-Fe_2O_3 [23], In_2O_3 [24], SnO_2 [25], and Fe_2O_4 [26] as well as doped-ZnO [18, 27, 28], and a summary of their results is listed in Table 4. From the results given in this work and Tables 3 and 4, it may be also deduced that ZnO is suitable to detect acetone vapor compared to the other metal oxide semiconductors.

To study the reproducibility of the ZnO thin film-based acetone sensor, the response of the sample was tested to 100 ppm of acetone at an operating temperature of 280 °C three times continuously. The results of this test are plotted in Fig. 8 and show a good reproducibility of the sample. The stability of the ZnO thin film-based acetone sensor was also investigated. To this end, the response of the sample was tested to 100 ppm of acetone at an operating temperature of 280 °C as a function of day in steps of 3 days for 21 days (Fig. 9). It can be seen that the response of the sample is decreased with the time lapse from 30 to 26 (namely, about 87% stability) and then found a steady state. The mentioned reduction of response may be

Table 3 Details of published works on acetone gas sensitivity of pure ZnO thin films prepared by various methods

Preparation method (morphology) (thickness)	Electrode	Substrate	Operating temp. (°C)	Detection limit (ppm)	Response	Res./Rec. time (s)	Refs.
Spray pyrolysis (granular) (250–300 nm)	Silver paste (on film)	Glass	300	100	~46%[a]	–	[18]
				25	~21%[a]		
Hydrothermal (flower like)	Au	Ceramic tube	300	100	18.6[b]	7/70	[8]
				50	~4[b]	–	
Hydrothermal (nanorod)	Pt (on substrate)	SiO_2/Si	300	100	30.4[b]	5/15	[19]
Hydrothermal (mesoporous hollow microspheres)	Au	Ceramic tube	390	100	7.7[b]	–	[20]
				10	2.6[b]	3/5	
RF sputtering (granular) (250 nm)	Pt/Ti (on substrate)	SiO_2/Si	400	30	10%[a]	–	[3]
Sol–gel spin coating (granular) (80 nm)	Ru (on film)	Alumina	200	100	5.76[b]	30/72	[21]
Electrospinning (nanoparticles)	Au	Ceramic tube	340	100	105[b]	11/18	[15]
				5	36[b]	–	
e-beam and post-annealing (nanosheet) (100 nm)	Au	SiO_2/Si	280	100	30[b]	6/18	This work
				10	2[b]	15/9	

[a] $\frac{R_a - R_g}{R_a} \times 100$

[b] (R_a/R_g), (~): values given in column 6 were obtained from the curves in the literature and may not be accurate

Table 4 Details of published works on acetone gas sensitivity of doped ZnO thin films and other metal oxide semiconductors

Material	Operating temp. (°C)	Concentration (ppm)	Response	Response time (s)	Refs.
WO_3	300	200	~108%[a]	237	[22]
Au-WO_3	300	200	~131%[a]	98	[22]
α-Fe_2O_3	300	100	10[b]	33	[23]
α-Fe_2O_3-Au	260	100	20[b]	26	[23]
In_2O_3	400	25	7[b]	–	[24]
SnO_2	240	50	6.5[b]	–	[25]
Ce-doped Fe_2O_4	200	2000	157%[a]	38	[26]
In-doped ZnO	300	100	96.8[c]	–	[18]
Sn-doped ZnO	300	400	131%[a]	–	[27]
Co-doped ZnO	360	100	16[a]	–	[28]

[a] $(R_a/R_g) \times 100$

[b] (R_a/R_g), (~)

[c] $\frac{R_a - R_g}{R_a} \times 100$: values given in column 5 were obtained from the curves in the literature and may not be accurate

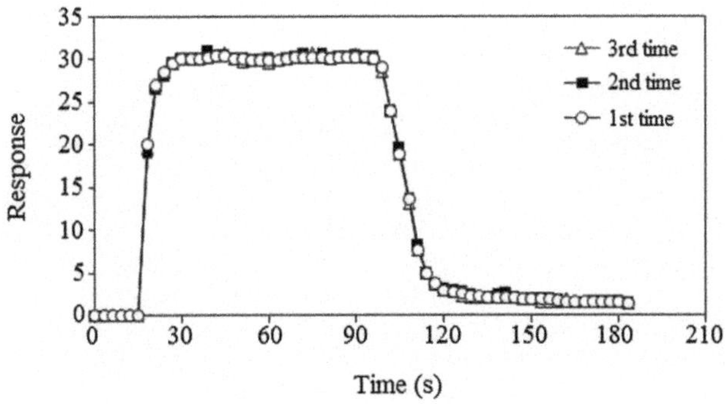

Fig. 8 Dynamic response of the ZnO thin film prepared in this work to acetone with 100 ppm concentration at the operating temperature of 280 °C for three times, a test for reproducibility of the sensor

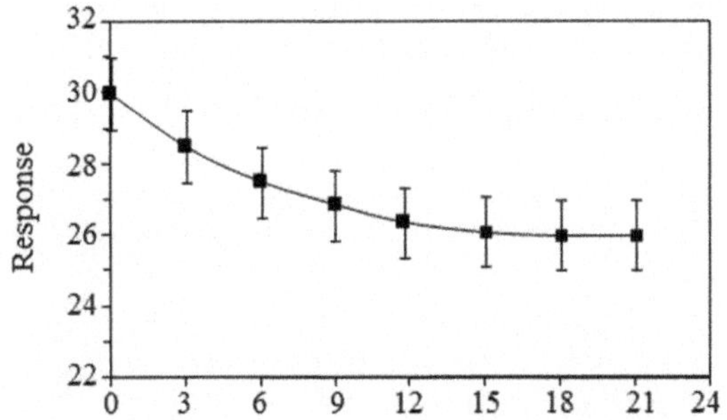

Fig. 9 Electrical response of the ZnO thin film-based acetone sensor prepared in this work as a test of stability of the sensor

attributed to the reaction of the film surface with its surroundings between and during the measurements, especially humidity [14].

It is also worth mentioning that the sensing property measurements in this research were carried out in dry air ,while the sensitivity of the ZnO-based sensor can be affected by relative humidity (RH). The effect of relative humidity on the sensitivity of the ZnO-based gas sensor can be also found in the literature [29–31].

Conclusion

ZnO thin films of 100 nm thickness with a hexagonal wurtzite crystallographic structure and (002) preferred orientation was prepared according to the knowledge obtained from the previous research, namely deposition by the e-beam evaporation method on the SiO_2/Si substrate and then post-annealing at 500 °C with an oxygen flow of 200 sccm for 60 min. The sensitivity and selectivity of the sample were optimized with respect to different VOC vapors and the operating temperature. The detection limit of the sample was also investigated in the range of 5–400 ppm. The sample showed the best sensitivity, selectivity as well as detection limit to the acetone vapor compared with the other vapors in the temperature range of 200–400 °C, whereas the mentioned parameters were optimum at the operating temperature of 280 °C. The ZnO thin film-based acetone sensor also showed a remarkable reproducibility and about 87% stability at the operating temperature of 280 °C.

Acknowledgements This work was carried out with the support of the Islamic Azad University, Central Tehran branch. The authors are grateful to Dr. Arman Toni for partial support of this work.

References

1. Bloemen, H.J.Th., Burn, J.: Chemistry and analysis of volatile organic compounds in the environment. Chapman & Hall (1993)
2. Boeglin, M.L., Wessels, D., Henshel, D.: An investigation of the relationship between air emissions of volatile organic compounds

and the incidence of cancer in Indiana counties. Environ. Res. **100**, 242–254 (2006)
3. Al-Hardan, N.H., Abdullah, M.J., Abdul Aziz, A., Ahmad, H., Low, L.Y.: ZnO thin films for VOC sensing applications. Vac. **85**, 101–106 (2010)
4. Tiwary, A., Colls, J.: Air Pollution: Measurement, Modelling and Mitigation. CRC, Boca Raton (2009)
5. Khojier, K., Savaloni, H., Amani, E.: Influence of annealing conditions on the crystallographic structure, chemical composition and luminescence of ZnO thin films. Appl. Surf. Sci. **289**, 564–570 (2014)
6. Teimoori, F., Khojier, K., Dehnavi, N.Z.: Investigation on the electrical and methane gas-sensing properties of ZnO thin films produced by different methods. J. Elec. Mater. **45**, 4881–4889 (2016)
7. Khojier, K., Savaloni, H.: Improving the H_2 gas sensitivity of ZnO thin films by modifying the annealing conditions. J. Elec. Mater. **44**, 3458–3464 (2015)
8. Peng, C., Guo, J., Yang, W., Shi, C., Liu, M., Zheng, Y., Xu, J., Chen, P., Huang, T., Yang, Y.: Synthesis of three-dimensional flower-like hierarchical ZnO nanostructure and its enhanced acetone gas sensing properties. J. Alloy. Compd. **654**, 371–378 (2016)
9. Muthukrishnan, K., Vanaraja, M., Boomadevi, S., Karn, R.K., Singh, V., Singh, P.K., Pandiyan, K.: Studies on acetone sensing characteristics of ZnO thin film prepared by sol-gel dip coating. J. Alloy. Compd. **673**, 138–143 (2016)
10. Luo, S., Shen, Y., Wu, Z., Cao, M., Gu, F., Wang, L.: Enhanced ethanol sensing performance of mesoporous Sn-doped ZnO. Mater. Sci. Semicond. Process. **41**, 535–543 (2016)
11. Wei, S., Wang, S., Zhang, Y., Zhou, M.: Different morphologies of ZnO and their ethanol sensing property. Sens. Actuators B **192**, 480–487 (2014)
12. Mirzaei, A., Leonardi, S.G., Neri, G.: Detection of hazardous volatile organic compounds (VOCs) by metal oxide nanostructures-based gas sensors: a review. Ceram. Int. **42**, 15119–15141 (2016)
13. Teimoori, F., Khojier, K., Dehnavi, N.Z.: On the dependence of H_2 gas sensitivity of ZnO thin films on film thickness. Procedia Mater. Sci. **11**, 474–479 (2015)
14. Zolghadr, S., Khojier, K., Kimiagar, S.: Study of sensitivity and selectivity of α-Fe_2O_3 thin films for different toxic gases and alcohols. Mater. Sci. Semicond. Process. **54**, 6–13 (2016)
15. Bian, H., Ma, S., Sun, A., Xu, X., Yang, G., Yan, S., Gao, J., Zhang, Z., Zhu, H.: Improvement of acetone gas sensing performance of ZnO nanoparticles. J. Alloy. Compd. **658**, 629–635 (2016)
16. Tian, S., Yang, F., Zeng, D., Xie, C.: Solution-processed gas sensors based on ZnO nanorods array with an exposed (0001) facet for enhanced gas-sensing properties. J. Phys. Chem. C **116**, 10586–10591 (2012)
17. Dean, J.A.: Lange's Handbook of Chemistry. McGraw-Hill, New York (1985)
18. Prajapati, C.S., Sahay, P.P.: Influence of In doping on the structural, optical and acetone sensing properties of ZnO nanoparticulate thin films. Mater. Sci. Semicond. Process. **16**, 200–210 (2013)
19. Zeng, Y., Zhang, T., Yuan, M., Kang, M., Lu, G., Wang, R., Fan, H., He, Y., Yang, H.: Growth and selective acetone detection based on ZnO nanorod arrays. Sens. Actuators B **143**, 93–98 (2009)
20. Li, Y., Li, D.L., Liu, J.C.: Optical and gas sensing properties of mesoporous hollow ZnO microspheres fabricated via a solvothermal method. Chin. Chem. Lett. **26**, 304–308 (2015)
21. Kakati, N., Jee, S.H., Kim, S.H., Oh, J.Y., Yoon, Y.S.: Thickness dependency of sol-gel derived ZnO thin films on gas sensing behaviors. Thin Solid Films **519**, 494–498 (2010)
22. Kim, S., Park, S., Park, S., Lee, C.: Acetone sensing of Au and Pd-decorated WO_3 nanorod sensors. Sens. Actuators B **209**, 180–185 (2015)
23. Kaneti, Y.V., Moriceau, J., Liu, M., Yuan, Y., Zakaria, Q., Jianga, X., Yu, A.: Hydrothermal synthesis of ternary α-Fe_2O_3–ZnO–Au nano composites with high gas-sensing performance. Sens. Actuators B **209**, 889–897 (2015)
24. Vomiero, A., Bianchi, S., Comini, E., Faglia, G., Ferroni, M., Poli, N., Sberveglieri, G.: In_2O_3 nano wires for gas sensors: morphology and sensing characterization. Thin Solid Films **515**, 8356–8359 (2007)
25. Mei, L., Deng, J., Yin, X., Zhang, M., Li, Q., Zhang, E., Xu, Z., Chen, L., Wang, T.: Ultra sensitive ethanol sensor based on 3D aloe-like SnO_2. Sens. Actuators B **166–167**, 7–11 (2012)
26. Khandekar, M.S., Tarwal, N.L., Mulla, I.S., Suryavanshi, S.S.: Nanocrystalline Ce doped $CoFe_2O_4$ as an acetone gas sensor. Ceram. Int. **40**, 447–452 (2014)
27. Li, X., Chang, Y., Long, Y.: Influence of Sn doping on ZnO sensing properties for ethanol and acetone. Mater. Sci. Eng., C **32**, 817–821 (2012)
28. Liu, L., Li, S., Zhuang, J., Wang, L., Zhang, J., Li, H., Li, H., Liu, Z., Han, Y., Jiang, X., Zhang, P.: Improved selective acetone sensing properties of Co-doped ZnO nanofibers by electrospinning. Sens. Actuators B **155**, 782–788 (2011)
29. Santra, S., Guha, P.K., Ali, S.Z., Hiralal, P., Unalan, H.E., Covington, J.A., Amaratunga, G.A.J., Milne, W.I., Gardner, J.W., Udrea, F.: ZnO nanowires grown on SOI CMOS substrate for ethanol sensing. Sens. Actuators B **146**, 559–565 (2010)
30. Santra, S., De Luca, A., Bhaumik, S., Ali, S.Z., Udrea, F., Gardner, J.W., Ray, S.K., Guha, P.K.: Dip pen nanolithography-deposited zinc oxide nanorods on a CMOS MEMS platform for ethanol sensing. RSC Adv. **5**, 47609–47616 (2015)
31. Rakshit, T., Santra, S., Manna, I., Ray, S.K.: Enhanced sensitivity and selectivity of brush-like SnO_2 nanowire/ZnO nanorod heterostructure based sensor for volatile organic compounds. RSC Adv. **4**, 36749–36756 (2014)

18

Patterns of deformations of Peregrine breather of order 3 and 4 solutions to the NLS equation with multi parameters

Pierre Gaillard[1] · Mickaël Gastineau[2]

Abstract In this article, one gives a classification of the solutions to the one dimensional nonlinear focusing Schrödinger equation (NLS) by considering the modulus of the solutions in the (x, t) plan in the cases of orders 3 and 4. For this, we use a representation of solutions to NLS equation as a quotient of two determinants by an exponential depending on t. This formulation gives in the case of the order 3 and 4, solutions with, respectively 4 and 6 parameters. With this method, beside Peregrine breathers, we construct all characteristic patterns for the modulus of solutions, like triangular configurations, ring and others.

Keywords NLS equation · Peregrine breathers · Rogue waves

Introduction

The rogue waves phenomenon currently exceed the strict framework of the study of ocean's waves [1–4] and play a significant role in other fields; in nonlinear optics [5, 6], Bose–Einstein condensate [7], superfluid helium [8], atmosphere [9], plasmas [10], capillary phenomena [11] and even finance [12]. In the following, we consider the one-dimensional focusing nonlinear equation of Schrödinger (NLS) to describe the phenomena of rogue waves. The first results concerning the NLS equation date from the Seventies. Precisely, in 1972 Zakharov and Shabat solved it using the inverse scattering method [13, 14]. The case of periodic and almost periodic algebro-geometric solutions to the focusing NLS equation was first constructed in 1976 by Its and Kotlyarov [15, 16]. The first quasi rational solutions of NLS equation were constructed in 1983 by Peregrine [17]. In 1986 Akhmediev, Eleonskii and Kulagin obtained the two-phase almost periodic solution to the NLS equation and obtained the first higher order analogue of the Peregrine breather [18–20]. Other analogues of the Peregrine breathers of order 3 and 4 were constructed in a series of articles by Akhmediev et al. [21–23] using Darboux transformations. The present paper presents multi-parametric families of quasi rational solutions of NLS of order N in terms of determinants of order $2N$ dependent on $2N - 2$ real parameters. The aim of this paper was to try to distinguish among all the possible configurations obtained by different choices of parameters, one those which have a characteristic in order to try to give a classification of these solutions.

✉ Pierre Gaillard
pierre.gaillard@u-bourgogne.fr

Mickaël Gastineau
gastineau@imcce.fr

[1] Université de Bourgogne, 9 Av. Alain Savary, Dijon, France

[2] IMCCE, Observatoire de Paris, PSL Research University, CNRS, Sorbonne Universités UPMC Univ, Paris 06 Univ. Lille 77 Av. Denfert-Rochereau, 75014 Paris, France

Expression of solu-tions of NLS equation in terms of a ratio of two determinants

We consider the focusing NLS equation

$$iv_t + v_{xx} + 2|v|^2 v = 0. \tag{1}$$

To solve this equation, we need to construct two types of functions $f_{j,k}$ and $g_{j,k}$ depending on many parameters. Because of the length of their expressions, one defines the functions $f_{\nu,\mu}$ and $g_{\nu,\mu}$ of argument A_ν and B_ν only in the appendix. We have already constructed solutions of equation NLS in terms of determinants of order $2N$ which we call solution of order N depending on $2N - 2$ real parameters. It is given in the following result [24–27]:

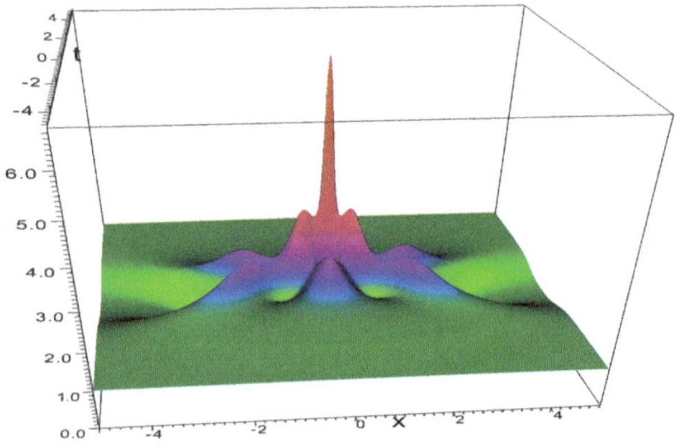

Fig. 1 Solution P_3 to NLS, N=3, $\tilde{a}_1 = \tilde{b}_1 = \tilde{a}_2 = \tilde{b}_2 = 0$

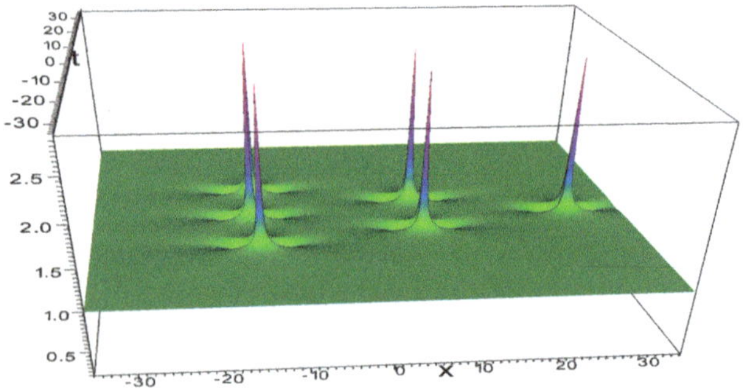

Fig. 2 Solution $1T6$ to NLS, N=3, $\tilde{a}_1 = 10^4$, $\tilde{b}_1 = 0$, $\tilde{a}_2 = 0$, $\tilde{b}_2 = 0$

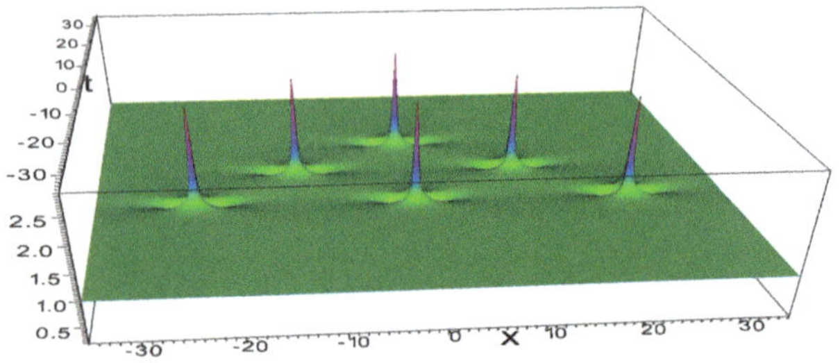

Fig. 3 Solution $1T6$ to NLS, N=3, $\tilde{a}_1 = 0$, $\tilde{b}_1 = 10^4$, $\tilde{a}_2 = 0$, $\tilde{b}_2 = 0$

Theorem 1 *The functions v defined by*

$$v(x,t) = \frac{\det((n_{jk})_{j,k\in[1,2N]})}{\det((d_{jk})_{j,k\in[1,2N]})} e^{2it-i\varphi} \tag{2}$$

are quasi-rational solution of the NLS Eq. (1) *depending on* $2N-2$ *parameters* $\tilde{a}_j$, $\tilde{b}_j$, $1 \leq j \leq N-1$, *where*

$$n_{j1} = f_{j,1}(x,t,0),$$
$$n_{jk} = \frac{\partial^{2k-2} f_{j,1}}{\partial \epsilon^{2k-2}}(x,t,0),$$
$$n_{jN+1} = f_{j,N+1}(x,t,0),$$
$$n_{jN+k} = \frac{\partial^{2k-2} f_{j,N+1}}{\partial \epsilon^{2k-2}}(x,t,0),$$

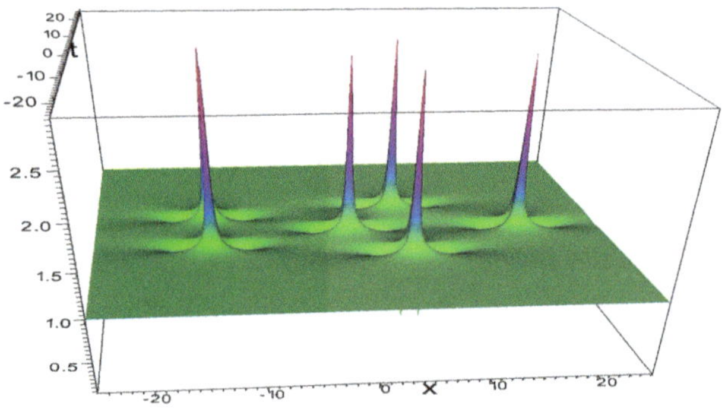

Fig. 4 Solution $1R5+1$ to NLS, N=3, $\tilde{a}_1 = 0$, $\tilde{b}_1 = 0$, $\tilde{a}_2 = 10^6$, $\tilde{b}_2 = 0$

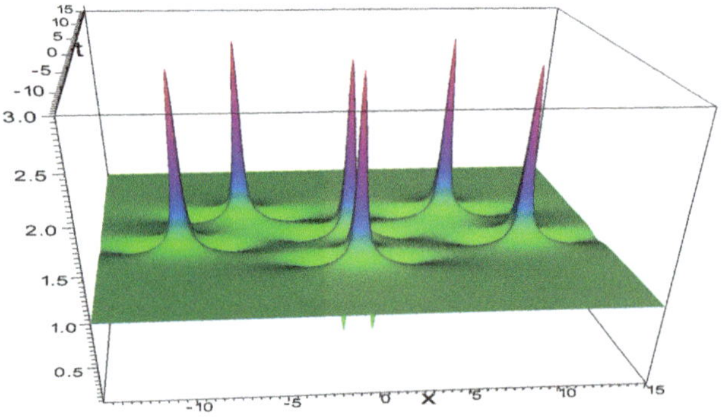

Fig. 5 Solution $1R5+1$ to NLS, N=3, $\tilde{a}_1 = 0$, $\tilde{b}_1 = 0$, $\tilde{a}_2 = 0$, $\tilde{b}_2 = 10^5$

$$d_{j1} = g_{j,1}(x,t,0),$$
$$d_{jk} = \frac{\partial^{2k-2} g_{j,1}}{\partial \epsilon^{2k-2}}(x,t,0),$$
$$d_{jN+1} = g_{j,N+1}(x,t,0),$$
$$d_{jN+k} = \frac{\partial^{2k-2} g_{j,N+1}}{\partial \epsilon^{2k-2}}(x,t,0),$$
$$2 \leq k \leq N,\ 1 \leq j \leq 2N \tag{3}$$

The functions f and g are defined in (9), (10), (11), (12).

Patterns of quasi rational solutions to the NLS equation

The solutions v_N to NLS Eq. (2) of order N depending on $2N-2$ parameters $\tilde{a}_j$, $\tilde{b}_j$ (for $1 \leq j \leq N-1$) have been already explicitly constructed and can be written as

$$v_N(x,t) = \frac{n(x,t)}{d(x,t)} \exp(2it)$$

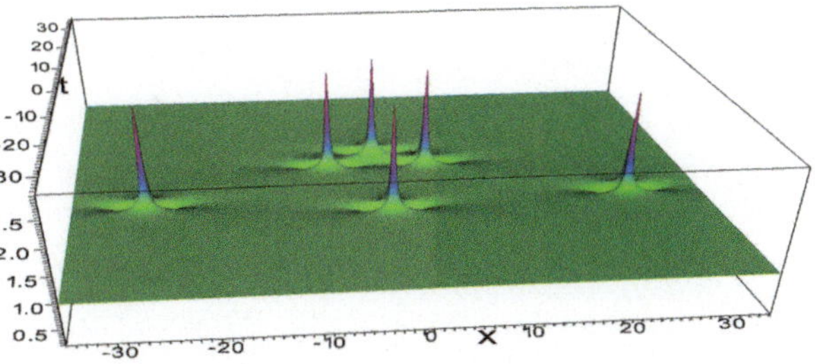

Fig. 6 Solution $1A3 + 1T3$ to NLS, N=3, $\tilde{a}_1 = 0$, $\tilde{b}_1 = 10^4$, $\tilde{a}_2 = 0$, $\tilde{b}_2 = 5 \times 10^6$

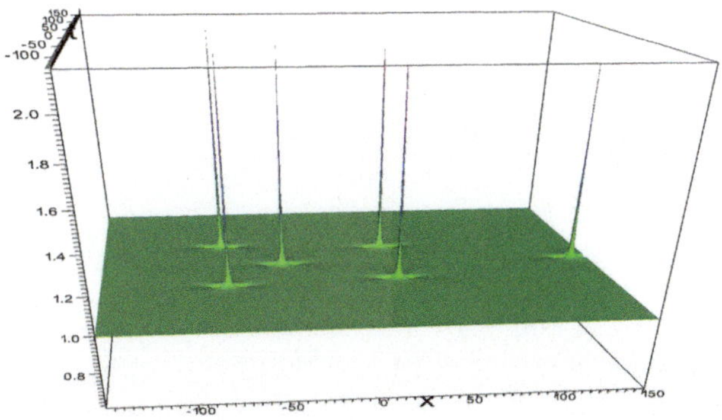

Fig. 7 Solution $1A3 + 1T3$ to NLS, N=3, $\tilde{a}_1 = 10^6$, $\tilde{b}_1 = 0$, $\tilde{a}_2 = 10^{10}$, $\tilde{b}_2 = 0$

$$= \left(1 - \alpha_N \frac{G_N(2x, 4t) + iH_N(2x, 4t)}{Q_N(2x, 4t)}\right) e^{2it}$$

with

$$G_N(X, T) = \sum_{k=0}^{N(N+1)} g_k(T)X^k,$$
$$H_N(X, T) = \sum_{k=0}^{N(N+1} h_k(T)X^k,$$
$$Q_N(X, T) = \sum_{k=0}^{N(N+1} q_k(T)X^k.$$

For order 3 these expressions can be found in [28]; in the case of order 4, they can be found in [29]. In the following, based on these analytic expressions, we give a classification of these solutions by means of patterns of their modulus in the plane (x; t).

Patterns of quasi-rational solutions of order 3 with 4 parameters

P_3 breather

If we choose all parameters equal to 0, $\tilde{a}_1 = \tilde{b}_1 = \ldots = \tilde{a}_{N-1} = \tilde{b}_{N-1} = 0$, we obtain the classical Peregrine breather given by

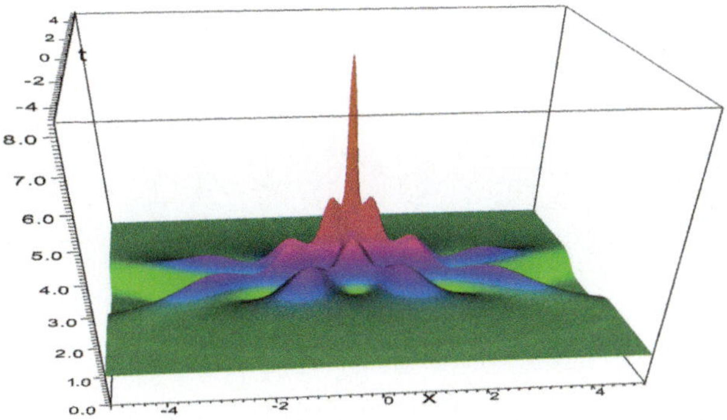

Fig. 8 Solution P_4 to NLS, N=4, $\tilde{a}_1 = \tilde{b}_1 = \tilde{a}_2 = \tilde{b}_2 = \tilde{a}_3 = \tilde{b}_3 = 0$

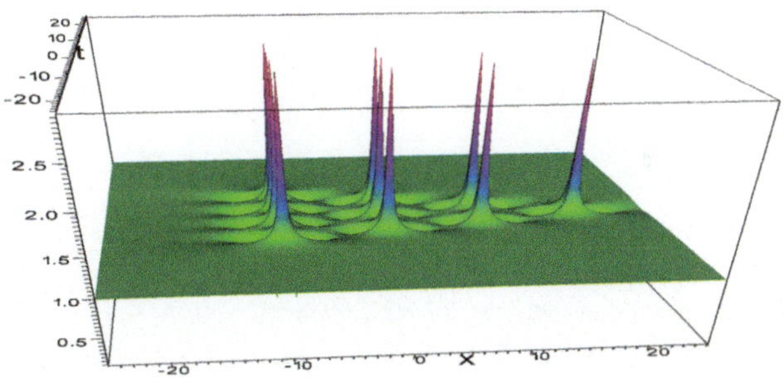

Fig. 9 Solution $1T10$ to NLS, N=4, $\tilde{a}_1 = 10^3$, $\tilde{b}_1 = 0$, $\tilde{a}_2 = 0$, $\tilde{b}_2 = 0$, $\tilde{a}_3 = 0$, $\tilde{b}_3 = 0$

Triangles

To shorten, the following notations are used: for example, the sequence $1A3 + 1T3$ means that the structure has one arc of 3 peaks and one triangle of 3 peaks. If we choose $\tilde{a}_1$ or $\tilde{b}_1$ not equal to 0 and all other parameters equal to 0, we obtain triangular configuration with 6 peaks.

Rings

If we choose $\tilde{a}_2$ or $\tilde{b}_2$ not equal to 0, all other parameters equal to 0, we obtain ring configuration with peaks.

Arcs

If we choose $\tilde{a}_1$ and $\tilde{a}_2$ not equal to 0 and all other parameters equal to 0 (and vice versa, $\tilde{b}_1$ and $\tilde{b}_2$ not equal to 0 and all other parameters equal to 0), we obtain deformed triangular configuration which we can call as arc structure.

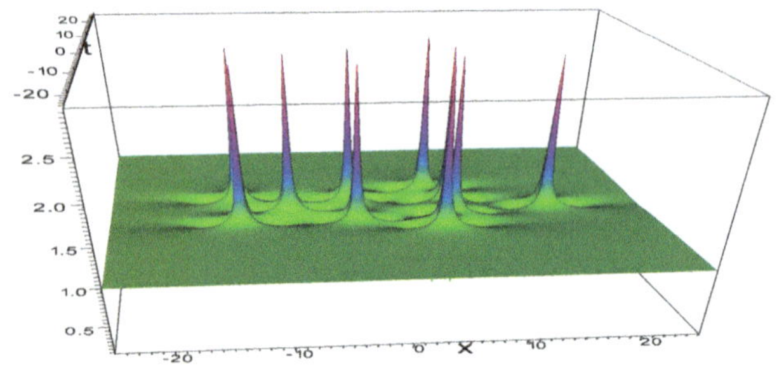

Fig. 10 Solution $2R5$ / 5 to NLS, N=4, $\tilde{a}_1 = 0$, $\tilde{b}_1 = 0$, $\tilde{a}_2 = 10^5$, $\tilde{b}_2 = 0$, $\tilde{a}_3 = 0$, $\tilde{b}_3 = 0$

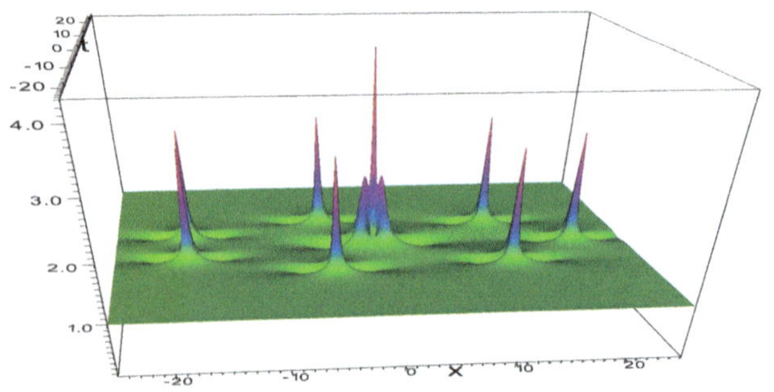

Fig. 11 Solution $1R7 + P_2$ to NLS, N=4, $\tilde{a}_1 = 0$, $\tilde{b}_1 = 0$, $\tilde{a}_2 = 0$, $\tilde{b}_2 = 0$, $\tilde{a}_3 = 10^8$, $\tilde{b}_3 = 0$

Patterns of quasi rational solutions of order 4 with 6 parameters

P_4 breather

If we choose all parameters equal to 0, $\tilde{a}_1 = \tilde{b}_1 = \ldots = \tilde{a}_{N-1} = \tilde{b}_{N-1} = 0$, we obtain the classical Peregrine breather given in the following figure.

Triangles

To shorten, we use the notations defined in the previous section. If we choose $\tilde{a}_1$ or $\tilde{b}_1$ not equal to 0 and all other parameters equal to 0, we obtain triangular configuration with 10 peaks.

Rings

If we choose $\tilde{a}_2$ or $\tilde{a}_3$ not equal to 0 and all other parameters equal to 0 (or vice versa $\tilde{b}_2$ or $\tilde{b}_3$ not equal to 0 and all other parameters equal to 0), we obtain ring configuration with 10 peaks.

Arcs

If we choose two parameters non equal to 0, $\tilde{a}_1$ and $\tilde{a}_2$, or $\tilde{a}_1$ and $\tilde{a}_3$ not equal to 0, or $\tilde{a}_2$ and $\tilde{a}_3$ and all other parameters equal to 0 (or vice versa for parameters b), we obtain arc configuration with 10 peaks.[1]

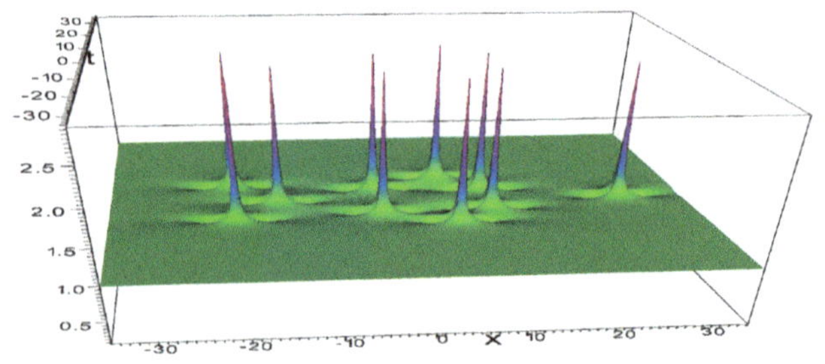

Fig. 12 Solution $2A3/4I + T3$ to NLS, N=4, $\tilde{a}_1 = 10^3$, $\tilde{b}_1 = 0$, $\tilde{a}_2 = 10^6$, $\tilde{b}_2 = 0$, $\tilde{a}_3 = 0$, $\tilde{b}_3 = 0$

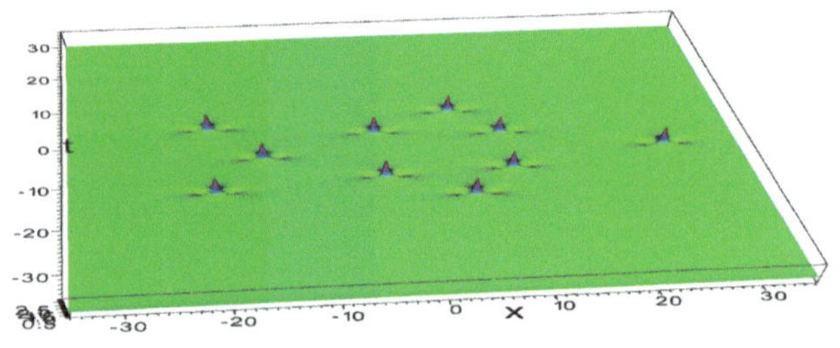

Fig. 13 Solution $2A3/4I + T3$ to NLS, N=4, $\tilde{a}_1 = 10^3$, $\tilde{b}_1 = 0$, $\tilde{a}_2 = 10^6$, $\tilde{b}_2 = 0$, $\tilde{a}_3 = 0$, $\tilde{b}_3 = 0$, sight top

Triangles inside rings

If we choose three parameters non equal to 0, $\tilde{a}_1$, $\tilde{a}_2$ and $\tilde{a}_3$ and all other parameters equal to 0 (or vice versa for parameters b), we obtain ring with inside triangle.

Conclusion

We have presented here patterns of modulus of solutions to the NLS focusing equation in the (x, t) plane. This study can be useful at the same time for hydrodynamics as well for nonlinear optics; many applications in these fields have been realized, as it can be seen in recent works of Chabchoub et al. [30] or Kibler et al. [31]. This study tries to bring all possible types of patterns of quasi-rational solutions to the NLS equation. We see that we can obtain 2^{N-1} different structures at the order N. Parameters a or b give the same type of structure. For $a_1 \neq 0$ (and other parameters equal to 0), we obtain triangular rogue wave; for $a_j \neq 0$ ($j \neq 1$ and other parameters equal to 0) we get ring rogue wave; in the other choices of parameters, we get in particular arc structures (or claw structure). This type of study has been realized in preceding works. Akhmediev

[1] In the following notations $2A4$ / $3I$, I meaning Reversed.

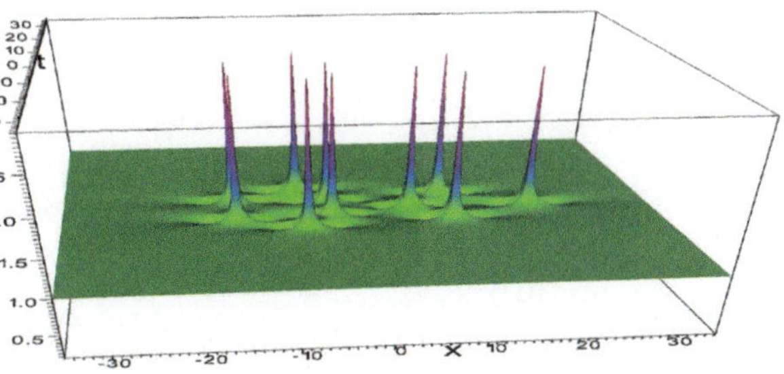

Fig. 14 Solution $2A4/3 + 1T3$ to NLS, N=4, $\tilde{a}_1 = 10^3$, $\tilde{b}_1 = 0$, $\tilde{a}_2 = 0$, $\tilde{b}_2 = 0$, $\tilde{a}_3 = 5 \times 10^7$, $\tilde{b}_3 = 0$

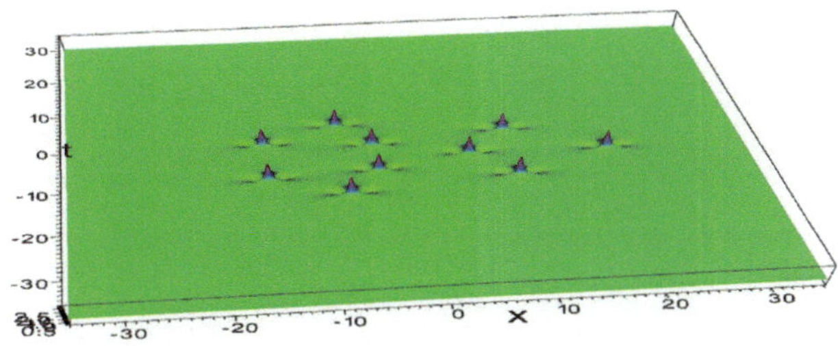

Fig. 15 Solution $2A4/3 + 1T3$ to NLS, N=4, $\tilde{a}_1 = 10^3$, $\tilde{b}_1 = 0$, $\tilde{a}_2 = 0$, $\tilde{b}_2 = 0$, $\tilde{a}_3 = 5 \times 10^7$, $\tilde{b}_3 = 0$, sight top

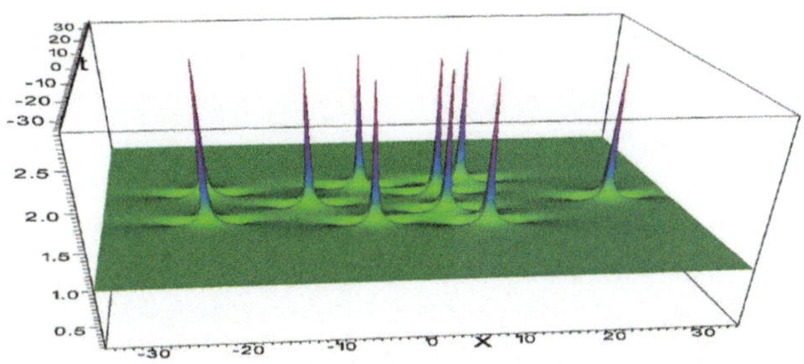

Fig. 16 Solution $2A3/4 + 1T3$ to NLS, N=4, $\tilde{a}_1 = 0$, $\tilde{b}_1 = 0$, $\tilde{a}_2 = 10^6$, $\tilde{b}_2 = 0$, $\tilde{a}_3 = 3 \times 10^8$, $\tilde{b}_3 = 0$

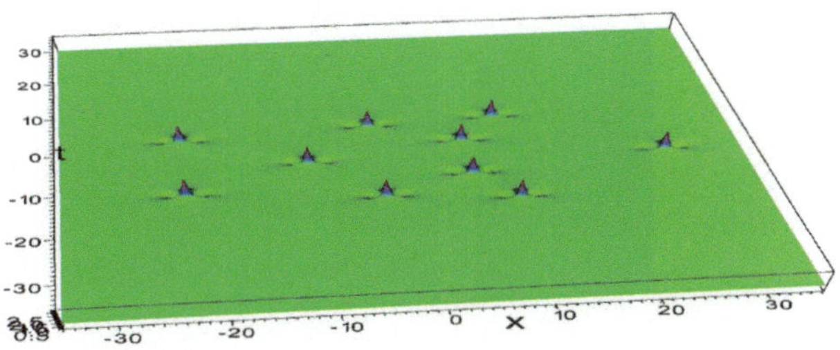

Fig. 17 Solution $2A3/4 + 1T3$ to NLS, N=4, $\tilde{a}_1 = 0$, $\tilde{b}_1 = 0$, $\tilde{a}_2 = 10^6$, $\tilde{b}_2 = 0$, $\tilde{a}_3 = 3 \times 10^8$, $\tilde{b}_3 = 0$, sight top

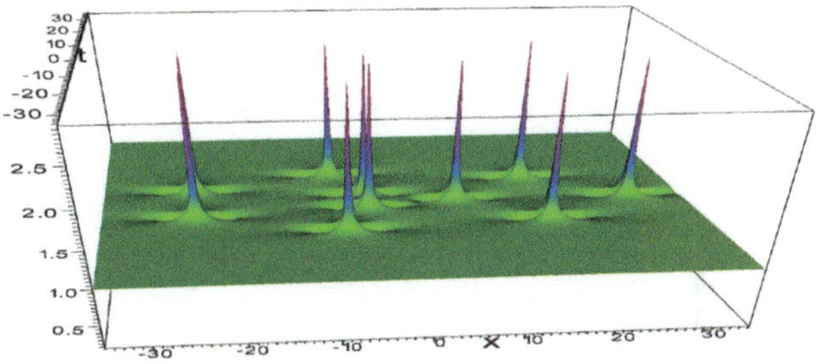

Fig. 18 Solution $1A7 + 1T3$ to NLS, N=4, $\tilde{a}_1 = 10^3$, $\tilde{b}_1 = 0$, $\tilde{a}_2 = 10^3$, $\tilde{b}_2 = 0$, $\tilde{a}_3 = 10^9$, $\tilde{b}_3 = 0$

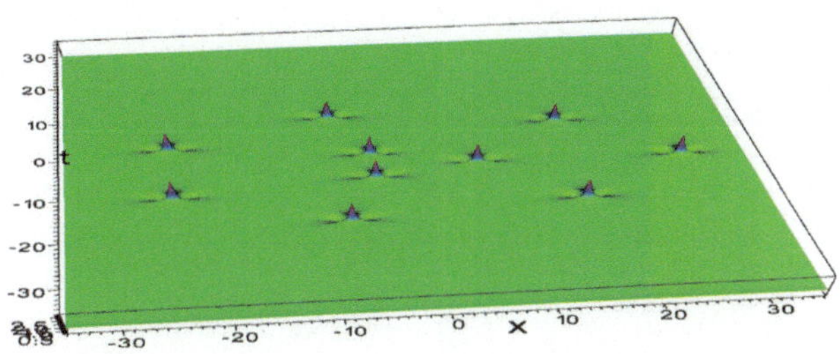

Fig. 19 Solution $1A7 + 1T3$ to NLS, N=4, $\tilde{a}_1 = 10^3$, $\tilde{b}_1 = 0$, $\tilde{a}_2 = 10^3$, $\tilde{b}_2 = 0$, $\tilde{a}_3 = 10^9$, $\tilde{b}_3 = 0$, sight top

et al, study the order $N = 2$ in [32], $N = 3$ in [33]; the case $N = 4$ was studied in particular ($N = 5$, 6 were also studied) in [34, 35] showing triangle and arc patterns; only one type of ring was presented. The extrapolation was done until the order $N = 9$ in [36]. Ohta and Yang [37] presented the study of the case cas $N = 3$ with rings and triangles. Recently, Ling and Zhao [38] presented the cases $N =$ 2, 3, 4 with rings, triangle and also claw structures.

In the present study, one sees appearing richer structures, in particular the appearance of a triangle of 3 peaks inside a ring of 7 peaks in the case of order $N = 4$; to the best of my knowledge, it is the first time that this configuration for order 4 is presented. In this way, we try to bring a better understanding to the hierarchy of NLS rogue wave solutions. It will be relevant to go on this study with higher orders.

Appendix: Parameters and functions

We consider the terms λ_ν satisfying the relations for $1 \leq j \leq N$

$$0 < \lambda_j < 1, \; \lambda_{N+j} = -\lambda_j,$$
$$\lambda_j = 1 - 2\epsilon^2 j^2,$$

with ϵ a small number intended to tend towards 0. The terms κ_ν, δ_ν, γ_ν are functions of the parameters λ_ν,

$1 \le v \le 2N$. They are given by the following equations, for $1 \le j \le N$:

$$\kappa_j = 2\sqrt{1-\lambda_j^2}, \quad \delta_j = \kappa_j \lambda_j,$$
$$\gamma_j = \sqrt{\frac{1-\lambda_j}{1+\lambda_j}}, \tag{4}$$

$$\kappa_{N+j} = \kappa_j, \quad \delta_{N+j} = -\delta_j,$$
$$\gamma_{N+j} = 1/\gamma_j. \tag{5}$$

The terms $x_{r,v}$ $r = 3, 1$ are defined by

$$x_{r,j} = (r-1)\ln\frac{\gamma_j - i}{\gamma_j + i},$$
$$x_{r,N+j} = (r-1)\ln\frac{\gamma_{N+j} - i}{\gamma_{N+j} + i}. \tag{6}$$

The parameters e_v are given by

$$e_j = ia_j - b_j, \quad e_{N+j} = ia_j + b_j, \tag{7}$$

where a_j and b_j are chosen in the form

$$a_j = \sum_{k=1}^{N-1} \tilde{a_k} \epsilon^{2k+1} j^{2k+1},$$
$$b_j = \sum_{k=1}^{N-1} \tilde{b_k} \epsilon^{2k+1} j^{2k+1}, \tag{8}$$
$$1 \le j \le N,$$

with $\tilde{a_j}$, $\tilde{b_j}$,, $1 \le j \le N-1$, $2N-2$, arbitrary real numbers.

The functions $f_{v,1}$ and $g_{v,1}$, $1 \le v \le N$ are defined by

$$f_{4j+1,1} = \gamma_k^{4j-1} \sin A_1,$$
$$f_{4j+2,1} = \gamma_k^{4j} \cos A_1,$$
$$f_{4j+3,1} = -\gamma_k^{4j+1} \sin A_1,$$
$$f_{4j+4,1} = -\gamma_k^{4j+2} \cos A_1, \tag{9}$$

$$f_{4j+1,N+1} = \gamma_k^{2N-4j-2} \cos A_{N+1},$$
$$f_{4j+2,N+1} = -\gamma_k^{2N-4j-3} \sin A_{N+1},$$
$$f_{4j+3,N+1} = -\gamma_k^{2N-4j-4} \cos A_{N+1},$$
$$f_{4j+4,N+1} = \gamma_k^{2N-4j-5} \sin A_{N+1}, \tag{10}$$

$$g_{4j+1,1} = \gamma_k^{4j-1} \sin B_1,$$
$$g_{4j+2,1} = \gamma_k^{4j} \cos B_1,$$
$$g_{4j+3,1} = -\gamma_k^{4j+1} \sin B_1,$$
$$g_{4j+4,1} = -\gamma_k^{4j+2} \cos B_1, \tag{11}$$

$$g_{4j+1,N+1} = \gamma_k^{2N-4j-2} \cos B_{N+1},$$
$$g_{4j+2,N+1} = -\gamma_k^{2N-4j-3} \sin B_{N+1},$$
$$g_{4j+3,N+1} = -\gamma_k^{2N-4j-4} \cos B_{N+1},$$
$$g_{4j+4,N+1} = \gamma_k^{2N-4j-5} \sin B_{N+1}. \tag{12}$$

The arguments A_v and B_v of these functions are defined by $1 \le v \le 2N$:

$$A_v = \kappa_v x/2 + i\delta_v t - ix_{3,v}/2 - ie_v/2,$$
$$B_v = \kappa_v x/2 + i\delta_v t - ix_{1,v}/2 - ie_v/2.$$

References

1. Kharif, C., Pelinovsky, E., Slunyaev, A.: Rogue waves in the Ocean. Springer (2009)
2. Akhmediev, N., Pelinovsky, E.: Discussion and debate: rogue waves - towards a unifying concept? Eur. Phys. Jour. Spec. Top. V **185** (2010)
3. Kharif, Ch., Pelinovsky, E.: Physical mechanisms of the rogue wave phenomenon. Eur. Jour. Mech. B Fluid **22**(6), 603–634 (2003)
4. Slunyaev, A., Didenkulova, I., Pelinovsky, E.: Rogue waters. Cont. Phys. **52**(6), 571–590 (2003)
5. Solli, D.R., Ropers, C., Koonath, P., Jalali, B.: Optical rogue waves. Nature **450**, 1054–1057 (2007)
6. Dudley, J.M., Genty, G., Eggleton, B.J.: Optical rogue waves. Opt. Express **16**, 3644 (2008)
7. Bludov, Y.V., Konotop, V.V., Akhmediev, N.: Matter rogue waves. Phys. Rev. A **80**(033610), 1–5 (2009)
8. Ganshin, A.N., Efimov, V.B., Kolmakov, G.V., Mezhov-Deglin, L.P., McClintok, P.V.E.: Observation of an inverse energy cascade in developed acoustic turbulence in superfluid helium. Phys. Rev. Lett. **101**, 065303 (2008)
9. Stenflo, L., Marklund, M.: Rogue waves in the atmosphere. J. Plasma Phys. **76**(3–4), 293–295 (2010)
10. Ruderman, M.S.: Freak waves in laboratory and space plasmas. Eur. Phys. J. Spec. Top. **185**, 57–66 (2010)
11. Shats, M., Punzman, H., Xia, H.: Matter rogue waves. Phys. Rev. Lett. **104**(104503), 1–5 (2010)
12. Yan, Z.Y.: Financial rogue waves. Commun. Theor. Phys. V **54**, 5, 947, 1-4, (2010)
13. Zakharov, V.E.: Stability of periodic waves of finite amplitude on a surface of a deep fluid. J. Appl. Tech. Phys **9**, 86–94 (1968)
14. Zakharov, V.E., Shabat, A.B.: Exact theory of two dimensional self focusing and one dimensinal self modulation of waves in nonlinear media. Sov. Phys. JETP **34**, 62–69 (1972)
15. Its, A.R., Kotlyarov, V.P.: Explicit expressions for the solutions of nonlinear Schrödinger equation. Dockl. Akad. Nauk. SSSR S A V **965**, 11 (1976)
16. Its, A.R., Rybin, A.V., Salle, M.A.: Exact integration of nonlinear Schrödinger equation. Teore. i Mat. Fiz. **74**(1), 29–45 (1988)
17. Peregrine, D.: Water waves, nonlinear Schrödinger equations and their solutions. J. Aust. Math. Soc. Ser. B **25**, 16–43 (1983)
18. Akhmediev, N., Eleonsky, V., Kulagin, N.: Generation of periodic trains of picosecond pulses in an optical fiber: exact solutions. Sov. Phys. JETP V **62**, 894–899 (1985)
19. Eleonskii, V., Krichever, I., Kulagin, N.: Rational multisoliton solutions to the NLS equation, Soviet Doklady 1986 sect. Math. Phys. **287**, 606–610 (1986)
20. Akhmediev, N., Eleonskii, V., Kulagin, N.: Exact first order solutions of the nonlinear Schrödinger equation. Theor. Math. Phys. **72**(2), 183–196 (1987)
21. Akhmediev, N., Ankiewicz, A., Soto-Crespo, J.M.: Rogue waves and rational solutions of nonlinear Schrödinger equation. Phys. Rev. E **80**(026601), 1–9 (2009)
22. Akhmediev, N., Ankiewicz, A., Clarkson, P.A.: Rogue waves, rational solutions, the patterns of their zeros and integral relations. J. Phys. A Math. Theor. **43**(122002), 1–9 (2010)

23. Chabchoub, A., Hoffmann, H., Onorato, M., Akhmediev, N.: Super rogue waves: observation of a higher-order breather in water waves. Phys. Rev. X **2**(011015), 1–6 (2012)
24. Gaillard, P.: Families of quasi-rational solutions of the NLS equation and multi-rogue waves. J. Phys. A Meth. Theor. **44**, 1–15 (2011)
25. Gaillard, P.: Wronskian representation of solutions of the NLS equation and higher Peregrine breathers. J. Math. Sci. Adv. Appl. **13**(2), 71–153 (2012)
26. Gaillard, P.: Degenerate determinant representation of solution of the NLS equation, higher Peregrine breathers and multi-rogue waves. J. Math. Phys. V **54**, 013504-1-32 (2013)
27. Gaillard, P.: Multi-parametric deformations of the Peregrine breather of order N solutions to the NLS equation and multi-rogue waves. Adv. Res. **4**, 346–364 (2015)
28. Gaillard, P.: Deformations of third order Peregrine breather solutions of the NLS equation with four parameters. Phys. Rev. E V **88**, 042903-1-9 (2013)
29. Gaillard, P.: Six-parameters deformations of fourth order Peregrine breather solutions of the NLS equation. J. Math. Phys. V **54**, 073519-1-22 (2013)
30. Chabchoub, A., Hoffmann, N.P., Akhmediev, N.: Rogue wave observation in a water wave tank, Phys. Rev. Lett. V **106**, 204502-1-4 (2011)
31. Kibler, B., Fatome, J., Finot, C., Millot, G., Dias, F., Genty, G., Akhmediev, N., Dudley, J.M.: The Peregrine soliton in nonlinear fibre optics. Nat. Phys. **6**, 790–795 (2010)
32. Ankiewicz, A., Kedziora, D.J., Akhmediev, N.: Rogue wave triplets. Phys Lett. A **375**, 2782–2785 (2011)
33. Kedziora, D.J., Ankiewicz, A., Akhmediev, N.: The phase patterns of higher-order rogue waves. J. Opt. V **15**, 064011-1-9 (2013)
34. Kedziora, D.J., Ankiewicz, A., Akhmediev, N.: Circular rogue wave clusters. Phys. Rev. E **84**(056611), 1–7 (2011)
35. Kedziora, D.J., Ankiewicz, A., Akhmediev, N.: Triangular rogue wave cascades. Phys. Rev. E **86**(056602), 1–9 (2012)
36. Kedziora, D.J., Ankiewicz, A., Akhmediev, N.: Classifying the hierarchy of the nonlinear Schrödinger equation rogue waves solutions. Phys. Rev. E V **88**, 013207-1-12 (2013)
37. Ohta, Y., Yang, J.: General high-order rogue waves and their dynamics in the nonlinear Schrödinger equation. Proc. R. Soc. A **468**(2142), 1716–1740 (2012)
38. Ling, L., Zhao, L.C.: Simple determinant representation for rogue waves of the nonlinear Shrodinger equation. Phys. Rev. E V **88**, 043201-1-9 (2013)

19

Formulation of atomic positions and carbon–carbon bond length in armchair graphene nanoribbons: an ab initio study

Mehran Balarastaghi[1] · **Vahid Ahmadi**[2]

Abstract In this paper, we investigate the atomic positions of single layer armchair graphene nanoribbon for two cases, with and without hydrogen-passivate edges, accurately and propose a formula which either removes the need of structural relaxation generally or decreases its time extremely (up to seven times). We also propose a general pattern (hyperbolic) for these positions. On the other hand, we show that edge effect influences several atoms near the edge not just one. These results can be used in software, which compute atomic positions and can increase their efficiency. In addition, we prove that the C–C bond distance depends on dimer number and differs in length and width directions, especially for narrow AGNRs. The maximum value of these differences is about 0.017 Å.

Keywords Armchair graphene nanoribbon (AGNR) · Structural relaxation · Atomic position · Dimer · C–C bond length

Introduction

Graphene is a one-atom-thick crystal of sp^2-bonded carbon atoms ordered in a two-dimensional (2D) hexagonal lattice [1]. Graphene and graphene nanoribbon (GNRs) have attracted much recent attention for their unique two-dimensional structures and physicochemical properties [2], as well as their wide potential applications in electronics [3, 4], photonics, and optoelectronics [5], energy storage and conversion [6], and chemical-bio sensing [7].

Regarding the broad usage of GNR in nano-electronics, physics and chemistry, this substance is under wide investigations at present. Most of these investigations are done using some software based on density functional theory (DFT), and they need structural relaxation of nanoribbons. It takes much time, especially if the length of nano-transistors being investigated is more than 10 nm. In this case, the number of atoms will increase. If the method for simulation of the structure is based on order-N3, the time it needs will be much more even for supercomputers. In the case of investigating multilayer graphene nanoribbon, the situation will deteriorate.

The idea is proposing a method having the dimer number of GNR and its length to acquire the structural relaxation and the exact position of atoms. In this case, either the general need for structural relaxation would be unnecessary or even by doing this, the time will be extremely decreased.

To use graphene nanoribbon as a channel, we should first determine the approximate position of atoms. Then these positions should be given to applications such as SIESTA, VASP and ATK [8–10] which change them using structural relaxation and spending a long time, until these atoms are placed in their precise positions. The running time for relaxation depends on several parameters: the accuracy of initial positions given to the software, the method used in the software for structural relaxation, the parameters that the user sets in the software, and the value of the requested accuracy to end run.

Except for the initial positions, the rest of the parameters are usually specified by the researcher at the beginning, and do not change [10–12]. Given that the initial position of the atoms is of

✉ Vahid Ahmadi
v_ahmadi@modares.ac.ir

1 Department of Electrical Engineering, Science and Research Branch, Islamic Azad University, Tehran, Iran

2 Faculty of Electrical and Computer Engineering, Tarbiat Modares University, Tehran, Iran

paramount importance in comparison to the rest of the parameters, it is under way in the present study. It is natural that if we relax an atomic structure and set the obtained positions to the software as new initial positions and repeat the structural relaxation, the run time would be reduced several times. This time for very large structures, can be reduced from months to days. So acquiring a way to determine the precise initial atomic positions sounds completely logical.

Researchers obtain these initial positions from some specific ways: downloading from the Internet, using simple formulas available in books and articles [13, 14], and the use of applications that calculate these positions for a number of particular structures such as carbon nanotube [15]. It is natural that the atomic structures found on the Internet are limited. Using the simple formula readily accessible in the books and articles, we can calculate the positions with low accuracy, and so much time is required to run structural relaxation.

Applications that are available in this field are also very limited. For example, Nano Tube Modeler software, which is used for CNT [15, 16]. Such applications use simple and ideal formulas to calculate the atomic positions too, which make the run time very long. So by determining the exact formulas for the positions in GNRs in this paper, researchers can utilize these formulas.

In all applications and simple formulas, C–C bond lengths are considered constant for all GNRs in different directions [13, 14]. As we calculate the exact atomic position for all atoms, it will be obvious that the D_{cc} (carbon-carbon bond length) is not equal for them. We consider the average value of middle atoms in all GNRs in two directions and compare them.

The paper is organized as follows: in Sect. 2, we present a basic formula for atomic positions for all carbon atoms in ideal (slab cut of graphene) and real (relaxed) AGNR. In Sect. 3, we derive all atomic positions in x and y directions and energy gap, and then present a generic formula for them and show the fact that the C–C bond lengths are different in two directions. Finally, Sect. 4 summarizes our findings.

Problem statement

In this paper, we present two groups of AGNR slabs. For the first group, all edge carbon atoms saturate with hydrogen atoms and for second group, they are bare (without hydrogen). The width of AGNR is defined by the number of dimer lines (N_y) and that of length is defined by the number of atoms in slab direction, as illustrated in Fig. 1. Since the band gaps of AGNRs depend on its width and exhibit three distinct families, $(3n + m)$ behavior [17, 18], we choose AGNRs with length $N_c = 6$ (number of carbon atoms in X direction), and width $N_y = 5$–30, representing the family behavior with the $n = 1$–10 and $m = 0$, 1 and 2. In addition, wider AGNRs with $N_y = 40$–42, are also calculated to examine the size-dependent effects, and to see some especial effects better.

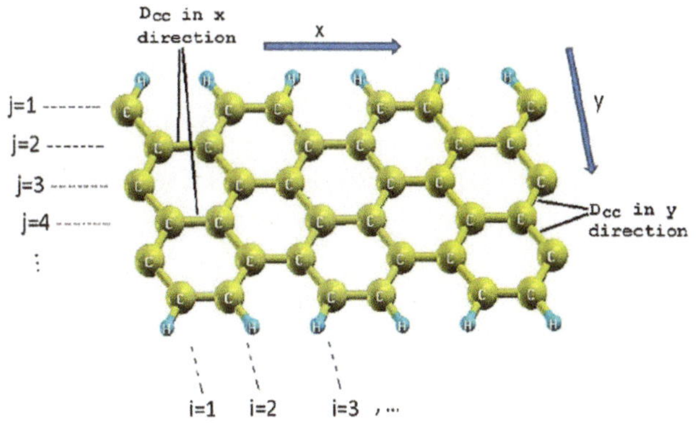

Fig. 1 An AGNR slab with its whole details

At first, we can consider an ideal graphene nanoribbon. The exact position of each atom is obtained by following equations [13, 14] (we show these positions in ideal states with capital letters):

$$X(i=1,j) = \begin{cases} C_1 & : j \text{ is Odd} \\ C_1 + \frac{Dcc}{2} & : j \text{ is Even} \end{cases} \quad (1)$$

$$X(i=2,j) = \begin{cases} C_1 + 2 \times Dcc & : j \text{ is Odd} \\ C_1 + 3 \times \frac{Dcc}{2} & : j \text{ is Even} \end{cases} \quad (2)$$

$$Y(i,j) = C_2 + (j-1)\frac{\sqrt{3}}{2}D_{cc} : \text{for all } i \,\&\, j, \quad (3)$$

$$X(i = \text{Odd}, j) = X(i = 1, j) + 3 \times N_x \times D_{cc} \quad (4)$$

$$X(i = \text{Even}, j) = X(i = 2, j) + 3 \times N_x \times D_{cc}, \quad (5)$$

where N_x is the number of unit cells in X direction (every two adjacent columns of atoms in X direction, is a unit cell), C_1 and C_2 are arbitrary constant numbers, and D_{cc} is C–C bond length. Now we are going to relax the various nanoribbons with different widths. The computed positions for these atoms are denoted by small letters. Let us call it the real value. In general, the positions in real and ideal states are not equal. We define this variation as:

$$x(\text{Real}) = X(\text{Ideal}) + \Delta x(i,j,N_y), \quad (6)$$

$$y(\text{Real}) = Y(\text{Ideal}) + \Delta y(i,j,N_y). \quad (7)$$

Now if we obtain the exact value of Δx and Δy, according to obtained ideal value of X and Y with formulas (1)–(5), we will have a general formula for the relaxed value of x and y, and the relaxation of nanoribbons will not be needed any more. Even if we need to relax the nanoribbons the time of relaxation will be decreased considerably.

Results and discussion

Electronic structures and geometry relaxations are calculated based on the DFT [19], using the Spanish initiative for electronic simulations with thousands of atoms package (SIESTA) [20, 21] with local-density approximation (LDA) [22]. A double-ζ plus polarization basis set is employed to describe the localized atomic orbitals and an energy cutoff for real-space mesh size is set to be 380 Ry. All nanostructure geometries are fully relaxed with a force tolerance, 0.01 eV/Å [11].

We do the structural relaxation using SIESTA for AGNRs with N_y from 5 to 30 in two states, with and without hydrogen for carbon atoms in edges. Then we use the output files of these simulations as input files in MATLAB and process all the output data of SIESTA by several programs written in MATLAB. According to reports by other researchers, it is obvious that energy bandgap diagram vs the dimer number is divided into three categories. The dimer number is shown as:

$$N_y = 3n + m, \tag{8}$$

where n is a natural number and m is 0, 1 or 2, other studies have shown that energy bandgap diagram depends on the value of m [23]. We also obtain the same diagrams in this research. Figures 2, and 3, show this diagram in two states, with and without hydrogen. These diagrams indicate that each branch of these diagrams can be shown with a specific formula:

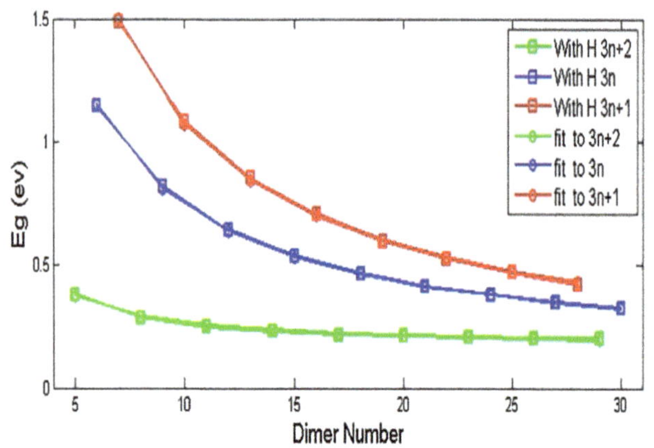

Fig. 2 Energy gap vs dimer number for AGNR with H

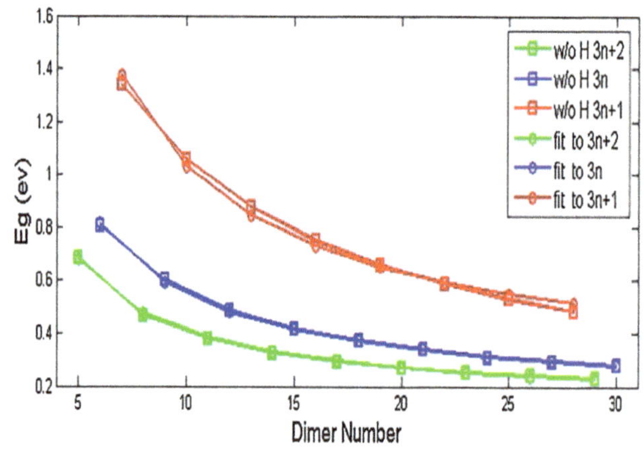

Fig. 3 Energy gap vs dimer number for AGNR without H

$$f(x) = A_0 + \frac{A_1}{x}, \tag{9}$$

where $f = E_g$ and $x = N_y$. As a result, we fit these diagrams with the obtained data using nonlinear regression. We obtain the coefficient A_0, A_1 and the mean square error (MSE) for various three branches of diagrams and in two cases, with and without hydrogen by listing them in Table 1. The diagrams related to nonlinear regression (as fitted to) of the data together with the data itself obtained in simulations are shown in Figs. 2, and 3. It can be figured out from these diagrams that the obtained data in simulations is perfectly coinciding with the proposed formula so that five out of six branches coincide in a way that almost no difference can be observed between simulation and nonlinear regression.

In Table 1, each column of data is related to a specific category of AGNRs, which has been shown by the value $m = 0$, 1, and 2. In MSE section of the table, it can be seen clearly that the value of error for $m = 1$, and for the case without hydrogen is more than the other considerably.

As a result, the diagram of simulation and nonlinear regression (for $m = 1$) in Fig. 3, are slightly different from each other. Although there exist some minor errors in one

Table 1 Energy gap in AGNRs vs m

	$m = 0$	$m = 1$	$m = 2$
With H			
A_0	0.1206	0.0741	0.1585
A_1	6.1866	9.9777	1.0715
MSE	3.7e−05	8.2e−05	4.1e−05
Without H			
A_0	0.1475	0.2286	0.1329
A_1	3.9968	8.0067	2.7275
MSE	2.9e−05	8.0e−04	2.1e−05

branch, simulation results show enough good matches with that of proposed formula.

Computing Δx

We take the idea from last paragraph and extend this regression to all parameters of GNRs, including Δx and Δy. The diagram of Δx based on the atom index in Y direction (j) is drawn in Figs. 4, and 5, for the two dimer numbers, 29 and 30 (odd and even). By looking over these figures it is observable that the diagram is odd for nanoribbons having even dimer number and vice versa. In addition, except for edge carbon atoms of nanoribbons (on both sides) which have the most variations, the diagram relating to the rest of atoms is zigzag shape. It means the amount of variations in X direction depends on atom index being odd or even, and this state looks normal considering the arrangement of atoms in graphene (or AGNR) unit cell which they are placed alternatively as atoms type 1 and 2.

Considering these points, we can write the Δx formula according atoms' index. It is enough to have the mean and peak to peak value of the zigzag. In this case the formulas can be written as:

$$\Delta x(i = \text{Odd},\ j = \text{Odd}) = \text{Av}(i = 1) - \frac{1}{2}\text{P2P}\,(i = 1), \tag{10}$$

$$\Delta x(i = \text{Odd},\ j = \text{Even}) = \text{Av}(i = 1) + \frac{1}{2}\text{P2P}\,(i = 1), \tag{11}$$

$$\Delta x(i = \text{Even},\ j = \text{Odd}) = \text{Av}(i = 2) + \frac{1}{2}\text{P2P}\,(i = 2), \tag{12}$$

$$\Delta x(i = \text{Even},\ j = \text{Even}) = \text{Av}(i = 2) - \frac{1}{2}\text{P2P}\,(i = 2), \tag{13}$$

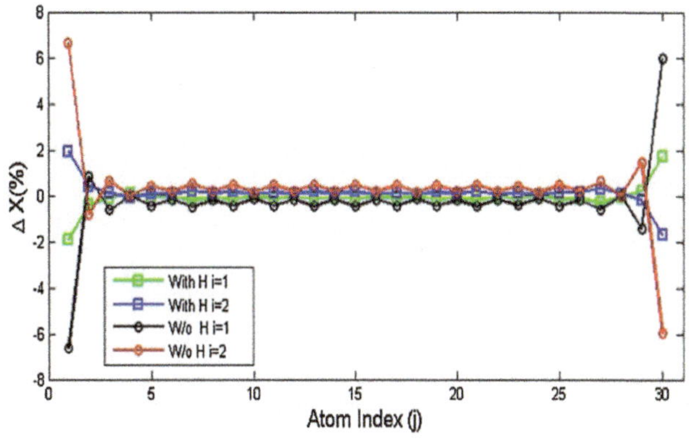

Fig. 5 Percentage of Δx for $N_y = 30$ vs atom index in Y direction (j) for two atom index in x direction ($i = 1$, $i = 2$)

where Av and P2P are the mean and peak to peak values of zigzags, respectively. The diagram relating to the mean value is depicted in Fig. 6. Moreover, that relating to the peak to peak is shown in Fig. 7, (both of them are plotted for nanoribbons without hydrogen). The same diagrams obtain for nanoribbons with hydrogen. Considering these diagrams, it can be seen that the hyperbolic formula (9) which is proposed for GNR's bandgap energy is also valid here. In this formula $f =$ Av or P2P and $x = N_y$. In this case, each diagram alone is divided into three categories (for three different values of m) and the nonlinear regression for each category.

The obtained coefficients and the MSE for each category are shown in Tables 2, and 3 for the mean and peak to peak of zigzag section, respectively, for two cases, with and without hydrogen.

Now we have to obtain the value of Δx of GNR for edge carbon atoms and atom near the edge. In addition, beside the edge carbon atom, the atoms near the edge are a little far from the ideal zigzag shape that we consider for central atoms. We have sketched the diagrams relating to Δx for edge carbon atom ($j = 1$) in Fig. 8, and for the second atom from the edge ($j = 2$) in Fig. 9, according to the

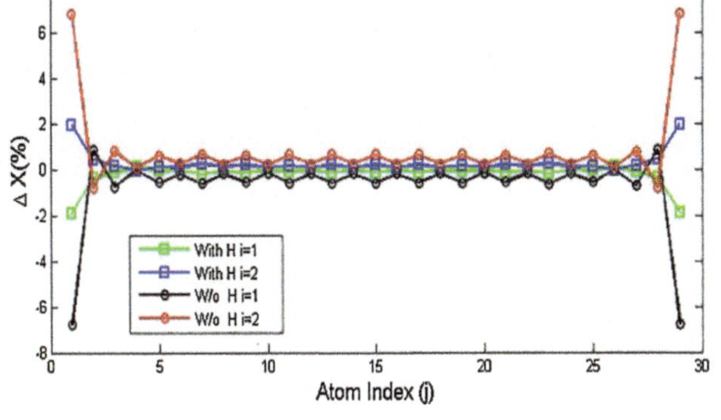

Fig. 4 Percentage of Δx for $N_y = 29$ vs atom index in Y direction (j) for two atom index in x direction ($i = 1$, $i = 2$)

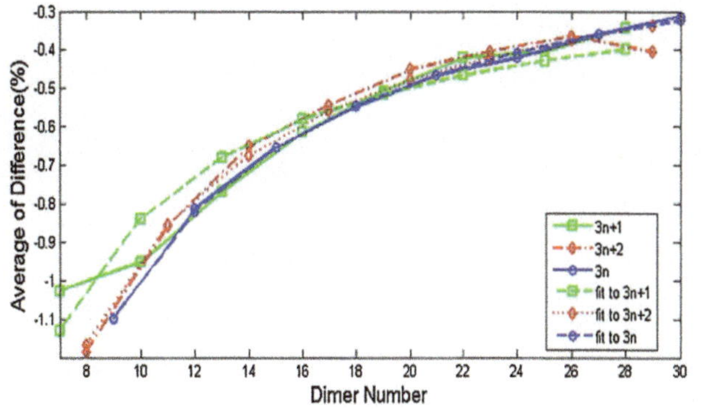

Fig. 6 Average of differences in percentage in X direction for AGNRs without H for all first atoms ($i = 1$) and their fitted curves

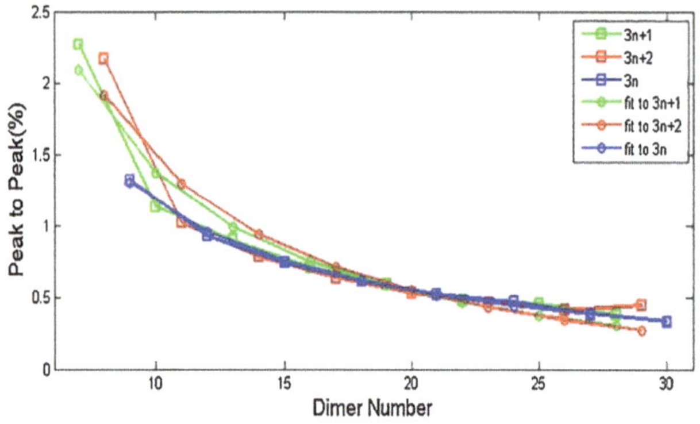

Fig. 7 Peak to Peak percentage of Δx for middle first atoms in Y direction ($i = 1$) for AGNRs without H vs dimer number

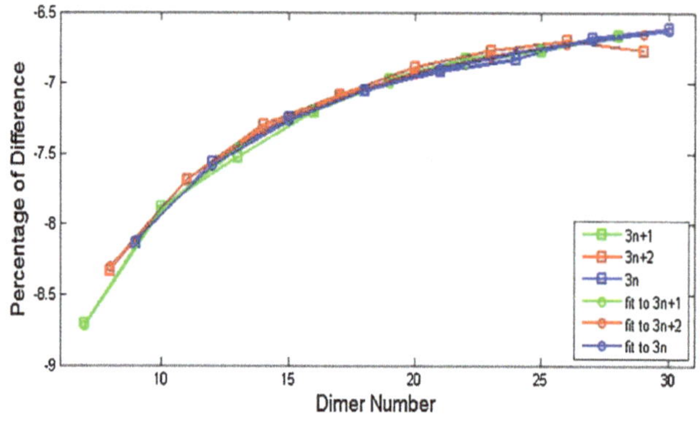

Fig. 8 Percentage of Δx for first carbon in Y direction ($j = 1$) for all AGNRs without H vs dimer number, and their fitted curves

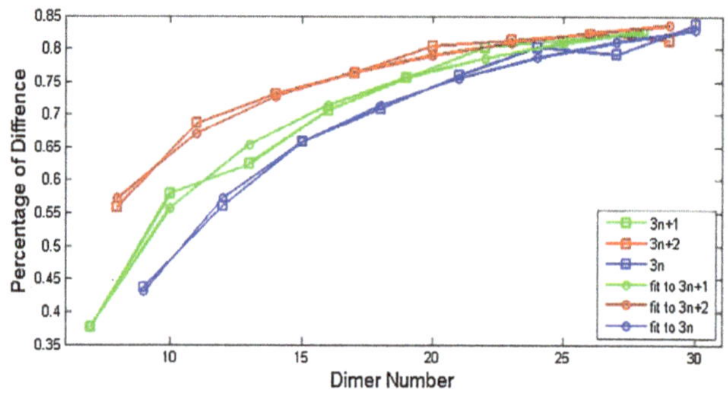

Fig. 9 Percentage of Δx for second carbon in Y direction ($j = 2$) for all AGNRs without H vs dimer number, and their fitted curves

Table 2 The average of Δx in middle carbons

	$m = 0$	$m = 1$	$m = 2$
With H			
A_0	−0.0082	−0.0374	0.0618
A_1	−2.7524	−2.5069	−4.1218
MSE	7.4e−08	3.3e−08	3.0e−07
Without H			
A_0	−0.0204	0.0075	−0.1548
A_1	−9.1735	−9.9429	−6.8449
MSE	1.2e−07	7.2e−09	6.4e−07

Table 3 Peak to Peak of Δx in middle carbons

	$m = 0$	$m = 1$	$m = 2$
With H			
A_0	0.0972	−0.0222	0.0613
A_1	1.5239	3.5494	2.2642
MSE	1.9e−07	8.3e−08	6.0e−08
Without H			
A_0	−0.3470	−0.0671	−0.2856
A_1	18.0746	12.3439	16.6161
MSE	3.3e−06	3.7e−08	1.8e−06

dimer number. Both of these figures are plotted for nanoribbons without hydrogen.

We have achieved all these diagrams for both cases, with and without hydrogen, and for four atoms near the edge ($j = 1, 2, 3, 4$), but considering the high number of figures, we do not draw them all. In all these diagrams, the general shape of hyperbolic can be seen. Thus, we use the formula (9). In this formula $f = \Delta x$ and $x = N_y$.

In Fig. 8, it can be seen that the diagrams are coinciding for all three categories. It means this diagram does not depend on m value, but as we can see in Fig. 9, those three categories are unlike each other. This manner is extendable for other atoms in the same way. It means that the various categories of the diagram coincide for odd atoms but not for even atoms. We use the general formula (9) again. In this formula $f = \Delta x$ and $x = N_y$. The various coefficients obtained from nonlinear regression for the first four atoms near the edge and for AGNRs with, and without hydrogen are listed in Tables 4 and 5, respectively. The points mentioned for odd and even edge atoms, are obvious in these tables.

Now by considering diagrams being odd or even, we have to write the formulas relating to second edge for first four atoms of the edge. These formulas are:

$$\Delta x(i, N_y - j) = \Delta x(i, j) : j = 1, 2, 3, 4 \text{ and } N_y = \text{Odd}, \tag{14}$$

$$\Delta x(i, N_y - j) = \text{Av}(i) - \Delta x(i, j) : j = 1, 2, 3, 4 \text{ and } N_y = \text{Even}. \tag{15}$$

Therefore, using Tables 2, 3, 4 and 5 in Eqs. (10)–(15), we compute the exact value of Δx for all atoms and for each dimer number. By inserting this value in Eq. (7), we obtain the real value of x.

Table 4 Δx of four first carbons in AGNRs with H

	$m = 0$	$m = 1$	$m = 2$
Carbon 1			
A_0	−1.7607	−1.7344	−1.6468
A_1	−5.1316	−5.5395	−7.3415
MSE	3.5e−08	3.0e−08	6.4e−07
Carbon 2			
A_0	−0.4238	−0.3344	−0.3516
A_1	−0.1986	−2.6628	−1.5550
MSE	3.6e−09	1.5e−08	7.1e−08
Carbon 3			
A_0	−0.0203	0.0647	0.0694
A_1	−3.7301	−5.6972	−5.3053
MSE	3.3e−08	3.6e−08	4.6e−07
Carbon 4			
A_0	0.1402	0.1537	0.1080
A_1	−1.6357	−2.2305	−0.7365
MSE	3.9e−09	8.6e−09	1.7e−08

Table 5 Δx of four first carbons in AGNRs without H

	$m = 0$	$m = 1$	$m = 2$
Carbon 1			
A_0	−6.0231	−5.9924	−5.9971
A_1	−18.244	−19.098	−19.027
MSE	3.3e−07	7.1e−08	1.1e−07
Carbon 2			
A_0	0.9351	0.9999	0.9759
A_1	−2.9057	−5.1337	−4.1839
MSE	1.8e−08	1.4e−08	2.9e−08
Carbon 3			
A_0	−0.1980	−0.1265	−0.2033
A_1	−13.489	−15.533	−13.714
MSE	2.8e−07	2.5e−08	8.5e−08
Carbon 4			
A_0	0.1357	0.1678	0.1086
A_1	−4.4469	−5.4186	−3.9314
MSE	2.0e−08	3.1e−09	3.0e−08

Computing Δy

Figure 10, shows **Δ***y* diagram in two cases, with and without hydrogen, according to atom index in *Y* direction (*j*) for GNR with dimer number 29 as an example. The basic difference between diagrams related to **Δ***x* and **Δ***y* is that the coefficients relevant to various indices in *X* direction (*i*) are completely coinciding in **Δ***y* diagrams while these coefficients in **Δ***x* diagrams are different.

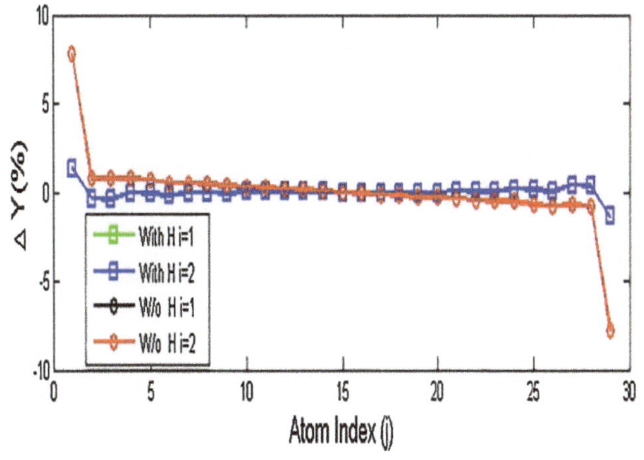

Fig. 10 Percentage of ΔY for $N_y = 29$ vs atom index in *Y* direction (*j*) with edge atoms

Now to look at the inner atoms more carefully, we eliminate two atoms from the edges of the diagram and focus on central atoms. This diagram has shown as an example for dimer number 30 in Fig. 11. Now suppose we divide atom's index in Y direction (*j*) into three categories means $j = 3 \times n + m_j$. Based on this figure it is obvious that the value of **Δ***y* depends on the categories (m_j). However, there is not much difference between categories (m_j) for the most dimer numbers. This point can be seen in Fig. 12, for dimer number 29. However, if we consider original figure in that two edge atoms are not eliminated (Fig. 10), the effect of m_j can be ignored.

Another important point is that if we eliminate the edge atoms, a linear shape can be seen for other atoms. Thus, we can consider **Δ***y* formula for central atoms as linear. For doing it, we use the linear regression in MATLAB to obtain the slope and offset values for each AGNR with various dimer numbers. As an example, the slope diagram is

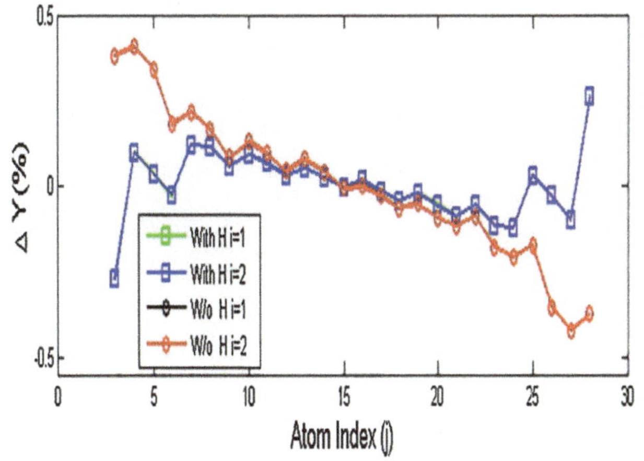

Fig. 11 Percentage of ΔY for $N_y = 30$ vs atom index in *Y* direction (*j*) without edge atoms

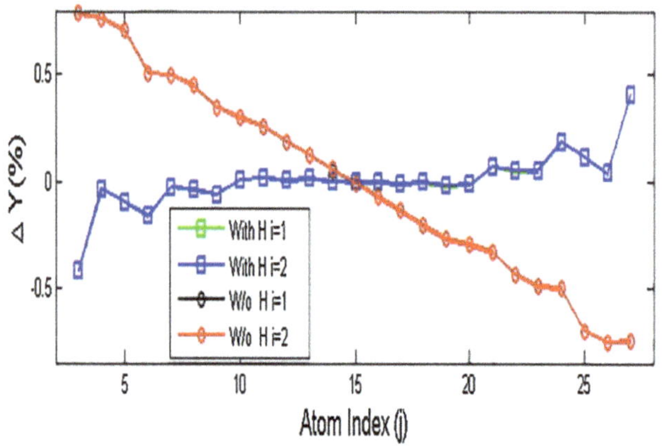

Fig. 12 Percentage of ΔY for $N_y = 29$ vs atom index in Y direction (j) without edge atoms

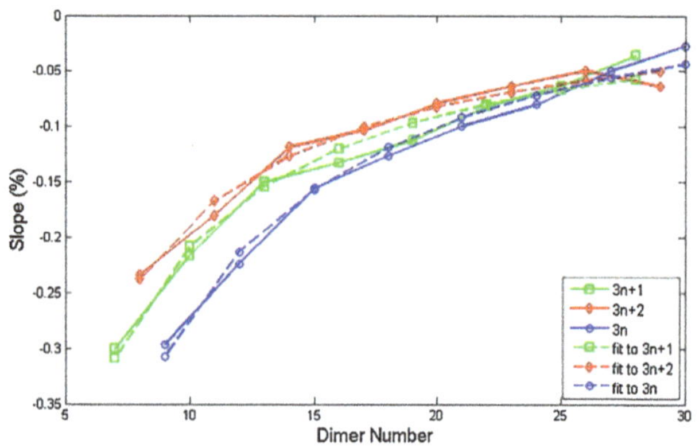

Fig. 13 Slope in the curve of D_{cc} difference in percentage in AGNRs without H in Y direction for middle atoms vs dimer number

depicted for AGNR without hydrogen in Fig. 13. As it can be seen easily, the diagram is hyperbolic and depends on m.

Considering this point, we use the general formula (9) and the linear regression. The obtained coefficients for the slopes and offsets of these lines have listed in Tables 6 and 7, respectively, in two cases, with and without hydrogen.

Thus, by having the values of the slope and offset, $\mathbf{\Delta y}$ formula can be written for central atoms as:

Table 6 The slope (**S**) in AGNRs in Y direction

	$m = 0$	$m = 1$	$m = 2$
With H			
A_0	0.0165	0.0140	0.0034
A_1	0.3108	0.0344	0.6737
MSE	1.0e−07	3.6e−08	7.9e−08
Without H			
A_0	0.0296	0.0667	0.0073
A_1	−2.2085	−3.3721	−2.1073
MSE	2.9e−08	2.4e−08	8.5e−08

Table 7 The offset (**O**) in AGNRs in Y direction

	$m = 0$	$m = 1$	$m = 2$
With H			
A_0	−0.2819	−0.1644	−0.2724
A_1	−0.8469	0.0477	−2.0991
MSE	8.0e−06	3.5e−06	9.4e−06
Without H			
A_0	0.5908	0.4035	0.7838
A_1	4.5999	11.5680	3.5510
MSE	3.2e−06	6.6e−06	6.7e−06

$$\Delta y(i,j) = S(N_y) \times j + O(N_y). \quad (16)$$

It can also be figured out from Figs. 11, and 12, that it is not only the edge atom is placed out of linear regression but also there are a couple of other atoms, which do the same slightly. As a result, the value of $\mathbf{\Delta y}$ for the first four edge atoms for AGNRs in two cases, without and with hydrogen, are listed in Tables 8 and 9, respectively. These values also follow the general hyperbolic shape. These values for ANGRs without hydrogen depend on m, but for AGNRs with hydrogen are independent of it.

As we have mentioned before, the diagram ΔY based on j (compared to the central atom of ANGR) is an odd function, thus for atoms near the second edge the following formula can be written:

$$\Delta y(i, N_y - j) = -\Delta y(i,j) : j = 1, 2, 3, 4. \quad (17)$$

Table 8 ΔY of four first carbons in AGNRs without H

	$m = 0$	$m = 1$	$m = 2$
Carbon 1			
A_0	7.3245	7.2128	7.4956
A_1	7.4807	13.1602	7.0796
MSE	1.8e−06	5.1e−06	3.2e−06
Carbon 2			
A_0	0.6364	0.6042	0.7627
A_1	−0.5570	2.3996	−0.9311
MSE	1.3e−06	4.1e−06	2.2e−06
Carbon 3			
A_0	0.6678	0.6608	0.8137
A_1	−1.5009	0.0079	−3.0802
MSE	1.2e−06	3.9e−06	2.4e−06
Carbon 4			
A_0	0.8059	0.7015	0.9006
A_1	−5.3071	−0.9615	−5.4591
MSE	1.1e−06	3.2e−06	2.1e−06

Table 9 ΔY of four first carbons in AGNRs with H

	A_0	A_1	MSE1
Carbon 1	1.3966	2.0813	1.0e−06
Carbon 2	−0.3637	1.3296	6.6e−07
Carbon 3	−0.4808	0.8744	6.3e−07
Carbon 4	−0.1334	0.5408	8.0e−07

Therefore, using Tables 6, 7, 8 and 9 in Eqs. (16) and (17), we compute the exact value of Δy for all atoms and for each dimer number. By inserting this value in Eq. (8), we obtain the real value of y. Using these equations, we repeat structural relaxations again. Comparing the calculation time required for new relaxations using the real positions and the ideal positions, we show that the calculation time is approximately reduced between 4 and 7 times. For example, this time for relaxation of AGNR with $N_y = 42$ reduced from 17 days to less than four days. This reduction is unbelievable when the channel length is more 10 nm or when the GNR is multilayer.

Difference between C–C bond length in two directions

If we look at the two Figs. 4, and 5 more carefully, we understand with a little change in D_{cc} in X direction, the mean value of zigzag shape can be equal to zero in both cases, $i = 1$ and $i = 2$. By summing this change to the last value of D_{cc}, we obtain its new value. Because this value is the mean value for all central atoms, we can name it the real value for D_{cc} in x direction. By writing Eq. (3) in two states before and after the change of D_{cc}, we can calculate the amount of change in D_{cc} accurately:

$$\Delta D_{ccx} = -\frac{4}{3}\mathrm{Av1}, \tag{18}$$

$$\mathrm{P2P}_2 = \mathrm{P2P}_1 + \frac{2}{3}\mathrm{Av1}, \tag{19}$$

where Av1 and $\mathrm{P2P}_1$ are mean value and peak to peak of the zigzag part of ΔX diagram in central atoms which their hyperbolic coefficients are listed in Tables 2 and 3, respectively. $\mathrm{P2P}_2$ is the amount of peak to peak after the changing of the D_{cc} obtained from the formula (18). The figures obtained from MATLAB programs have proven the improvements which we refuse to draw these figures because of their numbers and being repetitious.

Now, we consider D_{cc} in Y direction. If we look at the Figs. 10, 11, and 12, again, we can observe a slope in central atoms. This slope shows that the D_{cc} we considered in this direction, is different than the last one, and we can remove the slope by improving it. Using the Eq. (3) and considering the fact that slope in this diagram means the change of ΔY in two next atoms in Y direction, the following equation can be easily written as:

$$\Delta D_{ccy} = -\frac{2}{\sqrt{3}}S_1. \tag{20}$$

The parameter S_1 is the amount of obtained slope before applying the change in D_{cc} in Y direction which its hyperbolic coefficients are shown in Table 6. We achieved other diagrams using MATLAB programs for this state, which confirms the validity of our sayings. However, we refuse to draw these figures because of their numbers and being repetitious.

Now that we can have an estimation from existing differences in D_{cc} in two various directions, we compute both changes by using the formulas (18) and (20) for specific dimer number, for example, $N_y = 29$, for the state of AGNR without hydrogen. These values are approximately equal to:

$$\Delta D_{ccx} \approx \%0.4, \quad \Delta D_{ccy} \approx \%0.06. \tag{21}$$

We can see that the amount of change is different in two directions of AGNR. Regarding to the value of D_{cc} for graphene ($D_{cc} \approx 1.42$ Å), the difference between C–C bond lengths is approximately 0.005 Å. As another example, the value for $N_y = 5$, it is nearly 0.017 Å, and it reduces to zero when the dimer number goes to infinity.

Conclusions

We proposed a general hyperbolic formula, and we proved that the way the AGNRs bandgap energy is relatable, there are lots of other parameters, which can be shown by this formula. In this way, we showed that the atomic positions of all carbon atoms in AGNRs can be calculated by the formula, and so we can perform the analysis without the need to its structural relaxations. Using obtained equations, we did the structural relaxations for AGNRs with different dimer numbers again, and could reduce the analysis times, 4–7 times. We also showed that the edge effect would not be only for the first edge atom, and this effect can be observed in the couple of atoms close the edges.

We also showed that the C–C bond length is slightly different for each various directions. This difference is about 0.017 Å for AGNRs with small dimer numbers and is reduced to zero for AGNRs with large width (graphene).

References

1. Novoselov, K., Geim, A.K., Morozov, S., Jiang, D., Katsnelson, M., Grigorieva, I., Dubonos, S., Firsov, A.: Two-dimensional gas of massless Dirac fermions in graphene. Nature **438**(7065), 197–200 (2005)

2. Neto, A.C., Guinea, F., Peres, N., Novoselov, K.S., Geim, A.K.: The electronic properties of graphene. Rev. Mod. Phys. **81**(1), 109 (2009)
3. Chen, Z., Lin, Y., Rooks, M., Avouris, P.: Low-dimensional systems and nanostructures. Physica E **40**(2), 228–232 (2007)
4. Geim, A.K., Novoselov, K.S.: The rise of graphene. Nat. Mater. **6**(3), 183–191 (2007)
5. Bonaccorso, F., Sun, Z., Hasan, T., Ferrari, A.C.: Graphene photonics and optoelectronics. Nat Photon **4**(9), 611–622 (2010)
6. Hu, Y.H., Wang, H., Hu, B.: Thinnest two-dimensional nanomaterial—graphene for solar energy. Chemsuschem **3**(7), 782–796 (2010)
7. Zhang, Z., Liu, B., Hwang, K.-C., Gao, H.: Surface-adsorption-induced bending behaviors of graphene nanoribbons. Appl. Phys. Lett. **98**(12), 121909 (2011)
8. Vandescuren, M., Hermet, P., Meunier, V., Henrard, L., Lambin, P.: Theoretical study of the vibrational edge modes in graphene nanoribbons. Phys. Rev. B **78**(19), 195401 (2008)
9. Matthes, L., Hannewald, K., Bechstedt, F.: Ab initio investigation of graphene-based one-dimensional superlattices and their interfaces. Phys. Rev. B **86**(20), 205409 (2012)
10. Chauhan, S., Srivastava, P., Shrivastava, A.: Electronic and transport properties of boron and nitrogen doped graphene nanoribbons: an ab initio approach. Appl. Nanosci. **4**(4), 461–467 (2014). doi:10.1007/s13204-013-0220-2
11. Ren, H., Li, Q.-X., Luo, Y., Yang, J.: Graphene nanoribbon as a negative differential resistance device. Appl. Phys. Lett. **94**(17), 173110 (2009)
12. García-Fuente, A., Gallego, L.J., Vega, A.: Spin currents and filtering behavior in zigzag graphene nanoribbons with adsorbed molybdenum chains. J. Phys. Condens. Mater **27**(13), 135301 (2015)
13. Rosales, L., González, J.W.: Transport properties of two finite armchair graphene nanoribbons. Nanoscale Res. Lett. **8**(1), 1–6 (2013)
14. Ruseckas, J., Juzeliūnas, G., Zozoulenko, I.: Spectrum of π electrons in bilayer graphene nanoribbons and nanotubes: an analytical approach. Phys. Rev. B **83**(3), 035403 (2011)
15. Hesabi, M., Hesabi, M.: The interaction between carbon nanotube and skin anti-cancer drugs: a DFT and NBO approach. J Nanostruct. Chem. **3**(1), 1–6 (2013). doi:10.1186/2193-8865-3-22
16. Ganesh, E.: Single walled and multi walled carbon nanotube structure, synthesis and applications. Int. J. Innov. Technol. Explor. Eng. **2**(4), 311–320 (2013)
17. Son, Y.-W., Cohen, M.L., Louie, S.G.: Half-metallic graphene nanoribbons. Nature **446**(7133), 342 (2007)
18. Son, Y.-W., Cohen, M.L., Louie, S.G.: Energy gaps in graphene nanoribbons. Phys. Rev. Lett. **97**(21), 216803 (2006)
19. Afshar, M., Babaei, M., Kordbacheh, A.H.: First principles study on structural and magnetic properties of small and pure carbon clusters (C_n, $n = 2$–12). J. Theor. Appl. Phys. **8**(4), 59 (2013). doi:10.1007/s40094-014-0136-6
20. Ordejón, P., Artacho, E., Soler, J.M.: Self-consistent order-N density-functional calculations for very large systems. Phys. Rev. B **53**(16), R10441 (1996)
21. Soler, J.M., Artacho, E., Gale, J.D., García, A., Junquera, J., Ordejón, P., Sánchez-Portal, D.: The SIESTA method for ab initio order-N materials simulation. J. Phys. Condens. Matter **14**(11), 2745 (2002)
22. Perdew, J.P., Zunger, A.: Self-interaction correction to density-functional approximations for many-electron systems. Phys. Rev. B **23**(10), 5048 (1981)
23. Zhang, X., Kuo, J.-L., Gu, M., Bai, P., Sun, C.Q.: Graphene nanoribbon band-gap expansion: broken-bond-induced edge strain and quantum entrapment. Nanoscale **2**(10), 2160–2163 (2010)

On validity of paraxial theory for super-Gaussian laser beams propagating in a plasma

Lalita Devi[1] · Hitendra K. Malik[1]

Abstract In the present paper, we have investigated a situation where a high intensity laser beam passes through a gas and ionizes this gas by tunnel ionization. Here the electric field of the laser provides a sufficient velocity to the electrons to surpass the Coulomb barrier of the atom. Owing to the ionization the plasma density enhances which affects the laser beam propagation. The use of paraxial ray approximation theory for the present case of super-Gaussian lasers reveals the self-focusing of the beams and frequency upshifting. The predicted self-focusing of the laser beams is contrary to the expected outcome of defocusing of these beams in the plasma, indicating that the paraxial theory may not be valid for the case of super-Gaussian lasers even for the inclusion of most of the near axis region in the theory.

Keywords Tunnel ionization · Self-focusing · Super-Gaussian lasers · Frequency upshift · Paraxial ray approximation

Introduction

The interaction of high power laser pulse with gases and plasmas is a major field of research due to its applications in laser electron accelerator, laser driven fusion, and super-continuum generation [1–4]. For many of these applications, it is highly desirable that the laser pulse propagate extended distances (many Rayleigh lengths) at high intensity. The problem of spot size evolution, self-focusing and defocusing of the laser pulse in plasma are of great interest. Self-focusing is a non-linear optical process induced by a change in refractive index of material exposed to intense laser radiation. A medium whose refractive index increases with electric field intensity acts as focusing lens for a laser beam characterized by an initial transverse intensity gradient. There are studies where people have tried to enhance the Rayleigh length. Most of the studies have been conducted for the laser beams having Gaussian profile using paraxial ray approximation [5–8].

Sodha et al. [9] reviewed the self-focusing of electromagnetic beams in plasmas, referring primarily to steady state self-focusing. Esarey et al. [10] gave an extensive review of paraxial ray theory of self-focusing and self-guiding of short laser pulses in ionizing gases and plasmas. Fedosejevs et al. [11] reported experimental results on the competing processes of ionization-induced refraction and relativistic self-focusing. Liu and Tripathi [7] developed a theoretical framework to study the combined effects of the laser-frequency upshift, self-defocusing, and ring formation. The self-defocusing of the laser prepulse and laser guiding of the second laser pulse in an axially nonuniform plasma channel has also been studied by them [12]. The Ohmic and the ponderomotive nonlinearities at sub-picosecond pulse lengths are weak, as the time scale for ambipolar plasma diffusion is much longer. However, two other major nonlinearities are important. These arise due to tunnel ionization of atoms in the optical field of the laser and due to the relativistic dependence of electron mass on its velocity. The former nonlinearity is important when the laser field is comparable with the atomic Coulomb field.

However, the super-Gaussian lasers have recently attracted the researchers all over the world [13–17] for their

✉ Hitendra K. Malik
hkmalik@hotmail.com

1 Plasma Waves and Particle Acceleration Laboratory, Department of Physics, Indian Institute of Technology Delhi, New Delhi 110016, India

better extraction from saturated laser amplifiers, better nonlinear conversion for same peak intensity, and top-hat beams have a parabolic thermal lens which does not decrease beam quality. Due to this reason, it becomes vital to investigate the profile of super-Gaussian lasers when they tunnel ionize the gas and produce the plasma. Hence, in the present article, we have derived the coupled equations for the amplitude and phase of super-Gaussian laser beam and solve it numerically using initial conditions.

Coupled equations for amplitude and phase

Let us consider the propagation of a circularly polarized laser beam in a gas jet target. At $z = 0$, the laser field is given by

$$\vec{E} = (\hat{x} + i\hat{y})E_0(r)e^{-i\omega_0 t}, \tag{1}$$

where $E_0 = E_{00}\exp(-r^4/2r_0^4)$ for $t > 0$ and $E_0 = 0$ for $t < 0$.

The field of the laser ionizes the gas as it passes through it. The tunnel ionization of the gas gives rise to a changing plasma density n_p and plasma frequency ω_p, given by:

$$\frac{\partial\omega_\mathrm{P}^2}{\partial t} = \gamma_T(\omega_\mathrm{N}^2 - \omega_\mathrm{P}^2), \tag{2}$$

where

$$\gamma_\mathrm{T} = (\pi/2)^{1/2}(I_0/\hbar)(|E|/E_\mathrm{a})^{1/2}\exp(-E_\mathrm{a}/|E|), \tag{3}$$

is the rate of tunnel ionization by the laser electric field E, $\omega_\mathrm{N}^2 = 4\pi n_\mathrm{N}e^2/m$, n_N is the initial density of neutral atoms, $E_\mathrm{a} = (4/3)\sqrt{2m}(I_0)^{3/2}/e\hbar$ is the characteristic atomic field, I_0 is the ionization potential, and $h = 2\pi\,\hbar$ is the Planck's constant.

Inside the ionizing gas $(z > 0)$, $\vec{E} = (\hat{x} + i\hat{y})A(t,z,r)\exp(-i\varphi)$, where A is the slowly varying complex amplitude and $\varphi(t,z)$ is the fast phase of the wave. We define $\omega = \partial\varphi/\partial t$, $k = -\,\partial\varphi/\partial z$, $\omega_\mathrm{P0}^2 = \omega_\mathrm{P}^2(t,z,r=0)$. The wave equation for the laser pulse is written as:

$$\nabla^2\vec{E} - \frac{1}{c^2}\frac{\partial^2\vec{E}}{\partial t^2} = \frac{\omega_\mathrm{P}^2}{c^2}\vec{E}. \tag{4}$$

Substituting for $\vec{E}$ we obtain, to successive orders, in the WKB approximation,

$$\omega^2 = \omega_\mathrm{P0}^2 + k^2c^2, \tag{5}$$

$$2ik\frac{\partial A}{\partial z} + \frac{2i\omega}{c^2}\frac{\partial A}{\partial t} + \nabla_\perp^2 A + i\frac{\partial k}{\partial z}A + \frac{i}{c^2}\frac{\partial\omega}{\partial t}A = \frac{1}{c^2}\left[\omega_\mathrm{P}^2(t,z,r) - \omega_\mathrm{P0}^2(t,z)\right]A. \tag{6}$$

Differentiating Eq. (5) w.r.t. t, using $\partial k/\partial t = -\,\partial\omega/\partial z$, and defining $v_\mathrm{g} = c(1 - \omega_\mathrm{P0}^2/\omega^2)$, we obtain

$$\frac{\partial\omega^2}{\delta t} + v_\mathrm{g}\frac{\partial\omega^2}{\partial z} = \frac{\partial\omega_\mathrm{P0}^2}{\partial t}. \tag{7}$$

Combining the first and fourth, and second and fifth terms of Eq. (6) and defining the action amplitude $A_\mathrm{F} = (\omega/\omega_0)^{1/2}A$, $t' = t - z/c$, $z' = z$, and assuming $\omega_\mathrm{P}^2/\omega^2 \ll 1$, we can rewrite Eqs. (6) and (7) simply as

$$\frac{2i\omega}{c}\frac{\partial A_\mathrm{F}}{\partial z'} + \nabla_\perp^2 A_\mathrm{F} = \frac{\omega_\mathrm{P}^2 - \omega_\mathrm{P0}^2}{c^2}A_\mathrm{F}, \tag{8}$$

$$\frac{\partial\omega^2}{\partial z'} = (\gamma_\mathrm{T0}/c)(\omega_\mathrm{N}^2 - \omega_\mathrm{P0}^2), \tag{9}$$

together with $\gamma_\mathrm{T0} = \gamma_\mathrm{T}(r = 0)$. We consider cylindrically symmetric propagation and write $A_\mathrm{F} = A_\mathrm{F0}\exp(iQ)$, where $A_\mathrm{F0}(t', z', r)$ and $Q\,(t', z', r)$ are real. Separating the real and imaginary parts of Eq. (8), we obtain:

$$-\frac{2\omega}{c}\frac{\partial Q}{\partial z'}A_\mathrm{F0} + \frac{\partial^2 A_\mathrm{F0}}{\partial r^2} + \frac{1}{r}\frac{\partial A_\mathrm{F0}}{\partial r} - \left(\frac{\partial Q}{\partial r}\right)^2 A_\mathrm{F0} = \frac{\omega_\mathrm{P}^2 - \omega_\mathrm{P0}^2}{c^2}A_\mathrm{F0}, \tag{10}$$

$$\frac{\omega}{c}\frac{\partial A_\mathrm{F0}^2}{\partial z'} + \left(\frac{\partial^2 Q}{\partial r^2} + \frac{1}{r}\frac{\partial Q}{\partial r}\right)A_\mathrm{F0}^2 + \frac{\partial Q}{\partial r}\frac{\partial A_\mathrm{F0}^2}{\partial r} = 0. \tag{11}$$

To solve these equations, we expand ω_P^2, Q, γ_T in powers of r^2, as:

$$\omega_\mathrm{P}^2 = \omega_\mathrm{P0}^2 + \omega_\mathrm{P2}^2 r^2/r_0^2 + \omega_\mathrm{P4}^2 r^4/r_0^4 + \omega_\mathrm{P6}^2 r^6/r_0^6,$$

$$Q = Q_0 + Q_2 r^2/r_0^2 + Q_4 r^4/r_0^4 + Q_6 r^6/r_0^6,$$

$$\gamma_\mathrm{T} = \gamma_\mathrm{T0} + \gamma_\mathrm{T2}r^2/r_0^2 + \gamma_\mathrm{T4}r^4/r_0^4 + \gamma_\mathrm{T6}r^6/r_0^6,$$

i.e., in the non paraxial approximation. Since the electric field profile of the super-Gaussian laser beam is flat over a certain region away from the axis, we have included higher order terms (up to r^6) so that most of the region can be included. Keeping the analysis up to r^4 order terms, a significant error shall occur. However, the higher order terms shall have negligible impact in the case of Gaussian laser beams.

In this approximation, we write the laser intensity as:

$$A_{F0}^2 = \left(1 + \frac{a_2 r^2}{r_0^2 f_B^2} + \frac{a_4 r^4}{r_0^4 f_B^4} + \frac{a_6 r^6}{r_0^6 f_B^6}\right)\frac{E_{00}^2}{f_B^2}\exp\left(-\frac{r^4}{r_0^4 f_B^4}\right).$$

The coefficients involved in the expansion of γ_T can be determined using Eq. (3). On comparing the successive powers of r^2, we obtain the various equations:

$$\frac{\partial Q_0}{\partial\varsigma} = \frac{a_2}{\omega_\mathrm{SF} f_\mathrm{B}^2},$$

$$Q_2 = \left(\frac{\omega_{SF}}{2f_B}\right)\frac{\partial f_B}{\partial \varsigma},$$

$$\frac{\partial^2 f_B}{\partial \varsigma^2} + \frac{1}{\omega_{SF}}\frac{\partial \omega_{SF}}{\partial \varsigma}\frac{\partial f_B}{\partial \varsigma} = -\frac{\omega_{P2}^2 f_B r_0^2}{c^2\omega_{SF}^2} + \frac{\left(8a_4 - 8 - 3a_2^2\right)}{f_B^3\omega_{SF}^2},$$

$$\frac{\partial}{\partial \varsigma}\left(Q_4 f_B^4\right) = -\frac{\omega_{P4}^2 f_B^4 r_0^2}{c^2\omega_{SF}} + \frac{\left(2a_2^3 - 8a_2a_4 + 9a_6 - 2a_2\right)}{f_B^2\omega_{SF}},$$

$$\frac{\partial}{\partial \varsigma}\left(Q_6 f_B^6\right) = -\frac{\omega_{P6}^2 f_B^6 r_0^2}{2c^2\omega_{SF}} - \frac{8Q_4^2 f_B^6}{\omega_{SF}} + \frac{1}{2f_B^2\omega_{SF}}\left[a_2^2\left(18a_4 - 3a_2^2 + 4\right) - 12a_4^2 - 28a_2a_6 + 4 - 8a_4\right],$$

$$\frac{\partial a_2}{\partial \varsigma} = -\frac{16Q_4 f_B^2}{\omega_{SF}},$$

$$\frac{\partial a_4}{\partial \varsigma} = \frac{1}{\omega_{SF}}\left[-36Q_6 f_B^4 - 24a_2Q_4 f_B^2 + 4a_2^2Q_2\right],$$

$$\frac{\partial a_6}{\partial \varsigma} = -(-4 + 6a_4)f_B a_2\frac{\partial f_B}{\partial \varsigma} - \frac{8Q_4 f_B^2}{\omega_{SF}}\left(a_2^2 + 4a_4\right) - \frac{48Q_6 f_B^4 a_2}{\omega_{SF}},$$

$$\frac{\partial \omega_{P2}^2}{\partial \tau} = -g^{(0)}\omega_{P2}^2 + \frac{g^{(0)}a_2}{4f_B^2}(1 + 2g_1)\left(\omega_N^2 - \omega_{P0}^2\right),$$

$$\frac{\partial \omega_{P4}^2}{\partial \tau} = \frac{\gamma_4}{\gamma_{TO}^{(0)}}\left(\omega_N^2 - \omega_{P0}^2\right) - \frac{\gamma_2}{\gamma_{TO}^{(0)}}\omega_{P2}^2 - g^{(0)}\omega_{P4}^2,$$

$$\frac{\partial \omega_{P6}^2}{\partial \tau} = \frac{\gamma_6}{\gamma_{TO}^{(0)}}\left(\omega_N^2 - \omega_{P0}^2\right) - \frac{\gamma_4}{\gamma_{TO}^{(0)}}\omega_{P2}^2 - \frac{\gamma_2}{\gamma_{TO}^{(0)}}\omega_{P4}^2 - g^{(0)}\omega_{P6}^2$$

$$\frac{\partial \omega_{SF}}{\partial \varsigma} = \frac{\left(\omega_N^2 - \omega_{P0}^2\right)\gamma_{TO}^{(0)}g^{(0)}r_0^2}{2c^2\omega_{SF}\omega_0}.$$

The above coupled equations are written in the dimensionless quantities $\varsigma = z'/R_d$, $\tau = \gamma_{TO0}t'$, $\omega_{SF} = \omega/\omega_0$, and $R_d = (\omega_0/c)r_0^2$ is Rayleigh length.

Results and discussion

The coupled equations are solved by applying boundary conditions for an initially plane wave front. The initial and boundary conditions are at $\tau = 0, \omega_{P0}^2, \omega_{P2}^2, \omega_{P4}^2, \omega_{P6}^2 = 0$ for all ς and at $\varsigma = 0$, $f_B = 1, \frac{\partial f_B}{\partial \varsigma} = 0$, $\omega_{SF} = 1, a_2, a_4, a_6 = 0, Q_4, Q_6 = 0$ for all τ. Since the Keldysh parameter γ decides the nature of ionization and $\gamma \ll 1$ for the tunnel ionization, we select the laser and plasma parameters such that this condition is satisfied. The Keldysh parameter is given by $\gamma = \sqrt{\frac{E_I}{2U_P}} = \sqrt{\frac{\varepsilon_0 m_e c E_I \omega^2}{q^2 I}}$, where E_I is the ionization potential of atom, q is charge on the electron, I is the intensity of the laser beam and other symbols have their usual meanings.

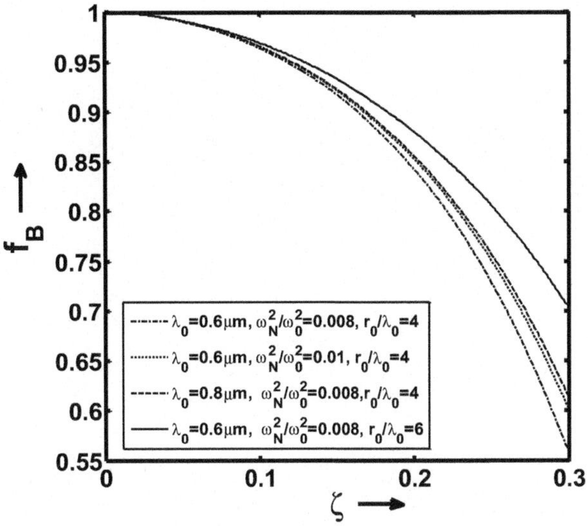

Fig. 1 Variation of beamwidth parameter f_B as a function of distance of propagation ς, for argon gas when intensity = 6.8×10^{19} W/m^2

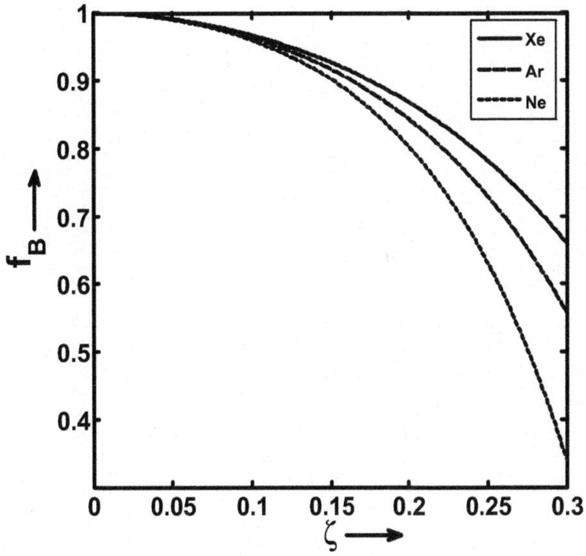

Fig. 2 Variation of beamwidth parameter f_B as a function of distance of propagation ς for different gases. The parameters are $r_0/\lambda_0 = 4$, $\omega_N^2/\omega_0^2 = 0.008$ and intensity = 6.8×10^{19} W/m^2

Figure 1 shows that the beam width parameter goes down with the distance travelled by the laser beam, indicating the self-focusing of the beams. The effect of self-focusing is found to be more for a laser beam of small spot size and smaller wavelength. On the other hand, self-focusing is found to be reduced as we increase the density of neutral atom in the gas.

Figure 2 shows the variation of beam width parameter for three different gases, namely xenon, argon and neon. Here one finds that the self-focusing of the laser in the case of neon gas is more as compared to the argon and xenon gases. On the other hand, this is observed that the lasers of much higher intensities encounter less significant effect of the focusing (see Fig. 3).

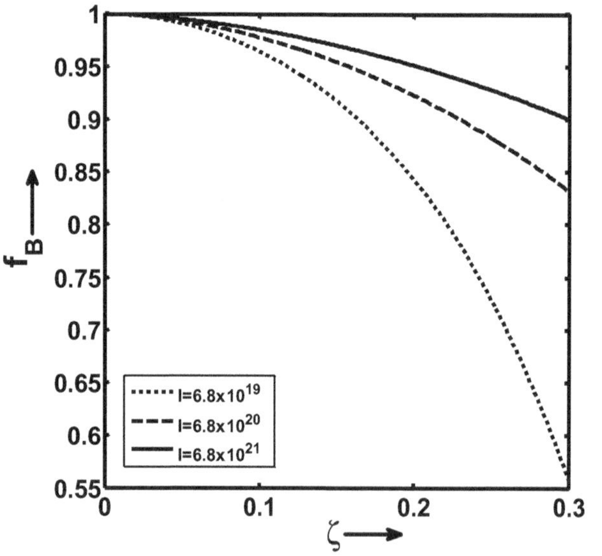

Fig. 3 Variation of beamwidth parameter f_B for different intensities of the laser beams, when $\tau = 0.3$, $r_0/\lambda_0 = 4$ and $\omega_N^2/\omega_0^2 = 0.008$ and the other parameters are the same as in Fig. 2. The intensity I in the legends carries the units W/m^2

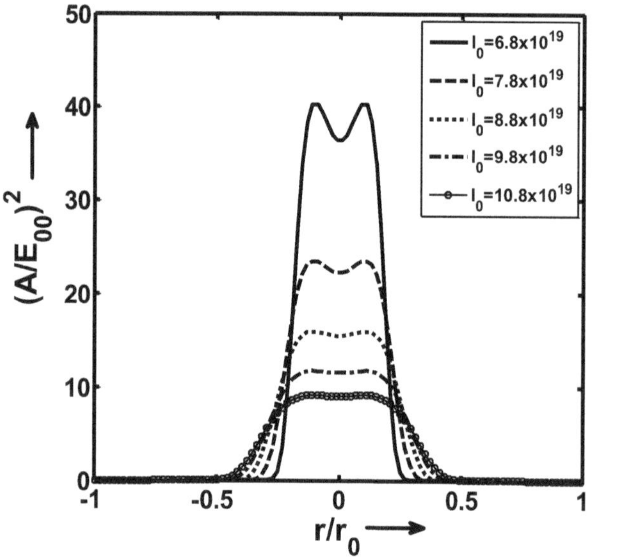

Fig. 5 Normalized intensity $(A/E_{00})^2$ pattern of the lasers as a function of r/r_0 for different intensities. This figure corresponds to Fig. 4. The intensity I_0 in the legends carries the units W/m^2

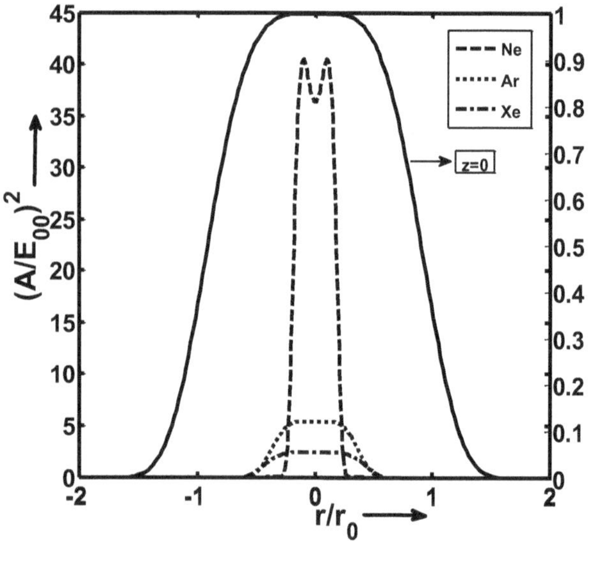

Fig. 4 Normalized intensity $(A/E_{00})^2$ pattern of the laser as a function of r/r_0 for different gases at $\varsigma = 0.3$, $\tau = 1.4$, $\omega_N^2/\omega_0^2 = 0.01$ and intensity $= 6.8 \times 10^{19}$ W/m^2

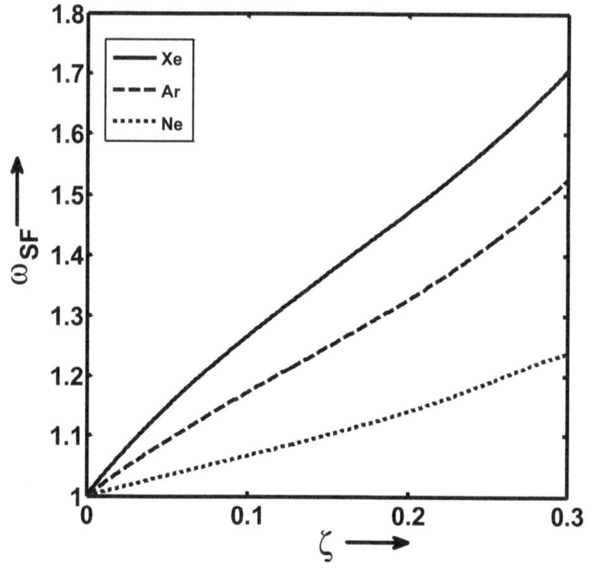

Fig. 6 Dependence of frequency ω_{SF} on the propagation distance ς for different gases, when $r_0/\lambda_0 = 4$ and intensity $= 6.8 \times 10^{19}$ W/m^2

We have plotted Fig. 4 for examining the change in intensity profile of the lasers when they propagate in the tunnel ionized plasma for different gases. The figure shows strong self-focusing of the lasers for all the gases. However, there is a significant change in the laser intensity profile when it ionizes the xenon gas and propagates in it. This effect is found to be prominent in the case of lasers carrying lower intensity (Fig. 5).

The theory used in the present article also predicts the frequency shifting of the lasers when they tunnel ionize the gas and propagate in the formed plasma. The change in the frequency with various parameters is shown in the forthcoming figures.

Figures 6, 7 and 8 show the variation of frequency of the laser when it propagates in the plasma. This is evident from these figures that the frequency gets upshifted during the

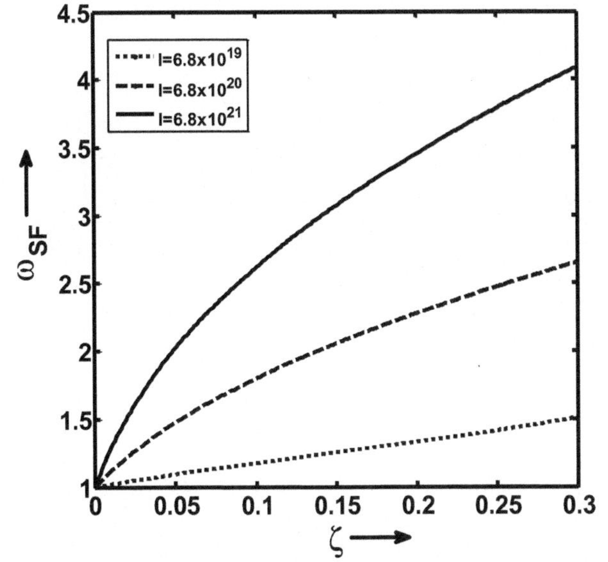

Fig. 7 Dependence of frequency ω_{SF} on the laser intensity, when $r_0/\lambda_0 = 4$ and laser propagates in argon gas

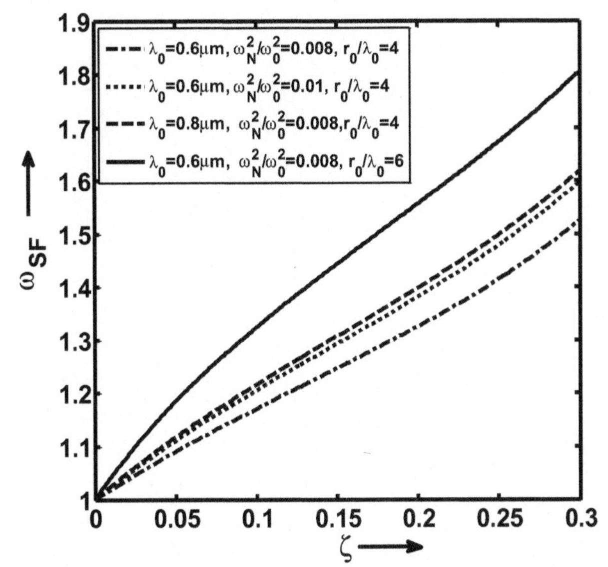

Fig. 8 Dependence of frequency ω_{SF} on the wavelength of the laser, spot size and density of neutrals, when intensity = 6.8×10^{19} W/m^2 and the other parameters are the same as in Fig. 2

propagation of the laser beam in the plasma. The highest shifting takes place in the case of xenon gas, whereas this is the least in the case of neon gas. The significant upshifting of the laser frequency takes place when the laser carries higher intensity (Fig. 7). Under this situation, less significant change in the laser profiles was also noticed. This is observed that the lasers with higher spot size undergo much significant changes in the frequency (Fig. 8); similar is the effect when there is a gas with larger number of neutrals. Another observation is that the lasers of larger wavelength undergo with the larger upshifting in its frequency.

Conclusions

We have considered super-Gaussian beam profile having maximum intensity on the axis and minimum intensity away from the axis. As the laser beam propagates in the gas to form the plasma, the plasma density should stay maximum on the axis which should result in minimum refractive index on the axis. Since the phase velocity of the lasers depends inversely on the refractive index, the phase velocity becomes maximum on the axis and falls down sharply on the edges. This would lead to a diverging wavefront, causing the laser beam to defocus. Kumar and Tripathi [6] observed the self-defocusing of Gaussian laser pulse based on the paraxial ray theory in tunnel ionized helium plasma. Liu and Tripathi [7] also observed the self-defocusing of Gaussian beam based on the paraxial ray theory in tunnel ionized neon plasma. However, we point out that the use of this theory in the case of super-Gaussian lasers beams provides an unphysical result of self-focusing of these beams. Hence, the paraxial ray approximation theory does not seem to be authentic in the case of super-Gaussian laser beams.

Acknowledgements One of the authors, Lalita Devi, would like to thank CSIR, Govt. of India for providing the financial support for this work.

References

1. Braun, A., Korn, G., Liu, X., Du, D., Squier, J., Mourou, G.: Self-channeling of high-peak-power femtosecond laser pulses in air. Opt. Lett. **20**, 73–75 (1995)
2. Esarey, E., Sprangle, P.: Generation of stimulated back scattered harmonic generation from intense laser interactions with beams and plasmas. Phys. Rev. A **45**, 5872–5882 (1992)
3. Malik, H.K.: Analytical calculation of wakefield generated by microwave pulses in a plasma filled waveguide for electron acceleration. J. Appl. Phys. **104**, 053308(1–7) (2008)
4. Deutsch, G.C., Furukawa, H., Mima, K., Murakami, M., Nishihara, K.: Interaction physics of the fast ignitor concept. Phys. Rev. Lett. **77**, 2483 (1996)
5. Faisal, M., Mishra, S.K., Verma, M.P., Sodha, M.S.: Ring formation in self-focusing of electromagnetic beams in plasmas. Phys. Plasmas **14**, 103103 (2007)
6. Kumar, N., Tripathi, V.K.: Non-paraxial theory of self-defocusing/focusing of a laser pulse in a multiple-ionizing gas. Appl. Phys. B **82**, 53–58 (2006)
7. Liu, C.S., Tripathi, V.K.: Laser frequency upshift, self defocusing, and ring formation in tunnel ionizing gases and plasmas. Phys. Plasmas **7**, 4360 (2000)
8. Singh, A., Singh, N.: Guiding of a laser beam in a collisionless magnetoplasma channel. J. Opt. Soc. Am. B **28**, 081844(1–7) (2011)
9. Sodha, M.S., Prasad, S., Tripathi, V.K.: Nonstationary self-focusing of a Gaussian pulse in a plasma. J. Appl. Phys. **46**, 637 (1975)

10. Esarey, E., Sprangle, P., Krall, J., Ting, A.: Self-focusing and guiding of short laser pulses in ionizing gases and plasmas. IEEE J. Quantum Electron. **33**, 1879 (1997)
11. Fedosejevs, R., Wang, X.F., Tsakiris, G.D.: Onset of relativistic self-focusing in high density gas jet targets. Phys. Rev. E **56**, 4615 (1997)
12. Liu, C.S., Tripathi, V.K.: Laser guiding in an axially nonuniform plasma channel. Phys. Plasmas **1**, 3100 (1994)
13. Singh, D., Malik, H.K.: THz generation by mixing of two super-Gaussian laser beams in collisional plasma. Phys. Plasmas **21**, 083105(1–5) (2014)
14. Malik, A.K., Malik, H.K.: Tuning and focusing of terahertz radiation by DC magnetic field in a laser beating process. IEEE J. Quantum Electron. **49**, 232–237 (2013)
15. Malik, H.K.: Terahertz radiation generation by lasers with remarkable efficiency in electron–positron plasma. Phys. Lett. A **379**, 2826–2829 (2015)
16. Malik, H.K.: Density bunch formation by microwave in a plasma-filled cylindrical waveguide. Europhys. Lett. **106**, 55002(1–5) (2014)
17. Malik, H.K., Malik, A.K.: Strong and collimated terahertz radiation by super-Gaussian lasers. Europhys. Lett. **100**, 45001(1–5) (2012)

21

Surface coupling effects on contact mechanics: contact area and interfacial separation between an elastic solid and a hard substrate with randomly rough, self-affine fractal surfaces

M. Feshanjerdi[1] · A. A. Masoudi[1] · M. Khorrami[1]

Abstract The objective of this study is to investigate both the contact area and the interfacial separation between two surfaces. Both surfaces are considered to be rough, one of them being elastic and the other one hard. The work is based on an extended version of Persson's model of contact mechanics to study the behavior of the contact area, the interfacial separation and the pressure distribution. The results are compared with the case merely the hard substrate is rough. It is seen that introducing a roughness in the elastic surface decreases the real contact, if the surfaces are uncorrelated. A positive (negative) correlation increases (decreases) the real contact. A reverse pattern occurs for the width of the pressure distribution, as well as the interfacial separation (at equal pressures).

Keywords Self-affine fractal · Cross-correlation · Surface effects

Introduction

All of the surfaces occurring in nature and industry are rough, provided they are observed with sufficiently high magnifications (small length scales) [1, 2]. So, for two contacting solid surfaces, microscopically, there are many non-contact regions (the interfacial separation), and microscopic contact occurs only at a fraction of the macroscopic contact. This fraction of real contact, as well as the interfacial separation, are affected by the roughness of the surfaces, and play important roles in the mechanical properties of the system. The area of real contact characterizes the frictional properties of the contact, as well as the strength of adhesion and the amount of wear [3–5]. Some other phenomena are affected by the interfacial separation, among which are the heat transfer, the contact resistivity, lubrication, and sealing [5–8]. The effect of the surface roughness on the area of real contact has been studied by two classes of analytical models. The first class involves multiasperity contact theories (originally formulated by Greenwood and Williamson (GW) [9–13]), where the contact between the surfaces is modeled as an ensemble of randomly distributed Hertzian contacts between the asperities. The second class is based on Persson's model of contact mechanics [1, 2], where the probability distribution of the contact pressure is shown to be governed by a diffusive process in terms of the magnification at which the interface is observed. Numerical studies [14–17] have shown that, in the case of non-adhesive contacts, when a flat elastic body is brought into contact with a rough surface, the real contact area increases proportional to the applied normal squeezing pressure (applied load). In [18], it has been shown that the GW-type theories predict linearity only for vanishingly small contact areas, corresponding to vanishingly small applied normal squeezing pressures. When the applied normal squeezing pressure is increased, the theoretical predictions rapidly deviate from the asymptotic linearity. This behavior is not seen in Persson's model, which predicts linearity between contact area and applied normal squeezing pressure for real contact values of up to about 15–20 percent of nominal contact area. This is in agreement with some experimental and numerical

✉ A. A. Masoudi
masoudi@alzahra.ac.ir

M. Feshanjerdi
m.feshanjerdi@alzahra.ac.ir

M. Khorrami
mamwad@mailaps.org

1 Department of Physics, Alzahra University, 1993891167 Tehran, Iran

results. Under full contact conditions, Persson's model is exact, but in the case of partial contact, some numerical results [16, 19] regarding non-adhesive contacts between rough surfaces indicate an underestimation of the contact area in Persson's model, while the results of the model still qualitatively agree with numerical calculations [20].

In all of these works, the area of real contact between a smooth elastic solid surface and a hard substrate with randomly rough surface has been studied. As stated earlier, however, there are essentially no surfaces which are smooth on atomic scales. Here, the elastic solid is assumed to have a rough surface as well, and the effect of roughness on the area of real contact and the interfacial separation is studied. The contacts are assumed to be frictionless and non-adhesive, and the roughness of both surfaces is assumed to be random. An extended version of Persson's model of contact mechanics is used to investigate the area of real contact, as well as the interfacial separation between two surfaces.

The outline of the paper is as follows. In Sect. 2, Persson's model of contact area and interfacial surface separation is reviewed. In Sect. 3, an extended version of Persson's model of contact mechanics is used to calculate the contact area and the interfacial surface separation for randomly rough elastic solids and hard substrates with randomly rough surfaces in contact with each other. Numerical results corresponding to the randomly rough self-affine fractal surfaces are presented in Sect. 4. Sections 3 and 4 contain the main results. The novelty, which is described in these sections, is the introduction of a second rough surface and the investigation of the effects of both surfaces being rough and also the effect of their correlation. Section 5 contains the concluding remarks.

The contact area, and the interfacial surface separation

The main topics of the Persson's model of contact mechanics are reviewed in this section. The model is based on the investigation of the interfacial at different magnifications ζ, [1, 2]. The system's saturation length is L, the length scale is λ, and the wave numbers q and q_L correspond to the length scale and the size of the system, respectively:

$$\begin{aligned} q &= \frac{2\pi}{\lambda}, \\ q_L &= \frac{2\pi}{L}, \\ \zeta &= \frac{L}{\lambda}, \\ &= \frac{q}{q_L}, \end{aligned} \tag{1}$$

The goal is to find the probability distribution $P(\sigma, \zeta)$ of the normal stress σ at the interface under the magnification ζ. The equation governing this distribution is shown to be a diffusion equation. For a non-adhesive contact, the boundary condition is [21]

$$P(\sigma = 0, \zeta) = 0, \tag{2}$$

and for the resulting probability distribution [14, 22]

$$P(\sigma, \zeta) = \frac{1}{2\,(\pi G)^{1/2}} \left\{ \exp\left[-\frac{(\sigma - p)^2}{4G} \right] - \exp\left[-\frac{(\sigma + p)^2}{4G} \right] \right\}, \tag{3}$$

where p is the nominal squeezing pressure and

$$G(q) = \frac{\pi}{4} \left(\frac{E}{1 - \upsilon^2} \right)^2 \int_{q_L}^{q} dq'\, q'^3\, C(q'), \tag{4}$$

with E and υ being the elastic modulus and the Poisson's ratio of the elastic block, respectively. $C(q)$ is the auto-spectral density function [23] of the hard randomly rough substrate. Denoting the actual (microscopic) and the nominal (macroscopic) contact areas by A and A_0, respectively, the relative contact area is

$$\begin{aligned} \frac{A}{A_0} &= \int_{0^+}^{\infty} d\sigma\, P(\sigma, \zeta), \\ &=: P(q). \end{aligned} \tag{5}$$

So,

$$\begin{aligned} \frac{A}{A_0} &= \frac{1}{(\pi G)^{1/2}} \int_0^p d\sigma\, \exp\left(-\frac{\sigma^2}{4G} \right), \\ &= \operatorname{erf}\left(\frac{p}{2\,G^{1/2}} \right). \end{aligned} \tag{6}$$

The error function can be approximated by a linear function of its argument, for the case the argument is small. The result is

$$\frac{A}{A_0} \approx \frac{p}{(\pi G)^{1/2}}, \quad p \ll G^{1/2}. \tag{7}$$

Regarding the interfacial surface separation, let us consider an elastic block squeezed against a flat hard surface. The separation between the average surface plane of the block and the average surface plane of the substrate is denoted by $\bar{u}$, which is nonnegative. Due to the external pressure p required to produce this separation, an elastic energy is stored in the block. Denoting this by U_{el}, one arrives at

$$\begin{aligned} U_{\mathrm{el}}(\bar{u}) &= \int_{\bar{u}}^{\infty} du\, A_0\, p(u), \\ p &= -\frac{1}{A_0} \frac{dU_{\mathrm{el}}}{du}. \end{aligned} \tag{8}$$

When the elastic block is soft and the hard substrate is rough, Persson's model results in the following for the elastic energy [2, 24–26].

$$U_{\rm el} \approx A_0 E^* \gamma \frac{\pi}{2} \int_{q_L}^{q_a} dq\, q^2 P(q)\, C(q), \tag{9}$$

$$E^* = \frac{E}{(1-\upsilon^2)}, \tag{10}$$

where q_a is the largest surface-roughness wave vector, and $C(q)$ is the auto-spectral density function of the hard randomly rough surface. In cases where the applied normal squeezing pressure p is small, the surface asperities do not fully penetrate the elastic block and only a partial contact is realized. So, the full contribution of the auto-spectral density function is not received by the elastic energy. In (9), this has been addressed through the factor $\gamma P(q)$, where $P(q)$ is the relative contact area for elastic nonadhesive contact and is given by [1, 21]

$$P(q) = \frac{1}{\sqrt{\pi}} \int_0^{S(q)p} dx \exp(-x^2). \tag{11}$$

$$S(q) = \frac{w(q)}{E^*}. \tag{12}$$

$$w(q) = \left[\pi \int_{q_L}^{q} dq'\, q'^3\, C(q')\right]^{-1/2}. \tag{13}$$

γ is less than one (but of the order one), when the squeezing pressures are small, and this factor takes into account the fact that the elastic energy stored in the contact region is less than the average elastic energy for full contact [22].

Substituting (9), (10), (11), and (13) in (8), after some calculations, it is shown that for nonadhesive interactions and small applied pressures, the relation between the average interfacial separation $\bar{u}$ and the small applied normal squeezing pressure p is [22, 24].

$$\begin{aligned} p &\approx \beta E^* \exp\left(-\frac{\bar{u}}{u_0}\right), \\ \bar{u} &\approx u_0 \log\left(\frac{\beta E^*}{p}\right), \end{aligned} \tag{14}$$

where

$$u_0 = \sqrt{\pi}\gamma \int_{q_L}^{q_a} dq\, q^2\, C(q)\, w(q), \tag{15}$$

$$\beta = \varepsilon \exp\left\{-\frac{\int_{q_L}^{q_a} dq\, q^2\, C(q)\, w(q) \log[w(q)]}{\int_{q_L}^{q_a} dq\, q^2\, C(q)\, w(q)}\right\}, \tag{16}$$

$$\begin{aligned} \varepsilon &= \exp\left[\int_0^{\infty} dx\, 2x \log x \exp(-x^2)\right], \\ &\approx 0.7493. \end{aligned} \tag{17}$$

An extension to the case of two randomly rough surfaces in contact with each other

As stated before, usually both of surfaces which are in contact with each other are rough. Here, Persson's model of contact mechanics is extended to such cases.

The contact area

The relation between the Fourier transform of the normal stress $\sigma_z(\boldsymbol{q})$ and that of the normal displacement field $u_z(\boldsymbol{q})$ of the surface of the elastic solid is [26]:

$$u_z(\boldsymbol{q}) = M_{zz}(\boldsymbol{q})\, \sigma_z(\boldsymbol{q}), \tag{18}$$

$$M_{zz}(\boldsymbol{q}) = -\frac{2(1-\upsilon^2)}{E q}. \tag{19}$$

When a randomly rough elastic surface is in contact with the randomly rough surface of a hard substrate, Eq. (18) still holds, but now, if the contact between the surfaces is full, the normal displacement field of the elastic solid is equal to the difference of the heights of the surfaces [27, 28]

$$u_z(\boldsymbol{q}) = h_2(\boldsymbol{q}) - h_1(\boldsymbol{q}), \tag{20}$$

so that

$$h_2(\boldsymbol{q}) - h_1(\boldsymbol{q}) = M_{zz}(\boldsymbol{q})\, \sigma_z(\boldsymbol{q}). \tag{21}$$

So, the relative contact area and pressure distribution at the interface ($\sigma > 0$) change. The change could be implemented through a substitution of the autocorrelation function C with the autocorrelation corresponding to $(h_2 - h_1)$, that is [27, 28]

$$C \to \left(C_1 + C_2 - 2\eta\sqrt{C_1 C_2}\right), \tag{22}$$

where $C_1(q)$ and $C_2(q)$ are the auto-spectral density functions for the rough substrate and the rough elastic block, respectively, and η is the coherence function [29, 30]. The surfaces are assumed to be homogeneous and isotropic, so (4) is transformed into:

$$\begin{aligned} G(q) = \frac{\pi}{4}\left(\frac{E}{1-\upsilon^2}\right)^2 \int_{q_L}^{q} dq'\, q'^3 \Big[C_1(q') + C_2(q') \\ -2\eta\sqrt{C_1(q')\, C_2(q')}\Big]. \end{aligned} \tag{23}$$

This new form of G is to be put in (6) and (7), to obtain the contact area. Here, η is considered to be a constant (independent of $\boldsymbol{q}$), and the results for this simple case are presented. Special cases are $\eta = 0$ (uncorrelated surfaces), $\eta = +1$ (completely positive correlated surfaces), or $\eta = -1$ (completely negative correlated surfaces).

The interfacial surface separation

When a randomly rough elastic solid is squeezed against a randomly rough surface of a hard substrate, the elastic energy is similar to (9), but with C substituted according to (22) [27, 28]:

$$U_{\mathrm{el}} \approx \frac{\pi E A_0 \gamma}{2(1-\upsilon^2)} \int_{q_L}^{q_a} \mathrm{d}q\, q^2 P(q) \left[C_1(q) + C_2(q) - 2\eta \sqrt{C_1(q)\, C_2(q)}\right]. \tag{24}$$

The relative contact area for elastic nonadhesive contact is obtained from (11) with

$$w(q) = \left(\pi \int_{q_L}^{q} \mathrm{d}q'\, q'^3 \left[C_1(q') + C_2(q') - 2\eta \sqrt{C_1(q')\, C_2(q')}\right]\right)^{-1/2}. \tag{25}$$

Using (11),

$$\frac{\partial P}{\partial \bar{u}} = \frac{2}{\sqrt{\pi}} S(q) \exp\{-[S(q)]^2 p^2\} \frac{\mathrm{d}p}{\mathrm{d}\bar{u}}. \tag{26}$$

Substituting (24) and (26) in (8) results in:

$$\mathrm{d}\bar{u} = -\sqrt{\pi}\gamma \int_{q_L}^{q_a} \mathrm{d}q\, q^2 \left[C_1(q) + C_2(q) - 2\eta \sqrt{C_1(q)\, C_2(q)}\right] \times w(q) \exp\left\{-\left[\frac{w(q)\, p}{E^*}\right]^2\right\} \frac{\mathrm{d}p}{p}. \tag{27}$$

Integrating this from $\bar{u} = 0$ (full contact, corresponding to $p = \infty$) to $\bar{u}$ gives:

$$\bar{u} = \sqrt{\pi}\gamma \int_{q_L}^{q_a} \mathrm{d}q\, q^2 \left[C_1(q) + C_2(q) - 2\eta \sqrt{C_1(q)\, C_2(q)}\right] \times w(q) \int_{p}^{\infty} \frac{\mathrm{d}p'}{p'} \exp\left\{-\left[\frac{w(q)\, p'}{E^*}\right]^2\right\}. \tag{28}$$

For very low squeezing pressures, one has

$$p \approx \beta E^* \exp\left(-\frac{\bar{u}}{u_0}\right), \tag{29}$$

$$u_0 = \sqrt{\pi}\gamma \int_{q_L}^{q_a} \mathrm{d}q\, q^2 \left[C_1(q) + C_2(q) - 2\eta \sqrt{C_1(q)\, C_2(q)}\right] w(q), \tag{30}$$

$$\beta = \varepsilon \exp \left\{-\frac{\int_{q_L}^{q_a} \mathrm{d}q\, q^2 \left[C_1(q) + C_2(q) - 2\eta \sqrt{C_1(q)\, C_2(q)}\right] w(q) \log[w(q)]}{\int_{q_L}^{q_a} \mathrm{d}q\, q^2 \left[C_1(q) + C_2(q) - 2\eta \sqrt{C_1(q)\, C_2(q)}\right] w(q)}\right\}, \tag{31}$$

with ε being obtained from (17).

Numerical results

In all cases, it is assumed that the randomly rough elastic solid surface and the randomly rough surface of the hard substrate are both self-affine fractal surfaces with the auto-spectral density functions described by [24]:

$$C(q) \approx \frac{H}{\pi} \left(\frac{h_{\mathrm{rms}}}{q_L}\right)^2 \left(\frac{q}{q_L}\right)^{-(2H+2)}, \tag{32}$$

where H is the Hurst exponent, and it is assumed that q_a is much larger than q_L, which usually is. For the hard substrate and the elastic block, these values have been used.

$$q_L = 2 \times 10^8\,\mathrm{m}^{-1}, \quad q_a = 4 \times 10^{10}\,\mathrm{m}^{-1}, \tag{33}$$

$$H_1 = 0.8, \quad h_{\mathrm{rms}\,1} = 1\,nm, \qquad H_2 = 0.7, \quad h_{\mathrm{rms}\,2} = 1\,nm, \tag{34}$$

$$E = 77.2\,\mathrm{GPa}, \quad \upsilon = 0.42. \tag{35}$$

Contact area

Figure 1 shows the contact area ratio A/A_0 calculated from (6) and (23), as a function of the normalized pressure p/E^* for the magnification $\zeta = 4$.

The pressure distribution calculated from (3) and (23), as a function of the normalized pressure σ/E^*, is shown in Fig. 2 for $\zeta = 4$ and for three different nominal pressures.

In these curves, the circle curve is for the case where only the substrate is rough. The asterisk, solid, and dotted curves correspond to the cases where both surfaces are rough and uncorrelated ($\eta = 0$), completely positively correlated ($\eta = 1$), and completely negatively correlated

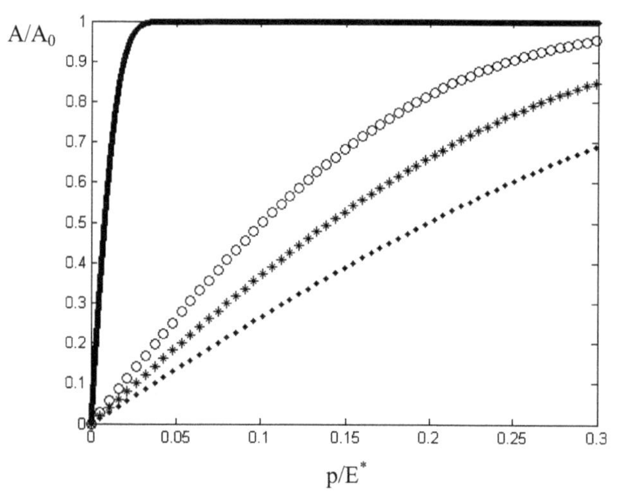

Fig. 1 The contact area ratio A/A_0 as a function of the normalized pressure p/E^* for the magnification $\zeta = 4$. The *circle curve* is for the case where only the substrate is rough. The *asterisk, solid*, and *dotted curves* correspond to the cases where both surfaces are rough and uncorrelated ($\eta = 0$), completely positively correlated ($\eta = 1$), and completely negatively correlated ($\eta = -1$), respectively

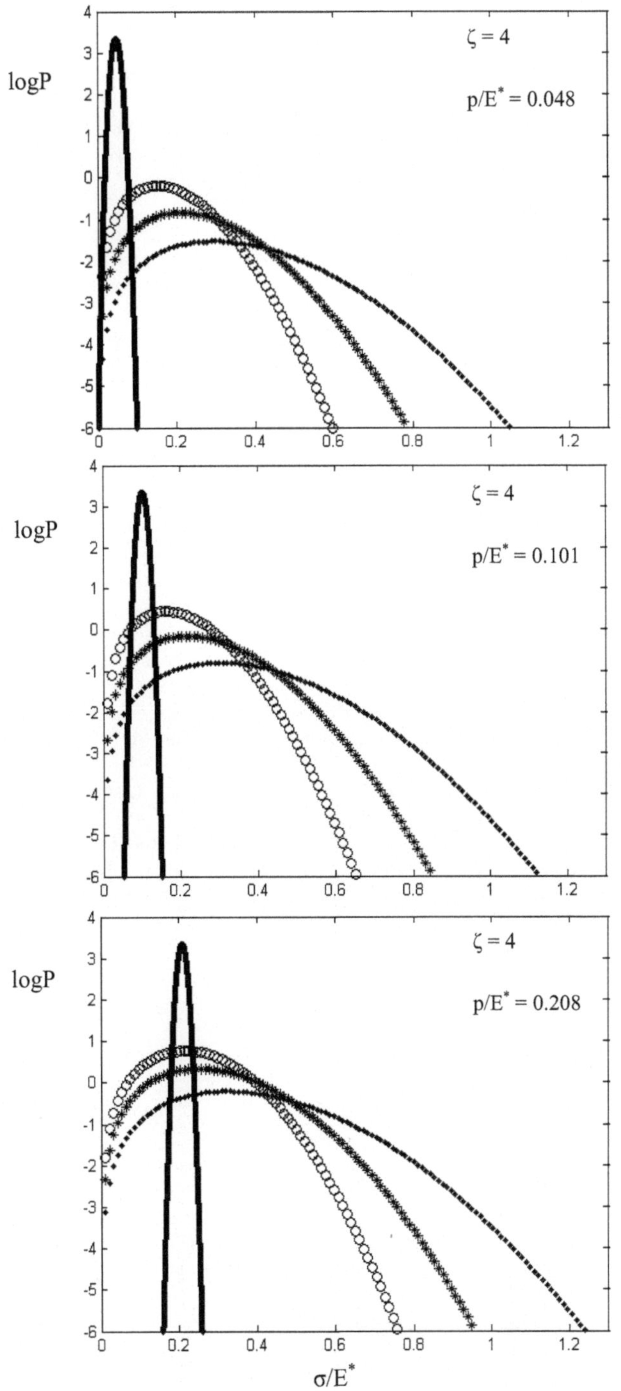

Fig. 2 The pressure distribution as a function of the normalized pressure σ/E^* for the magnification $\zeta = 4$ and for three different nominal pressures. The *circle curve* is for the case where only the substrate is rough. The *asterisk*, *solid*, and *dotted curves* correspond to the case where both surfaces are rough and uncorrelated ($\eta = 0$), completely positively correlated ($\eta = 1$), and completely negatively correlated ($\eta = -1$), respectively

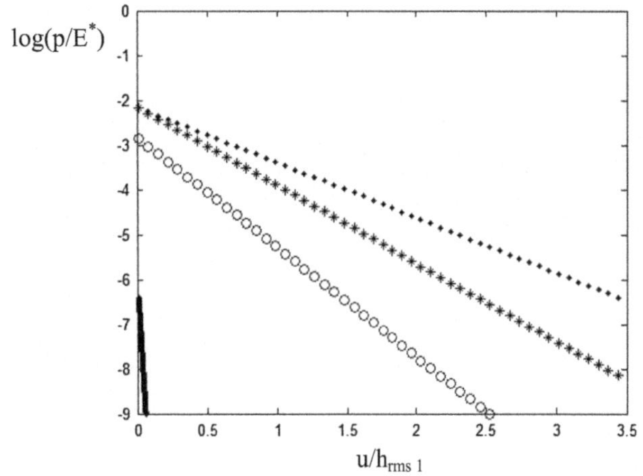

Fig. 3 The logarithm of the normalized average pressure p/E^*, as a function of the separation $\bar{u}$ between the average plane of the substrate and the average plane of the lower surface of the elastic block for the magnification $\zeta = 4$. The *circle curve* is for the case where only the substrate is rough. The *asterisk*, *solid*, and *dotted curves* correspond to the case where both surfaces are rough and uncorrelated ($\eta = 0$), completely positively correlated ($\eta = 1$), and completely negatively correlated ($\eta = -1$), respectively

($\eta = -1$), respectively. Figure 2 shows that both surfaces being rough, but uncorrelated, results in a decrease in the contact area, compared to the case of only one rough surface. If both surfaces are rough, and they are correlated, depending on the sign of the correlation, an increase or decrease in the values of the contact area is resulted, compared to the case of two uncorrelated surfaces and the case of only one rough surface. A positive correlation ($\eta = 1$) increases the contact area between the two surfaces, so that the pressure distribution vanishes in a smaller normalized pressure, as seen from Fig. 2. For a negative correlation $\eta = -1$, however, the contact area is decreased compared to the case of uncorrelated surfaces and the case of only rough surface, so that the pressure distribution vanishes in larger pressure. It is seen that when both surfaces are rough but uncorrelated, the width of the pressure distribution is larger compared to the case of only one rough surface. A positive (negative) correlation results in a decrease (an increase) of the width of the pressure distribution.

Interfacial surface separation

The logarithm of the normalized average pressure p/E^*, as a function of the separation $\bar{u}$ between the average plane of the substrate and the average plane of the lower surface of the elastic block is shown in Figs. 3 and 4. Figure 3 corresponds to the magnification $\zeta = 4$, and Fig. 4 corresponds to the magnification $\zeta = 200$. In these curves, the circle curve is for the case where only the substrate is rough. The asterisk, solid, and dotted curves correspond to the cases where both surfaces are rough and uncorrelated ($\eta = 0$), completely positively correlated ($\eta = 1$), and completely negatively correlated ($\eta = -1$), respectively. It is seen that the interfacial separation at a fixed pressure is larger when

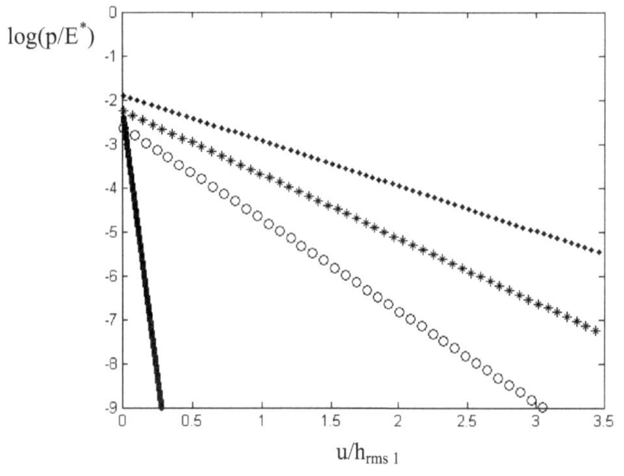

Fig. 4 The logarithm of the normalized average pressure p/E^{*}, as a function of the separation $\bar{u}$ between the average plane of the substrate and the average plane of the lower surface of the elastic block for the magnification $\zeta = 200$. The *circle curve* is for the case where only the substrate is rough. The *asterisk*, *solid*, and *dotted curves* correspond to the cases where both surfaces are rough and uncorrelated ($\eta = 0$), completely positively correlated ($\eta = 1$), and completely negatively correlated ($\eta = -1$), respectively

both surfaces are rough but uncorrelated. A positive (negative) correlation decreases (increases) the interfacial separation.

Concluding remarks

An extension of Persson's model of contact mechanics was used to study the contact area and the interfacial separation, when the elastic solid and the hard substrate are both rough. It was seen that when the two surfaces are rough but uncorrelated, the real contact decreases compared to the case where only the substrate is rough, effectively the roughness has been increased. It was also shown that when the surfaces are correlated, a positive correlation increases the real contact area compared to the case of no correlation, while a negative correlation decreases the real contact area compared to the case of no correlation. A reverse pattern is seen for the width of the pressure distribution, as well as the interfacial separation (at equal pressures): making both surfaces rough but uncorrelated increases these, and a positive (negative) correlation results in a decrease (an increase) in these.

Acknowledgements This work was supported by the research council of the Alzahra University.

References

1. Persson, B.N.J.: Theory of rubber friction and contact mechanics. J. Chem. Phys. **115**, 3840 (2001)
2. Persson, B.N.J.: Adhesion between an elastic body and a randomly rough hard surface. Eur. Phys. J. E. **8**, 385 (2002)
3. Bowden, F.P., Tabor, D.: Friction and Lubrication of Solids. Wiley, New York (1956)
4. Johnson, K.L.: Contact Mechanics. Cambridge University Press, Cambridge (1985)
5. Persson, B.N.J.: Sliding Friction: Physical Principles and Applications, 2nd edn. Springer, Berlin (2000)
6. Volokitin, A.I., Persson, B.N.J.: Near-field radiative heat transfer and noncontact friction. Rev. Mod. Phys. **79**, 1291 (2007)
7. Rabinowicz, E.: Friction and Wear of Materials, 2nd edn. Wiley, New York (1995)
8. Patir, N., Cheng, H.S.: An average flow model for determining effects of three dimensional roughness onpartial hydrodynamic lubrication. J. Lubr. Technol. **100**, 12 (1978)
9. Greenwood, J.A., Williamson, J.B.P.: Contact of nominally flat surfaces. Proc. R. Soc. Lond. A **295**, 300 (1966)
10. Bush, A.W., Gibson, R.D., Thomas, T.R.: The elastic contact of a rough surface. Wear **35**, 87 (1975)
11. Thomas, T.R.: Rough Surfaces, vol. 8. Longman Group Limited, London (1982)
12. Greenwood, J.A.: A simplified elliptic model of rough surface contact. Wear **261**, 191 (2006)
13. Carbone, G.: A slightly corrected Greenwood and Williamson model predicts asymptotic linearity between contact area and load. J. Mech. Phys. Solids **57**, 1093 (2009)
14. Yang, C., Tartaglino, U., Persson, B.N.J.: A multiscale molecular dynamics approach to contact mechanics. Eur. Phys. J. E **19**, 47 (2006)
15. Borri-Brunetto, M., Chiaia, B., Ciavarella, M.: Incipient sliding of rough surfaces in contact: a multiscale numerical analysis. Comput. Methods Appl. Mech. Eng. **190**, 6053 (2001)
16. Hyun, S., Pei, L., Molinari, J.F., Robbins, M.O.: Finite-element analysis of contact between elastic self-affine surfaces. Phys. Rev. E **70**, 026117 (2004)
17. Campana, C.: Using Green's function molecular dynamics to rationalize the success of asperity models when describing the contact between self-affine surfaces. Phys. Rev. E **78**, 026110 (2008)
18. Carbone, G., Bottiglione, F.: Asperity contact theories: do they predict linearity between contact area and load?. J. Mech. Phys. Solids **56**, 2555 (2008)
19. Yang, C., Persson, B.N.J.: Molecular dynamics study of contact mechanics: contact area and interfacial separation from small to full contact. Phys. Rev. Lett. **100**, 024303 (2008)
20. Campana, C., Mušer, M.H., Robbins, M.O.: Elastic contact between self-affine surfaces: comparison of numerical stress and contact correlation fucntions with analytic predictions. J. Phys.: Condens. Mater. **20**, 354013 (2008)
21. Persson, B.N.J., Bucher, F., Chiaia, B.: Elastic contact between randomly rough surfaces: comparison of theory with numerical results. Phys. Rev. B **65**, 184106 (2002)
22. Yang, C., Persson, B.N.J.: Contact mechanics: contact area and interfacial separation from small contact to full contact. J. Phys.: Condens. Matter. **20**, 215214 (2008)
23. Persson, B.N.J., Albohr, O., Tartaglino, U., Volokitin, A.I., Tosatti, E.: On the nature of surface roughness with application to contact mechanics, sealing, rubber friction and adhesion. J. Phys.: Condens. Matter. **17**, R1 (2005)
24. Persson, B.N.J.: Relation between interfacial separation and load: a general theory of contact mechanics. Phys. Rev. Lett. **99**, 125502 (2007)
25. Persson, B.N.J.: Contact mechanics for randomly rough surfaces. Surf. Sci. Rep. **61**, 201 (2006)

26. Persson, B.N.J., Tosatti, E.: The effect of surface roughness on the adhesion of elastic solids. J. Chem. Phys. **115**, 5597 (2001)
27. Feshanjerdi, M., Masoudi, A.A., Khorrami, M.: Surface coupling effects on the adhesive contact between an elastic solid and a hard substrate with randomly rough, self-affine fractal surfaces. J. Stat. Mech. **2015**(2), P02018 (2015)
28. Feshanjerdi, M., Khorrami, M., Masoudi, A.A.: The hysteretic contribution of friction for the polished rubber on the concrete surface. Appl. Surf. Sci **394**, 528 (2017)
29. Bendat, J.S., Piersol, A.G.: Random Data: Analysis and Measurement Procedures, 3rd edn. Wiley, NewYork (2000)
30. Palasantzas, G., Barnas, J., Hosson, J.T.M.D.: Correlated roughness effects on electrical conductivity of quantum wires. J. Appl. Phys. **89**, 8002 (2001)

22

Elastic and thermodynamical properties of cubic (3*C*) silicon carbide under high pressure and high temperature

Dinesh Varshney[1] · S. Shriya[1] · M. Varshney[2] · N. Singh[3] · R. Khenata[4]

Abstract Pressure-dependent first-order phase transition, mechanical, elastic, and thermodynamical properties of cubic zinc blende to rock-salt structures in 3*C* silicon carbide (SiC) are presented. An effective interatomic interaction potential for SiC is formulated. The potential for SiC incorporates long-range Coulomb, charge transfer interactions, covalency effect, Hafemeister and Flygare type short-range overlap repulsion extended up to the second-neighbour ions, van der Waals interactions and zero point energy effects. The developed potential including many body non-central forces validates the Cauchy discrepancy successfully to explain the high-pressure structural transition, and associated volume collapse. The 3*C* SiC ceramics lattice infers mechanical stiffening, thermal softening, and ductile (brittle) nature from the pressure (temperature) dependent elastic constants behaviour. To our knowledge, these are the first quantitative theoretical predictions of the pressure and temperature dependence of mechanical and thermodynamical properties explicitly the mechanical stiffening, thermally softening, and brittle/ductile nature of 3*C* SiC and still awaits experimental confirmations.

Keywords Carbide · High pressure · Elastic properties · Mechanical properties · Thermal expansion

Introduction

Silicon carbide, (SiC), a high quality technical grade ceramics, possesses wide energy band gap, low density, high strength, low thermal expansion, high thermal conductivity, high hardness, high melting point, large bulk modulus, low dielectric constant, high elastic modulus, excellent thermal shock resistance, and superior chemical inertness. The IV–IV SiC compound possesses tetrahedral of C and Si atoms with strong bonds in the crystal lattice and the availability of wide variety of its polytypes with unique structural and electronic properties. SiC has been a subject of immense interest as it is a very hard and strong ceramics with its application in requiring high endurance, such as car brakes and clutches and ceramics plates in bulletproof vests. The high thermal conductivity coupled with low thermal expansion and high strength gives this material exceptional thermal shock resistant qualities. Ceramics SiC with little or no grain boundary impurities maintain strength at very high temperatures with no strength loss [1, 2].

The structural transition, mechanical and elastic properties under low and high pressures of SiC have attracted much interest in both experimental [3, 4] and theoretical [5–15] investigations to elucidate the pressure-dependent behaviour of polytype [cubic (3*C*), hexagonal (6*H*), and rhombohedral (15*R*)] SiC. The energy-dispersive X-ray

✉ Dinesh Varshney
vdinesh33@rediffmail.com

[1] Materials Science Laboratory, School of Physics, Vigyan Bhavan, Devi Ahilya University, Khandwa Road Campus, Indore 452001, India

[2] Department of Physics, M. B. Khalsa College, Indore 452002, India

[3] Department of Physics, Ranchi College, Ranchi, Jharkhand 834008, India

[4] Laboratoire de Physique Quantique et de Modélisation Mathématique (LPQ3M), Département de Technologie, Université de Mascara, 29000 Mascara, Algeria

diffraction with a diamond anvil cell has identified structural transformation in *3C* SiC [zinc-blende (ZB) (*B*3) structure to the rock-salt (*B*1) structure $\approx$ 100 GPa] with a volume collapse of about 20.3 % [3]. The transition is reversible and the zinc-blende phase is recovered below 35 GPa upon decompression. Furthermore, *6H* polytype SiC is found to be stable up to about 90.0 GPa. Later, shock compression experiments on *6H* SiC show a first-order phase transition into a sixfold coordinated rock-salt structure around 105 $\pm$ 4 GPa with a volume reduction of about 15 $\pm$ 3 % [4].

Structural and thermal stability as well as high-pressure behaviour of 3*C*-SiC has been described both by ab initio [5–7, 11, 14, 16] and molecular dynamics simulations [8, 10, 15]. Based on ab initio density functional calculations with the local-density approximation (LDA) show that the transition pressure of 3*C*-SiC is around 60 GPa [ZB to rock salt (RS)] [5–7, 11–14, 16]. Following Perdew–Wang generalized gradient approximation (GGA) for the exchange-correlation potential and the Troullier–Martins pseudopotentials; the transition pressure of SiC at about 63 GPa is predicted [6, 12, 13]. Using the Troullier–Martins pseudopotentials and the LDA, the phase-transition pressure of 100 GPa is also documented [7, 14]. It is noted that first-principles LDA calculations underestimate critical pressure for structural phase transition.

The constant-pressure molecular dynamics (MD) simulation retraces the reversible phase transformation [3*C* to RS] in SiC [8]. The first-principle calculations clearly demonstrate the structural transformation of SiC from a fourfold coordinated structure to sixfold coordinated structure under pressure. The phase transition from the zinc-blende phase to the RS phase is associated with a cubic to only one intermediate state as monoclinic unit cell transformation. Later on, the structural phase transition and mechanical properties of SiC from the ZB structure to the RS structure under pressure are investigated in detail by the first-principles plane-wave pseudopotential density functional theory method [9]. The results on the high-pressure elastic constants illustrate that the ZB structure SiC is found unstable when the applied pressure is larger than 126.6 GPa consistent with the experimental data and the molecular dynamics (MD) simulation results.

The molecular dynamics with effective interatomic interaction potential for SiC incorporating two-body and three-body covalent interactions is also proposed. The covalent characteristics SiC are described by the three-body potential using modified Stillinger–Weber form. The molecular dynamics method with the developed interaction potential is employed to investigate the structural and dynamical properties of crystalline 3*C*, amorphous, and liquid states of SiC for several densities and temperatures [10, 15]. The phase stability of SiC under high pressure and behaviour of elastic constants with temperature is worth for microscopic understanding as well as technological applications.

The quantum computations based on density functional theory (DFT) as the full-relativistic version of the full-potential augmented plane-wave plus local orbitals method (FP $-$ APW $+$ lo) are powerful techniques and have the advantage of elucidating the ground state properties not only for small atomic systems but also for large molecules [17]. Density functional theory with approximate local and semilocal density functionals with nonlocal and long-range Coulomb interactions are effective for dense molecules and materials as well as short-range interactions for soft matter, van der Waals complexes, and biomolecules [18].

The quantum mechanical calculations suggest a favourable phase transition into a RS structure with different transition pressure for 3*C* SiC. However, these underestimate critical pressure for structural phase transition. We thus aimed at computing the pressure-induced phase transition of 3*C* SiC by formulating an effective interatomic potential. Note that the determination of ground state properties of IV–IV compounds with complex bonds based on lattice dynamical models is not easy. One needs to take care of experimental data with high accuracy and precision that accounts for low degree of freedom. The object of the proposed investigation is to develop a differential model as well to integrate the observed effect enabling useful prediction with reduced input experimental parameters.

The lattice dynamical models are useful in yielding both qualitative and quantitative information also with suggestive parameterization of the materials parameter. The density functional quantum calculations need precisely structural information as atomic positions and space groups. It also cares the value of volume around the experimental volume of the system. The successive iteration is thus made to determine the total energy corresponding to this volume. The thermal equation of state is needed to determine the ground state structural properties. It includes the lattice parameters within stable structure, the bulk modulus and pressure derivatives of elastic constants. The transition pressure is obtained by the common tangent between the two pressure–volume curves.

While exploring the ground state and anharmonic properties of simple and complex molecules, two major methods, one based on analytical form of cohesion with effective physical understanding and other based on expensive and time taking computational methods, are progressive. In the analytical models with two body interactions, the force constant deduced does not validate the Cauchy discrepancy. Incorporating many body interactions in the interaction potential with different cohesive energy forms is seen as the relevant potential which validates the Cauchy discrepancy. It is thus the major objective to seek the importance of both charge transfer interactions and covalent nature apart from zero point energy effects of group IV–IV cubic SiC in the interaction potential.

While discussing the mechanical and thermodynamical properties, the lattice dynamical model based on charge transfer interactions has been found successful [19–26]. The overlap repulsion potential (extended up to second-neighbour ions) is based on Hafemeister and Flygare [27]. The short-range interactions as the induced charge dipole–dipole and charge dipole–quadruple (van der Waals) interaction are found to be successful to determine the cohesion in several alkaline-earth solids [28]. As far as the pressure (temperature) induced mechanical and elastic properties as hardness, ductile nature, mechanical stiffening and thermal softening of SiC ceramics is concerned, these are probably the first quantitative theoretical prediction of its kind and still awaits experimental confirmations. Deduced aggregate Young's modulus, compression and shear wave velocities are in agreement with the observed values.

The opportunities offered by SiC ceramics have, therefore, motivated large research efforts which in turn have addressed only structural transitions and elastic, thermal and thermodynamical properties are still lagging. The proposed studies are organized in the following sequence. We first discuss the key assumptions to develop an effective interatomic potential between a pair of ions. We then support them by physical arguments for ceramics SiC followed by an equation of state in "Computational methodology" section. The Slater–Kirkwood variational method is employed to determine the induced charge dipole–dipole and charge dipole–quadruple (van der Waals) coefficients keeping in mind that both ions (Si and C) are polarizable. In sequence, we compute phase-transition pressures, the second-order aggregate elastic constants within the Shell model. Thus, the importance of long-range Coulomb with charge transfer interactions, covalent nature of bonds, charge dipole–dipole and charge dipole–quadruple (van der Waals) interaction, and the short-range overlap repulsive interaction up to second-neighbour ions is validated. In "Results and discussion" section, we discuss the various elastic properties.

As regard the ceramics 3*C* SiC, a number of works have been made on the structural stability of high-pressure phases but the pressure and temperature-dependent properties are sparse as normalized volume, aggregate second-order elastic constants C_{ij}, Bulk modulus B_T, Cauchy discrepancy in second-order elastic constant Δ_1^2, second-order elastic constant anisotropy γ_1^2, melting temperature T_M, aggregate third-order elastic constants C_{ijk}, Cauchy discrepancy Δ_i^3 in third-order elastic constant, third-order elastic anisotropy γ_i^3, isotropic shear modulus G_H, Voigt's shear modulus G_V, Reuss's shear modulus G_R, Young's modulus E, Poisson's ratio ν and Pugh ratio ϕ (=B_T/G_H) leading to ductility (brittleness), Lamé's constant λ and μ, and elastic wave velocity v_l and v_s.

The anharmonic effects on SiC lattice are further studied by investigating the Gruneisen parameter γ_G, isothermal compressibility β, Debye temperature θ_D, hardness H_V, heat capacity C_v, and thermal expansion coefficient $\alpha_{th.exp.}$ either in ZB or RS structures. The relevant expressions are documented in Appendix 1 and 2. We also provide a comparison of computed values with observed experimental and other theoretical studies. The major conclusions are presented in "Concluding remarks" section.

Computational methodology

To study the phase stability and the aggregate elastic constants of SiC under high pressures, we formulate an interatomic potential. The proposed interaction potential is based on the following assumptions: variations in force constants of Si and C are small, the short-range interactions between Si and C atoms are effective up to their second-neighbour ions, and harmonic elastic forces are viable for a pair of atoms without any internal strains within the crystal. Application of mechanical pressure as external variable causes an increase in the overlap of adjacent ions in a crystal. The result is a charge transfer takes place between the overlapping electron shells of Si and C atoms. The transferred charges interact with neighbouring charges around the lattice via Coulomb's law. Also, the chemical bonds in SiC are both ionic and covalent in nature and the application of pressure causes change in crystal structure (first-order structural phase transition), volume collapse and elastic properties.

The proposed interatomic potential is thus a route to discuss the structural transitions, mechanical properties in particularly, about stiffness, softeness, ductile, brittle nature, elastic constants anisotropy, compression and shear elastic wave velocity. In addition, the potential is predictive to explain the validity of non-central forces and anharmonicity from thermodynamical properties namely Debye temperature, heat capacity and thermal expansion coefficient of 3*C* SiC. At zero pressure and variable temperature, the underlined effective interatomic potential also describes some of the temperature-induced thermal and thermodynamical properties of 3*C* SiC.

The effective interatomic potential at ambient pressure invokes an isolated stable phase when the crystal free energy is minimized for the specified thermodynamic conditions. For this purpose, if the variables such as temperature, pressure or magnetic field applied on the crystal are altered, either the free energy changes smoothly and continuously or discontinuously. Variations in free energy cause structural phase transition. The SiC ceramics transform from their initial *B*3 to *B*1 structure under pressure.

The minimum Gibbs's free energy, G, infers the stability of a particular structure. Herein, Gibbs's free energy, $G = U + PV - TS$. The notations are: U is internal energy, which at 0 K is the cohesive energy, S is the vibrational entropy at absolute temperature T, pressure P and volume V. The thermodynamically stable phase at a given pressure P and at zero temperature is the one with lowest enthalpy. The thermodynamical potential is thus the Helmholtz free energy (H).

The interatomic potential with pressure or temperature as thermodynamical variable requires the estimation of the Gibbs's free energies for ZnS ($B3$) phase and NaCl ($B1$) phase. We use Born equation: $G_{B3}(r) = U_{B3}(r) + 3.08Pr^3$ for ZnS ($B3$) phase and $G_{B1}(r') = U_{B1}(r') + 2Pr'^3$ for NaCl ($B1$) phase. At phase-transition pressure P and at zero temperature: $G_{B1} = G_{B3}$ [29]. The notations U_{B3} (r) infer the total potential energy of ZnS ($B3$) phase and U_{B1} (r') as total potential energy for the RS ($B1$) phase. The unit cell volumes are V_{B3} (=3.08r^3) and V_{B1} (=2 r'^3). The nearest neighbour distance is r (r') in ZB (RS) structure. The total potential energy for ZnS ($B3$) and NaCl ($B1$) phases is

$$\begin{aligned}U_{B3} = &(-\alpha_M Ze^2/r)[Z + 2nf(r)] - Cr^{-6} - Dr^{-8}\\ &+ nb\beta_{ij}\exp\left[(r_i + r_j - r_{ij})/\rho\right]\\ &+ (n'b/2)\left[\beta_{ii}\exp\left((2r_i - kr_{ij})/\rho\right)\right.\\ &\left.+\beta_{jj}\exp\left((2r_j - kr_{ij})/\rho\right)\right] + \left[\{\hbar<\omega^2>^{1/2}/2\right] \qquad (1)\end{aligned}$$

$$\begin{aligned}U_{B1} = &(-\alpha'_M Ze^2/r')[Z+2mf(r')] - Cr'^{-6} - Dr'^{-8}\\ &+mb\beta_{ij}\exp\left[\left(r_i+r_j-r'_{ij}\right)/\rho\right]\\ &+(m'b/2)\left[\beta_{ii}\exp\left(\left(2r_i-k'r'_{ii}\right)/\rho\right)\right.\\ &\left.+\beta_{jj}\exp\left(\left(2r_j-k'r'_{jj}\right)/\rho\right)\right]+\left[\{\hbar<\omega^2>^{1/2}/2\}\right] \qquad (2)\end{aligned}$$

Due to complex nature of bonds in SiC, the ionic charge for Si and C atom cannot be determined uniquely. The calculation of the Madelung energy is thus modified by incorporating the covalency effects [10, 15, 30, 31]. The charge in above equations is thus written incorporating the polarization of a spherical shaped dielectric in displacing the constituent positive ions. The charge transfer interactions caused by the deformation of the electron shells of the overlapping ions and the covalency effects are the major attributes of long-range Coulomb effects. The IV–IV semiconducting compound contains covalent bonds so that some electrons are distributed over the region between neighbouring atoms; in such situation the interaction energies are attributed from the contribution of charge dipole-dipole and charge dipole-quadruple terms. The induced charge dipole–dipole and charge dipole–quadruple (van der Waals) interactions are the third and fourth terms which are the short-range vdW attractive potential energies.

The fifth and sixth terms are the short-range overlap repulsive energies. This is due to the overlap repulsion between ij, ii and jj ions. The Madelung constants for $B3$ ($B1$) phases are represented by α_m (α_m'). β_{ij} symbolized for Pauling coefficient and is defined as $\beta_{ij} = 1 + (Z_i/n_i) + (Z_j/n_j)$ with $Z_i(Z_j)$ and $n_i(n_j)$ as the valence and the number of electrons in the outermost orbit. The numbers of the nearest unlike n (=4) and like n' (=6) neighbours are for $B3$ (ZnS). Similarly, numbers of the nearest unlike m (=6) and like m' (=6) are for $B1$ (NaCl) structure. The Ze is being the ionic charge, k (k') is the structure factor for $B3$ ($B1$) structures, and b (ρ) is the hardness (range) parameters. We denote the nearest neighbour ion separations as r (r')for $B3$ ($B1$) structures.

The last term in Eqs. 1 and 2 is the lowest possible energy of the system and is due to the zero point energy. Here, $<\omega^2>^{1/2}$ (=$k_B\theta_D/\hbar$) is the mean square frequency related to the Debye temperature θ_D. The Debye temperature can be known either from Heat capacity measurements or from the Bulk modulus value using $\theta_D = (\hbar/k_B)\sqrt{(5r_0B_T/\mu)}$. Herein, r_0, B and μ are the equilibrium distance, Bulk modulus and reduced mass of the compounds. Henceforth, model potential for ground state incorporates the attractive, repulsive and zero point energy.

We use the variational approach to deduce the overall vdW coefficients C (charge dipole–dipole) and D (charge dipole–quadruple) [32]. The short-range vdW coefficients due to induced charge dipole–dipole and charge dipole–quadruple interactions caused by Si atom and C atom are

$$c_{ij} = \frac{3}{2}\frac{e\hbar}{\sqrt{m_e}}\alpha_i\alpha_j\left[\left(\frac{\alpha_i}{N_i}\right)^{1/2} + \left(\frac{\alpha_j}{N_j}\right)^{1/2}\right]^{-1}, \qquad (3)$$

$$d_{ij} = \frac{27}{8}\frac{\hbar^2}{m_e}\alpha_i\alpha_j\left[\left(\frac{\alpha_i}{N_i}\right)^{1/2} + \left(\frac{\alpha_j}{N_j}\right)^{1/2}\right]^2\left[\left(\frac{\alpha_i}{N_i}\right) + \frac{20}{3}\left(\frac{\alpha_i\alpha_j}{N_iN_j}\right)^{1/2} + \left(\frac{\alpha_j}{N_j}\right)\right]^{-1}. \qquad (4)$$

$$c_{ii} = \frac{3}{2}\frac{e\hbar}{\sqrt{m_e}}\alpha_i\alpha_i\left[\left(\frac{\alpha_i}{N_i}\right)^{1/2} + \left(\frac{\alpha_i}{N_i}\right)^{1/2}\right]^{-1}, \qquad (5)$$

$$d_{ii} = \frac{27}{8}\frac{\hbar^2}{m_e}\alpha_i\alpha_i\left[\left(\frac{\alpha_i}{N_i}\right)^{1/2} + \left(\frac{\alpha_i}{N_i}\right)^{1/2}\right]^2\left[\left(\frac{\alpha_i}{N_i}\right) + \frac{20}{3}\left(\frac{\alpha_i\alpha_i}{N_iN_i}\right)^{1/2} + \left(\frac{\alpha_i}{N_i}\right)\right]^{-1} \qquad (6)$$

$$c_{jj} = \frac{3}{2}\frac{e\hbar}{\sqrt{m_e}}\alpha_j\alpha_j\left[\left(\frac{\alpha_j}{N_j}\right)^{1/2} + \left(\frac{\alpha_j}{N_j}\right)^{1/2}\right]^{-1}, \qquad (7)$$

$$d_{jj} = \frac{27}{8}\frac{\hbar^2}{m_e}\alpha_j\alpha_j\left[\left(\frac{\alpha_j}{N_j}\right)^{1/2} + \left(\frac{\alpha_j}{N_j}\right)^{1/2}\right]^2\left[\left(\frac{\alpha_j}{N_j}\right) + \frac{20}{3}\left(\frac{\alpha_j\alpha_j}{N_jN_j}\right)^{1/2} + \left(\frac{\alpha_j}{N_j}\right)\right]^{-1}. \qquad (8)$$

In the above equations, the notations: m_e, e and Z are mass of the electron, charge and valence of the constituent metallic element, respectively. The symbols α_i,and α_j represent the polarizabilities of ith and jth ion, respectively. The effective number of electrons responsible for polarization is symbolized by N_i and N_j. The lattice sums S_{ij} and T_{ij} enable one to compute the overall vdW coefficients C and D in terms of c and d values determined from Eqs. 1 to 8. The lattice sums S_{ij}, and T_{ij} are expressed as [28]:

$$C = c_{ij}S_{ij} + c_{ii}S_{ii} + c_{jj}S_{jj} \tag{9}$$

$$D = d_{ij}T_{ij} + d_{ii}T_{ii} + d_{jj}T_{jj} \tag{10}$$

SiC is a tetrahedrally coordinated covalent material and the complex chemical bonding corroborate both ionic and covalent nature. The Coulomb interaction between ions of Si and C atoms leads to charge transfer interactions. Apart from this, the covalent character of bond bending and stretching also needs to be incorporated in the potential. Thus, the second term in Eqs. 1 and 2 is an algebraic sum of non-central many body forces as the charge transfer force parameter and the force parameter arise due to covalent nature *i.e.* $f(\mathrm{r}) = f_{cti} + f_{cov}$. The charge transfer between ions of Si and C atoms is denoted in terms of a force parameter f_{cti} and is expressed as [27, 33, 34]:

$$f_{cti} = f_0 \exp(-r/\rho) \tag{11}$$

Here, r_i (r_j) is the ionic radii of $i(j)$ ion.

The complex chemical bonding in IV–IV semiconducting compounds infers SiC as partially ionic and partially covalent in bonding. The attractive forces due to covalent nature thus modify the charge and are now the effective charge. The polarization effects originate from changes in covalency due to Si–Si, Si–C, and C–C interacting electric fields. The covalency term in the interaction potential is thus expressed as [30, 31]:

$$f_{\mathrm{cov}}(r) = \frac{4e^2V_{sp\sigma}^2}{r_0E_{\mathrm{g}}^3} \tag{12}$$

Here, $V_{sp\sigma}$ represents the transfer matrix element between the outermost p orbital and the lowest excited of s state. The transfer energy of electron from p to the s orbital is denoted as E_{g}. The effective charge e_S^* of SiC is related with the number of electrons transferred to the unoccupied orbitals from its surrounding nearest neighbour. The electron density is thus $n_c = 1 - e_S^*/e$. Thus, in SiC for overlap distortion effect $e_S^* \neq e$. The transfer matrix element $V_{sp\sigma}$ and the transfer energy E_{g} are related to electron density as $n_c/12 \cong V_{sp\sigma}^2/E_{\mathrm{g}}^2$. The effective charge e_S^* is thus

$$\frac{V_{sp\sigma}^2}{E_{\mathrm{g}}^2} = \frac{1 - e_s^*}{12} \tag{13}$$

The transfer energy E_{g} is

$$E_{\mathrm{g}} = E - I + (2\alpha - 1)\frac{e^2}{r} \tag{14}$$

Here, E is the electron affinity for C and I is the ionisation potential of constituent atom.

The Szigeti effective charge e_S^* $(=Ze)^*$ is written in terms of the optical static dielectric constant ε_0 and the high frequency dielectric constant ε_∞ as [19–26]:

$$e_s^{*2} = \frac{9\mu\omega_{TO}^2(\varepsilon_0 - \varepsilon_\infty)}{4\pi N_k(\varepsilon_\infty + 2)^2} \tag{15}$$

and

$$\frac{e_s^{*2}}{e^2} = \frac{9V\mu\omega_{TO}^2(\varepsilon_0 - \varepsilon_\infty)}{4\pi e^2(\varepsilon_\infty + 2)^2} \tag{16}$$

The symbol μ is the reduced mass, N_k is the number of atoms present per unit cell volume i.e. $N_k = 1/V$, ω_{TO} is the long wavelength transverse optical phonon frequency. Thus, for 3C SiC e_S^* deviates from e and is attributed to covalent nature of Si–Si, Si–C, and C–C bonds.

Usually materials document a transition when the thermodynamical potential relevant to the given ensemble of the lower pressure phase equals that for some other structure in the absence of any barrier. The low pressure phase becomes the stable phase above this coexistence pressure. After determining the stable phase we also compute the higher order elastic constants, their pressure derivatives and anisotropy. The Appendix 1 illustrates the essential equations for the higher order elastic constants and their pressure derivatives. With these understanding of interatomic potential in SiC, we have four material parameters, namely, modified ionic charge, hardness, range, force parameter [Z_m, b, ρ, $f(\mathrm{r})$]. Values of them can be deduced from equilibrium conditions [35–42].

Results and discussion

The application of pressure, temperature and magnetic field probably transforms materials from one structure to another. The relative stability of two crystal structures requires an extremely accurate prediction. The interatomic interaction potential with charge transfer interactions caused by ions of Si and C atom and covalent nature of Si–Si, Si–C, and C–C bonds are effective in studying the structural phase transitions and elastic properties of tetrahedrally coordinated ceramics 3C SiC. We evaluate the phase transition pressure by computing the Gibbs free energy $G = U + PV - TS$ for the ZB and RS phases. We note that the Gibbs free energy is thus the enthalpy H $(=U + PV)$ at $T = 0$ K.

While doing high-pressure experiments, the huge applied pressure causes a reduction of the material volume.

The temperature variations during the experiments will normally produce much smaller changes in the relative stabilities of different phases. Thus, the Gibbs free energy at zero temperature, which is the enthalpy H, is measured. The thermodynamically stable phase of crystal at 0 K and at ambient pressure P is the one with the lowest enthalpy. Thus, the zero-temperature theoretical calculations are valid with experiment. In a situation when temperature variations are large during experiment for certain materials the effects of finite temperature may be significant. With these assumptions, we investigate structural and elastic properties of SiC in an ordered way.

The thermodynamical potential G or H in SiC is computed involving modified ionic charge, hardness, range and charge transfer parameters [Z_m, b, ρ, $f(r)$] as [35–42]:

$$\left|\frac{dU(r)}{dr}\right|_{r=r_0} = 0 \qquad (17)$$

Also, the bulk modulus (B_T):

$$\left|\frac{d^2U(r)}{dr^2}\right|_{r=r_0} = (9kr_0)^{-1}B_T \qquad (18)$$

We first deduce vdW coefficients C and D involved in expressions (1) and (2) from the Slater–Kirkwood variational method [32], for 3C SiC ceramics material parameters, and are enlisted in Table 1. The charge dipole–dipole and charge dipole–quadruple vdW coefficients are influenced by electronic polarizabilities. The polarizability values have been obtained from least-squares fit of experimental refraction data using additive rule and a Lorentz factor of $4\pi/3$ [43, 44]. We consider that the SiC to be partially ionic and covalent to discuss their structural, mechanical, elastic and thermodynamical properties in a systematic manner.

As a next step, we use the experimental data on lattice constant (a) [45], the bulk modulus (B_T) [46], ionic (Ze), effective charge (e_s^*) and the second-order aggregate elastic constant C_{12} (C_{44}) [47] for determining the material parameters. Deduced values of hardness (b), range (ρ) parameter, and non-central many body forces arising due to charge transfer (f_{cti}) and covalency (f_{cov}) for 3C SiC ceramics are illustrated in Table 1. The effective charge e_S^* depends on the values of optical dielectric constant ε_s and the high frequency dielectric constant ε_∞. The value of long wave length transverse optical phonon frequency ω_{TO} is taken from [48] to have the effective charge e_S^* and hence the covalency contribution.

We then minimize the Gibbs's free energies $G_{B3}(r)$ and $G_{B1}(r')$ for the equilibrium interatomic spacing (r) and (r') to determine structural phase transition of SiC. Table 1 shows the optimized values of equilibrium interatomic spacing in $B3$ and $B1$ phases. The Gibbs's free energy $G_{B3}(r)$ [$G_{B1}(r')$] as functions of pressure (P) for SiC is discerned in Fig. 1a. At zero pressure, the Gibb's free energy for SiC in $B3$ crystal phase is more negative. Thus, at zero pressure SiC in $B3$ phase is thermodynamically and mechanically stable, while the $B1$ is not. On the other hand, above the phase transition pressure ($P_T = 90$ GPa), the Gibb's free energy for $B1$ system becomes more negative than $B3$ phase, implying $B1$ phase will be more stable. The cohesive energy per particle is obtained as 6.301 eV for 3C SiC which is consistent with earlier experimental value of 6.34 eV [49]; 7.415 eV from ab initio density functional calculations and molecular dynamics method yields 6.3410868 eV for SiC [50].

The phase stability of cubic 3C SiC under high pressures is essentially based on material parameters namely hardness (b), range (ρ), non-central many body forces as charge transfer force (f_{cti}) and covalency parameter (f_{cov}). These are obtained from the experimental data. We comment that the available data based on one kind of experiment depend on the conditions of measurement. Henceforth, while developing a theory, one faces certain complications and one need to find suitable data that varies from technique to technique. For SiC, we pay special attention while

Table 1 Estimated and input crystal data: vdW coefficients [c_{ii}, c_{ij}, c_{jj}, C, d_{ii}, d_{ij}, d_{jj}, D], lattice constant (a_0), bulk modulus (B_T), second-order elastic constant C_{12} (C_{44}), optimized value of ionic radii r_i (r_j), hardness (b), range (ρ), charge transfer parameter f (r), equilibrium distance: r_O ($B3$); r_O' ($B1$), Gibbs's free energy: $G_{B3}(r)$; $G_{B1}(r)$

Input parameters	SiC
c_{ii} (10^{-60} erg cm^6)	28.76
c_{ij} (10^{-60} erg cm^6)	0.71
c_{jj} (10^{-60} erg cm^6)	0.047
C (10^{-60} erg cm^6)	14.07
d_{ii} (10^{-76} erg cm^8)	14.284
d_{ij} (10^{-76} erg cm^8)	0.297
d_{jj} (10^{-76} erg cm^8)	0.002718
D (10^{-76} erg cm^8)	3.019
a_0 (Å)	4.36 [45]
B_T (GPa)	227.0 [46]
C_{12} (GPa)	142.0 [47]
C_{44} (GPa)	256.0 [47]
r_i (Å)	0.42
r_j (Å)	1.162
b (10^{-12} erg)	7.512
ρ (10^{-9} cm)	3.29
f (r) (10^{-3})	5.267
Equilibrium distance (Å) r_O ($B3$)	1.89
Equilibrium distance (Å) r_O' ($B1$)	2.11
Gibb's free energy (kJ mol^{-1}) $G_{B3}(r)$	−4570
Gibb's free energy (kJ mol^{-1}) $G_{B1}(r)$	−4498

Fig. 1 Variation of Gibb's free energy for *B*3 and *B*1 phases with pressure and normalized volume with pressure and temperature

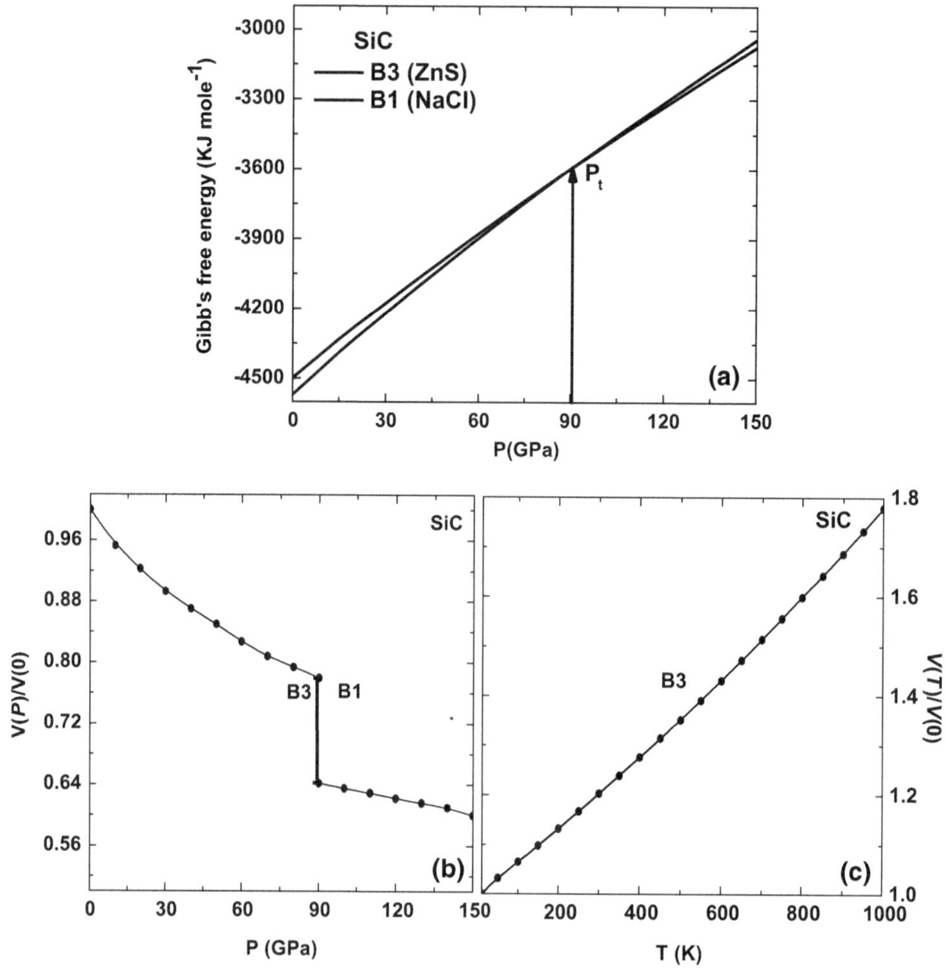

formulating interatomic potential to ensure whether long range or short-range interactions are at the origin of the structural transition. The many body force parameter as charge transfer *f*(r) is positive as seen in Table 1. It is attributed to the fact that the charge transfer parameter is computed from the difference of second-order elastic constants C_{12} and C_{44}. The Cauchy energy $C_{12} - C_{44}$ is negative for SiC. Thus, consistent results lead to a test of the validity of assumptions made in model potential for SiC.

SiC shows a crystallographic transition from *B*3 to *B*1 in certain pressure range. Table 2 illustrates the computed phase-transition pressure (P_T) and compared with available experimental data [3, 4] and theoretical results [5–12, 14–16]. P_T for SiC is consistent with the experimental and other reported values and is attributed to proper formulation of interatomic potential, which considers the various interactions, explicitly the non-central many body forces as charge transfer interactions and covalency effects, as well as use of materials parameter based on experimental data. We may comment that any computational technique has its own limitations related to the chosen materials basic parameters, basic sets, as well the accuracy and precisions used apart from the approximations laid in the method. Needless to suggest that there is always a variation in estimated parameters by each technique.

The values of relative volumes associated with various compressions are estimated from [51]

$$\frac{V_P}{V_0} = \left(1 + \frac{B'}{B_0}P\right)^{-1/B'}, \tag{19}$$

Here, V_O (B_O) is the cell volume (bulk modulus) of SiC at ambient conditions and V_P is at finite pressure. The symbol B' is the pressure derivative of the bulk modulus. Figure 1b documents the estimated value of pressure-dependent radius for both *B*3 and *B*1 structures, the curve of volume collapse (V_P/V_0) with pressure to show phase diagram for 3*C* SiC. The phase diagram will let us estimate the magnitude of the discontinuity in volume at the transition pressure. The value of relative volumes is shown in Table 2. It is also compared with various experimental [3, 4] and other theoretical works [5–9, 11–14, 16].

Table 2 Calculated transition pressure P_T (GPa), volume collapse (%), aggregate second-order elastic constants (C_{11}, C_{12} and C_{44}), aggregate bulk modulus (B_T), (all are in 10^{10} Nm^{-2}) and pressure derivatives of SOECs (dB_T/dP, dC_{44}/dP and dC_S/dP) for Silicon Carbide in $B3$ phase

Property	Present	Expt.	FP LMTO	LDF	LDA	MD	DFT
Transition pressure P_T	90.0	100 [3],			66 ± 5 [5, 11]	100 [8]	66.6 [16]
		105 ± 4 [4]			63 [6]	90 [10, 15]	100 [7, 14]
							75.4 [9]
Volume collapse	13.8	20.3 [3]			18.5 [5, 11]	21 [8]	18.5 [16]
		15 ± 3 [4]					12 [7, 14]
							18 [9]
C_{11}	37.11	39.0 [47]	42.0 [51]		44.9 [62, 64]	39.0 [8]	41.51 [9]
		36.3 [56]	35.23 [58]		43.6 [63]	39.0 [10, 15]	37.1 [56]
C_{12}	22.34	14.2 [47]	12.6 [51]		14.6 [62, 64, 65]	14.4 [8]	13.19 [9]
		15.4 [56]	14.04 [58]		12.0 [63]	14.26 [10, 15]	16.9 [56]
C_{44}	27.93	25.6 [47]	28.7 [51]		25.6 [62, 64, 65]	17.9 [8]	26.54 [9]
		14.9 [56]	23.29 [58]		25.5 [63]	19.11 [10, 15]	17.6 [56]
B_T	27.30	22.7 [46]	22.3 [51]	24.9 [34]	21.2 [5, 11]	22.5 [8]	23.5 [7, 14]
		22.4 [59]	21.1 [58]	22.5 [61]	22.5 [62–65]	22.52 [10, 15]	22.71 [9]
		22.5 [60]					20.0 [57]
							22.5 [56]
C_S	7.38						
dB_T/dP	5.188	3.57 [46]	3.8 [51]	3.2 [62, 64, 65]	3.7 [5, 11]	5.5 [10, 15]	3.79 [9]
							7.3 [57]
dC_{44}/dP	5.198						
dC_S/dP	−0.0716						

Compressions in SiC at higher pressure indicate the mechanical stiffening of lattice.

Figure 1c discerns the variation of V_T/V_0 as functions of temperature in $B3$ phase. Here, V_T symbolizes the volume at various temperatures and V_0 at zero temperature and zero pressure volumes, respectively. A steep increase in the ratio V_T/V_0 with increasing temperature infers expansion of SiC lattice and is susceptible to temperature. On the other hand, SiC is compressed at higher pressures as shown in Fig. 1b. Henceforth, SiC lattice is thermally softened and mechanical stiffened. The normalized volume V_T/V_0 dependences on temperature are not known for SiC, but the present behaviour is consistent with available experimental [52] and theoretical [53] data on Li_2O.

The response of any material that undergoes stress, deforms and then recovers and returns to its original shape after stress ceases leads to the determination of elastic properties. The elastic properties are vital in generating information about the binding characteristic between adjacent atomic planes, anisotropic character of binding and structural stability. Apart from the structural stability of SiC in ZnS ($B3$) and NaCl ($B1$) structures, we now compute the aggregate elastic constants at normal and under hydrostatic pressure. Deduced values are documented in Table 2.

Using the stress–strain coefficients, one determines the second-order aggregate elastic constants C_{ij} under hydrostatic pressure with respect to finite strain. Also, proper parametrization of Coulomb, non-central many body forces, overlap repulsion, van der Waals interactions and zero point energy terms are essentially required.

The cubical symmetry of SiC dealt with three independent elastic constants C_{ij}. C_{11} is a response of resistance to deformation by a stress applied on (1,0,0) plane with polarization in the direction <100>. C_{11} probes elasticity in length and a longitudinal strain produces a change in C_{11}. C_{44} refers to the measurement of resistance to deformation with respect to a shearing stress applied across the (100) plane with polarization in the <010> direction. C_{12} and C_{44} are related to the elasticity in shape, which is a shear constant. A transverse strain causes a change in shape without a change in volume and hence C_{12} and C_{44} are less sensitive of pressure as compared to C_{11}.

The variation of three independent second-order aggregate elastic constants (SOECs): C_{11}, C_{12}, and C_{44} with external pressure for SiC in $B3$ and $B1$ phase is first discussed. As seen from Fig. 2a that C_{11} and C_{12} increase with increase in pressure in both $B3$ and $B1$ phases. Also, C_{44} decreases with the increase of pressure away from zero till the phase-transition pressures and then increases in $B1$

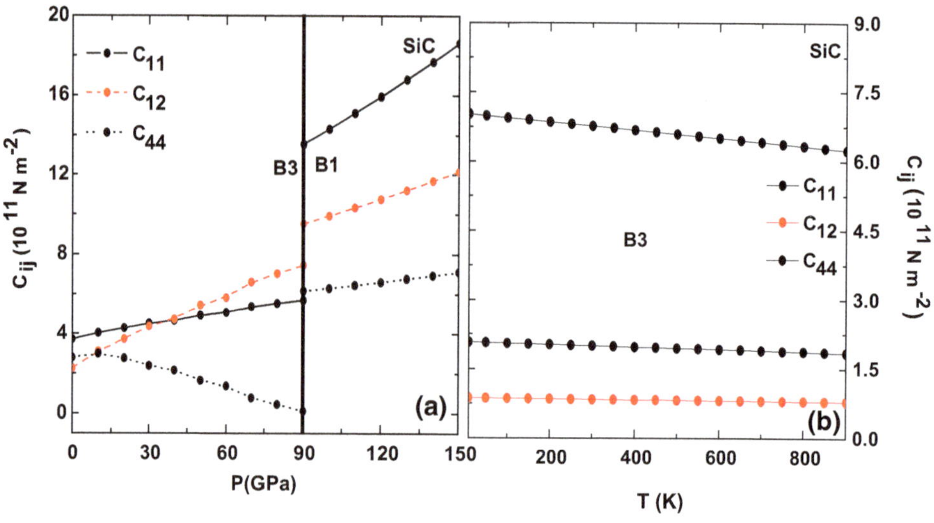

Fig. 2 Variation of aggregate second-order elastic constants (C_{ij}) with pressure and temperature

phase. Similar observations have earlier been reported in SiC [9, 10, 15]. A crossover of C_{12} and C_{44} in CaS has also been reported [54]. At phase transition pressures, SiC has witnessed a discontinuity in aggregate second-order elastic constants C_{ij}, which identifies the first-order phase transition. Thus, the proposed interaction potential incorporating charge transfer interactions ions of Si and C atom and covalency effects caused by Si–Si, Si–C, and C–C bonds consistently explains the high-pressure elastic behaviour.

The variations in C_{ij} with temperatures (T) for 3C SiC ceramics are plotted in Fig. 2b. It can be seen that the aggregate elastic constants C_{ij} (T) decrease linearly with the temperature in ZB phase. We note that the pressure dependence of aggregate elastic constants C_{ij} (P) documents an increasing trend (please see Fig. 2a). The physical interpretation of temperature dependence of C_{ij} showed that (a) values of C_{11} decrease more steeply with enhancing temperature, (b) C_{12}, and C_{44}, are less sensitive to temperature for 3C SiC ceramics, (c) C_{11} is remarkably larger than C_{12}, and C_{44}, and (d) values of all aggregate elastic constants C_{ij} are influenced by temperature dependence indicating that anharmonicity is substantial. Deduced values of C_{ij} with temperatures (T) are documented in Table 3 along with the available data on SiC at room temperature [55]. From the second-order aggregate elastic constants pressure and temperature-dependent behaviour, we comment that SiC lattice is mechanical stiffened and thermally softened.

Born criterion for a lattice to be mechanically stable infers that the elastic energy density must be a positive definite quadratic function of strain. The principal minors (alternatively the eigenvalues) of the elastic constant matrix should all be positive at ambient conditions. The mechanical stability conditions for a crystal suggest that elastic constants of a cubic crystal are as follows [29],

$$B_T = (C_{11} + 2C_{12})/3 > 0, \tag{20}$$

$$C_{11}, C_{44} > 0, \tag{21}$$

and

$$C_S = (C_{11} - C_{12})/2 > 0. \tag{22}$$

Here, C_{ij} are conventional aggregate elastic constants and B_T is bulk modulus. We represent, C_{44} and C_S as the shear and tetragonal moduli of a cubic crystal.

Table 2 illustrates the computed values of bulk modulus (B_T), shear moduli (C_{44}) and tetragonal moduli (C_S) which validates the elastic stability criteria for 3C SiC in $B3$ phase. The second-order elastic constants critically depend upon pressure leading to $C_{12} - C_{44} \neq 0$. The mechanically stable phases for cubic crystal satisfy the Born criteria: $C_{12} - C_{44} > 0$. The validity of above is readily seen by referring to Eqs. 51 and 52 for C_{12} and C_{44}. The equilibrium condition leads to $B_1 + B_2 = -1.261Z_m^2$ with emphasis on charge transfer interactions as well as covalency effects. For optimized values of r_i (r_j) the Cauchy discrepancy $C_{12} - C_{44}$ is nonzero at zero pressure and at zero temperature. It is also valid when the many body noncentral forces are not involved in long-range forces. The short-range and long-range effects are naturally of similar order of magnitude. This is due to the fact C_{ij} are calculated at optimized values of equilibrium distances rather than at experimental values.

Table 2 illustrates the calculated values of pressure derivatives of aggregate second-order elastic constants (dB_T/dP, dC_{44}/dP and dC_S/dP). These are compared with available experimental [46, 47, 56] and theoretical studies [8–10, 15, 46, 57–65]. For mechanical stability, the shear elastic constant C_{44} is nonzero and is known by combining mechanical stability with minimum energy conditions. The high-pressure stability also suggests that the stable phase of

Table 3 Calculated second-order elastic constant Cauchy discrepancy (Δ_1^2), anisotropy parameter (γ_i^2), isotropic shear modulus (G_H), Voigt's shear modulus (G_V), Reuss's shear modulus (G_R), Young's modulus (E), Poisson ratio (ν), compressibility (β) and Gruneisen parameter (γ_G) of Silicon Carbide in $B3$ phase at zero pressure

Property	Present	Expt.	FP LMTO	LDF	LDA	MD	DFT
$\gamma_1^2(10^{10}\ Nm^{-2})$	−0.736						
$\Delta_1^2\ (10^{10}\ Nm^{-2})$	−5.588						
$G_H\ (10^{10}\ Nm^{-2})$	16.47	19.2 [60]	21.9 [50]			12.37 [10, 15]	14.1 [56]
			16.9 [58]				
$G_V\ (10^{10}\ Nm^{-2})$	19.71		23.1 [50]				14.6 [56]
			18.2 [58]				
$G_R\ (10^{10}\ Nm^{-2})$	13.22		20.8 [50]				13.6 [56]
			15.7 [58]				
$E\ (10^{10}\ Nm^{-2})$	41.22	44.8 [60]	49.60 [50]		56.7 [62, 64, 65]	31.36 [10, 15]	35.2 [56]
			40.1 [58]		55.0 [62, 64, 65]		
ν	0.249	0.267 [47]	0.146 [50]			0.268 [10, 15]	1.0 [5, 11]
		0.168 [60]	0.201 [58]				0.259 [56]
$\beta\ (10^{-11}\ Pa^{-1})$	0.024						0.1518 [5, 11]
γ_G	1.015			1.01 [61]	1.12 [5, 11]		

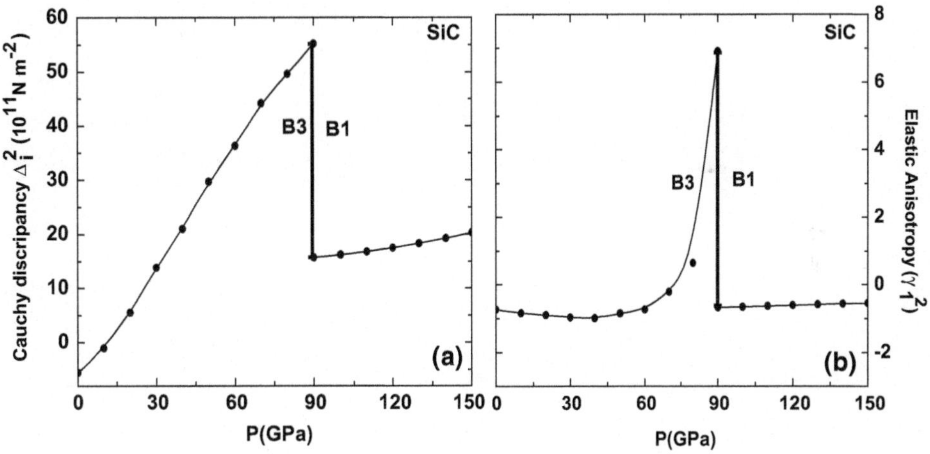

Fig. 3 Variation of Cauchy discrepancy (Δ_1^2) and elastic anisotropy (γ_1^2) in second-order elastic constant (C_{ij}) with pressure

the crystal possesses the lowest potential energy among the mechanically stable lattices [66].

The elasticity in 3*C* SiC is thus probed by a non-central many body force potential which assumes that the interatomic forces have a certain shape and directionality. The Cauchy discrepancy is defined as $\Delta_1^2 = C_{12} - C_{44} - 2P$. Here, Δ_1^2 is a measure of the contribution from the non-central many body force. However, for pure central interatomic potentials, Cauchy relation is $C_{12} = C_{44} + 2P$.

At zero pressure, the Cauchy discrepancy (Δ_1^2) in SiC is about $-5.588 \times 10^{10}\ Nm^{-2}$. The Δ_1^2 further enhances on increasing the pressure in both phases as depicted in Fig. 3a. In SiC, larger deviation of Δ_1^2 essentially points to the importance of the many body non-central (charge transfer and covalency) interaction in the interatomic potential and anharmonic effects are substantial at high pressures. The strength of non-central many body forces incorporating charge transfer interactions and covalency effects is witnessed by significant deviation in Δ_1^2 at different pressures not only in ZB but also in RS structure, although weak. The importance of many body non-central forces and anharmonic effects is further explored by analysing the higher order elastic constants explicitly the third-order elastic constants. Usually, the anharmonic effects are noticeable at high pressure as reflected from elastic constants behaviour.

The anisotropy in second-order elastic constants is reflected from geophysical activities of various materials and alloys. The anisotropic parameter γ is unity for isotropic elasticity. As far as cubic crystal is concerned, although it is isotropic in structure, it has elastic anisotropy other than unity. This is a consequence of fourth rank tensor property of elasticity.

We define elastic anisotropic parameter γ_1^2 in terms of aggregate C_{ij} as [67]:

$$\gamma_1^2 = \frac{C_{11} - C_{12} - 2C_{44}}{2C_{44}} \tag{23}$$

The pressure dependence of the elastic anisotropic parameter γ_1^2 in SiC is shown in Fig. 3b. It is evident that γ_1^2 in SiC is insensitive below transition pressure and also at low pressures. A jump has been noted at P_T (=90 GPa) inferring first-order structural phase transition. Furthermore, γ_1^2 in SiC remains unaltered for $B1$ phase for higher pressures. The value of anisotropic parameter γ_1^2 for SiC is given in Table 4 at $T = 0$ K and $P = 0$ GPa.

During mechanical processing, explicitly in fabrication the melting ranges of materials and alloys are substantial. The usage of an alloy in the applications as the success of the melting and casting operations depends on the correct selection of temperature. Once solidified and primary processed (rolling or forging), the melting temperature has little significance to designers, engineers and users. The melting temperature influences elevated temperature properties, such as creep strength, but the researchers have limited interest. The pressure dependence of the melting temperature: $T_M = 553\ \text{K} + 5.91C_{11}\ \text{K GPa}^{-1}$ for SiC in $B3$ and $B1$ phase is discerned in Fig. 4. It is noticed that T_M enhances with increased pressure or in other words the resistance to deformation by a stress increases. It is noticed that at zero pressure, the melting temperature of SiC is 2746 K consistent with reported value of about 3100 ± 40 K [1, 2]. At P_T (=90 GPa), its value is about 4000 K which enhances further in $B1$ phase. An increase in T_M with variations in pressure infers the hardening or stiffening of the lattice. Higher melting temperature symbolizes higher shear modulus (G), and Young's modulus (E) values that we shall see later on. The data on its melting under high pressure are very limited and extremely contradictory, which does not allow one to make any conclusions about congruent or incongruent melting behaviour as well as the slope of the melting curve of SiC.

Table 4 Calculated aggregate second-order elastic constants (C_{11}, C_{12} and C_{44}) and aggregate bulk modulus (B_T), second-order elastic constant anisotropy parameter (γ_i^2), isotropic shear modulus (G_H), Voigt's shear modulus (G_V), Reuss's shear modulus (G_R), Young's modulus (E), Poisson ratio (ν), First and second Lame constant (λ, μ), longitudinal (v_l), shear (v_s) and average elastic wave velocity (v_m), Kleinman parameter (ξ) and thermodynamical properties: Debye temperature (θ_D) of silicon carbide in $B3$ phase at temperature of about 300 K

Materials parameter	SiC
C_{11} (10^{10} Nm^{-2})	67.65, 35.23 [49]
C_{12} (10^{10} Nm^{-2})	8.31, 14.04 [49]
C_{44} (10^{10} Nm^{-2})	19.99, 23.29 [49]
B_T (10^{10} Nm^{-2})	28.1
C_S (10^{10} Nm^{-2})	29.7
G_H (10^{10} Nm^{-2})	23.43
G_V (10^{10} Nm^{-2})	23.86
G_R (10^{10} Nm^{-2})	22.99
E (10^{10} Nm^{-2})	54.9, 43.7 [71]
ν	0.174, 0.167 [71]
λ (10^{10} Nm^{-2})	12.47
μ (10^{10} Nm^{-2})	23.43
v_l (ms^{-1})	12,440, 12,182 [71]
v_s (ms^{-1})	7819, 7701 [71]
v_m (ms^{-1})	1824
ξ	0.274
θ_D (K)	708.60

SiC is the only compound in the Si(IV)–C(IV) binary system and is obtained by electromelting high purity silica sand with petroleum coke, also of good quality. This melting takes place at high temperature about 2473 K and requires a large quantity of energy to produce. It also requires energy for it to dissociate into about 2/3 Si and 1/3 C in the induction furnace [1, 2]. The better quality raw materials produce better quality SiC, which is lower in nitrogen, sulphur, hydrogen and other trace elements. Figure 4 shows the temperature dependence of the melting temperature for SiC estimated from the C_{11} elastic constant as discussed previously. At room temperature its value is about 4550 K which drops and is 4200 K at 1000 K in $B3$ phase. The suppressed T_M with increased temperature indicates that there is a decrease in the resistance to deformation by a stress induced due to temperature. The suppressed T_M infers the weakening of the lattice as a result of thermal softening. Usually, SiC does not melt, it actually dissolves since its melting point is about 2973 K. Its behaviour in the molten metal is similar to sugar dissolving in coffee. This aspect is very important for the use of SiC.

For cubic lattice, three second-order elastic constants and the six non-vanishing third-order elastic constants are obtained from crystal geometry. The anharmonicity of a crystal lattice is successfully probed in terms of higher order elastic constants. The third-order terms in the strain variables are deduced from derivatives of elastic energy (please see Appendix for both ZB $B3$ and RS $B1$ phases). For SiC, the third-order aggregate elastic constants C_{111}, C_{112}, C_{166}, C_{144}, and C_{456}, are negative and only C_{123} is positive at $P = 0$ GPa i.e., in $B3$ phase. We note that no such efforts have been made in the past for third-order elastic constants of SiC. Thus, deduced information on pressure-dependent C_{ijk} will serve as a guide line for future

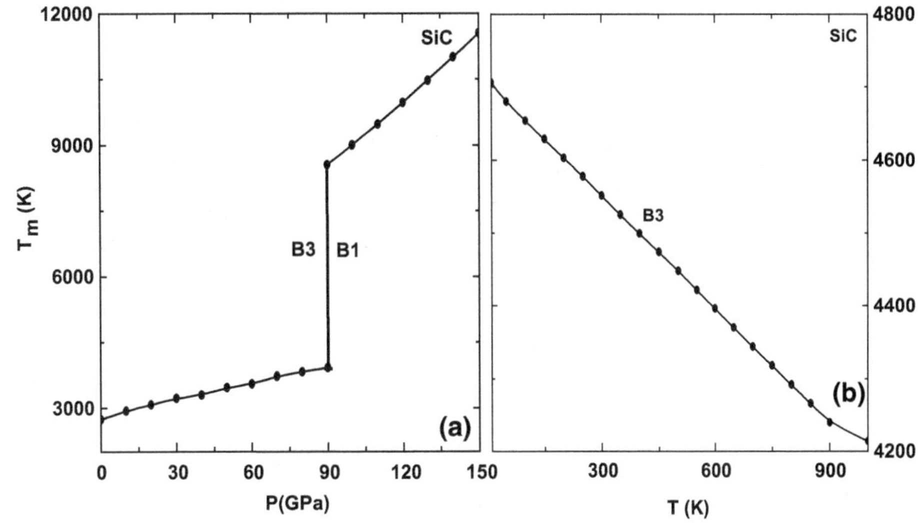

Fig. 4 Variation of melting temperature (T_M) with pressure and temperature

research and its application in materials technology. This information is useful as the structural strengths of a material are successfully known from the knowledge of elastic constants. The microstructures developed on the applied pressure and temperature can be known from third-order elastic constants variations.

The variation of aggregate third-order elastic constants (TOECs) with pressure for SiC in $B3$ and $B1$ phase is sketched in Fig. 5. For SiC in cubic phase, the C_{144}, and C_{166} increases with enhancing pressure. Other TOECs as C_{456}, C_{112}, C_{111}, and C_{123} infer a decreasing trend. We note that C_{144} is remarkably larger as compared to other TOECs. Furthermore, aggregate elastic constants C_{ijk} inferring the anharmonic effects are influenced by application of pressure in SiC. In the present interatomic potential, C_{ijk} are affected by the inclusion of second-nearest-neighbour interaction and are influenced by the short-range interactions (please see Eqs. 60–65 in the Appendix). Also, many body non-central forces as long-range Coulomb, charge transfer interactions and covalency are effective in SiC. We comment that pressure-dependent C_{ijk} behaviour can have a direct means to describe the interatomic forces at high pressure. It successfully cares the short-range forces, and a balance between long range and short-range forces.

As a next step, we intend to study the anisotropy among C_{ijk}. The equilibrium condition in the interatomic potential infers $B_1 + B_2 = -1.261\ Z_m^2$. The Cauchy discrepancy Δ_i^3 among third-order elastic constants are: (a) $\Delta_1^3 = C_{112} - C_{166} - 2P$; (b) $\Delta_2^3 = C_{123} - C_{456} - 2P$; (c) $\Delta_3^3 = C_{144} - C_{456} - 2P$; and (d) $\Delta_4^3 = C_{123} - C_{144} - 2P$. The C_{ijk} at zero pressure i.e., in $B3$ phase are influenced by contributions from both long-range and short-range interactions. Henceforth, Δ_i^3 is an indicator of the contribution from the non-central many body force. Figure 5a shows the variation of Δ_i^3 as functions of pressure. The significant deviation in Δ_i^3 is a natural consequence of the non-central many body forces as charge transfer interactions ions of Si and C atom and covalency effects caused by Si–Si, Si–C, and C–C bonds apart from short-range interactions as the induced charge dipole–dipole and charge dipole–quadruple (van der Waals) interaction and the overlap repulsion. At zero pressure, Δ_2^3 and Δ_3^3 are positive, while to that Δ_1^3 and Δ_4^3 are negative in $B3$ phase. A growth in Δ_2^3 and Δ_3^3 and decay in with enhanced pressure is witnessed with enhanced pressure till phase transition pressure. However, all Δ_i^3 is negative in $B1$ phase indicating the importance of non-central many body forces and anharmonic effects in SiC ceramics. As far as we know, there are no experimental results available for Cauchy discrepancy Δ_i^3 in SiC.

For cubic crystal structures, the aggregate third-order elastic constants C_{ijk} discern three anisotropy coefficients and three isotropic coefficients. It is useful to express linear combinations of the anisotropy coefficients and dividing them by the isotropic coefficients [68–70]. The anisotropy coefficients γ_i^3 are as follows:

$$\gamma_1^3 = \frac{3C_{112} - C_{111} - 12C_{144} + 12C_{166} - 16C_{456} - 2C_{123}}{2C_{123}} \tag{24}$$

$$\gamma_2^3 = \frac{C_{111} - C_{123} - 2C_{144}}{2C_{144}} \tag{25}$$

$$\gamma_3^3 = \frac{C_{166} - C_{144} - 2C_{456}}{2C_{456}} \tag{26}$$

The pressure dependence of the elastic anisotropic parameter γ_i^3 for both phases is sketched in Fig. 5b. The elastic anisotropy (γ_2^3 and γ_3^3) in $B3$ phase is less sensitive. On the other hand, γ_1^3 shows variation with increase in pressure in $B3$ phase. The anisotropy factor γ_1^3 is negative and shows a

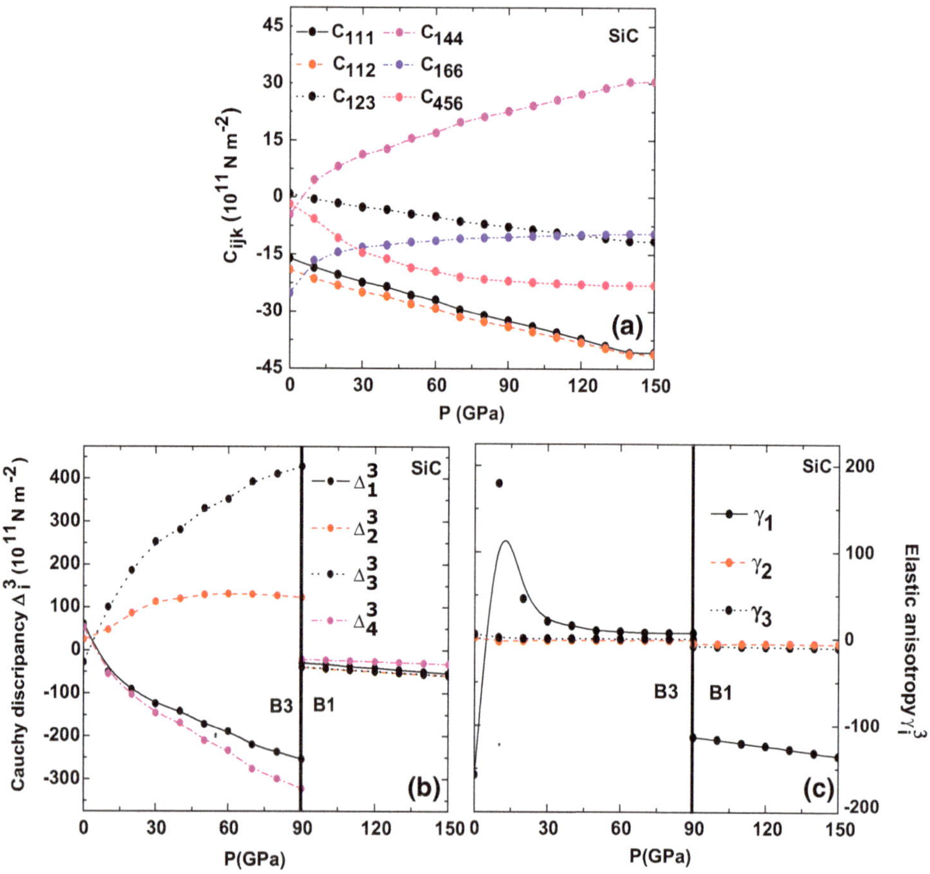

Fig. 5 Variation of aggregate third-order elastic constants (C_{ijk}) with pressure, Cauchy discrepancy (Δ_i^3) and elastic anisotropy (γ_i^3) in third-order elastic constant with pressure

decreasing trend with pressure in $B1$ phase at higher pressures. Values of γ_i^3 are given in Table 5 at zero temperature and pressure. As far as we know, there are no experimental and theoretical results available for comparison.

The strength and hardness are key issues for materials useful in technological applications. The mechanical strength and hardness of SiC can be known once elastic constants are known. We now determine elastic moduli at various pressures of SiC which are of substantial importance in engineering, geophysical and seismological application. The elastic properties are important in probing the bonding characteristic between adjacent atomic planes and the anisotropic character of the solid. Crystals leading to elasticity under application of pressure identify the response of a crystal under external strain. This valuable information is characterized by isotropic shear modulus (G_H), and Young's modulus (E). We note that hardness of polycrystalline materials is inhibited in elastic response as G_H and E.

The resistance to volume change and resistance to reversible deformations upon applied pressures are accounted by Voigt–Reuss–Hill approximation using G_H and E following [68–70]:

$$G_H = \frac{G_V + G_R}{2} \tag{27}$$

Herein, G_V (G_R) is Voigt's (Reuss's) shear modulus corresponding to the upper (lower) bound of G_H values. The G_V (G_R) is obtained from C_{ij} assuming uniform strain throughout the sample as

$$G_V = \frac{C_{11} - C_{12} + 3C_{44}}{5} \tag{28}$$

and

$$\frac{5}{G_R} = \frac{4}{(C_{11} - C_{12})} + \frac{3}{C_{44}} \tag{29}$$

Hardness of materials is usually measured in terms of isotropic shear modulus (G_H), and Young's modulus (E) or bulk modulus (B_T). The bulk modulus infers resistance to volume change by applied pressure. The isotropic shear modulus (G_H) measures the resistance to reversible deformations upon shear stress. In such a situation, G_H is a better probe to measure hardness than the bulk modulus (B_T). Table 4 illustrates the calculated isotropic shear modulus (G_H), Voigt's shear modulus (G_V) and Reuss's shear modulus (G_R) for SiC ceramics at zero temperature and

Table 5 Calculated aggregate third-order elastic constant (C_{ijk}), third-order elastic constant anisotropy parameter (γ_i^3), First and second Lame constant (λ, μ), longitudinal (v_l), shear (v_s) and average elastic wave velocity (v_m), Kleinman parameter (ξ) thermodynamical properties: force constant (f), Reststrahlen frequency (v_0), Debye temperature (θ_D), and average elastic constant (C), of Silicon Carbide in $B3$ phase at zero pressure

Materials parameter	SiC
C_{111} (10^{10} Nm^{-2})	−160.70
C_{112} (10^{10} Nm^{-2})	−189.70
C_{123} (10^{10} Nm^{-2})	8.33
C_{144} (10^{10} Nm^{-2})	−44.85
C_{166} (10^{10} Nm^{-2})	−250.90
C_{456} (10^{10} Nm^{-2})	−17.38
γ_1^3	−157.283
γ_2^3	1.207
γ_3^3	4.927
λ (10^{10} Nm^{-2})	16.29
μ (10^{10} Nm^{-2})	16.47
v_l (ms^{-1})	12,400
v_s (ms^{-1})	7172
v_m (ms^{-1})	1688
ξ	0.709
f (10^5 dyne cm^{-1})	7.187
v_0 (10^{12} Hz)	6.867
θ_D (K)	611.758
$C_{av.}$ (10^{10} Nm^{-2})	22.82

pressure along with a comparison with the available theoretical results [10, 15, 50, 56, 58, 60].

A decreasing and then increasing nature of the isotropic shear modulus G_H, G_V and G_R is witnessed in $B3$ phase of SiC. On the other hand, a steep increase in G_H, G_V and G_R is seen with enhanced pressure in $B1$ phase at higher pressures as plotted in Fig. 6a. The explanation of the above characteristics lies in a fact that both G_V and G_R are influenced by aggregate elastic constant C_{44}. Thus, G_V decreases as C_{44} decreases with enhanced pressure in ZnS phase. On the other hand, Reuss's shear modulus (G_R) pressure-dependent behaviour is integrated by combination of $C_{11} - C_{12}$ pressure-dependent behaviour as well as to that of C_{44} pressure-dependent behaviour. G_R initially decreases and then starts increasing at about 30 GPa and is attributed to steep decrease in C_{44}. We end up by stating that above transition pressure an increase in G_H, G_V and G_R support our earlier prediction about mechanical stiffening of lattice.

The high temperature studies of materials at ambient pressure lead to the performance of a material in terms of (a) the understanding of vibrational anharmonicity that is associated with the relative interplay of long-range and short-range potential energy function, (b) thermal response in terms of softening or hardening and (c) the elastic behaviour of materials as the elasticity, extensibility, acoustic transmission velocity, Debye temperature, specific heat, and thermal conductivity.

The high temperature investigations cause laboratory difficulties and structural changes make the phenomenon more amenable to interpretation. Figure 6b displays the G_H, G_V and G_R behaviour of 3C SiC ceramics as functions of temperature (at zero pressure). We note that the isotropic shear modulus of SiC is decreasing with enhanced temperature. The steep decrease of G_H, G_V, and G_R is in accordance with aggregate second-order elastic constant C_{ij} behaviour with temperature. Suppressed G_H, G_V and G_R as functions of temperature infer the weakening of the lattice as a result of thermal softening.

The tensile modulus as Young's modulus (E) further elaborates the stiffness property. E is also defined in terms of Reuss's shear modulus (G_H), and bulk modulus (B_T):

$$E = \frac{9G_H B_T}{G_H + 3B_T} \tag{30}$$

The tensile strength as Young's modulus (E) for 3C SiC is illustrated in Table 4 for ZnS ($B3$) phase along with a comparison with the reported data [10, 15, 50, 55, 56, 60, 62, 64, 65]. The model calculations presented here lead to E value of about 411 GPa at zero pressure. It is known that the material is stiffer if its Young's modulus is high [E of steel, graphene and diamond is ~200, 1000 and 1220 GPa]. As inferred SiC is less stiff as compare to graphene and diamond. The pressure dependence of the Young's modulus (E), of SiC is sketched in Fig. 7a. A decreasing trend of E in $B3$ phase infers the weakening of tensile strength till 90 GPa and an increasing trend of E in $B1$ phase essentially identifies more stiffening. Looking to the wide usage of SiC in gas turbines, heat exchangers, ceramics fans, radar, microwave, solar cell, and high-voltage devices, these properties are worth investigating.

Figure 7b illustrates the temperature dependence of the tensile strength as Young's modulus (E) for 3C SiC ceramics. A decreasing trend is inferred with increase in temperature. Matsumoto and researchers have reported the Young's modulus and Poisson's ratio of SiC ceramics at temperatures >1400 °C using laser ultrasonics coupled with Fabry-Pérot interferometry as well ultrasonic pulse method [71]. It is reported that E is about 438 GPa at $T = 273$ K and shows a decreasing behaviour with increasing temperature. The model calculations presented here lead to a value of about 549 GPa at room temperature which is comparable to reported E of 437 GPa for 3C SiC ceramics [71]. A reduction of E with temperature is a signature of bond weakening or thermal softening.

We now make a comparison of tensile strength E (P) and E (T) of 3C SiC ceramics. It is noted that

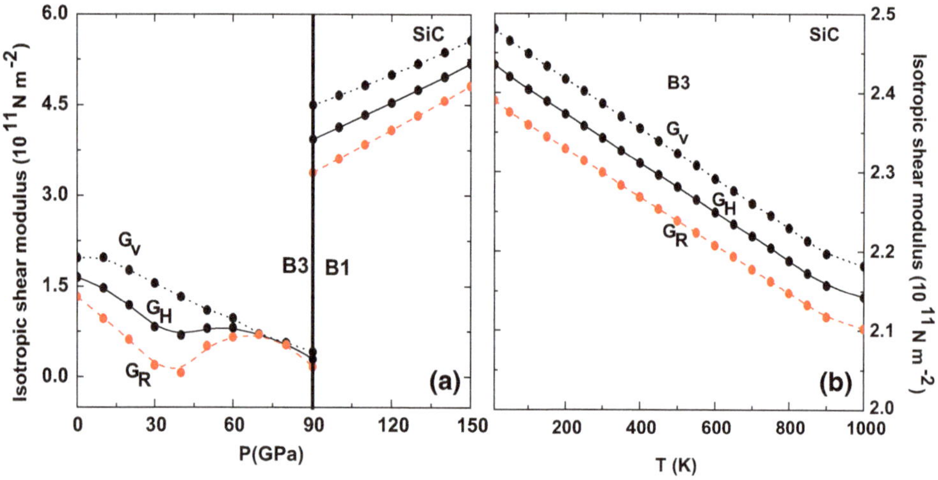

Fig. 6 Variation of Isotropic shear modulus (G_H, G_V, and G_R) with pressure and temperature

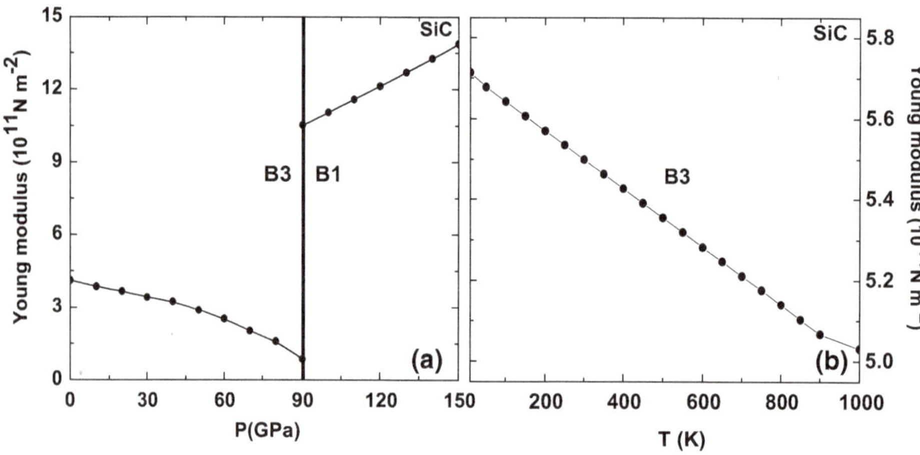

Fig. 7 Variation of Young's modulus (E) with pressure and temperature

E (P) values are smaller than E (T) for SiC at low pressures and temperatures. It implies that E (P) and E (T) are although susceptible to external variables as pressure and temperature, but it is their Si and C ions in 3*C* SiC ceramics that makes SiC lattice to be mechanical hard due to bond strengthening and thermal soft due to bond weakening. It is worth commenting that while tailoring the composites based on SiC, one should seek its performance in terms of tensile strength.

Another interest in 3*C* SiC is to probe the ductile and brittle nature of Silicon and Carbon-based alloys are important and predicted from the knowledge of second-order elastic constants. Ductile materials as Steel and aluminium sustain large strains before rupture, while to that brittle materials as glass and cast iron fractured at lower strains. For materials design and advances in metallurgy as well composite technology, the materials response for applied pressures and temperatures are often gauzed by stress–strain characteristics. Composites are predictive as depending on the external variable (pressure and temperature), the response of the constituent element is tailored in terms of ductile or brittle nature. Also ductile element in composites may become brittle as the pressure or temperature is increased or decreased. Once the pressure dependence of shear modulus (G_H), and Young's modulus (E) or bulk modulus (B_T) is known, it is worth investigating ductile and brittle nature of SiC.

The shear modulus (G_H,) measures the resistance to plastic deformation, while the bulk modulus (B_T) probes the resistance to fracture. An empirical relation in terms of the ratio of these moduli is known to differentiate ductile and brittle nature. In accordance with the Pugh's ratio ϕ (=B_T/G_H) > 1.75, the material sustains large strains before rupture *i.e.*, the ductile response. On the other hand, for $\phi \leq 1.75$ the material is brittle. Empirically, 1.75 is a critical value that separates ductile and brittle response of materials [72].

In Fig. 8a, we have shown the Pugh ratio ϕ as functions of pressure. Note that SiC is ductile in ZnS phase (at zero as well at low pressures). Brittle nature is noticed at higher pressures in ZnS phase. Furthermore, brittle response of SiC is also noticed in rock salt structure. Hence, ceramics SiC is ductile/brittle irrespective of any structural transformations. To our knowledge, perovskite are ductile/brittle that is a test for its deformation before fracture. In materials performance ductility is a powerful probe for quality control. It provides a means to assess the level of impurities and proper processing of a material. Hence, we classify 3*C* SiC as brittle material at higher pressures and ductile at zero pressure on the basis of Pugh ratio. The magnitude of elastic constants and shear moduli agrees with ductility degree of SiC in both *B*3 and *B*1 phase.

The Poisson's ratio is another measure to differentiate the ductility and brittleness of materials [73]. The critical value of Poisson's ratio (in terms of B_T and G_H) v is 0.33 that separates ductile and brittle nature of any material. If $v > 0.33$, the material is ductile, and for $v < 0.33$, the brittle response of material is observed such as ceramics. However, Poisson's ratio lies in between -1.0 and 0.5 which are the lower and upper bounds. The lower bound is a signature of the response of the materials that do not change its shape and the upper bound indicates that the volume is unchanged. The Poisson's ratio v in terms of bulk modulus B_T and the shear modulus G_H is expressed as [68–70],

$$v = \frac{1}{2}\left[3\frac{B_T}{G_{\mathrm{H}}} - 2\right]\left[3\frac{B_T}{G_{\mathrm{H}}} + 1\right]^{-1} \quad (31)$$

Figure 8a summarizes the pressure-dependent results of Poisson's ratio v. SiC in ZnS phase documents mixed behaviour. At zero pressure ($P = 0$ GPa), the value of v is about 0.25 for SiC. At lower pressures, SiC is brittle, above 15 GPa and till phase transition pressure, ductile nature of SiC is observed. Above P_T (=90 GPa), the value of v is about 0.34 for *B*1 phase. Above P_T, at higher pressures, in the RS structure, SiC remains brittle with increasing pressure. Deduced value of v is in good agreement with available experimental and theoretical results [5, 10, 11, 15, 47, 50, 56, 59, 63]. As per definition of Poisson's ratio, 3*C* SiC is brittle. It should be pointed out here that both Pugh (SiC as ductile) and Poisson's ratio (SiC as brittle) give contradictory results. The two empirical rules only differ on the exact border between the two types of behaviour. In view of this, the pressure-dependent variations of SiC allow it to consider as a borderline case between the classes of ductile and brittle materials. We note the ceramics are brittle, but perovskites are ductile.

The Poisson's ratio v behaviour as a function of temperature is sketched in Fig. 8b. The Poisson's ratio v is independent of temperature and is consistent with the earlier measured ν by laser ultrasonics method [71]. Beginning from zero temperature and at high temperatures in ZB phase, $\nu \approx 0.174$ is obtained for 3*C* SiC. The temperature-dependent Poisson's ratio reflects brittle nature of SiC, while to that a borderline case between the classes of ductile and brittle materials is known from pressure dependence. Poisson's ratio is thus an effective indicator to control the level of impurities and processing of SiC ceramics. For covalent materials, v is small ($v \sim 0.1$), whereas for metallic materials v is typically 0.33. It is worth mentioning that agreement with experimental and theoretical data is not fortuitous, but it is attributed to proper parametrization and formulation of potential with non-central many body forces as charge transfer interactions ions of Si and C atom and covalency effects caused by Si–Si, Si–C, and C–C bonds apart from short-range interactions as the induced charge dipole–dipole and charge dipole–quadruple (van der Waals) interaction and the overlap repulsion.

Usually, the elastic moduli describe only reversible response of a material to small strain near equilibrium. The intrinsic strength of a material reflects permanent plastic deformation at large shear strain. Consequently, to further understand the behaviour of SiC in terms of Vickers hardness:$H_V = 2 \quad (\phi^2 G_H)^{0.585} - 3$. Here, $\phi = B_T/G_H$. Figure 9a illustrates the theoretical Vickers hardness as a function of pressure. It is clear from the plot that the Vickers hardness H_V decreases in *B*3 phase and then increases in *B*1 phase with increase in pressure, which

Fig. 8 Variation of Poisson's ratio (v) and Pugh's modulus ratio (Φ) with pressure and temperature

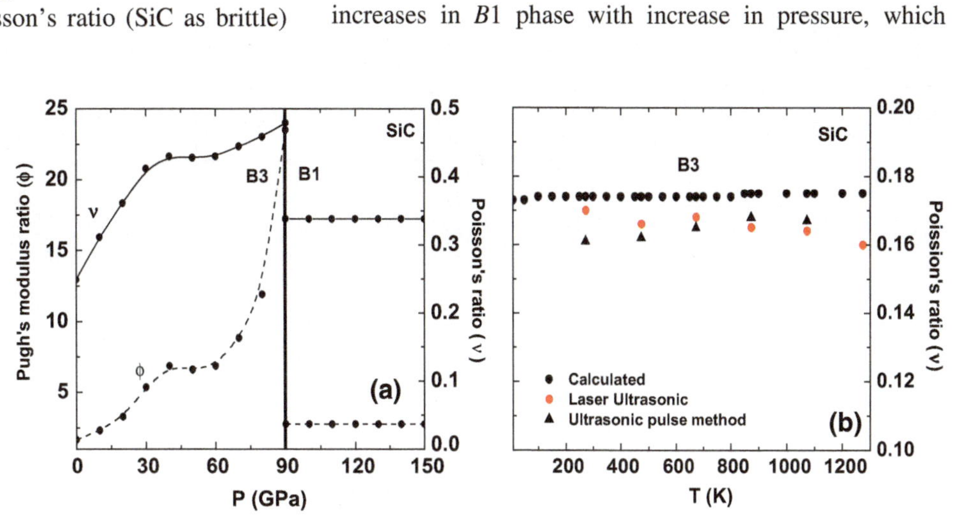

indicates that SiC becomes hard under pressure implying its good mechanical properties. Figure 9b shows Vickers hardness as a function of temperature for SiC. It is clear from the plot that the Vickers hardness H_V decreases in $B1$ phase, which indicates that SiC gets softened with enhanced temperature. Apart from the elastic anisotropy of crystals, the hardness is important to discuss their properties because it is highly correlated with the possibility of inducing microcracks in materials.

SiCs are promising materials with wide range of applicability's with effective mechanical properties. Usually, materials elastic properties are a source of valuable information where materials mechanics is significant as the knowledge of deformational characteristics of materials is essential in engineering design and construction of effective structures. Having, understood the materials elastic behaviour using Bulk modulus (B_T), shear modulus (G_H, G_V and G_R), and Young modulus (E), in the following, we will use our calculated elastic constants to discuss the compressional and shear wave velocity in ceramics SiC. The velocities of the longitudinal v_l and shear waves v_s are known from the Lamé's constants, λ and μ. The compressional wave with velocity v_l propagates back and forth in a crystal. The shear waves with velocity v_s go up and down.

The Lamé's constants are related to Young's modulus (E) and Poisson's ratio (v) [correlating the bulk modulus B_T and the shear modulus G_H]. The first Lamé's constant (λ) measures the compressibility of the material. The second Lamé's constant (μ) infers its shear stiffness [68–70]. The Lamé's constants (λ and μ) are expressed as:

$$\lambda = \frac{vE}{(1+v)(1-2v)} \tag{32}$$

$$\mu = \frac{E}{2(1+v)} \tag{33}$$

Figure 10a displays the pressure dependence of the first and second Lamé's constants (λ, μ). Starting from zero pressure and at high pressures, both Lamé's constants (λ, μ) are positive. Note that the Lamé's constant (λ) can be negative; however, for most materials it is also positive. The second Lamé's constant (μ) is positive. An increasing trend of compressibility (λ) of SiC is witnessed in terms of λ as seen in both ZB and RS structures and is attributed to mechanical hardening of lattice. The decreasing trend in $B3$ phase and increasing trend in $B1$ phase of shear modulus (G_H, G_V and G_R) results in a decreasing behaviour of shear stiffness (μ) in $B3$ phase and then enhanced shear stiffness in $B1$ phase. Note that λ and μ together constitute a parameterization of the elastic moduli for homogeneous isotropic media. The values of pressure-dependent Lamé's constants (λ, μ) are documented in Table 5. Deduced values of Lamé's constants (λ, μ) could not be compared due to lack of data on SiC. Figure 10b summarizes the temperature-dependent behaviour of Lamé's constants (λ, μ), for $3C$ SiC. It is noticed that both Lamé's constants (λ, μ) are decreasing with increasing temperature. Thus, both compressibility and shear stiffness showed decreasing trend with temperature variations and identifies once again the thermal softening of SiC.

The Lamé's constants (λ, μ) determine the longitudinal (shear) wave velocity as

$$v_l = \left[\frac{\lambda + 2\mu}{\rho}\right]^{\frac{1}{2}} \tag{34}$$

$$v_s = \left[\frac{\mu}{\rho}\right]^{\frac{1}{2}} \tag{35}$$

Here, ρ is the density. The average wave velocity v_m has been approximately given by

$$v_m = \left[\frac{1}{3}\left(\frac{2}{v_t^3} + \frac{1}{v_s^3}\right)\right]^{-\frac{1}{3}} \tag{36}$$

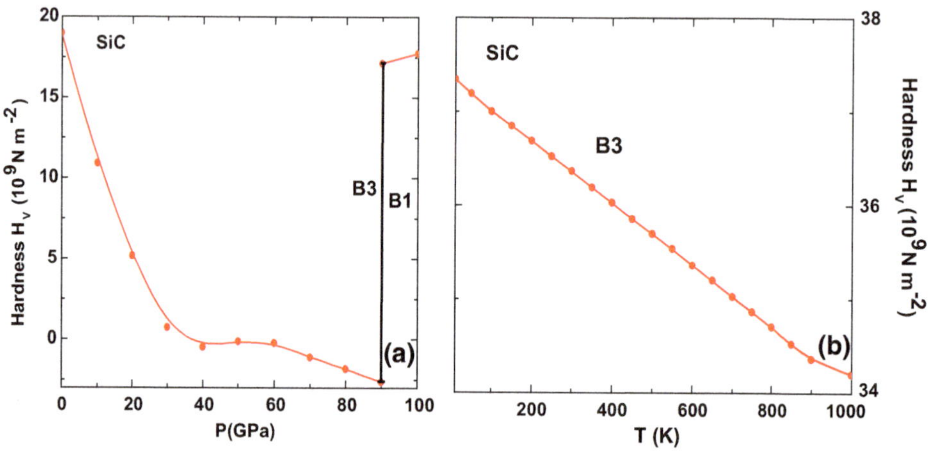

Fig. 9 Variation of Hardness (H_V) as a function of pressure and temperature

The compressional longitudinal (shear) wave velocity thus depends on density of the material as well on Bulk modulus (B_T), shear modulus (G_H, G_V and G_R), and Young modulus (E). We note that Bulk modulus (B_T) has a strong dependence on the density of the material, Young modulus (E) displays a weak dependence, while the shear modulus (G_H) is independent of density.

Figure 11a displays the pressure dependence of the longitudinal (shear) wave velocity in SiC. In ZB phase, beginning from zero pressures and at high pressures, compressional wave velocity v_l increases while to that shear wave velocity v_s decreases. Both compression and shear waves increase with enhanced pressure in $B1$ structure. Deduced values of longitudinal, shear and average elastic wave velocities propagating in SiC are illustrated in Table 5 at zero temperature and pressure. Figure 11b represents the temperature dependence of the longitudinal (shear) velocity, respectively. It is noticed that both longitudinal (shear) wave velocity decreases in $B3$ phase with enhanced temperature. The values of the longitudinal, shear and average elastic wave velocities propagating in $3C$ SiC ceramics are documented in Table 4 at room temperature. Deduced values of wave velocities are in good agreement with the measured values by laser ultrasonics method [71]. The high temperature behaviour of longitudinal (shear) wave velocity for SiC can be considered as predictive studies as they cannot be compared due to unavailability of high temperature data.

The Navier's equation is also used to determine the longitudinal and the shear wave velocity [74, 75]. These are written in terms of Reuss's shear modulus (G_H), and bulk modulus (B_T) as

$$v_l = \left[\frac{3B_T + 4G_H}{3\rho}\right]^{\frac{1}{2}} \quad (37)$$

$$v_s = \left[\frac{G_H}{\rho}\right]^{\frac{1}{2}} \quad (38)$$

Elastic, plastic and molten state properties with pressure as controlling variable are useful for tailoring composites. The Lamé's constants (λ, μ) are of substantial interest for plastic materials. Referring to equations 32 and 33, we note that as the Poisson's ratio (ν) increases, the Lamé's constants (λ, μ) numerically approach the bulk modulus (E). For fluids, the Reuss's shear modulus (G_H) vanishes as the viscosity of the fluid approaches zero. The above is

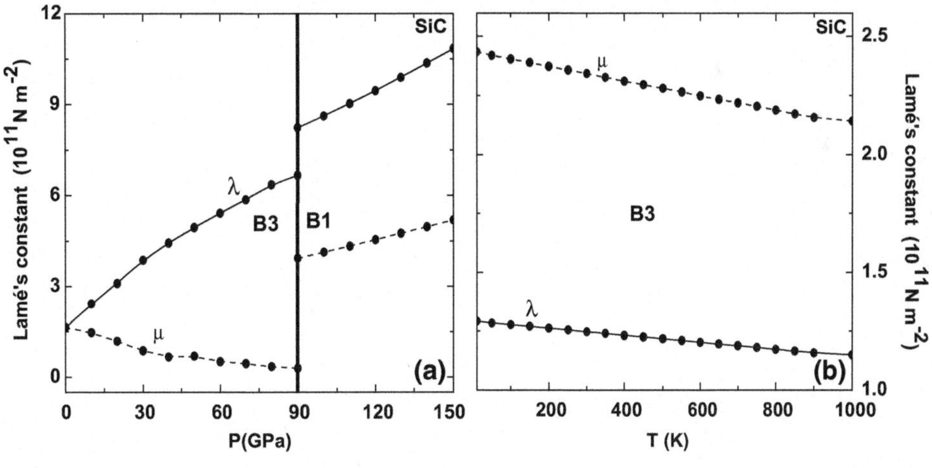

Fig. 10 Variation of Lamé's constant (λ, μ) with pressure and temperature

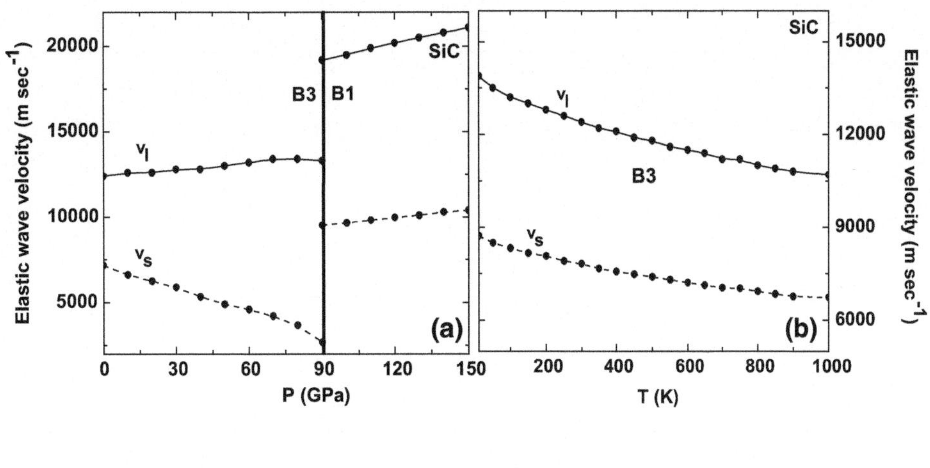

Fig. 11 Variation of elastic wave velocity v_l and v_s with pressure and temperature

important in context of the present computation and can also be cross-checked from the relation between bulk modulus and Reuss's shear modulus: $B_T = \lambda + 2\ G_H/3$. Furthermore, G_H approaches zero for fluids and hence the Poisson's ratio is ~ 0.5.

The response to deformations against bond bending or bond-angle distortion is relevant for materials with promising technological applications and also a test to validate the many body non-central potential that we dealt with. The aggregate elastic constants infer the elastic properties of material that undergo stress, deform and then recover after returns to its original shape after stress ceases. The nature of elastic constants in solids holds a great importance to elucidate the microscopic nature interatomic bonding, equations of state, and vibrational density of states. The above can be understood in terms of Kleinman parameter, ξ, which describes the relative positions of the cation and anion sub-lattices under volume-conserving strain distortions for which positions are not fixed by symmetry [76, 77]. We have explored its applicability to ceramics SiC using [78].

$$\xi = \frac{C_{11} + 8C_{12}}{7C_{11} + 2C_{12}} \tag{39}$$

Deduced value of $\xi \approx 0.709$ is documented in Table 5 for $3C$ SiC at zero temperature and pressure. The value of $\xi \approx 0.274$ is obtained at room temperature and is reported in Table 4 for $3C$ SiC. A low value of ξ implies a large resistance against bond bending or bond-angle distortion and vice versa [79]. Thus, SiC shows resistance against bond bending or bond-angle distortion and hence is of immense use in heat exchangers and ceramics fans. As far as we know, there is no experimental result available for Kleinman parameter and is considered as a prediction study. For both ZB and RS structures, the knowledge of elastic constants at variable pressure is worth investigating aimed for practical applications related to the mechanical properties of a solid: load deflection, thermo elastic stress, internal strain, elastic wave velocities, and fracture toughness.

Physical properties as thermal expansion, heat capacity and Grüneisen parameter can be explained with higher order terms of the interaction potential. The thermal process in terms of heat conduction is hindered if one considers a solid to be perfectly harmonic and thermal conductivity will be infinitely large. Thus, the role of anharmonic effects or phonon decay is important as enhanced pressures as well temperatures allows a change in volume or dimensions. The pressure-dependent calculations of elastic constants will provide a measure of the accuracy of the calculation of forces in SiC as well the mechanical stiffening or hardening attributed to Si–Si, C–C, and Si–C bond compression and bond strengthening due to lattice vibration.

We express the molecular force in the absence of the Lorentz effective field [19–26].

$$f = \frac{1}{3}\left[\frac{d^2}{dr^2} U_{SR}(r) - \frac{2}{r_0}\frac{d}{dr} U_{SR}(r)\right]_{r=r_0}, \tag{40}$$

The force constant is a function of the second-order derivatives of the short-range (SR) overlap repulsive potential and as well the charge dipole–dipole and charge dipole–quadruple van der Waals potential. The Reststrahlen frequency is obtained from the force constant using

$$\upsilon_0 = \frac{1}{2\pi}\left[\frac{f}{\mu}\right]^{1/2}, \tag{41}$$

Here, μ is the reduced mass of SiC.

The Grüneisen parameter γ_G is a ratio of second and first-order derivatives of the potentials enable us to discuss the anharmonic effects in a crystal. We express γ_G as

$$\gamma_G = -\frac{r_0}{6}\left[\frac{U'''(r_0)}{U''(r_0)}\right] \tag{42}$$

The pressure dependence of Grüneisen parameter is sketched in Fig. 12a for $3C$ SiC in ZnS ($B3$) and RS ($B1$) phases. A linear decrease of γ_G with pressure in both phases is observed. The Grüneisen constant jump [γ_G ($B3 - B1$)] at P_T is about 7 % in $3C$ SiC ceramics. Deduced γ_G value of about 1.015 for $3C$ SiC in ZnS ($B3$) phase is in good agreement with available theoretical results [5, 11, 61]. On the other hand, Fig. 12b discerns temperature dependence of Grüneisen parameter with a value of about 0.98 at 1000 K. For most of the solids, Grüneisen parameter ranges from 1.5 to 2.5. As Grüneisen constant γ_G behaviour is influenced by ratio of second and first-order derivatives of the potentials, we may thus comment that anharmonicity is significant in SiC ceramics. Note that these results are further validated by inelastic neutron scattering measurements to probe phonon frequencies as a function of the crystal volume.

The isothermal compressibility (β) is known from second-order derivative of the potential as

$$\beta = \left[\frac{r_0^2}{9V}\{U''(r_0)\}\right]^{-1} \tag{43}$$

The isothermal compressibility (β) is another interesting thermodynamical property which invokes the bond compression or strengthening on application of pressure. With this motivation, we have computed and sketched isothermal compressibility (β) with pressure in Fig. 13a. For SiC, at zero pressure, β is about 0.024 which is comparable with available theoretical result of 0.1518 [5, 11]. Note that the lattice of SiC is stiffened with increased pressure in $B3$ phase. At phase transition pressure it gets softened. Away from phase transition pressure and at higher pressures i.e., in $B1$ phase β again becomes stiff. The mechanical stiffening is reflected from all elastic response of SiC. It is attributed to Si–Si, C–C, and Si–C bond compression and bond strengthening due to lattice vibration in both $B3$ and $B1$ phase. We comment that the compressibility (β) at zero pressure is reduced by about 1 % of that at transition pressure P_T in $B3$ phase in SiC. However, the magnitude of β at transition pressure P_T is about 0.3 % of that at higher pressures ($\sim$150 GPa). We identify that SiC documents giant lattice softening at the boundary of $B3$–$B1$ phase transition.

The compressibility is useful in engineering applications as it measures the performance of a material explicitly the elasticity, extensibility, and thermal conduction. The isothermal compressibility (β) as functions of temperature is shown in Fig. 13b. It is seen that β displays a steep increase in $B3$ phase. The compressibility is thus tunable with applied pressure and temperature stimuli and plays a vital role in tailoring materials and composites.

We further compute Debye temperature (θ_D) to shed further light on mechanical stiffened and thermal softened characteristics of SiC following [80–84]:

$$\theta_D^3 = \frac{3.15}{8\pi}\left(\frac{h}{k_B}\right)^3\left(\frac{r}{M}\right)^{\frac{3}{2}}\left(\frac{C_{11}-C_{12}}{2}\right)^{\frac{1}{2}}\left(\frac{C_{11}+C_{12}+2C_{44}}{2}\right)^{\frac{1}{2}}C_{44}^{\frac{1}{2}}, \tag{44}$$

where, M is the acoustic mass of SiC. The notations h and k_B are the Planck and Boltzmann constants.

The Debye temperature as functions of pressure is plotted in Fig. 14a for SiC. At zero pressure, θ_D is about 611 K. It is noticed that with enhanced pressure, θ_D decreases in $B3$ phase for SiC. The Debye temperature from $B3$ to $B1$ phase is jumped by 655 K at P_T. On the other hand, θ_D showed an increase in $B1$ phase. Deduced value of θ_D is listed in Table 5. Suppressed θ_D in $B1$ phase at higher pressure indicates the mechanical stiffening of lattice and giant softening at P_T and above pressures.

Debye temperature is a function of temperature and its value varies from technique to technique as well as depends on the sample quality. Usually, a standard deviation in θ_D of about 15 K is agreeable. The change in the

Fig. 12 Varaition of Gruneisen parameter (γ_G) with pressure and temperature

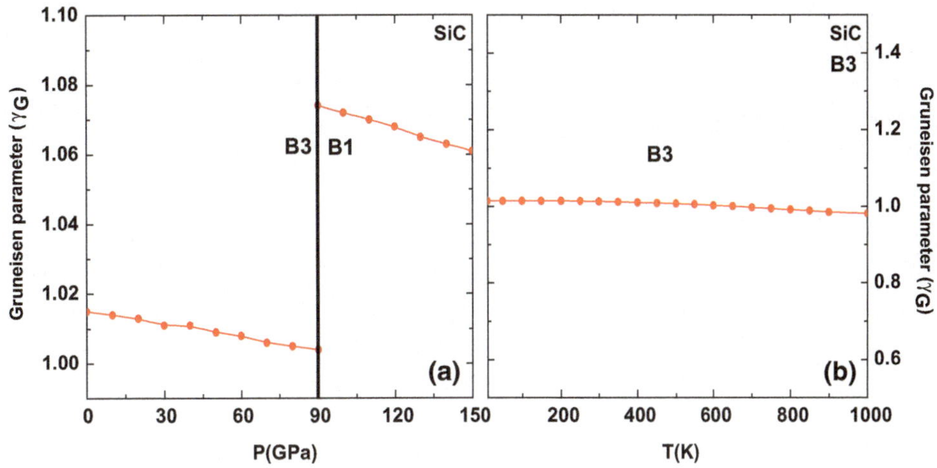

Fig. 13 Variation of isothermal compressibility (β) with pressure and temperature

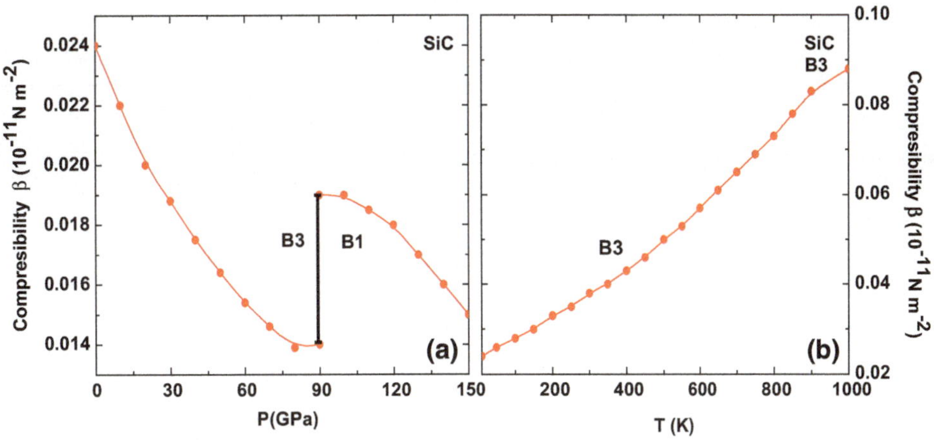

Fig. 14 Variation of Debye temperature (θ_D) as a function of pressure and temperature

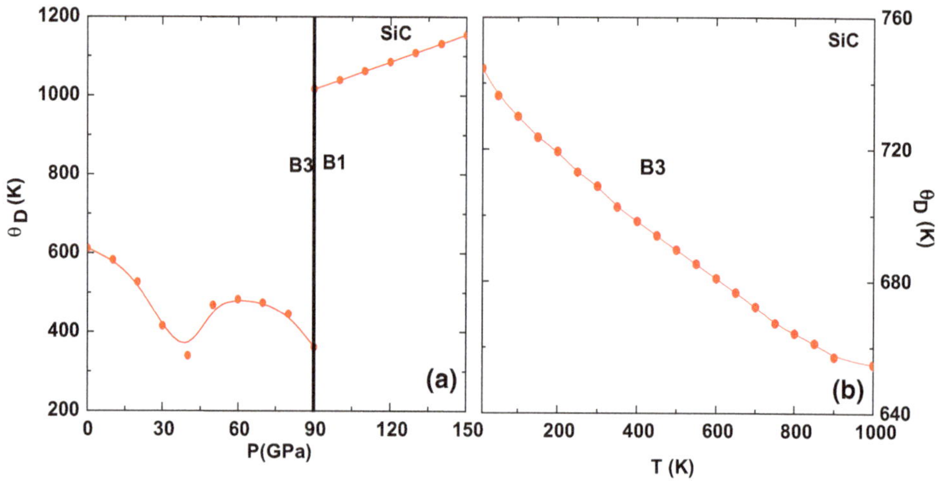

force constants induced by pressure decreases θ_D in $B3$ phase and after transition pressure it starts increasing which drives the system effectively towards the softening of lattice with increasing pressure.

Figure 14b displays the variation of the Debye temperature as functions of temperature at zero pressure. Starting from zero temperature, θ_D decreases rapidly. On the other hand, θ_D enhances with applied pressures at zero temperature. It is worth commenting from high pressure and high temperature Debye temperature behaviour that (a) the pressure-dependent Debye temperature infers the mechanical stiffened bulk modulus due to Si–Si, C–C, and Si–C bond compression and bond strengthening due to lattice vibration and (b) the thermal softening of bulk modulus results from bond expansion and bond weakening due to thermal stress in 3C SiC ceramics in ZB structure.

A comparison of values of $\theta_D(P)$ and $\theta_D(T)$ at zero pressure and zero temperature results: $\theta_D(T) > \theta_D(P)$ implying the susceptibility of θ_D with temperature. It is worth to comment that $\theta_D(P)$ and $\theta_D(T)$ behave differently with applied pressure and temperatures in $3C$ SiC ceramics. This is attributed to the fact vibrational spectrum of SiC lattice is controlled by aggregate elastic constants C_{ij} behaviour. Thus, the understanding of Debye temperature behaviour of a material not only provides essential features of the vibrational spectrum but is also mandatory for technological and engineering applications. This quantity is useful as a reference for future experimental studies.

The cumulative effect of both pressure and temperature led us to define aggregate elastic constants C_{av} as

$$C_{\mathrm{av.}} = \left(\frac{8\pi}{3.15}\right)^{\frac{2}{3}} \left(\frac{k_B}{h}\right)^2 \left(\frac{M}{r}\right) \theta_D^2, \tag{45}$$

This is useful once Debye temperature at zero pressure is known from experiments. Despite various investigations on mechanical and thermodynamical properties, basic properties controlled by elastic constants related with high temperature behaviours are lacking.

In order to further explore the role of anharmonic effects in terms of heat capacity at constant volume C_v and thermal expansion coefficient α, we express the vibration term $A_{\mathrm{vib.}}$ [85, 86]:

$$A_{\mathrm{vib.}}(\theta_D, T) = nk_B T\left[\frac{9\theta_D}{8T} + 3\ln\left\{1 - \exp\left(-\frac{\theta_D}{T}\right)\right\} - D(\theta_D/T)\right] \tag{46}$$

The non-equilibrium Gibbs function, $G^*(V; P, T)$, is minimized with respect to volume V as

$$\left[\frac{\partial G^*(V;P,T)}{\partial V}\right]_{P,T} = 0 \tag{47}$$

From Eq. (47), we determine heat capacity at constant volume C_v as

$$C_v = 3nk_B\left[4D\left(\frac{\theta_D}{T}\right) - \frac{3\theta_D/T}{e^{\theta_D/T} - 1}\right] \tag{48}$$

Figure 15a documents the variations of heat capacity at constant volume, C_v, with pressure P for both $B3$ and $B1$ phase of SiC at $T = 600$, 800, 1000 and 1200 K. The normalized heat capacity is $[C_v(P) - C_v(0)]/C_v(0)$, where $C_v(P)$ and $C_v(0)$ are heat capacity at any pressure P and at zero pressure. Starting from zero pressure, normalized heat capacity initially increases and with further enhanced pressure it decreases abruptly in $B1$ phase. This behaviour essentially points to the fact that the SiC lattice vibrations energy are controlled by both pressure as well temperature. Note that for higher temperatures $T \to \theta_D$, the variation in heat capacity with pressure is weak in $B1$ phase. Also, at P_T, reduced jump in between ZB and RS structures can be seen as compared to low temperatures.

The Debye temperature calculated from elastic constants must have a close resemblance to that measured from

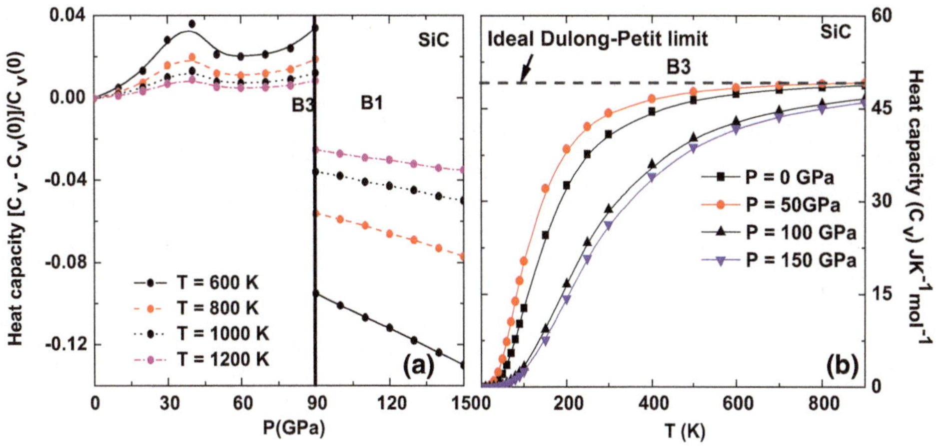

Fig. 15 Variation of heat capacity (C_v) with pressure at different temperatures and with temperature at different pressures

specific heat only at low temperatures. It is instructive to study the temperature-dependent behaviour of heat capacity. Figure 15b illustrates the heat capacity at constant volume, C_v, behaviour as functions of temperature for various pressures [0, 50, 100, 150 GPa] within the framework of quasi-harmonic model for 3*C* SiC ceramics. It can be seen that below room temperature (300 K), C_v increases very rapidly with the temperature at all pressures. Above room temperature, C_v increases slowly with the temperature. It almost approaches a constant ideal gas limit, the Dulong–Petit limit, $C_v(T) = 3R$, at higher temperatures as well at all pressures for SiC ceramics.

We further determine thermal expansion coefficient $\alpha_{\text{th.exp.}}$ using Eq. (47) as

$$\alpha_{\text{th.exp.}} = \frac{\gamma C_v}{B_T V} \quad (49)$$

Apart from the pressure dependence of heat capacity at constant volume C_v, the Gruneisen parameter and Bulk modulus are also needed to elucidate the thermal expansion coefficient ($\alpha_{\text{th.exp.}}$). It is a measure of any alteration in phonon frequency depending on the lattice's expansion or contraction in volume as a result of temperature variation. Figure 16a illustrates the pressure dependence of $\alpha_{\text{th.exp.}}$ for 3*C* SiC in both ZB and RS i.e., *B*3 and *B*1 phase. A nonlinear decrease in $\alpha_{\text{th.exp.}}$, is thus witnessed with pressure in both ZnS and NaCl phase. The decrease is more rapid in ZB structure while a slow decrease is documented in RS structure. We note that at P_T, the thermal expansion coefficient $\alpha_{\text{th.exp.}}$ is suppressed by 26–22 % in SiC, at temperatures 600, 800, 1000, and 1200 K, respectively.

The variations of $\alpha_{\text{th.exp.}}$ as functions of temperature at various pressures have been plotted in Fig. 16b for 3*C* SiC ceramics. This figure shows that at low temperatures ($T < 250$ K), $\alpha_{\text{th.exp.}}$ enhances rapidly with temperature at $P = 0$ and 50 GPa. At high temperatures ($T > 250$ K), a sharp increase of $\alpha_{th.exp}$ is witnessed at all pressures. This figure also suggests that with enhanced pressure, the increase of $\alpha_{\text{th.exp.}}$ with temperature becomes smaller. In other words, the slope of $\alpha_{\text{th.exp.}}$ gradually decreases at higher temperatures at all pressures except at $P = 0$ GPa. Returning back to pressure dependence of thermal expansion coefficient ($\alpha_{\text{th.exp.}}$), we note a rapid decrease in ZB structure with the increase of pressure. A comparison of $\alpha_{\text{th.exp.}}(P)$ and $\alpha_{\text{th.exp.}}(T)$ reveals that $\alpha_{\text{th.exp.}}(P)$ values are smaller than that $\alpha_{\text{th.exp.}}(T)$ for SiC under same pressure and temperature. Thus, $\alpha_{\text{th.exp.}}(P)$ and $\alpha_{\text{th.exp.}}(T)$ are differently sensitive to the Si and C ions in 3*C* SiC ceramics as SiC lattice is mechanical hard due to bond strengthening and thermal soft due to bond weakening.

Concluding remarks

The present study addresses for the first time, the high pressure and high temperature-dependent structural, elastic, and thermodynamical studies in ZB and RS structure of 3*C* SiC ceramics. We have formulated an interatomic pairwise potential that incorporates the long-range Coulomb with charge transfer interactions, covalent nature of bonds, zero point energy effects and the short-range interactions as charge dipole–dipole and charge dipole–quadruple (van der Waals), as well overlap repulsive interaction up to second-neighbour ions.

From the knowledge of Gibbs's free energies in ZB and RS structure, we determine the pressure-induced first-order structural phase transition of SiC at about 90 GPa. As a next step, we determine the cohesive energy as 6.301 eV for 3*C* SiC ceramics and the volume collapse is about 13.8 %. Compressions in SiC at higher pressure indicate the mechanical stiffening of lattice. The phase transition pressure and volume collapse are consistent with earlier observations. We emphasize that agreement with experimental and theoretical data is not fortuitous, but it is

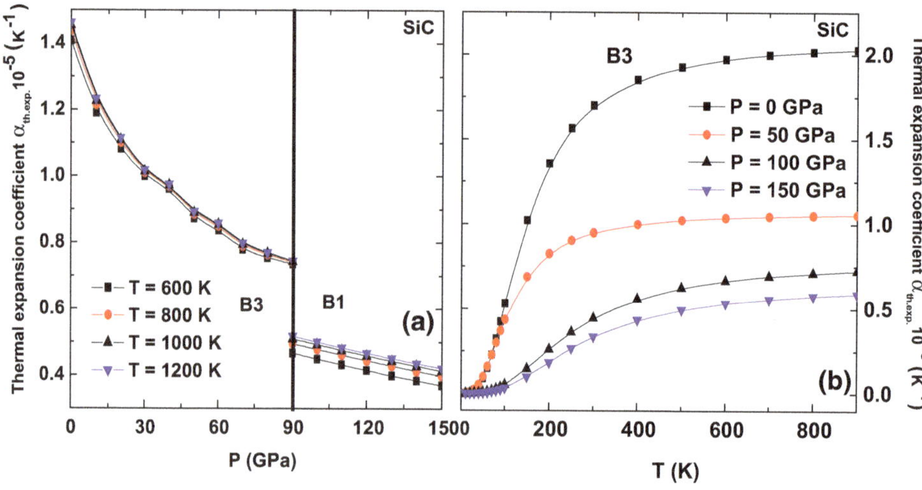

Fig. 16 Variation of the thermal expansion coefficient ($\alpha_{th.exp.}$) with pressure at different temperatures and with temperature at different pressures

attributed to proper parametrization and formulation of potential with non-central many body forces in terms of the screening of the effective Coulomb potential through modified ionic charge.

Furthermore, we make effort to determine the second-order aggregate elastic constants C_{ij} under applied pressure and temperature with respect to finite strain. Once pressure and temperature dependence of C_{ij} is known, Cauchy discrepancy and elastic anisotropy in second-order elastic constants, melting temperature, third-order elastic constants C_{ijk}, Cauchy discrepancy and anisotropy in third-order elastic constants, isotropic shear moduli as G_H, G_V, and G_R, Young's modulus E, Poisson's ratio v, Pugh's ratio ϕ and Vicker's hardness H_V to discuss ductile/brittle nature and mechanical stiffening/thermal softening of SiC lattice. In continuity, Lamé's constant (λ, μ), longitudinal (shear) wave velocity to enumerate the compressibility and shear stiffness of the material, Grüneisen constant, Debye temperature, isothermal compressibility, heat capacity and thermal expansion coefficient to shed light on anharmonicity of 3*C* SiC ceramics. We comment that incorporation of charge transfer interactions, covalent contribution and quantum effects leads to nonzero value of Cauchy discrepancy ($C_{12} - C_{44} \neq 0$) and is a corner stone of the proposed interatomic potential.

From the present investigations on 3*C* SiC ceramics, we draw the following conclusions:

(a) The volume collapse (V_P/V_0) in terms of compressions in SiC at higher pressure indicates the mechanical stiffening of lattice. The expansion of SiC lattice is inferred from steep increase in V_T/V_0 and is attributed to thermal softening of SiC lattice.

(b) Larger deviation in Cauchy discrepancy Δ_1^2 emphasizes the importance of the many body non-central (charge transfer and covalency) interaction and substantial anharmonic effects at high pressures.

(c) Elastic anisotropic parameter γ_1^2 in cubic SiC is negative and largely uninfluenced by application of pressure.

(d) Melting temperature (T_M), Vicker's hardness (H_V), shear modulus (G_H), Young's modulus (E) and bulk modulus (B_T) increase with enhanced pressure showing the hardening or stiffening of the lattice, and suppressed T_M, G_H, E and B_T variations in temperature suggest the weakening of the lattice as a result of thermal softening of SiC lattice.

(e) From the Pugh's ratio (ϕ) we classify 3*C* SiC as ductile material while to that Poisson's ratio suggest its brittle nature. The two empirical rules only differ on the exact border between the two types of behaviour and SiC is a borderline case between the classes of ductile and brittle materials.

(f) Grüneisen parameter (γ_G) Debye temperature (θ_D), isothermal compressibility (β), heat capacity at constant volume (C_v) and thermal expansion coefficient ($\alpha_{th.exp.}$) probe the importance of anharmonicity in SiC at heigh pressures and temperatures.

(g) Elastic properties showed $T_M(P) < T_M(T)$, $H_V(P) < H_V(T)$, $E(P) < E(T)$, $E(P) < E(T)$, $\theta_D(P) < \theta_D(T)$ and $\alpha_{th.exp.}(P) < \alpha_{th.exp.}(T)$ infers that these are differently sensitive to the Si and C ions as SiC lattice is mechanical hard due to bond strengthening and thermal soft due to bond weakening.

To an end, an interatomic pairwise potential for cubic SiC incorporating the non-central many body forces as long-range Coulomb with charge transfer interactions, covalent nature of bonds (due to Si–Si, Si–C, and C–C interacting electric fields), quantum effects, charge dipole–dipole and charge dipole–quadruple (van der Waals), and the short-range interactions as overlap repulsion up to second-neighbour ions successfully explains the pressure and temperature induced structural, elastic and thermodynamical properties of 3*C* SiC ceramics consistent with available results.

Appendix 1

The relevant expressions for the aggregate second-order elastic constants (SOECs), and the pressure derivatives of SOECs are expressed for zinc-blende structure as

$$C_{11} = L\left[0.2477Z(Z+8f(r_0)) + \frac{1}{3}(A_1+2B_1) + \frac{1}{2}(A_2+B_2) + 5.8243Zaf'(r_0)\right], \quad (50)$$

$$C_{12} = L\left[-2.6458Z(Z+8f(r_0)) + \frac{1}{3}(A_1-4B_1) + \frac{1}{4}(A_2-5B_2) + 5.8243Zaf'(r_0)\right], \quad (51)$$

$$C_{44} = L\left[-0.123Z(Z+8f(r_0)) + \frac{1}{3}(A_1+2B_1) + \frac{1}{4}(A_2+3B_2) - \frac{1}{3}\nabla(-7.53912Z(Z+8f(r_0)) + A_1 - B_1)\right] \quad (52)$$

and

$$B_1 + B_2 = -1.261Z[Z+8f(r)]. \quad (53)$$

Thus, second-order elastic constants difference lead to finite value at Cauchy pressure:

$$C_{12} - C_{44} \neq 0 \quad (54)$$

Henceforth, polarizability of the ions has effect on the elastic constants. It should be noted that if charge transfer mechanism is not taken into account, Cauchy relation: $C_{12} - C_{44} = 0$, The Cauchy violations ($C_{12} \neq C_{44}$) is seen by several crystals due to anisotropy in the electron distribution or angle bending. Here, C_{11} represents a measure of resistance to deformation by applied stress and C_{44} represents the measure of resistance to deformation with respect to applied shearing stress. The elastic constants C_{12} and C_{44} are related to the elasticity in shape, which is a shear constant.

The pressure derivatives of second-order elastic constants under hydrostatic pressure *P* are obtained in the form

$$3\Omega\frac{dB_T}{dp} = -\big[20.1788Z(Z+8f(r_0)) - 3(A_1+A_2) + 4(B_1+B_2) + 3(C_1+C_2) - 104.8433Zaf'(r_0) + 22.7008Za^2f''(r_0)\big] \quad (55)$$

$$2\Omega\frac{d\sigma}{dp} = -\left[-11.5756Z(Z+8f(r_0)) + 2(A_1-2B_1) + \frac{3}{2}A_2 - \frac{7}{2}B_2 + \frac{1}{4}C_2 + 37.5220Zaf'(r_0)\right], \quad (56)$$

and

$$\Omega\frac{dC_{44}}{dp} = -\left[0.4952Z(Z+8f(r_0)) + \frac{1}{3}(A_1-4B_1+C_1)r + \frac{1}{4}(2A_2-6B_2+C_2) + 4.9667Zaf'(r_0) + 2.522Za^2f''(r_0)\right] + \nabla\left[-17.5913Z(Z+8f(r_0)) + A_1 - B_2 - \frac{2}{3}C_{11} + 40.6461Zaf'(r_0) - 5.044Za^2f''(r_0)\right] + \nabla^2\left[3.1416Z(Z+8f(r_0) + \frac{2}{3}(A_1-B_1) + \frac{C_1}{3} - 15.9412Zaf'(r_0) + 8.8052Za^2f''(r_0)\right] \quad (57)$$

The notations are

$$\Omega = -5.0440g + A_1 + A_2 - 2(B_1+B_2) + 17.4730Zg_1 \quad (58)$$

$$\nabla = \left[\frac{-7.5391Z(Z+8f(r_0)) + (A_1-B_1)}{-3.141Z(Z+8f(r_0)) + (A_1+2B_1) + 21.765Zaf'(r_0)}\right], \quad (59)$$

Finally, the anharmonic third-order elastic constants (TOECs) for zinc-blende structure are as follows:

$$C_{111} = \frac{e^2}{4a^4}\left[0.5184g + \frac{1}{9}(C_1-6B_1-3A_1) + \frac{1}{4}(C_2-B_2-3A_2) - 2(B_1+B_1) - 9.9326Zg_1 + 2.522Zg_2\right], \quad (60)$$

$$C_{112} = \frac{e^2}{4a^4}\left[0.3828g + \frac{1}{9}(C_1+3B_1-3A_1) + \frac{1}{8}(C_2+3B_2-3A_2) - 11.642Zg_1 + 2.522Zg_2\right], \quad (61)$$

$$C_{123} = \frac{e^2}{4a^4}\left[6.1585g + \frac{1}{9}(C_1+3B_1-3A_1) - 12.5060Zg_1 + 2.5220Zg_2\right], \quad (62)$$

$$C_{144} = \frac{e^2}{4a^4}\left\{6.1585g + \frac{1}{9}(C_1+3B_1-3A_1) - 4.1681Zg_1 + .8407zg_2 + {} + \nabla\left[-3.3507g - \frac{2}{9}C_1 + 13.5486Zg_1 - 1.681Zg_2\right] + \nabla^2\left[-1.5637g + \frac{2}{3}(A_1-B_1) + \frac{C_1}{9} - 5.3138Zg_1 + 2.9350Zg_2\right]\right\} \quad (63)$$

$$C_{166} = \frac{e^2}{4a^4}\left\{ -2.1392g + \frac{1}{9}(C_1 - 6B_1 - 3A_1) + \frac{1}{8}(C_2 - 5B_2 - 3A_2) - (B_1 - B_2) - 4.168Zg_1 + .8407Zg_2 + \nabla\left[-8.3768g + \frac{2}{3}(A_1 - A_2) - \frac{2}{9}C_1 + 13.5486Zg_1 - 1.6813Zg_2\right] + \nabla^2\left[2.3527g + \frac{C_1}{9} - 5.3138Zg_1 + 2.9350Zg_2\right] \right. \tag{64}$$

$$C_{456} = \frac{e^2}{4a^4}\left\{4.897g + \frac{1}{9}(C_1 - 6B_1 - 3A_1) - B_2 + \nabla\left[-5.0261g - \frac{1}{9}C_1\right] + \nabla^2\left[7.0580g + \frac{1}{3}C_1\right] + \nabla^3\left[-4.8008g + \frac{1}{3}(A_1 - B_1) - \frac{1}{9}C_1\right]\right\} \tag{65}$$

In view of equilibrium condition, the third-order elastic constants difference also lead to finite value at Cauchy pressure inferring that polarizability of the ions has effect on the elastic constants:

$$C_{112} - C_{166} \neq 0 \tag{66}$$

$$C_{123} - C_{456} \neq 0 \tag{67}$$

$$C_{144} - C_{456} \neq 0 \tag{68}$$

$$C_{123} - C_{144} \neq 0 \tag{69}$$

Furthermore, the third-order elastic constants satisfy the identity:

$$C_{123} + 2C_{456} - 3C_{144} = 0 \tag{70}$$

Various symbols appear in the earlier expressions are associated with the crystal energy and have been defined below:

$$A_1 = A_{ij} = \frac{4a^3}{e^2}\left(\frac{d^2}{dr^2}V_{ij}(r)\right)_{r=a}, \tag{71}$$

$$A_2 = A_{ii} = A_{jj} = \frac{4a^3}{e^2}\left(\frac{d^2}{dr^2}V_{ii}(r) + \frac{d^2}{dr^2}V_{jj}(r)\right)_{r=\sqrt{3}a/4}, \tag{72}$$

$$B_1 = B_{ij} = \frac{4a^2}{e^2}\left(\frac{d}{dr}V_{ij}(r)\right)_{r=a}, \tag{73}$$

$$B_2 = B_{ii} = B_{jj} = \frac{4a^2}{e^2}\left(\frac{d}{dr}V_{ii}(r) + \frac{d}{dr}V_{jj}(r)\right)_{r=\sqrt{3}a/4}, \tag{74}$$

$$C_1 = C_{ij} = \frac{4a^4}{e^2}\left(\frac{d^3}{dr^3}V_{ij}(r)\right)_{r=a}, \tag{75}$$

$$C_2 = C_{ii} = C_{jj} = \frac{4a^4}{e^2}\left(\frac{d^3}{dr^3}V_{ii}(r) + \frac{d^3}{dr^3}V_{jj}(r)\right)_{r=a}, \tag{76}$$

$$g = Z + 8f(r) \tag{77}$$

$$g_1 = r_o df(r) \tag{78}$$

$$g(2) = r_o ddf(r) \tag{79}$$

Similarly, the expressions for the second-order elastic constants for RS structure crystals are as follows:

$$C_{11} = \frac{e^2}{4a^4}\left\{-5.112Z_m^2 + A_1 + \frac{A_2 + B_2}{2} + 9.3204Z(af'(r))\right\} \tag{80}$$

$$C_{12} = \frac{e^2}{4a^4}\left\{0.226Z_m^2 - B_1 + \frac{A_2 - 5B_2}{4} + 9.3204Z(af'(r))\right\} \tag{81}$$

$$C_{44} = \frac{e^2}{4a^4}\left\{2.556Z_m^2 + B_1 + \frac{A_2 + 3B_2}{4}\right\} \tag{82}$$

and

$$B_1 + B_2 = -1.165Z[Z + 12f(r)]. \tag{83}$$

For non-central many body forces as we deal with the second-order elastic constants difference lead to finite value at Cauchy pressure:

$$C_{12} - C_{44} \neq 0 \tag{84}$$

The pressure derivatives of second-order aggregate elastic constants under hydrostatic pressure P are obtained in the form

$$3\Omega\frac{dB_T}{dp} = -\{13.980Z_m^2 + C_1 - 3A_1 + C_2 - 3A_2 - 167.7648Z(af'(r)) + 41.9420Z(a^2f''(r))\} \tag{85}$$

$$2\Omega\frac{d\sigma}{dp} = -\left\{23.682Z_m^2 + C_1 + \frac{C_2 + 6A_2 - 6B_2}{4} - 50.0752Z(af'(r)) + 13.9808Z(a^2f''(r))\right\} \tag{86}$$

$$\Omega\frac{dC_{44}}{dp} = -\left\{-11.389Z_m^2 + A_1 - 3B_1 + \frac{C_2 + 2A_2 - 10B_2}{4} + 44.6528Z(af'(r))\right\} \quad (87)$$

The notation Ω is

$$\Omega = -2.330Z_m^2 + A_1 + A_2 + 27.9612Z(af'(r)) \quad (88)$$

Finally, the anharmonic third-order elastic constants for RS structure are as follows:

$$C_{111} = \frac{e^2}{4a^4}\left\{37.556Z_m^2 + C_1 - 3A_1 - \frac{3A_2 + 9B_2 - C_2}{4} - 89.305Z(af'(r)) + 13.980Z(a^2f''(r))\right\} \quad (89)$$

$$C_{112} = \frac{e^2}{4a^4}\left\{-4.836Z_m^2 - \frac{3A_2 + 3B_2 - C_2}{8} - 18.640Z(af'(r)) + 4.66Z(a^2f''(r))\right\} \quad (90)$$

$$C_{123} = \frac{e^2}{4a^4}\left\{2.717Z_m^2 + 16.692Z(af'(r))\right\} \quad (91)$$

$$C_{144} = \frac{e^2}{4a^4}\left\{2.717Z_m^2 + 5.564Z(af'(r))\right\} \quad (92)$$

$$C_{166} = \frac{e^2}{4a^4}\left\{-4.836Z_m^2 - 2(B_1 + B_2) - \frac{3A_2 - 3B_2 - C_2}{8} + 5.564Z(af'(r))\right\} \quad (93)$$

$$C_{456} = \frac{e^2}{4a^4}\left\{2.717Z_m^2 - (B_1 + B_2)\right\} \quad (94)$$

Various symbols appear in the above expressions are associated with the crystal energy and have the following form in RS structure

$$A_1 = \frac{8a^3}{e^2}\left[\frac{d^2}{dr^2}V_{ij}(r)\right]_{r=a}, \quad (95)$$

$$A_2 = \frac{16a^3}{e^2}\left[\frac{d^2}{dr^2}V_{ii}(r) + \frac{d^2}{dr^2}V_{jj}(r)\right]_{r=\sqrt{2}a}, \quad (96)$$

$$B_1 = \frac{8a^3}{e^2}\left[\frac{1}{r}\frac{d}{dr}V_{ij}(r)\right]_{r=a}, \quad (97)$$

$$B_2 = \frac{16a^3}{e^2}\left[\frac{1}{r}\frac{d}{dr}V_{ii}(r) + \frac{1}{r}\frac{d}{dr}V_{jj}(r)\right]_{r=\sqrt{2}a}, \quad (98)$$

$$C_1 = \frac{8a^3}{e^2}\left[r\frac{d^3}{dr^3}V_{ij}(r)\right]_{r=a}, \quad (99)$$

$$C_2 = \frac{16a^3}{e^2}\left[r\frac{d^3}{dr^3}V_{ii}(r) + \frac{d^3}{dr^3}V_{jj}(r)\right]_{r=\sqrt{2}a} \quad (100)$$

in terms of the short-range energy

$$V_{ij}(r) = \sum_{ij} b\beta_{ij}\exp\left(\frac{r_i + r_j - r_{ij}}{\rho}\right) - \sum_{ij}\frac{C}{r_{ij}^6} - \sum_{ij}\frac{D}{r_{ij}^8} \quad (101)$$

The short-range interaction (SR) energy is expressed in terms of the overlap repulsion (first term) and the induced charge dipole–dipole and charge dipole–quadruple (van der Waals) [second and third terms], respectively.

Appendix 2

The temperature-dependent elastic, thermal and thermodynamical properties of SiC are now studied within the quasi-harmonic Debye approximation. The Helmholtz free energy at temperature below the melting point of SiC ceramics in the quasi-harmonic model is applicable to evaluate equation of state and thus the elastic properties. The non-equilibrium Gibbs's free energy function, $G^*(V; P, T)$ within the quasi-harmonic approximation is $G = U + PV + A_{vib.}[\theta_D(V); T]$.

In order to determine the anharmonic effects on the elastic constants at higher temperatures, we use the equation of state (EOS) as:

$$P = -\frac{dU}{dV} + T\alpha_v B_T \quad (102)$$

Here, P is the pressure, T the temperature, U the total potential energy, α_v the volume thermal expansion coefficient, B_T the isothermal bulk modulus. The second term in Eq. (106) is the thermal phonon pressure. The temperature-dependent second-order elastic constants C_{ij} are derived from the dynamical matrix of modified Rigid Shell model [19–26, 35–42] and the method of long waves as

$$C_{11} = 1.573T\alpha_v B_T + \frac{e^2}{4a^4}\left\{\frac{1}{3}A_1 - 0.9B_1 - 1.573B_2 + 8.9558(af'(r))\right\} \quad (103)$$

$$C_{12} = -1.786T\alpha_v B_T + \frac{e^2}{4a^4}\left\{\frac{1}{3}A_1 + 1.45B_1 + 1.786B_2 + 8.9558(af'(r))\right\} \quad (104)$$

$$C_{44} = -0.786T\alpha_v B_T + \frac{e^2}{4a^4}\left\{\frac{1}{3}A_1 + 1.45B_1 + 1.786B_2 - \frac{1}{3}\Delta\right\} \quad (105)$$

with

$$\Delta = \left[A_1 + 6.77B_1 + 7.78B_2 - 7.77\left\{T\alpha_v B_T \frac{4a^4}{e^2}\right\}\{(A_1 + 2B_1) + 3(A_2 + 2B_2)\}^{-1}\right] \quad (106)$$

and

$$T\alpha_v B_T = \frac{e^2}{4a^4}[1.9395Z[Z + 8f(r_0)] + (B_1 + B_2)] \quad (107)$$

Herein, the potential energy incorporates the long-range Coulomb with charge transfer interactions, covalent nature of bonds, and short-range overlap repulsive interaction up to second-neighbour as well charge dipole–dipole and charge dipole–quadruple (van der Waals) interaction.

We use equilibrium condition to obtain temperature dependence of hardness and range parameters as

$$r\left|\frac{dU(r)}{dr}\right|_{r=r_0} = 3T\alpha_v VB_T \quad (108)$$

Here, V is the unit cell volume of SiC.

References

1. Levinshteæin, M.E., Rumyantsev, S.L., Shur, M.: Properties of advanced semiconductor materials: GaN, AlN, InN, BN, SiC, SiGe. Wiley, New York (2001)
2. Parfenova, I.I.: Substitutional 3d impurities in cubic silicon carbide. Semiconductors **38**, 189 (2004)
3. Yoshida, M., Onodera, A., Ueno Takemura, M.K., Shimomura, O.: Pressure-induced phase transition in SiC. Phys. Rev. B **48**, 10587 (1993)
4. Sekine, T., Kobayashi, T.: Shock compression of 6*H* polytype SiC to 160 GPa. Phys. Rev. B **55**, 803 (1997)
5. Chang, K.J., Cohen, M.L.: Ab initio pseudopotential study of structural and high-pressure properties of SiC. Phys. Rev. B **35**, 8196 (1987)
6. Miao, M.S., Lambrecht, W.R.L.: Unified path for high-pressure transitions of SiC polytypes to the rocksalt structure. Phys. Rev. B **68**, 92103 (2003)
7. Durandurdu, M., Drabold, D.A.: *Ab initio* simulations of the structural phase transformation of 2*H*-SiC at high pressure. Phys. Rev. B **75**, 235204 (2007)
8. Shimojo, F., Ebbsj, O.I., Lalia, R., Nakano, A., Rino, J.P., Vashista, P.: Molecular dynamics simulation of structural transformation in silicon carbide under pressure. Phys. Rev. Lett. **84**, 3338 (2000)
9. Yu-Ping, L., Duan-Wei, H., Jun, Z., Xiang Dong, Y.: First-principles study of pressure-induced phase transition in silicon carbide. Phys. B **403**, 3543 (2008)
10. Vashishta, P., Kalia, R.K., Nakano, A.: Interaction potential for silicon carbide: A molecular dynamics study of elastic constants and vibrational density of states for crystalline and amorphous silicon carbide. J. Appl. Phys. **101**, 103515 (2007)
11. Cheong, B.H.K., Chang, J., Cohen, M.L.: Pressure dependences of band gaps and optical-phonon frequency in cubic SiC. Phys. Rev. B **44**, 1053 (1991)
12. Miao, M.S., Prikhodko, M., Lambrecht, W.R.L.: Changes of the geometry and band structure of SiC along the orthorhombic high-pressure transition path between the zinc-blende and rocksalt structures. Phys. Rev. B **66**, 64107 (2002)
13. Miao, M.S., Prikhodko, M., Lambrecht, W.R.L.: Orthorhombic intermediate state in the zinc blende to rocksalt transformation path of SiC at high pressure. Phys. Rev. Lett. **88**, 18960 (2002)
14. Durandurdu, M.: Pressure-induced phase transition of SiC. J. Phys. Condens. Matter **16**, 4411–4417 (2004)
15. Stillinger, F.H., Weber, T.A.: Computer-simulation of local order in condensed phases of silicon. Phys. Rev. B **31**, 5262 (1985)
16. Karch, K., Bechstedt, F., Pavone, P., Strauch, D.: Pressure-dependent properties of SiC polytypes. Phys. Rev. B **53**, 13400 (1996)
17. Oganov, A.R. (ed.): Modern methods of crystal structure prediction (2011) Wiley VCH Verlag GmbH and Co. KGaA, Weinheim
18. Dion, M., Rydberg, H., Schroder, E., Langreth, D.C., Lundqvist, B.I.: Van der Waals density functional for general geometries. Phys. Rev. Lett. **92**, 246401 (2004)
19. Varshney, D., Kaurav, N., Sharma, P., Shah, S., Singh, R.K.: Structural phase transition and elastic properties of ZnSe at high pressure. Phase Transit **77**, 1075 (2004)
20. Varshney, D., Sharma, P., Kaurav, N., Shah, S., Singh, R.K.: Study of elastic properties and their pressure dependence of semi magnetic semiconductors. J. Phys. Soc. **74**, 382 (2005)
21. Varshney, D., Kaurav, N., Kinge, R., Singh, R.K.: B1-B2 structural phase transition and elastic properties of UX (X = S, Se and Te) compounds at high pressure. J. Phys. Condens. Matter **19**, 236204 (2007)
22. Varshney, D., Sharma, U., Kaurav, N.: Structural phase transformation and mechanical properties of semimagnetic semiconductors $Zn_{1-x}M_xSe$ (M = Mn, Fe and Cd). J. Phys. Condens. Matter **20**, 075204 (2008)
23. Varshney, D., Joshi, G., Kaurav, N., Singh, R.K.: Structural phase transition (zincblende–rocksalt) and elastic properties in AlY (Y = N, P and As) compounds: pressure-induced effects. J. Phys. Chem. Solids **70**, 451 (2009)
24. Varshney, D., Joshi, G.: High-pressure structural phase transition and elastic properties of Ga_{1-x} In_xAs semiconducting compounds. Eur. Phys. J. B **70**, 523 (2009)
25. Varshney, D., Joshi, G., Varshney, M., Shriya, S.: Pressure induced structural phase transition and elastic properties in BSb, AlSb, GaSb and InSb compounds. Phys. B **405**, 1663 (2010)
26. Varshney, D., Joshi, G., Varshney, M., Shriya, S.: Pressure dependent elastic and structural (B 3–B1) properties of Ga based monopnictides. J. Alloys Compd. **495**, 23 (2010)
27. Hafemeister, D.W., Flygare, W.H.: Outer-shell overlap integral as a function of distance for halogen–halogen, halogen–alkali, and alkali–alkali ions in the alkali halide lattices. J. Chem. Phys. Soc **43**, 795 (1965)
28. Tosi, M.P.: Cohesion of ionic solids in the born model. Solid State Phys. **16**, 1 (1964)
29. Born, M., Huang, K.: Dynamical Theory of Crystal Lattices. Clarendon, Oxford (1956)
30. Motida, K.: Szigeti charge and its correlation with hyperine coupling constant of doped Mn^{2+} ion in divalent metal compounds. J. Phys. Soc. Jpn. **49**, 213 (1980)
31. Motida, K.: Effect of covalency on phonon dispersion relations in NaCl type alkali halide crystal. J. Phys. Soc. Jpn. **55**, 1636 (1986)
32. Slater, J.C., Kirkwood, J.G.: The vander Waals forces in gases. Phys. Rev. **37**, 682 (1931)
33. Lowdin, P.O.: Quantum theory of cohesive properties of solids. Adv. Phys. **5**, 1 (1956)
34. Lundqvist, S.O.: Three body potential for alkali halides. Ark. Fys. **12**, 263 (1957)
35. Varshney, D., Joshi, G., Varshney, M., Shriya, S.: Pressure induced mechanical properties of boron-based pnictides. Solid State Sci. **12**, 864 (2010)

36. Varshney, D., Dodiya, N., Shaikh, M.W.: Structural properties and electrical resistivity of Na-substituted lanthanum manganites: $La_{1-x}Na_xMnO_{3+y}$ ($x = 0.1$, 0.125 and 0.15). J. Alloys Compd. **509**, 7447 (2011)
37. Varshney, D., Choudhary, D., Shaikh, M.W., Khan, E.: Electrical resistivity behaviour of sodium substituted manganites: electron–phonon, electron–electron and electron–magnon interactions. Eur. Phys. J. B **76**, 327 (2010)
38. Varshney, D., Kaurav, N., Kinge, R., Singh, R.K.: High-pressure induced structural phase transition in alkaline earth CaX (X = S, Se and Te) semiconductors: NaCl-type (*B*1) to CsCl-type (*B*2). J. Alloys Compd. **484**, 239 (2009)
39. Varshney, D., Kaurav, N., Kinge, R., Singh, R.K.: Pressure induced phase transition (B1–B2) and elastic properties in alkaline earth BaX (X = S, Se and Te) chalcogenides. Phase Transit. **81**, 1 (2008)
40. Varshney, D., Kaurav, N., Kinge, R., Singh, R.K.: High pressure structural (B1–B2) phase transition and elastic properties of II–VI semiconducting Sr chalcogens. Comput. Mater. Sci. **41**, 529 (2008)
41. Varshney, D., Kaurav, N., Sharma, U., Singh, R.K.: Pressure induced phase transition and elastic properties of Y and Sc antimonides. J. Alloys Compd. **448**, 250 (2008)
42. Varshney, D., Kaurav, N., Kinge, R., Shah, S., Singh, R.K.: Study of elastic properties and their pressure dependence of lanthanum monochalcogenides. High Press. Res. **25**, 145 (2005)
43. Tessman, J.R., Kahn, A.H., Shockley, W.: Electronic polarizabilities of ion in crystals. Phys. Rev. **92**, 890 (1953)
44. Shannon, R.D.: Dielectric polarizabilities of ions in oxides and fluorides. J. Appl. Phys. **73**, 348 (1993)
45. Weast, R.C. (ed.): Handbook of Chemistry and Physics, 63rd edn. CRC, BocaRaton (1982)
46. Aleksandrov, I.V., Goncharov, A.F., Stishov, S.M., Yakovenko.: Equation of state and Raman scattering in cubic BN and SiC at high pressure. EV, Pis'ma h.Eksp. Teor. Fiz. **50,** 116 (1989) (JETP Lett. **50**, 127 (1989))
47. Feldman, D.W., Parker, H., Choyke, W., Patric, L.: Phonon dispersion curves by Raman scattering in SiC polytypes 3C, 4H, 6H, 15R, and 21 R. Phys. Rev. **173**, 787 (1968)
48. Jun, L., Vohra, Y.K.: Raman modes of 6H polytype of silicon carbide to ultrahigh pressures: a comparison with silicon and diamond. Phys. Rev. Lett **72**, 4105 (1994)
49. Wagman, D.D., Evans, W.H., Parker, V.B., Halow, E., Baily, S.M., Shumm, R.H.: Selected values of chemical thermodynamics properties, tables for the first thirty four elements in the standard order of arrangement, Natl. Bur. Std. (U.S.) Tech. Note No. 270-3 (U.S. GPO, Washington, DC, 1968)
50. Lambrecht, W.R.L., Segall, B., Methfessel, M., van Schilfgaarde, M.: Calculated elastic constants and deformation potentials of cubic SiC. Phys. Rev. B **44**, 3685 (1991)
51. Murnaghan, F.D.: The compressibility of media under extreme, pressures. Proc. Natl. Acad. Sci. U. S. A. **3**, 244 (1944)
52. Goel, P., Choudhury, N., Chaplot, S.L.: Superionic behavior of lithium oxide Li_2O: a lattice dynamics and molecular dynamics study. Phys. Rev. B **70**, 174307 (2004)
53. Xiao-Feng, Li, Xiang-Rong, Chen, Chuan-Min, Meng, Guang-Fu, Ji: Ab initio calculations of elastic constants and thermodynamic properties of Li_2O for high temperatures and pressures. Solid State Commun. **139**, 197–200 (2006)
54. Yun-Dong, G., Ze-Jin, Y., Qing-He, G., Zi-Jiang, L., Wei, D.: The phase transition, and elastic and thermodynamic properties of CaS derived from first-principles calculations. J. Phys. Condens. Matter **20**, 115203 (2008)
55. Kunc, K., Balkanski, M.: Nusimovici MA Lattice dynamics of several A *N*B8–*N* compounds having the zincblende structure II. Numerical calculations. Phys. Status Solidi (b) **72**, 229 (1975)
56. Lee, D.H., Joannopoulos, J.D.: Simple scheme for deriving atomic force constants. Application to SiC. Phys. Rev. Lett. **48**, 1846 (1982)
57. Denteneer, P.J.H., van Haeringen, W.: Ground-state properties of polytypes of silicon carbide. Phys. Rev. B **33**, 2831 (1986)
58. Tolpygo, K. B. Optical, elastic and piezoelectric properties of ionic and covalent with the ZnS type lattice. F. Tverd. Tela (Leningrad) 1960; **2:**2655 Sov. Phys. Solid State 1961; **2:**2367
59. Yean Jr, D.H., Riter, J.R.: Estimates of isothermal bulk moduli for group iva crystals with the zinc blende structure. J. Phys. Chem. Solids **32**, 653 (1971)
60. Carnahan, R.D.J.: Elastic properties of silicon carbide. Am. Ceram. Soc. **51**, 223 (1968)
61. Churcher, N., Kunc, K., Heine, V.: Calculated ground-state properties of silicon carbide. J. Phys. C **19**, 4413 (1986)
62. Weber, W.J., Yu, N., Wang, L.M., Hess, N.J.: Temperature and dose dependence of ion-beam-induced amorphization in α-SiC. J. Nucl. Mater. **244**, 258 (1997)
63. Tang, M., Yip, S.: Lattice instability in β-SiC and simulation of brittle fracture. J. Appl. Phys. **76**, 2719 (1994)
64. Weber, W.J., Yu, N., Wang, L.M., Hess, N.: Structure and properties of ion-beam-modified (6H) silicon carbide. J. Mater. Sci. Eng. A **253**, 62 (1998)
65. Gao, F., Weber, W.J.: Mechanical properties and elastic constants due to damage accumulation and amorphization in SiC. Phys. Rev. B **69**, 224108 (2004)
66. Vukcevich, M.R.: The elastic constants of cubic crystals with covalent and partially covalent bonds. Phys. Stat. Solidi (b) **54**, 435 (1972)
67. Barsch, G.R.: Relation between third-order elastic constants of single crystals and polycrystals. J. Appl. Phys. **39**, 3780 (1968)
68. Hill, R.: The elastic behavior of a crystalline aggregate. Proc. Phys. Soc. (Lond.) **65A**, 349 (1952)
69. Voigt, W.: Lehrbuch der Kristallphysik. Teubner, Leipzig (1928)
70. Reuss, A.: Calculation of the flow limits of mixed crystals on the basis of the plasticity of monocrystals. Angew Z. Math. Mech. **9**, 49 (1929)
71. Takahiro, M., Tetsuro, N., Yasuaki, N., Katsuhiro, K., Tastuya, Y., Hidetoshi, N., Satoshi, N.: Measurement of high-temperature elastic properties of ceramics using a laser ultrasonic method. J. Am. Ceram. Soc. **84**, 1521–1525 (2001)
72. Pugh, S.F.: Relations between the elastic moduli and the plastic properties of polycrystalline pure metals. Philos. Mag. **45**, 823 (1954)
73. Frantsevich, N., Voronov, F.F., Bokuta, S.A.: In: Frantsevich, I. N. (ed.) Elastic Constants and Elastic Moduli of Metals and Insulators Handbook, vol. 60. Naukova Dumka, Kiev (1983)
74. Schreiber, E., Anderson, O.L., Soga, N.: Elastic constants and Their Measurements. McGraw-Hill, New York (1973)
75. Harrison, W.A.: Electronic Structure and Properties of Solids. Dover, New York (1989)
76. Bouhemadou, A., Khenata, R., Kharoubi, M., Seddik, T., Reshak, A.H., Al-Douri, Y.: FP-APW+ lo calculations of the elastic properties in zinc-blende III-P compounds under pressure effects. Comput. Mater. Sci **45**, 474 (2009)
77. Maachou, A., Aboura, H., Amrani, B., Khenata, R., Bin Omran, S., Varshney, D.: Structural stabilities, elastic and thermodynamic properties of scandium chalcogenides via first-principles calculations. Comput. Mater. Sci. **50**, 3123 (2011)
78. Kleinman, L.: Deformation potentials in silicon. I. Uniaxial strain. Phys. Rev. **128**, 2614 (1962)
79. Kim, K., Lambrecht, W.R.L., Segal, B.: Electronic structure of GaN with strain and phonon distortions. Phys. Rev. B **50**, 1502 (1994)
80. Blackman, M. The specific heat of solids. Proc. R. Soc. Lond. A 1942; **181:**58; *ibid* 1937; **159:** 416; 1935; **149:**126; 1935; **148:**384; 1935; **148:**365

81. Blackman, M.: The specific heat of solids. Proc. R. Soc. Lond. A **159**, 416 (1937)
82. Blackman, M.: The specific heat of solids. Proc. R. Soc. Lond. A **149**, 126 (1935)
83. Blackman, M.: The specific heat of solids. Proc. R. Soc. Lond. A **148**, 384 (1935)
84. Blackman, M.: The specific heat of solids. Proc. R. Soc. Lond. A **148**, 365 (1935)
85. Gopal, E.S.R.: Specific heats at low temperatures. Plenum Press, New York (1966)
86. Tari, A.: The Specific Heat of Matter at Low Temperatures. Imperial College Press, London (2003)

23

Magnetized plasma sheath with two positive ions where collision frequencies have a power law dependency on ions velocities

S. Farhad Masoudi[1] **· Mansor Khoramabadi**[2]

Abstract We study the dynamics of collisional magnetized plasma sheath with two species of positive ions by using the plasma fluid model. The basic equations of the fluid model are solved numerically where the sheath is in the external magnetic field and the elastic collision between ions and neutrals has been taken into account. In our model, we assume that the collisional momentum transferring cross section has a power law dependency on ion flow velocity. Our analysis demonstrates that the sheath dynamics are sensitive to the power law dependency, especially for the ion with greater density.

Keywords Magnetized plasma sheath · Multi-component plasma · Ion–neutral collision

Introduction

The understanding of sheath formation of a plasma in contact with an absorbing wall is one of the oldest problems in plasma physics. However, the study of the plasma sheath is still one of the interests in plasma [1–9], since it plays an important role in many plasma applications such as processing and fabrication of semiconductor devices [10], modification of the chemical and physical properties of the materials [11], film deposition [12, 13], etching [14, 15], ion extraction [16], sputtering [17, 18], and modal and discrete focusing [19, 20].

In the sheath region, the ions are accelerated by the electric field and implanted on the absorbing wall. Therefore, understanding the dynamic properties of the plasma sheath is imperative for proper control over the implantation processing. Hence, many theoretical and experimental researches have been developed to study the plasma sheath structure [21–26].

External magnetic field and collisions between ions and neutrals are two important components that can affect the plasma sheath dynamics. Many works have investigated the collisional and magnetized plasma sheath containing ion and electron. Besides the works studying two-component plasma sheath composed of ion and electron, the investigation of multi-component plasma sheath has recently drawn considerable attention [22–39]. For example, experimental investigation of plasma sheath containing two positive ions can be found in Refs. [34–37]. References [34] and [35] have investigated the plasma sheath containing two positive ions, Ar^+ and He^+, experimentally, using dc hot-filament multidipole plasma system. Experimental study of two-ion species argon–xenon plasma can be found in Refs. [36, 37] and [36, 37].

Besides experimental investigations, there exist related theoretical studies on multi-component plasma sheath. For instance, the plasma including hot electrons and multiply charged ions has been investigated in Ref. [38]. In Ref. [39], the effect of positive ion temperature on magnetized electronegative sheath in low-pressure argon–oxygen plasma mixture has been studied. Multi-component non-magnetized and magnetized plasmas containing electron and two positive ions have been analyzed in Refs. [26] and [32], respectively.

Investigation of the effect of collisions between ions and neutrals on plasma sheath dynamics is more important in magnetized plasma sheath, since it can diminish the effects

✉ S. Farhad Masoudi
masoudi@kntu.ac.ir

1 Department of Physics, K.N. Toosi University of Technology, P.O. Box 15875-4416, Tehran, Iran

2 Department of Physics, Boroujerd Branch, Islamic Azad University, Boroujerd, Iran

of magnetic force on sheath dynamics. For example, Refs. [40] and [33] have reported this effect in two-component and multi-component plasma sheaths, respectively.

In Ref. [33], the effects of ion–neutral collision on the characteristics of a magnetized plasma sheath composed of two species of positive ions have been investigated. Three assumptions were taken into consideration in Ref. [33] as follows: (1) the ion–neutral collision cross section is constant, (2) the ion–neutral collision frequency depends on ion velocity in the cathode direction, and (3) it has been assumed that the ions enter into the sheath vertically. However, the momentum transfer cross section of ion–neutral collision is generally a function of velocity (see chapter 3 of Ref. [41]). Although two specific cases have been investigated in literature (constant mean free path and constant ion mobility), however, the dependency of ion–neutral collision cross section on ion velocity depends on plasma characteristics such as pressure, ions in plasma and the range of their energies.

In general, the ion–neutral collision cross section σ and frequency ν have a power law dependency on the ion speed v, as follows [42, 43]:

$$\sigma = \sigma_s\left(\frac{v}{c_s}\right)^p, \tag{1}$$

$$\nu = n_n\sigma_s c_s\left(\frac{v}{c_s}\right)^{p+1} = n_n v\sigma_s\left(\frac{v}{c_s}\right)^p = n_n v\sigma, \tag{2}$$

where n_n is the neutral gas density, c_s is the ion acoustic velocity, σ_s is the ion collision cross section measured at ion acoustic velocity, and 'p' is a dimensionless parameter that ranges from 0 to -1. For example, the plasma sheath at low and medium pressures is examined by constant cross section ($p = 0$); the constant collision frequency for momentum transfer ($p = -1$) is considered for higher pressures, and for argon plasma in the range of 3–300 eV, p is -0.25. Many works have investigated the effects of power law dependency on two-component or electronegative plasma sheath such as Refs. [43–45]. From here, we name the power law dependency of collision cross section to ion velocity as collisional regime between ions and neutrals.

In this article, we modified the results of Ref. [33] by considering the power law dependency between collision cross section and ion velocity. Also, in our model, the collisions frequency depends on ion velocity and not on its vertical component. It leads to increasing the effects of collision. Although these two cases approximately lead to the same ion dynamics in two-component magnetized plasma sheath [46], our study shows that they lead to different results in multi-component magnetized plasma sheath. In addition, we consider oblique entrance of the ions into the sheath as a result of magnetic presheath.

The layout of the article is as follows. In "Fluid model and basic equations", we present the model and its equations for studying the magnetized collisional plasma sheath composed of electron and two positive ions with different masses. "Fluid model and basic equations" ends with the presentation of the normalized form of the basic equations. In "Numerical results and discussion", we report our numerical results and analyze their implications. "Conclusions" concludes the article by a summary and outlook of our finding.

Fluid model and basic equations

In the presence of a magnetic field, the ions enter into the sheath with an oblique entrance angle as a result of a magnetic presheath. To find the entrance angle of ions into the sheath, it is necessary to solve the fluid equations in the presheath exactly. However, to avoid complex mathematical techniques, the fluid equations in the plasma sheath can be investigated by considering the Bohm criterion at the sheath boundary. Following some theoretical works used in this approach (e.g., [47–52]), we consider a magnetized plasma sheath having one-dimensional coordinate space and three-dimensional velocity space. Also, we consider a Cartesian coordinate where the z-direction is chosen as the depth direction (normal to the wall). The plasma sheath boundary is located at $z = 0$, and the external magnetic field is embedded in the x–z plane and makes θ angle with respect to the z-direction. As a result of the magnetic field, the plasma-wall transition consists of a magnetic presheath followed by a Debye sheath. In this article, we study the Debye sheath in which the neutrality is broken.

We assume that the plasma consists of thermal electrons and two positive cold ions with different masses. We suppose that the energy of the electrons is sufficiently low. It enables us to neglect the ionization in the plasma sheath. We explain the basic equations of the model using the following variables:

Variables	Local value	Boundary value	Normalized form
Electron density	n_e	n_{e0}	$N_e = n_e/n_{e0}$
Electrostatic potential	ϕ	ϕ_0	$\eta = -\phi/(k_B T_e/e)$
Ion velocity	$\mathbf{v}_1 = (v_{1x}, v_{1y}, v_{1z})$	$\mathbf{v}_{10} = (v_{1x0}, v_{1y0}, v_{1z0})$	$\mathbf{u}_1 = \mathbf{v}_1/c_{s1}$
	$\mathbf{v}_2 = (v_{2x}, v_{2y}, v_{2z})$	$\mathbf{v}_{20} = (v_{2x0}, v_{2y0}, v_{2z0})$	$\mathbf{u}_2 = \mathbf{v}_2/c_{s1}$
Ion speed	v_1	v_{10}	$u_1 = v_1/c_{s1}$
	v_2	v_{20}	$u_2 = v_2/c_{s1}$
Ion density	n_1	n_{10}	$N_1 = n_1/n_{e0}$
	n_2	n_{20}	$N_2 = n_2/n_{e0}$
Depth direction	z	0	$\xi = z/\lambda_D$

where T_e is the electron temperature, $c_{sj} = (k_B T_e/m_j)^{0.5}$ is the ion acoustic speed for the j-th ion (j = 1 and 2), m_j is the mass of the j-th ion, $\lambda_D = (\varepsilon_0 k_B T_e/n_{e0}e^2)^{0.5}$ is the electron Debye length and k_B is the Boltzmann constant. Using these definitions, the basic equations of the plasma sheath according to the fluid model are as follows:

Continuity equation of j-th ion:

$$\vec{\nabla}.(n_j\vec{v}_j) = 0. \tag{3}$$

Motion equation of the j-th ion:

$$m_j(\vec{v}_j.\vec{\nabla})\vec{v}_j = -e\vec{\nabla}\varphi + e\vec{v}_j \times \vec{B} - m_j v_j \vec{v}_j, \tag{4}$$

where v_j is the ion–neutral collision frequency of the j-th ion.

Poisson equation:

$$\nabla^2\varphi = \frac{e}{\varepsilon_0}(n_e - n_1 - n_2). \tag{5}$$

In addition, the quasi-neutral condition at the sheath edge leads to

$$n_{e0} - n_{10} - n_{20} = 0. \tag{6}$$

These equations are completed by Boltzmann relation for electron density. Compared with ions, the electrons would be accelerated to high energy very quickly as they are light and very mobile. Hence, they leave behind a large ion charge and so forces on them are closely in balance. Neglecting the drift terms in the force balance equation and considering the limit of $m_e \to 0$ leads to Boltzmann relation for the electrons as follows [53]:

$$n_e = n_{e0}\exp(e\varphi/k_B T_e). \tag{7}$$

In the next section, we solve these basic equations numerically to study the sheath dynamics. For numerical solving of the equations, we assume that the plasma sheath variables vary only in the z-direction. Based upon this assumption, the dimensionless forms of the plasma sheath equations are as follows:

$$N_e = \exp(-\eta), \tag{8}$$

$$N_1 u_{1z} = \frac{M_1}{1+\delta}, \quad N_2 u_{2z} = \frac{\delta M_2}{1+\delta}, \tag{9}$$

$$u_{1z}\partial_\xi u_{1x} = \gamma_1 \cos\theta\, u_{1y} - \alpha_1 u_1^{p_1+1} u_{1x}, \tag{10}$$

$$u_{1z}\partial_\xi u_{1y} = \gamma_1 \sin\theta\, u_{1z} - \gamma_1 \cos\theta\, u_{1x} - \alpha_1 u_1^{p_1+1} u_{1y}, \tag{11}$$

$$u_{1z}\partial_\xi u_{1z} = \partial_\xi \eta - \gamma_1 \sin\theta\, u_{1y} - \alpha_1 u_1^{p_1+1} u_{1z}, \tag{12}$$

$$u_{2z}\partial_\xi u_{2x} = \mu\gamma_2 \cos\theta\, u_{2y} - \mu^{-p_2}\alpha_2 u_2^{p_2+1} u_{2x}, \tag{13}$$

$$u_{2z}\partial_\xi u_{2y} = \mu\gamma_2 \sin\theta\, u_{2z} - \mu\gamma_2 \cos\theta\, u_{2x} - \mu^{-p_2}\alpha_2 u_2^{p_2+1} u_{2y}, \tag{14}$$

$$u_{2z}\partial_\xi u_{2z} = \mu^2\,\partial_\xi \eta - \mu\gamma_2 \sin\theta\, u_{2y} - \mu^{-p_2}\alpha_2 u_2^{p_2+1} u_{2z}, \tag{15}$$

$$\partial_\xi^2 \eta = N_1 + N_2 - N_e, \tag{16}$$

where $M_j = u_{jz0} = v_{jz0}/c_{s1}$, $\delta = n_{20}/n_{10}$, $\mu = (m_1/m_2)^{0.5}$, $\gamma_j = (\varepsilon_0/n_{e0}m_j)^{0.5}B_0$ and $\alpha_j = n_n\lambda_D\sigma_{sj}$ for j = 1 and 2. α and $\gamma_2 = \mu\gamma_1$ are the dimensionless parameters that specify the strength of the drag and magnetic forces in the plasma sheath equations, respectively. Also, p_j is the p value for the j-th ion (see Eqs. 1 and 2).

Numerical results and discussion

This section includes the numerical solution of the normalized equations and discussion on the sheath dynamics for different dependency of collision frequency on the ions velocities. At first, we use a numerical example for investigating the effect of collisional regime on plasma sheath dynamics. The dimensionless parameters in our numerical example are as follows: $\mu = 0.1$, $\alpha_1 = 0.15$, $\alpha_2 = 0.45$, $\delta = 0.1$, $\gamma_1 = 5$ and $\theta = 40°$. Also, the boundary values have been considered as $\phi_O = 0$, $u_{10} = (0.3, 0.3, 1)$ and $\boldsymbol{u}_{20} = (0.1, 0.1, 1)$ and the electrostatic field at the sheath edge is taken to be an infinitesimal value to prevent the divergence of the numerical results. The boundary values of the ion velocity have been adopted, so that the Bohm criterion in the multi-component plasma is satisfied [8, 54]. Considering the non-zero components of the velocities parallel to the wall enables us to investigate the oblique entrance of the ions into the sheath.

Figure 1 shows the components of the velocity of lighter ion for constant mean free path ($p_1 = 0$) and constant collision frequency ($p_1 = -1$).

The figure shows the ion gyration movement because of the external magnetic field. As it is seen, the ion gyro radius decreases by increasing the distance from the sheath edge. Also, the decrease of ion gyro radius for $p_1 = -1$ is smaller than that of $p_1 = 0$. In other words, the drag force diminishes the helical movement of the lighter ion, and the diminution of the ion helical movement for constant mean free path is much stronger than that of constant collision frequency. It means that the drag force conquers the magnetic force quickly in constant collision mean free path regime, but it dominates slowly in constant collision frequency regime. These effects can be investigated by studying the effect of 'p' value on the fluctuations of the ion flow velocity in depth direction.

Figure 2 shows the velocity of lighter ion in the z-direction versus the distance from the sheath edge.

As seen, the amplitude of the fluctuations of the ion velocity decreases by penetrating into the sheath and increasing the drag force. Also, the results show that the ion velocity decreases by increasing the p_1 value from -1 to 0. From the results, the sheath dynamics for $p_1 = -1$ is approximately the same as the collisionless magnetized plasma sheath. Our numerical results showed that the velocity of the lighter ion is approximately independent of the p_2 value. Therefore, in a multi-component plasma sheath where the density of the heavier ion is a small percent of the lighter ion, the dynamics of the lighter ion is independent of the collisional regime of the heavier ion. The characteristics of the lighter ion in the plasma sheath depend on its collisional regime. Results shows that the oscillations of the velocity of the lighter ion and their amplitudes depend on the collisional regime of the lighter ion: for low-pressure plasma, the plasma sheath dynamics is approximately the same as that of the collisionless plasma sheath; however, by increasing the pressure, the collisions between ions and stationary neutrals tend to damp the oscillations and gradually diminish their amplitudes. Thus, the effects of magnetic field on plasma increase if the law dependency of the collision regime is satisfied by p values smaller than zero.

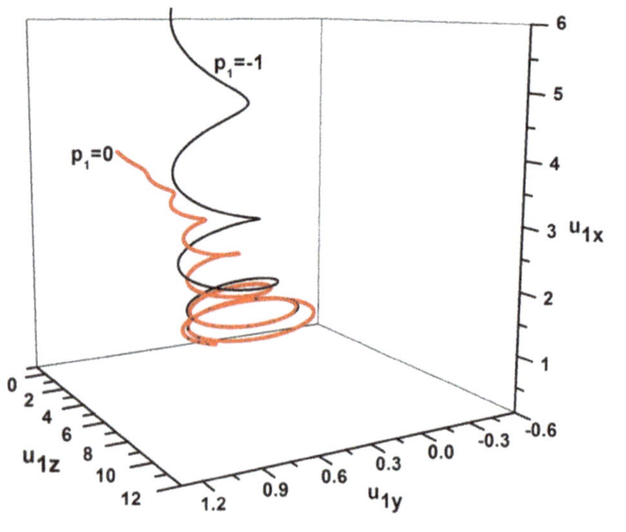

Fig. 1 The ion helical movement of the lighter ion in two collisional regimes: constant frequency ($p_1 = -1$) and constant mean free path ($p_1 = 0$)

Figure 3 shows the effects of p_1 and p_2 values on the velocity of the heavier ion in the z-direction. From the results, u_{2z} increases by increasing the p_1 value (unlike the lighter ion velocity). It means that increasing the p_2 value tends to increase the collision effect and decrease the heavier ion velocity in the depth direction; however increasing the p_1 value tends to increase the heavier ion velocity. Nonetheless, as the effect of the collisional regime of heavier ion is stronger than that of the lighter ion, the ion velocity of the heavier ion decreases on increase of pressure.

Figure 4 indicates the effects of p_1 and p_2 values on the sheath electrostatic potential. As seen, the electrostatic potential decreases by decreasing the p_1 and p_2 values. The results show that the effect of the p_1 value is stronger than that of the p_2 value. The reason for this behavior can be explained by using the effect of the p_1 and p_2 values on the net positive space charge.

Figure 5 shows the net positive space charge versus the depth for different dependencies of ion collision frequency on the ion velocity. The fluctuations of the net positive space charge are due to the fluctuations of the lighter ion velocity in the z-direction, which leads to the fluctuations of the density distribution of the lighter ion. From the results, if the collisional regime of the plasma sheath is near to constant collision cross-sectional regime, the effects

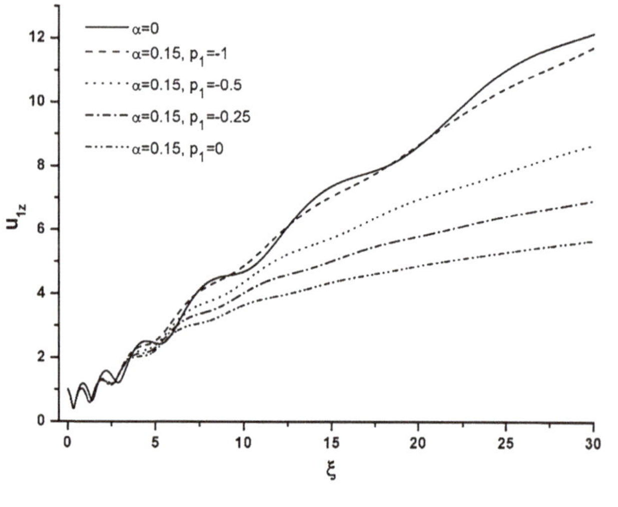

Fig. 2 The lighter ion velocity in the depth direction for different values of p_1 where $p_2 = 0$

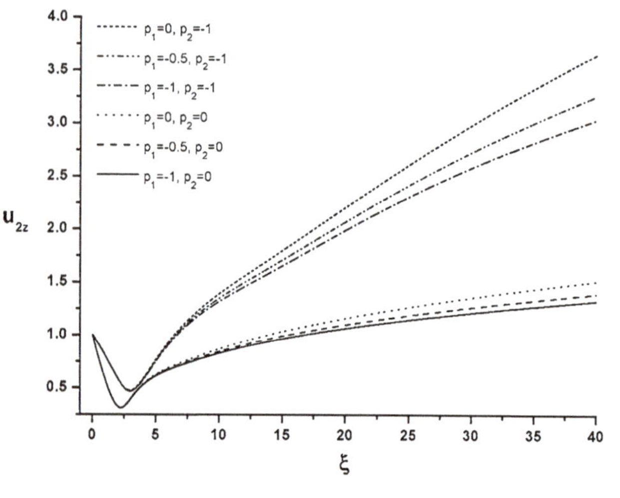

Fig. 3 The heavier ion velocity in the depth direction for different values of p_1 and p_2, where $\delta = 0.1$

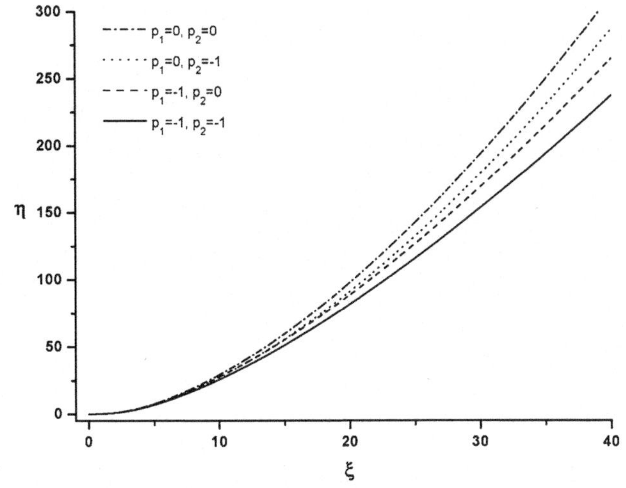

Fig. 4 The electrostatic potential versus the depth direction for different values of p_1 and p_2 where $\delta = 0.1$

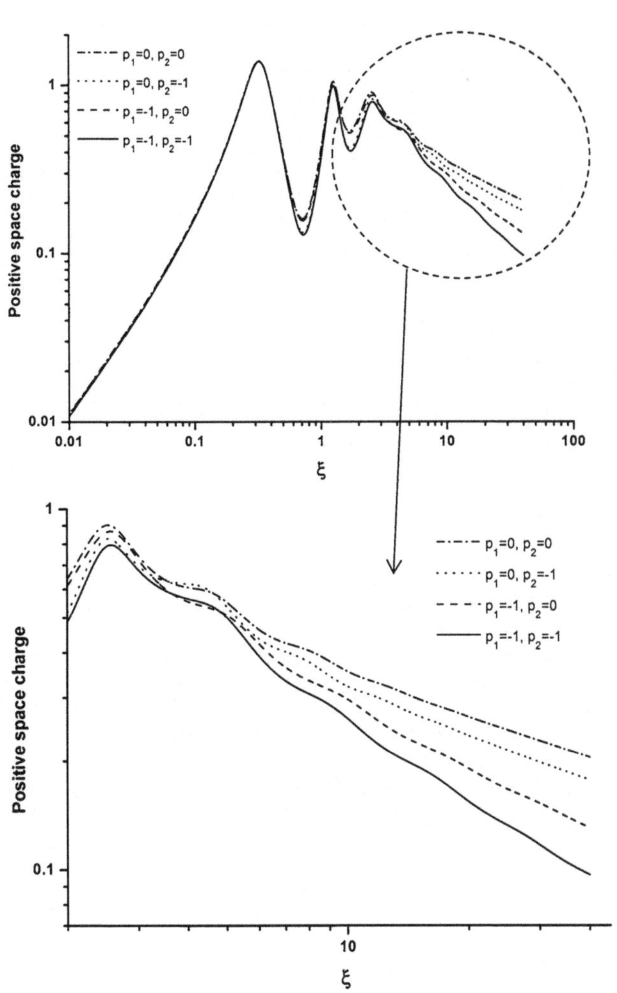

Fig. 5 The net positive space charge in the plasma sheath versus the depth direction for different dependencies of collision frequency on the ion velocity

of collisions on the sheath structure increase (compared with constant collision frequency regime), which tends to decrease the ion velocity and gathering of ions. Also, the effect of the collisional regime of the lighter ion on the net positive space charge and electrostatic potential is more than that of the heavier ion.

The results of the numerical example show that the collisional regime between ions and neutrals has significant effects on plasma sheath dynamics, such as the energy and densities of the ions and the electrostatic potential. In the next subsection, we calculate the sheath dynamics of Ar^+–He^+ plasma considering the real collisional regime and compare the results with specific cases: constant collision cross section and constant collision frequency.

Specific example: He–Ar plasma

Consider a multi-component plasma containing two ion species, Ar^+–He^+, with the same density. Here, $m_1 = 4$ and $m_2 = 40$, so $\mu = 0.1$. Also, assume that the plasma is in B = 4 mT external magnetic field, making an angle of $\theta = 40°$ with respect to the depth direction, neutral gas density $n_g = 3 \times 10^{14}$ cm^{-3} in 10 mTorr pressure at room temperature KT = 0.02 eV, electron density $n_{e0} = 10^9$ cm^{-3} and $\sigma_{2s} = 3\sigma_{1s} = 5 \times 10^{-15}$ cm^2 [41]. Therefore, the dimensionless parameters are $\gamma_I = 5$ and $\alpha_1 = \alpha_2/3 = 0.15$. For argon in the range of 3–300 eV, the power law is most closely satisfied by -0.25, so $p_2 = -0.25$. However, for helium ions streaming through a low-pressure plasma sheath, the cross section is more closely approximated by the constant mean free path, $p_1 = 0$. Figure 6 shows the normalized kinetic energy of argon ions and compares the result with the results of Ref. [33] in which $p_1 = p_2 = 0$. The result shows that the argon ion energy is not the same as the constant collision cross-sectional regime. Consider that the most closely satisfied p_2 value for argon ion tends to increase of the energies of argon ions and small decrease of that of helium ions. As the results show, the difference between the real collisional regime and constant collision cross-sectional regime increases by increase of the ions' distance from the sheath edge. Therefore, the most closely satisfied dependency of collision cross section to the ion velocity should be considered in the investigation of the multi-component plasma sheath.

Conclusions

We studied the sheath dynamics containing two positive cold ions in the presence of an external magnetic field. We considered a collisional plasma sheath where the collision frequencies of both ions have a power law dependency on the ions' flow speeds. The numerical results showed that

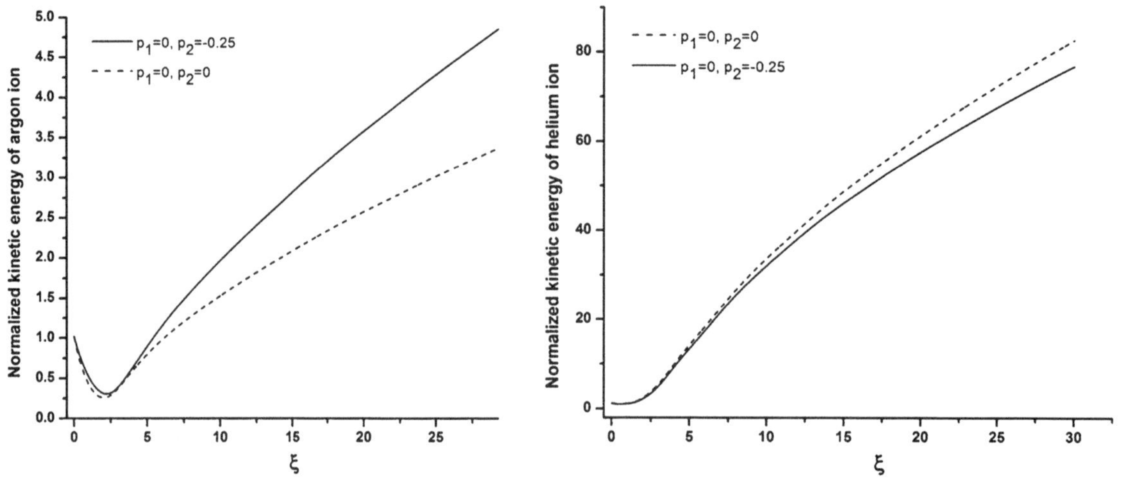

Fig. 6 Normalized kinetic energy of argon and helium ions in the Ar^+–He^+ plasma sheath where the densities of the ions are the same

near the sheath edge, the sheath dynamics are independent of the collision regime. However, by increasing the distance from the sheath edge and increasing the ions' velocity in the cathode direction, the effects of the collision regime on the sheath dynamics are as follows:

- The velocity of the lighter ion in the depth direction decreases by increasing the p_1 value and is independent of the p_2 value. In other words, the maximum effect of drag force on the lighter ion velocity corresponds to the case where the lighter ion is in the constant collisional mean free path regime.
- Maximum effect of drag force on the heavier ion velocity corresponds to $p_1 = -1$ and $p_2 = 0$.
- The effect of collision regime of lighter (heavier) ion on some sheath dynamics such as electrostatic potential and net positive space charge is more than that of heavier (lighter) ion for $\delta < 1$ ($\delta > 1$).
- When the occupation fraction of the heavier and lighter ions in the sheath is approximately the same ($\delta \sim 1$), two ions can be considered as two independent ions. Hence, the maximum effect of drag force on the sheath dynamics corresponds to the case that both ions are in the constant mean free path regime ($p_1 = p_2 = 0$). Also, the effect of the collisional regime of the heavier ion on the sheath dynamics increases by increasing the δ value.
- The sheath dynamics is approximately the same as collisionless magnetized plasma sheath when the collision frequencies of the heavier and lighter ions are constant.

References

1. Riemann, K.U.: J. Phys. D Appl. Phys. **24**, 493–518 (1991)
2. Franklin, R.N.: J. Phys. D Appl. Phys. **36**, R309–R320 (2003)
3. Riemann, K.U., Seebacher, J., Tskhakaya Sr, D.D., Kuhn, S.: Plasma Phys. Control. Fusion **47**, 1949 (2005)
4. Franklin, R.N.: J. Phys. D **37**, 1342 (2004)
5. Riemann, K.U.: Plasma Sour. Sci. Technol. **18**, 014006 (2009)
6. Vranjes, J., Pandey, B.P., Tanaka, M.Y., Poedts, S.: Phys. Plasmas **15**, 123505 (2008)
7. Stokes, J.D.E., Samarian, A.A., Vladimirov, S.V.: Phys. Rev. E **78**, 036402 (2008)
8. Masoudi, S.F., Jafari, Gh, Shorakaee, H.: Vacuum **83**, 1031–1035 (2009)
9. Khoramabadi, M., Ghomi, H., Kant Shukla, P.: J. Appl. Phys. **109**, 073307 (2011)
10. Cheung, N.W.: Mater. Chem. Phys. **46**, 132 (1996)
11. Conrad, J.R.: J. Appl. Phys. **62**, 777 (1987)
12. Anders, A.: Handbook of Plasma Immersion Ion Implantation. Wiley, New York (2000)
13. Flender, U., Wiesemann-K, K.: Plasma Chem. Plasma Process. **15**, 123 (1995)
14. Donnelly, V.M., Kornblit, A.: J. Vaccum Sci. Tech. A **31**, 050825 (2013)
15. Lan, Ch., Peng, Y.F., Yang, Z., Long, J.D.: Plasma Sci. Technol **15**, 945 (2013)
16. Chao-Hui, L., Yu-Fei, P., Zhen, Y., Ji-Dong, L.: Chins. Phys. C **37**, 057002 (2013)
17. Han, Jeon G.: J. Phys. D Appl. Phys. **42**, 043001 (2009)
18. Yang, Y.Z., Kim, K.H., Ong, J.L.: Biomaterials **26**, 327–337 (2005)
19. Stamate, E., Sugai, H.: Phys. Rev. Lett. **94**, 125004 (2005)
20. Stamate, E.: Plasma Phys. Control. Fusion **54**, 124048 (2012)
21. Kim, G.H., Hershkowitz, N., Diebold, D.A., Cho, M.-H.: Phys. Plasmas **2**, 3222 (1995)
22. Chengsen, L., Dezhen, W.: Surf. Coat. Technol. **171**, 119 (2003)
23. Sheridan, T.E., Alport, M.J.: Appl. Phys. Lett. **64**, 1783 (1994)
24. Pandey, B.P., Samarian, A., Vladimirov, S.V.: Plasma Phys. Control. Fusion **50**, 10 (2008)
25. Kostov, K.G., Barrosob, J.J., Ueda, M.: Surf. Coat. Technol. **201**, 8398 (2007)
26. Franklin, R.N.: J. Phys. D Appl. Phys. **36**, 1806 (2003)
27. Valentini, H.B., Herrmann, F.: J. Phys. D **29**, 1175 (1996)
28. Benilov, M.S.: J. Phys. D **29**, 364 (1996)
29. Franklin, R.N.: Plasma Sources Sci. Technol. **10**, 162 (2001)
30. Severn, G., Wang, X., Ko, E., Hershkowitz, N.: Phys. Rev. Lett. **90**, 145001 (2003)

31. Franklin, R.N.: J. Phys. D **36**, 34 (2003)
32. Hatami, M.M., Niknam, A.R., Shokri, B., Ghomi, H.: Phys. Plasmas **15**, 053508 (2008)
33. Hatami, M.M., Shokri, B., Niknam, A.R.: J. Phys. D Appl. Phys. **42**, 025204 (2009)
34. Hala, M.A., Hershkowitz, N.: Rev. Sci. Instrum. **72**, 2279 (2001)
35. Severn, G.D., Wang, X., Ko, E., Hershkowitz, N.: Phys. Rev. Lett. **90**, 145001 (2003)
36. Lee, D., Severn, G., Oksuz, L., Hershkowitz, N.: J. Phys. D Appl. Phys. **39**, 5230–5235 (2006)
37. Lee, Dongsoo, Hershkowitz, Noah, Severn, Greg D.: Appl. Phys. Lett. **91**, 041505 (2007)
38. Khoramabadi, M., Ghomi, H., Shukla, P.K.: J. Plasma Phys. **79**, 267–271 (2013)
39. Shaw, A.K., Kar, S., Goswami, K.S.: Phys. Plasmas **19**, 102108 (2012)
40. Masoudi, S.F.: J. Phys. D Appl. Phys. **40**, 6641–6645 (2007)
41. Lieberman, M.A., Lichtenberg, A.J.: Principle of Plasma Discharges and Materials Processing, 2nd edn. Wiley, Hoboken (2005)
42. Sheridan, T.E., Goree, J.: Phys. Fluids B **3**, 2796 (1991)
43. Liu, J., Wang, Z., Wang, X.: Phys. Plasmas **10**, 3032 (2003)
44. Yasserian, K., Aslaninejad, M.: Eur. Phys. J. D **67**, 161 (2013)
45. Moulick, R., Mahanta, M.K., Goswami, K.S.: Phys. Plasmas **20**, 094501 (2013)
46. Masoudi, S.F.: Vacuum **81**, 871–874 (2007)
47. Hatami, M.M.: Phys. Plasmas **20**(1), 013509 (2013)
48. Yankun, J., Xiu, Z., Huiping, L., Minghui, Q.: Plasma Sci. Technol. **13**(5), 519 (2011)
49. Zou, X., Qiu, M., Liu, H., Zhang, L., Liu, J., Gong, Y.: Vacuum **83**(1), 205–208 (2009)
50. Gan, Bao-Xia, Chen, Yin-Hua: Chin. Phys. Lett. **24**(7), 2003–2005 (2007)
51. Shaw, A.K., Kar, S., Goswami, K.S.: Phys. Plasmas **19**(10), 102108 (2012)
52. Yasserian, K., Aslaninejad, M.: Phys. Plasmas **19**(7), 073507 (2012)
53. Chen, F.: Introduction to Plasma Physics and Controlled Fusion. Plenum, New York (1984)
54. Riemann, K.-U.: IEEE Trans. Plasma Sci. **23**, 709 (1995)

24

Surface modification of Raw and Frit glazes by non-thermal helium plasma jet

M. Ghasemi[1] · F. Sohbatzadeh[1,2] · S. Mirzanejhad[1,2]

Abstract In this study, non-thermal atmospheric pressure plasma jet (APPJ) was utilized to improve the adhesion of Raw and Frit glazes. These glazes are widely used in industry to make chinaware, decorative dishes and tiles applied at wall and floor. As they should be painted before use, increasing their adhesive properties leads to a better paint durability. Electrical and optical characteristics of the plasma jet are investigated to optimize for efficient treatment. Contact angle measurement and surface energy calculation demonstrate a drastic increase after the plasma treatment indicating wettability and paintability enhancement. Moreover, atomic force microscopy and X-ray photoelectron spectroscopy analyses were performed on the specimens to explore the influence of helium plasma jet on the physical and chemical properties of the glazes, microscopically. AFM analysis reveals surface etching resulted from the bombardment of the solid surfaces by the APPJ using helium fed gas. The process aims to enhance adhesive properties of glaze surfaces.

Keywords Adhesion · Plasma jet · Surface modification

✉ F. Sohbatzadeh
f.sohbat@umz.ac.ir

1 Department of Atomic and Molecular Physics, Faculty of Basic Sciences, University of Mazandaran, 47416-95447 Babolsar, Iran

2 Nano and Biotechnology Research Group, Faculty of Basic Sciences, University of Mazandaran, 47416-95447 Babolsar, Iran

Introduction

Some materials such as polymers, textiles and metals, which are widely used in industrial and medical applications, have low surface free energy. This limitation restricts them as it leads to poor wettability, poor adhesion and poor printability, so they require surface modification to be more efficient. Among conventional methods of surface modification, atmospheric pressure plasma processing has attracted great interests due to its numerous advantages over the traditional processes [1, 2]. Atmospheric pressure plasma surface activation is an environmentally friendly method as it does not require the use of water and chemicals which leads to pollutants and corresponding cost for effluent treatment [3]. Furthermore, non-thermal plasma is a suitable choice for heat-sensitive materials and there is no need for vacuum equipment which increases the cost. Production of a large variety of chemically active functional groups makes it fast, effective and versatile technique for surface activation.

Noh et al. [2] employed O_2 APPJ as a second step of treatment of Polytetrafluoroethylene (PTFE) after chemical treatment. PTFE is extensively utilized in synthetic vascular surgery and medical devices because of its chemical and mechanical stability, but it is limited due to lack of haemocompatibility. APPJ is a suitable choice as it could improve cellular interaction property and tailor the biological response to the implant. Sorrentino and Carrino demonstrated that an oxygen cold plasma treatment could improve wettability and adhesion of Al2024 surface which is an important alloy at aeronautical application. It should be painted to prevent corrosion which is caused by aeronautical critical environmental condition. They also investigated the influence of different parameters of plasma on the durability of paint on its surface [4]. The plasma processing results in physical and chemical modification in scale of several nanometres (first few molecular layers) while maintaining the properties of the bulk unchanged [3]. There are many investigations, in which plasma was utilized to improve surface properties of materials such as polymers, textiles and etc. Their physical and chemical properties after plasma treatment were examined in detail as well. But there are a few studies investigated the effect of plasma on the properties of non-carbonic base surfaces and needs more investigations [1–10]. The aim of this paper is to utilize atmospheric pressure helium plasma jet on Raw and Frit glaze surfaces to improve their paintability and explore the effect of the plasma on their surface properties which seems not fully understood so far. Helium, in comparison with other gases, could be considered the most convenient gas working at atmospheric pressure, as it allows stabilizing homogeneous glow discharges at lower sustaining voltages and inter-electrode gaps. It is also very suitable for surface treatment because of low degradation effect and high properties of cross linking and functionalizing on to the surface [8, 10]. We used helium plasma as it is capable of maintaining simultaneous stability and reactivity, while in other inert gases chemical reactivity is obtained by adding a small amount of reactive gases at the expense of stability and increasing the temperature which may cause non-uniform processing.

Materials and methods

The atmospheric pressure plasma jet used in this study was single electrode configuration based on DBD principle of operation. The image of the cold helium plasma jet is presented in Fig. 1. The plasma tube consists of 10 mm capillary quartz of inner and outer diameter of 1 and 3 mm, respectively, around which a ring cupper electrode of 10 mm length and 50 μm thickness was wrapped 10 mm from its end. 0.5–1 standard litre per minute (SLM) of helium gas with 99.999 % purity was flourished inside the capillary tube. A home-built AC sinusoidal power supply at the frequency of 10 and 19 kHz was applied to ignite the plasma jet.

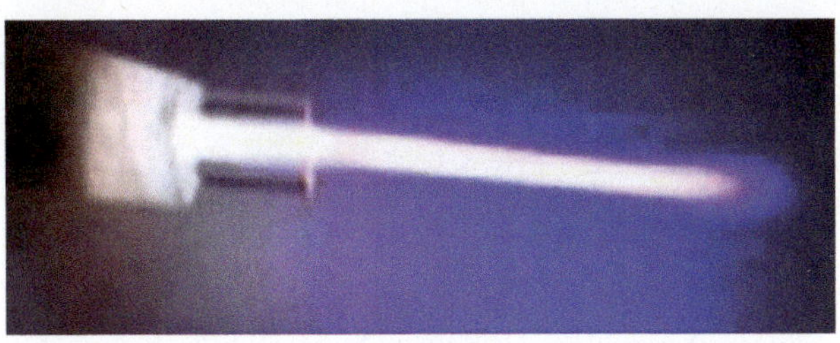

Fig. 1 Image of the cold helium plasma jet at atmospheric pressure utilized for surface activation of glazes

To explore the plasma capabilities, V-I characteristics and optical emission spectra (OES) were investigated. Applied voltage and jet current were measured using a high voltage probe (Tektronix P6015 A) and a current probe (Tektronix TCPA-300). Optical emission spectroscopy was used to detect plasma effective species produced by helium APPJ responsible for plasma processing. We used S100 spectrometer (Solar laser Systems) which covers 200–1100 nm with the resolution of 1 nm. Microscopic properties of the glaze surfaces were analysed using atomic force microscopy (Quto Probe, Cp, USA, Contact), and X-ray photoelectron spectroscopy (Aluminium Monochromated XPS hϑ – 1486.6 eV) was used to reveal the surface chemical change caused by the plasma jet.

Voltage and current curves are shown in Fig. 2. There is at least one current spike every half cycle of the voltage which occurs at fixed positions of each cycle. It is known as bullet mode which is suitable for surface treatment as it provides uniform treatment [11, 12].

Two types of glazes selected to improve their paintability are Raw and Frit glazes. These glazes are widely used in industry to make chinaware, decorative dishes and tiles utilized at wall and floor. As both of them need to be painted before use, the activation of their surfaces improves their paintability (paint absorption) and enhances their efficiency (paint durability is an important factor of a good earthenware or tile). Table 1 represents the compositions of Raw and Frit glazes and their percentages. Raw and Frit glazes were cut into pieces of 8 × 8 mm^2 and 9 × 11 mm^2, respectively, to be able to use for XPS and AFM analyses. Before the plasma treatment, the specimens were cleaned to remove their surface contaminations. To prepare the samples, they were cleaned ultrasonically in isopropanol for 1 min, then rinsed with fresh isopropanol again and dried out under nitrogen flow. After cleaning, the specimens were treated by the cold helium APPJ. AC sinusoidal high voltage power supply at the frequency of

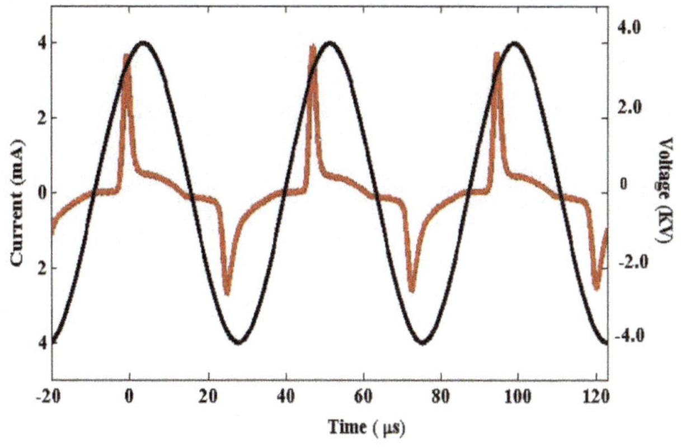

Fig. 2 Voltage and current curves measured at downstream electrode at the frequency of 19 kHz and applied voltage of 4 kV

Table 1 Compositions of Frit and Raw glazes

Oxide (%)	CaO	Na_2O	ZnO	K_2O	Al_2O_3	Fe_2O_3	SiO_2	B_2O_3	TiO_2	ZrO_2	MgO	BaO
Frit	9.03	0.19	10.69	2.23	6.74	0.40	49.1	5.80	0.03	13.95	2.84	–
Raw	8.45	0.92	6.13	–	12.54	0.36	61.01	–	0.14	–	3.01	5.77

10 and 19 kHz was used to ignite the plasma. The applied voltages of 4 and 14 kV were used for Raw and Frit glazes, respectively. 0.5 SLM helium gas was passed through the plasma tube, and glazes were exposed to the plasma jet for 1 min at the distance of 3 mm from the capillary end. It is worth noting that 1-min treatment is not a long time with regard to the size of the glazes and the jet cross section.

Plasma optical emission

Optical emission spectroscopy is a powerful technique to determine optical characteristics of the plasma. Identification of the plasma reactive species responsible for relevant application enables one to enhance plasma efficiency by making it optimum. The optical fibre probe was placed 3 mm below the exit of the quartz tube where the helium flow is mixed with the ambient air to produce reactive species. Figure 3 demonstrates the optical emission spectral lines of helium plasma jet with the key lines labelled. Colliding of energetic electrons produced at the discharge tube with the background helium atoms causes the production of excited and metastable helium atoms which play an important role initiating many chemical reactions inside the plasma tube [11]:

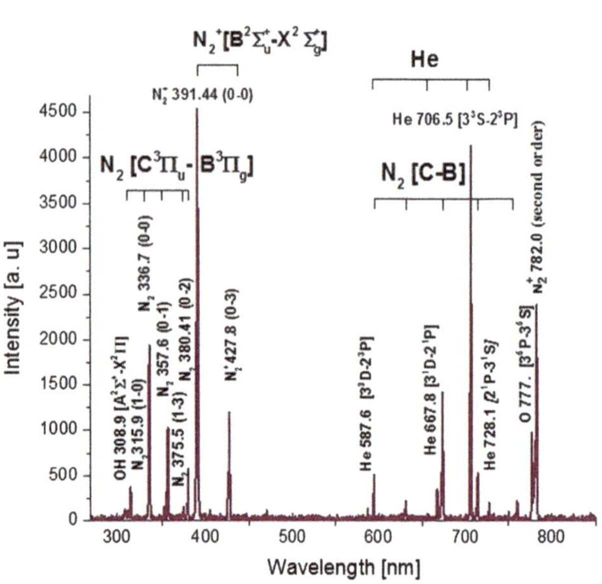

Fig. 3 Optical emission lines of helium plasma jet at 3 mm below the exit of the quartz tube. The *spectral lines* and their transition levels are labelled in the spectrum

$$\mathrm{He} + \mathrm{e} \rightarrow \mathrm{He}^*(2\mathrm{p}) + \mathrm{e} \tag{1}$$

$$\mathrm{He} + \mathrm{e} \rightarrow \mathrm{He}^*(2\mathrm{S}) + \mathrm{e} \tag{2}$$

Helium line of 706 nm [$3^3S \rightarrow 2^3P$] with the threshold energy of 22.7 eV is an indicator of the energetic electron [13] which is the second intense spectral line at the plasma jet emission spectrum shown in Fig. 3. The other helium atomic lines observed in Fig. 3 are 587.6 ($3^3D \rightarrow 2^3P$), 667.8 ($3^1D \rightarrow 2^1P$) and 728.1 ($3^1S \rightarrow 2^1P$). A helium atom naturally has two metastable levels, which are 2^1S_0 (singlet) and 2^3S_1 (triplet) with the energies of 19.8 and 20.6 eV, respectively [8, 13]. Helium metastable lines are reported at He^* 501.6 nm ($3^1P \rightarrow 2^1S$) and He_2^* 388.86 ($3^3P_2 \rightarrow 2^3S_1$) [13]. Highly energetic helium metastable atoms are able to ionize the nitrogen through the penning ionization due to the interaction with the ambient air [11].

$$\mathrm{He}^* + N_2 \rightarrow rN_2^+\left(B^2\sum\nolimits_u^+\right) + (1-r)N_2^+\left(X^2\sum\nolimits_g^+\right) + \mathrm{He} + e \tag{3}$$

$$\mathrm{He}_2^* + N_2 \rightarrow rN_2^+\left(B^2\sum\nolimits_u^+\right) + (1-r)N_2^+\left(X^2\sum\nolimits_g^+\right) + 2\mathrm{He} + e \tag{4}$$

It is well known that nitrogen molecules are very effective in quenching the helium metastables [17]. Rapid penning reaction reduces the effective lifetime of helium metastables and produces excited nitrogen species [11]. N_2 second positive system ($C^3\prod_u \rightarrow B^3\prod_g$) is observed at helium spectrum (Fig. 3) due to the existence of air at the two wavelength intervals of (315–381) and (590–770) in which the intense peaks are observed at the wavelengths of 315.9 (1–0), 336.7 (0–0), 357.6 (0–1), 375.5 (1–3) and 380.41 (0–2). Moreover, radiative dissociation, which takes place after recombination process, could lead to the decrease of He_2^* population according to the following reactions [15]

$$\mathrm{He}^* + 2\mathrm{He} \rightarrow 2\mathrm{He}^* + \mathrm{He} \tag{5}$$

$$\mathrm{He}_2^* \rightarrow 2\mathrm{He} + h\vartheta \tag{6}$$

It is possible that nitrogen second positive emission is capable of producing a great amount of seed electron for subsequent discharge events. It is reported that nitrogen has a crucial role in producing N_4^+ by penning interaction between two nitrogen molecules to maintain a stable glow discharge [14].

$$N_2\left(a^1\sum\nolimits_u^-\right) + N_2\left(A^3\sum\nolimits_u^+\right) \rightarrow N_4^+ + e \quad (7)$$

$$N_2\left(a^1\sum\nolimits_u^-\right) + N_2\left(A^1\sum\nolimits_u^-\right) \rightarrow N_4^+ + e \quad (8)$$

There are also several sources of electron such as the following reactions [15, 16]:

$$He^* + He^* \rightarrow He^+ + He + e \quad (9)$$

$$He_2^* + N_2 \rightarrow N_2^+ + 2He + e \quad (10)$$

$$He_2^* + He_2^* \rightarrow He^+ + 2He + e \quad (11)$$

Other positive ions are also produced in contact with nitrogen such as He^+, He_2^+, N^+, N_3^+ and N_4^+. Two intense spectral lines of N_2^+ first negative system ($\sum_u^+ \rightarrow X^2\sum_g^+$) at the wavelengths of 391.44 and 427.8 nm and a spectral line of N_2^+ second-order system at the wavelength of 782.0 are the other prominent productions of the discharge. N_2^+ is the dominant ion in the plasma jet reported in both numerical and experimental investigations [14–17]. N_2^+ line at 391.4 nm, which is well known as an indicator of helium metastable presence, is the intense line in the jet spectrum presented in Fig. 3. Oxygen $3^5P \rightarrow 3^5S$ at 777.4 nm and $OH[A^2\sum^+ \rightarrow X^2\prod]$ at 308 nm are the other discharge productions both present at the result of water vapour dissociation [15]. They are also labelled in Fig. 3.

$$H_2O + e \rightarrow OH + H + e \quad (12)$$

$$O_2 + e \rightarrow O + O + e \quad (13)$$

Three-body collisions occurred in the plasma state can lead to the quenching of atomic oxygen and reducing the oxygen emission intensity.

$$He + O + O \rightarrow He + O_2 \quad (14)$$

$$He + O_2 + O \rightarrow He + O_3 \quad (15)$$

Formation of such intense reactive species in helium as an inert gas is beneficial for atmospheric pressure plasma application especially plasma processing. Dr Walsh et al. demonstrated that in bullet mode of helium plasma, the nitrogen emission is in the direction of the plasma plume extension towards the downstream electrode. The plasma bullet front is largely made of high-energy electrons with a cloud of positive nitrogen ions located at a short distance behind the electron region [11].

Wettability measurement

Surface reactivity is characterized by water wettability which describes the ability of a liquid to spread over and penetrate into a surface and is measured by the contact angle between the liquid and the surface. In general, contact angle analysis is a surface-sensitive technique which allows the wetting properties and surface energy of the investigated sample to be measured. In this study, deionised distilled water droplet of 0.1 micro litre dispersed on the glaze surfaces; the pictures were captured as quickly as possible after placing the droplets on the surfaces. Dino Lite Digital Microscope (200X-ANMO Electronics Co.) was used to capture the image and the contact angles were measured using its software. The results of averaged contact angle measurements and their standard deviations are shown in Table 2. It should be noted that the averaged results were obtained after 15 times repeated tests. Similar measurements were repeated for Frit glaze. Figure 4 shows the image of water droplet on the Frit glaze surface before (4a) and after (4b) the cold APPJ treatment.

Surface free energy (SFE) indicates chemical reactivity and compatibility with adhesives, paints and inks. It is one of the thermodynamic quantities describing the equilibrium state of atoms in the surface layer of materials. In other words, it reflects the imbalance of intermolecular interactions at the phase boundary of two mediums and defines as a work necessary to create a new surface unit, at equilibrium separation of two phases in reversible isothermal processes. SFE is an indicator of adhesive properties which plays an important role in the processes of adhesion [1, 18]. The methods of calculation of SFE is from the contact angle measurements and based on Young equation which was derived from the equilibrium condition of forces representing surface tensions at the contact point of three phases of solid, liquid and gas. The energy equivalent of the Young equation is as follows [1, 18]:

$$\gamma_L(\cos\theta) = \gamma_S - \gamma_{SL}, \quad (16)$$

where γ_S denotes the SFE of solid in vacuum, γ_{SL} is the surface tension on the solid–liquid phase boundary, γ_L

Table 2 Contact angles of water and diiodomethane dispersed on the Frit and Raw glazes

Glaze	Liquid	θ_{nt}	Sdev	θ_t	Sdev
Frit	Water	87.78	3.86	12.33	4.41
	Diiodomethane	51.40	2.78	59.37	3.46
Raw	Water	94.66	3.83	11.44	4.82
	Diiodomethane	53.72	6.67	63.70	2.82

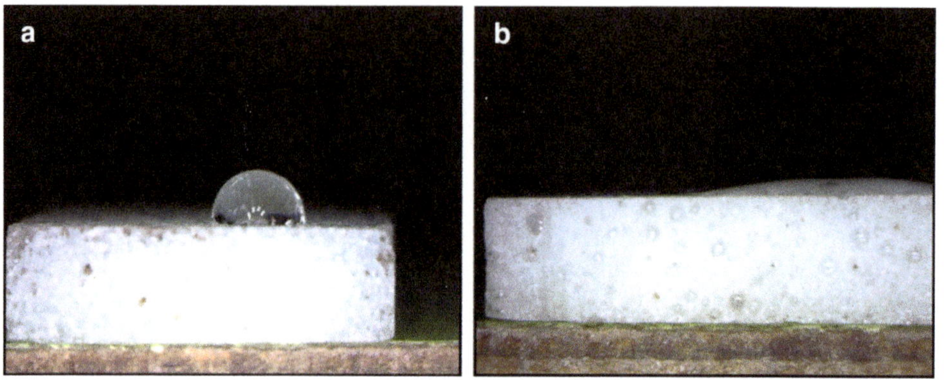

Fig. 4 The picture of water droplet on the Frit glaze surface before (**a**) and after (**b**) the plasma jet treatment

represents the SFE of the measured liquid, and θ is the contact angle that should be measured. In this paper, we used the Owens–Wendt method which considered two components of dispersion and polar SFE. It is based on Berthelot hypothesis which considered the geometric mean of intermolecular interactions within each substance as the interactions present in their surface layers. Combining this equation with Young's equation, the following geometric mean equation is obtained [18]

$$\gamma_L(1+\cos\theta) = 2\left(\sqrt{\gamma_s^d\gamma_L^d} + \sqrt{\gamma_s^p\gamma_L^p}\right) \tag{17}$$

Note that the total SFE is the sum of the two dispersive and polar components [18]

$$\gamma_s = \gamma_s^d + \gamma_s^p \tag{18}$$

Because of the presence of the polar term, the number of liquids required to calculate the solid surface components is two of known surface tension [1, 18–20]. One should be polar and the other should be non-polar. Distilled water with the surface energy of 72.8 mN/m ($\gamma_w^d = 21.8$, $\gamma_w^p = 51.0$) and diiodomethane (Merck Schuchardt OHG, 85662 Hohenbrunn, Germany) with the surface energy of 50.8 mN/m ($\gamma_d^d = 50.8$, $\gamma_d^p = 0.0$) were chosen as the two polar and non-polar liquids to measure the SFE of the glazes [20]. The contact angles of these two liquids with Raw and Frit glazes are shown in Table 2, and the results of their SFE are shown in Table 3.

Table 3 Surface free energy of Frit and Raw glazes

Glaze	γ_S^P	γ_S^D	γ_{total}
Frit (non-treated)	2.3	33.5	35.8
Frit (treated)	43.0	29.1	72.1
Raw (non-treated)	1.0	32.2	33.2
Raw (treated)	45.3	26.5	71.8

Morphology and surface chemical analyses

The physical modification of Frit glaze surface was investigated by atomic force microscopy technique (AFM). It will provide information about the morphology of the surface. Figure 5 represents the AFM image of Frit glaze surface at the scale of 5 μm^2 before (left side) and after the plasma treatment (right side). At the first glance, morphology of the Frit glaze surface changed significantly due to the bombardment of plasma ions, electrons and reactive species. The surface seems to be smoother after the plasma treatment. The data show a reduction in average roughness from 30.9 to 19.5. Roughness is defined as the normalized standard deviation calculated from the local heights (Z_{x-y}) and the average height (Z_{av}) determined over all x − y coordinates (N) measured in the AFM image. It should be noted that the result is the averaged number of data obtained from 6 different parts of the surface. It seems that the plasma removes a non-uniform layer in the scale of nanometre from the surface which made the surface smother with lower averaged roughness. It is likely etching.

X-ray photoelectron spectroscopy (XPS) reveals the surface chemistry changes of the surface after the plasma treatment. In Fig. 6, the chemical compositions of Raw glaze before and after plasma treatment are shown. Oxygen 1S was the intense peak observed in the XPS spectrum.

Results and discussion

In this contribution, we used non-thermal helium plasma jet at atmospheric pressure to improve the adhesive properties of Raw and Frit glazes. The cold helium plasma jet used in this study is convenient for surface activation, as it is inexpensive and easy to use. We explored the electrical and optical characteristics of the plasma jet to achieve efficient treatment. Current measurement shows a stable mode of

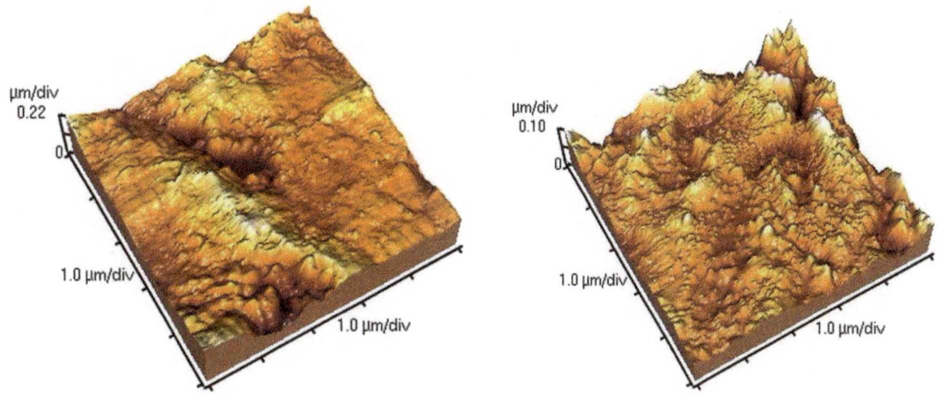

Fig. 5 AFM images of Frit glaze at the scale of 5 µm × 5 µm before (*left side*) and after the plasma jet treatment (*right side*)

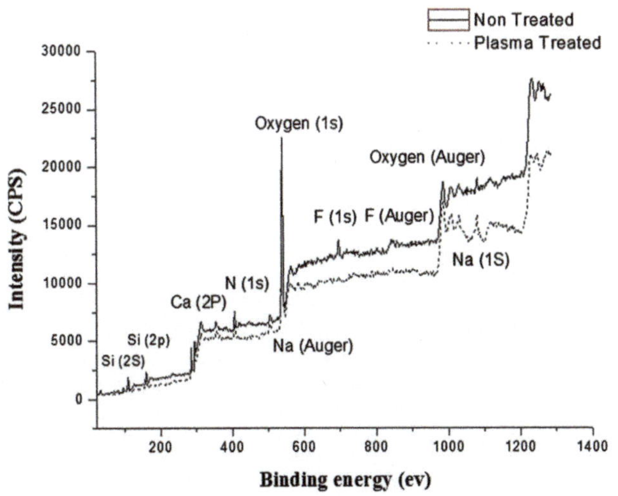

Fig. 6 XPS spectrum of Raw glaze before (*solid line*) compared with the case of plasma treatment (*dot line*)

helium plasma jet which could result in uniform treatment. Several reactive groups such as N_2 second positive, N_2^+ first negative and second-order systems, atomic oxygen, helium atomic lines and OH line were observed at the helium emission spectrum. The indicator of the energetic metastable helium and energetic electrons, N_2^+ 391.4 nm and He 706 nm, respectively, are the first and second intense lines detected in the jet spectrum. The presence of the dominant well known helium plasma lines, N_2^+ and other reactive species of helium plasma already mentioned represents the plasma capability for surface modification.

Wettability and hydrophilicity of the specimens were examined by comparison of contact angles measured before and after the plasma treatment. The results and their standard deviations are shown in Table 2 as every contact angle reported is the averaged number of 15 times repeated tests. Water contact angles of both glazes are significantly decreased after treatment. Contact angle of 94.66° for Raw glaze introduces it as a very hydrophobic surface before treatment, while reduction to 11.44° could represent a huge improvement of the surface wettability and change into hydrophilic surface. Similar water contact angle reduction was measured for Frit glaze with negligible difference.

SFEs were calculated from contact angle of two, polar and non-polar, known liquids. Since it reflects chemical reactivity and compatibility with adhesives, paints and inks, it could be a good criteria of adhesive properties improvement result in better stick (attach or joint) to paint. SFE of the both glazes increased after plasma treatment which is attributed to the polar component of SFE resulted from plasma treatment. AFM and XPS analyses were performed to reveal the physical and chemical changes which led to increase in the contact angle and SFE of the glazes and consequently resulted in surface activation. Figure 5 demonstrates a significant change of the surface morphology due to the plasma treatment. The average roughness decreased and the surface seems to be smoother. AFM analyses, reported as a result of treatment of polymers and textiles by helium or mixture of helium and nitrogen, revealed the creation of granular structure on the surface at nanoscale domain and the change of surface roughness after the plasma treatment [3, 5, 8, 9, 21]. Examining the AFM images of different parts of surface reveals that, although the effect of plasma at different parts seems to be different apparently, all of them show that the plasma jet removes a nanoscale non-uniform layer from the surface and that a non-uniform etching process occurred. In other words, it is no matter whether the surface is composed of a small number of high peaks and valleys or a large number of them with moderate height. In both cases, the plasma removes a non-uniform layer. Consequently, the surface average roughness decreases by increasing the number of peaks and valley with lower height.

XPS analysis demonstrates the surface chemistry as shown in Fig. 6. Oxygen 1S was the intense peak observed in the XPS spectrum. The ratio of O/Si increased from 11.95 to 13.61 after the plasma treatment which probably resulted in wettability improvement. Other prominent peaks observed such as Si (2S), Si (2P), N (1S), Ca (2P),

Na (Auger), Oxygen (Auger), F (1S), F (Auger) and Na (1S) are labelled in Fig. 6. Moreover, the intensity level of the spectrum was decreased after plasma treatment, and the reduction does not happen at the same rate for different species.

Acknowledgments The authors would like to thank Mr. Hossein Ghassae, associate professor of Iranian University of Science and Technology, to make the glazes, Dr. Jun-Seok Oh for his kind help, Mr. Dave Atkinson for cutting the specimens, Prof. Piter Weightman and Mr. Paul Unsworth for XPS test from the University of Liverpool and also Mrs. Sahar Tayebtaher for AFM test from Materials and Energy Research Center of Iran.

Reference

1. Awaja, F., Gilbert, M., Kelly, G., Fox, B., Pigram, P.J.: Adhesion of polymers. Prog. Polym. Sci. **34**(9), 948–968 (2009)
2. Noh, J.H., Baik, H.K., Noh, I., Park, J.C., Lee, I.S.: Surface modification of polytetrafluoroethylene using atmospheric pressure plasma jet for medical application. Surface Coat. Technol. **201**(9), 5097–5101 (2007)
3. Geyter, N.De, Morent, R., Leys, C., Gengembre, L., Payen, E., Vlierberghe, S.V., Schacht, E.: DBD treatment of polyethylene terephthalate: atmospheric versus medium pressure treatment. Surface Coat. Technol. **202**(13), 3000–3010 (2008)
4. Sorrentino, L., Carrino, L.: Influence of process parameters of oxygen cold plasma treatment on wettability ageing time of 2024 aluminum alloy. Int. J. Adhes. Adhes. **29**(2), 136–143 (2009)
5. Noeske, M., Degenhardt, J., Strudthoff, S., Lommatzsch, U.: Plasma jet treatment of five polymers at atmospheric pressure: surface, modifications and the relevance for adhesion. Int. J. Adhes. Adhes. **24**, 171–177 (2004)
6. Olszyna, A.R., Sokolowska, A., Kulakowska-pawlak, B., Dora, J.: Surface modification of Al_2O_3-Cr ceramic by non-isothermal N_2 plasma treatment at atmospheric pressure. Plasma Proces. Polym. **4**(1), S507–S511 (2007)
7. Sun, D., Stylios, G.K.: Fabric surface properties affected by low temperature plasma treatment. J. Mater. Process. Tech. **173**, 172–177 (2006)
8. Borcia, G., Chiper, A., Rusu, I.: Using a He + N_2 dielectric barrier discharge for the modification of polymer surface properties. Plasm. Sour. Sci. Technol. **15**(4), 849–857 (2006)
9. Wang, C.X., Ren, Y., Qiu, Y.P.: Penetration depth of atmospheric pressure plasma surface modification into multiple layers of polyester fabrics. Surface Coat. Technol. **202**, 77–83 (2007)
10. Massines, F., Rabehi, A., Decomps, P., Ben Gadri, R., Segur, P., Mayoux, C.: Study of a glow discharge at atmospheric pressure controlled by dielectric barrier. J. Appl. Phys. **83**(6), 2950–2957 (1998)
11. Walsh, J.L., Iza, F., Janson, N.B., Law, V.J., Kong, M.G.: Three distinct modes in a cold atmospheric pressure plasma jet. J. Phys. D: Appl. Phys. **43**, Art. ID 075201, pp. 14 (2010)
12. Ghasemi, M., Olszewski, P., Bradley, J.W., Walsh, J.L.: Interaction of multiple plasma plumes in an atmospheric pressure plasma jet array. J. Phys. D: Appl. Phys. **46**, Art. ID 052001, pp. 6 (2013)
13. Baldwin, K.G.H.: Metastable helium: atom optics with nanogrenades. Contemp. Phys. **46**(2), 105–120 (2005)
14. Gherardi, N., Gouda, G., Gat, E., Ricard, A., Massines, F.: Transition from glow silent discharge to micro-discharges in nitrogen gas. Plasm. Sour. Sci. Technol. **9**, 340–346 (2000)
15. Yuan, X., Raja, L.L.: Computational study of capacitively coupled high-pressure glow discharges in helium. IEEE Trans. Plasma Sci. **31**(4), 495–503 (2003)
16. Deloche, R., Monchicourt, P., Cheret, M., Lambert, F.: High-pressure helium afterglow at room temperature. Phys. Rev. A. **13**(3), 1140–1176 (1976)
17. Anghel, S.D., Simon, A., Radu, A.I., Hidi, I.J.: Nucl. Instrum. Methods Phys. Res. B **267**, 430–433 (2009)
18. Rudawska, A., Jacniacka, E.: Analysis for determining surface free energy uncertainty with the Owens-Wendt method. Int. J. Adhes. Adhes. **29**(4), 451–457 (2009)
19. Sohbatzadeh, F., Mirzanejhad, S., Ghasemi, M., Talebzadeh, M.: Characterization of a non-thermal plasma torch in streamer mode and its effect on polyvinyl chloride and silicone rubber surfaces. J. Electrost. **71**, 875–881 (2013)
20. Jung, M.H., Choi, H.S.: Photoresist Etching Using Ar/O_2 and He/O_2 atmospheric pressure plasma. Thin Solid Films **515**, 2295–2302 (2006)
21. Morent, R., De Geyter, N., Verschuren, J., Clerck, K.De, Kiekens, P., Leys, C.: Non-thermal plasma treatment of textiles. Surface Coat. Technol. **202**(14), 3427–3449 (2008)

25

Investigation of electromagnetic soliton in the Cairns–Tsallis model for plasma

Shabnam Rostampooran[1] · Sharooz Saviz[1]

Abstract The nonlinear Schrödinger equation (NLS) that describes the propagation of high intensity laser pulse through plasma is obtained by employing the multiple scales technique. One of the arresting solution for NLS equation is soliton like envelope for vector potential that is called electromagnetic soliton. The type and amplitude of electromagnetic soliton (EM) depends on the distribution function of plasma's particles. In this paper, distribution function of electrons obey the Cairns–Tsallis model and ions are assumed as stationary background. There are two flexible parameters, affect on the formation of EM soliton. By variation of nonextensive and nonthermal parameters, bright soliton could convert to dark one or versus. Due to positive kinetic energy, there are the limited region for nonextensive and nonthermal parameters as $q > 0.6$ and $0 < \alpha < 0.25$. The variation of EM soliton's amplitude is discussed analytically.

Keywords Electromagnetic soliton · Weakly relativistic plasma · Cairns–Tsallis model · Fluid equation · Nonlinear interaction · Multiple scales technique

Introduction

Soliton is a localized structure which can propagate in medium without diffraction spreading. Ion acoustic soliton, electrostatic soliton and electromagnetic soliton (EM) are different types of solitons that could be created in plasma medium [1–4]. EM soliton is one of the spectacular phenomena due to the nonlinear interaction between high intensity laser pulse and plasma. The electromagnetic solitons have reach variety applications such as laser fusion, plasma-based particle accelerators and etc. [5, 6]. This type of soliton is a result of various physical effects that betide in the propagation of strong laser pulse through plasma, including relativistic mass change of electrons, alteration of plasma density due to ponderomotive force and dispersion effects. The electromagnetic solitons were investigated by Kozlov et al. [7]. In the theoretical aspects, EM solitons are assumed as coupling between modulated laser pulse and electron plasma wave that have been studied by many different researchers continuously [8, 9]. For discussion on the formation and features of EM soliton, the Maxwell and fluid equations should be solved. A multiple scales technique is used to solve the fluid-Maxwell equations in cold plasma [10–14]. Kuehl and Zhang [15] have expressed the creation of bright and dark EM soliton in weakly relativistic approximation. Also, Borhanian et al. [16] employed same technique for investigation of EM solitons in magnetized plasma.

The presence of energetic particles in plasma, is an inherent factor in many space and laboratory evidences [17, 18]. In this case, the distribution function of plasma's particles don't obey Maxwell–Boltzmann distribution function [19]. It is obvious, existence and feature of nonlinear phenomena in plasma tightly depend on properties of plasma and distribution function of particles.

Observations by Viking spacecraft [20] and Freja satellite [21] indicated on existence of electrostatic solitary structure in magnetosphere which couldn't be expressed by Maxwell distribution function.

The nonthermal distribution function for plasma's particles of was proposed by Cairns et al. This distribution

✉ Sharooz Saviz
shahrooz.saviz@srbiau.ac.ir

[1] Laser Laboratory, Plasma Physics Research Center, Science and Research Branch, Islamic Azad University, Tehran, Iran

function could express the presentation of rarefactive ion sound solitons very similar to those observed by Freja satellite and Viking spacecraft [22]. In the nonthermal distribution function, population of the nonthermal particles are shown by α which could vary between 0 and 1. In the case $\alpha \to 0$, the Maxwellian distribution function is recovered. Many researchers have assumed the Cairns distribution function in plasma's model and investigated the phenomena in the presence of these particles [23–25]. In an attempt to generalize the Boltzmann–Gibbs (BG) entropy, Tsallis proposed the nonextensive statistical mechanic to describe the systems with long interaction [26, 27], as usually happen in astrophysics and plasma physics. Nonextensive statistic has been used for describing various phenomena in plasma such as dissipative optical lattices [28], plasma wave propagation [29, 30]. The foremost character of Tsallis distribution function, is q parameter which stands on the degree of nonextensivity. The nonextensive parameter has two separate states. For $-1 < q < 1$, particles cover all velocity. While in $q > 1$, the distribution function has a cutoff on the maximum permitted value for velocity of the particles, given by $v_{\max} = \sqrt{2/(q-1)}v_T$. In which, $v_T^2 = 2kT/m$ is thermal velocity of the plasma particle, T and m is the temperature and mass of the particle respectively. In the case $q < -1$, distribution function is unnormalized.

Comparison of Maxwellian and nonextensive distribution function demonstrate, for $q > 1$, high energy states are more likely in the Maxwellian distribution function. Although, for $-1 < q < 1$ high energy status are more probable in the nonextensive distribution function. In condition, $q \to 1$, Tsallis distribution function converts to Maxwell–Boltzmann distribution function.

Ion acoustic solitary wave was investigated in a two component plasma with nonextensive electrons by Tribeche et al. [31]. They realized this model of plasma could explain both rarefactive and compressive solitons. A recent study by them has proposed a hybrid Cairns–Tsallis distribution function, which purports to offer enhanced parametric flexibility in modeling nonthermal plasmas. Whereas such a two-parameter demonstration of the distribution function could be useful in fitting to a wider range of experimental plasmas. Subsequently, Amour et al. [32] applied this distribution to the study of acoustic solitary. Motivated by these efforts to explain various phenomena in the presence of non-Maxwellian particles, we investigate the electromagnetic soliton in non-Maxwellian plasma.

In the present paper we analyze the circularly polarized intense EM wave propagating in a weakly relativistic plasma. The electrons of plasma obey the mixed Cairns–Tsallis distribution function. The ions of plasma are assumed to be stationary. The relevant nonlinear Schrödinger equation is introduced. Roles of mixed electrons on the existence of bright and dark solitons and amplitude of them were discussed in detail.

The layout of this article goes as follows; following the introduction in "Introduction" section, we present the basic equations describing the dynamics of the nonlinear interaction of laser and plasma. We use the reductive perturbation method to derive the nonlinear Schrödinger (NLS) equation. In "Results and discussion" section, numerical results and discussion are presented and finally "Conclusion" section is devoted to conclusion.

Model description

We start with description of the propagation of a circularly polarized electromagnetic pulse in a weakly relativistic plasma in x direction. In this paper, we consider collisionless, unmagnetized, two component plasma, and ion is assumed as immoble singly charged positive particle. Then, we employ the momentum and continuity equations for electrons as Eqs. (1)–(2) respectively. Equations (3) and (4) are giving an expressions for the Poisson's equation and electromagnetic wave equation in the Coulomb gauge

$$m_e N_e\left(\frac{\partial}{\partial t} + v_e \cdot \nabla\right) v_e = -eN_e E - \frac{e}{c} N_e v_e \times B - \nabla \cdot P_e \tag{1}$$

$$\frac{\partial N_e}{\partial t} + \frac{\partial}{\partial x}(N_e v_e) = 0 \tag{2}$$

$$\frac{\partial^2 \phi}{\partial x^2} = 4\pi e(N_e - N_i) \tag{3}$$

$$\frac{\partial^2 \boldsymbol{A}}{\partial x^2} - \frac{1}{c^2}\frac{\partial^2 \boldsymbol{A}}{\partial t^2} - \frac{1}{c}\frac{\partial}{\partial t}\nabla\phi = \frac{4\pi e}{c}(N_e v_e - N_i v_i) \tag{4}$$

with one-dimensional approximation in which $\partial/\partial y = \partial/\partial z = 0$, set of the normalized hydrodynamic equations (continuity and momentum) for electrons of plasma and Maxwell's equations for the scalar and vector potentials, ϕ and A, can be written as

$$n_e\left(\frac{\partial}{\partial t} + u_e \frac{\partial}{\partial x}\right)(\gamma u_e) = n_e \frac{\partial \phi}{\partial x} - \frac{n_e}{2\gamma}\frac{\partial A_\perp^2}{\partial x} - \frac{1}{m_0 N_0 c^2}\frac{\partial P_e}{\partial x} \tag{5}$$

$$u_{e\perp} = \frac{A_\perp}{\gamma_e} \tag{6}$$

$$\frac{\partial n_e}{\partial t} + \frac{\partial}{\partial x}(n_e u_e) = 0 \tag{7}$$

$$\frac{\partial^2 \phi}{\partial x^2} = n_e - 1 \tag{8}$$

$$\frac{\partial^2 A_\perp}{\partial x^2} - \frac{\partial^2 A_\perp}{\partial t^2} = \frac{n_e A_\perp}{\gamma} y \tag{9}$$

Here γ is the relativistic factor.

$$\gamma = \sqrt{(1 + A_\perp^2)/(1 - u_e^2)} \tag{10}$$

In writing the above equations, length, time, velocity, scalar and vector potential and density are normalized over c/ω_{pe0}, ω_{pe0}^{-1}, c, $m_0 c^2/e$ and N_0, respectively; $\omega_{pe0} = \sqrt{4\pi N_0 e^2/m_0}$ is the electron plasma frequency. m_0 is the rest mass of electron and N_0 is the unperturbed electron background density.

We also rewrite momentum equation in parallel direction by using following expression

$$\frac{\partial P_e}{\partial x} = \left(\frac{\mathrm{d}P_e/\mathrm{d}\phi}{\mathrm{d}N_e/\mathrm{d}\phi}\right)\frac{\partial N_e}{\partial x} \equiv \left(\frac{\mathrm{d}P_e}{\mathrm{d}N_e}\right)_{\phi=0}\frac{\partial N_e}{\partial x} = m_0 c_{se}^2 \frac{\partial N_e}{\partial x} \tag{11}$$

where

$$c_{se}^2 = \left[\frac{1}{m_0}\left(\frac{\mathrm{d}P_e}{\mathrm{d}N_e}\right)_{\phi=0}\right] \tag{12}$$

As it is mentioned, in our plasma model, the electrons distribution function of plasma obey the Cairns–Tsallis distribution function as follow [31, 32]

$$f_e(v_x) = C_{q,\alpha}\left(1 + \alpha\frac{v_x^4}{v_{te}^4}\right)\left\{1 - (q-1)\frac{v_x^2}{2v_{te}^2}\right\}^{\frac{1}{q-1}} \tag{13}$$

where $C_{q,\alpha}$ is the constant of normalization which depends on q and α

$$C_{q,\alpha} = N_0\sqrt{\frac{m_e}{2\pi T_e}}\frac{\Gamma\left(\frac{1}{1-q}\right)(1-q)^{5/2}}{\Gamma\left(\frac{1}{1-q} - \frac{5}{2}\right)\left[3\alpha + \left(\frac{1}{1-q} - \frac{3}{2}\right)\left(\frac{1}{1-q} - \frac{5}{2}\right)(1-q)^2\right]} \quad \text{for } -1<q<1$$

and

$$C_{q,\alpha} = N_0\sqrt{\frac{m_e}{2\pi T_e}}\frac{\Gamma\left(\frac{1}{1-q} + \frac{3}{2}\right)(q-1)^{5/2}\left(\frac{1}{1-q} + \frac{3}{2}\right)\left(\frac{1}{1-q} + \frac{5}{2}\right)}{\Gamma\left(\frac{1}{1-q} + 1\right)\left[3\alpha + \left(\frac{1}{1-q} - \frac{3}{2}\right)\left(\frac{1}{1-q} - \frac{5}{2}\right)(1-q)^2\right]} \quad \text{for } q>1 \tag{14}$$

Here α is the parameter indicating the proportion of nonthermal electrons and q implicates nonextensive parameter, and Γ represents the standard Gamma function.

To derive the number density expression in the Cairns–Tsallis distribution function, Eqs. (13)–(14) are employed with replacing v_x^2 by $v_x^2 - (2e\varphi/m_e)$ in two ranges, $q > 1$ and $-1 < q < 1$. Then

$$N_e(\phi) = \begin{cases} \int_{-v_{\max}}^{v_{\max}} f_e(v_x)\mathrm{d}v_x & q>1 \\ \int_{-\infty}^{+\infty} f_e(v_x)\mathrm{d}v_x & -1<q<1 \end{cases}$$
$$= N_{e0}\left\{1 + (q-1)\frac{e\phi}{kT_e}\right\}^{\frac{1}{q-1}+\frac{1}{2}} \times \left\{1 + A\left(\frac{e\phi}{kT_e}\right) + B\left(\frac{e\phi}{kT_e}\right)^2\right\} \tag{15}$$

where the confidences of A and B are expressed as

$$A = -16q\alpha/(3 - 14q + 15q^2 + 12\alpha) \tag{16}$$

$$B = 16(2q-1)q\alpha/(3 - 14q + 15q^2 + 12\alpha) \tag{17}$$

It is noted in the case $q \to 1$, nonextensive density converts to pure Cairns et al. (nonthermal) as follow

$$N_e(\phi) = N_{e0}\left\{1 + A\left(\frac{e\phi}{kT_e}\right) + B\left(\frac{e\phi}{kT_e}\right)^2\right\}\exp\left(\frac{e\phi}{kT_e}\right) \tag{18}$$

Beside, in the limit $\alpha = 0$, Tsallis pure density is recovered as

$$N_e(\phi) = N_{e0}\left\{1 + (q-1)\frac{e\phi}{kT_e}\right\}^{\frac{1}{q-1}+\frac{1}{2}} \tag{19}$$

it is useful to determine the integral $\langle v_x^2\rangle = \int v_x^2 f_e(v_x)\mathrm{d}v_x$ over all permitted velocity as follow

$$v_x^2 = \begin{cases} \int_{-v_{\max}}^{v_{\max}} v_x^2 f_e(v_x)\mathrm{d}v_x & q>1 \\ \int_{-\infty}^{+\infty} v_x^2 f_e(v_x)\mathrm{d}v_x & -1<q<1 \end{cases}$$
$$= \frac{2(35q^2 - 46q + 60\alpha + 15)}{(7q-5)[12\alpha + (5q-3)(3q-1)]} N_0 v_t^2 \tag{20}$$

it is noted that the kinetic energy should be positive. So, the acceptable range for q is narrowed to the area that mean value of square velocity to be positive. Moreover, it is well known that the Cairns distribution for $\alpha > 0.25$ presents unstable behavior as it develops side wings, possibly leading to a kinetic instability. One would not expect stable nonlinear structures such as solitons to be supported by such a linearly unstable situation. That implies a need to introduce a cutoff in α governed by this consideration [33]. Indeed Williams et al. [3] demonstrated that the region for nonextensive parameter in the case $-1 < q < 1$, is restricted to $0.6 < q < 1$.

Using of Eq. (20), the pressure term in Eq. (5) is determined as

$$P_e = \frac{m_0}{3}\int v^2 f_e(v)\mathrm{d}v$$
$$= \frac{2(35q^2 - 46q + 60\alpha + 15)}{3(7q-5)[12\alpha + (5q-3)(3q-1)]} N_0 k T_e \tag{21}$$

By replacing Eqs. (10), (21) in the momentum equation, Eq. (5) is written in the following model

$$n_e\frac{\partial}{\partial t}(\gamma u_e) = n_e\frac{\partial}{\partial x}(\phi - \gamma) - \frac{c_{se}^2}{c^2}\frac{\partial n_e}{\partial x} \tag{22}$$

Now we apply the multiple scales technique [34] to treat Eqs. (7)–(10) and (22). According to this method, the amplitude of all wave harmonics such as density, velocity, scalar and vector potential will be assumed to have an envelope with the slower space/time evolution which are distinguish from the fast carrier wave (phase) dynamics. Let $S = (n, u, \Phi, \mathbf{A})$ is given by

$$S = S^{(0)} + \sum_{n=-\infty}^{n} \varepsilon^n S^{(n)} \tag{23}$$

where $S^{(0)} = (1, 0, 0, 0)$ indicates the equilibrium state of the system and ε is a smallness parameter which provides evolution equations for different harmonic amplitude in successive orders ε^n. We suppose $S^{(n)}$, the perturbed state, contains fast and slow part as

$$S^{(n)} = \sum_{l=-\infty}^{\infty} S_l^{(n)}(X_{m\geq 1}, T_{m\geq 1})\exp(il(kx - \omega t)) \tag{24}$$

The fast variables of state depend on the phase $\psi = kx - \omega t$, which k and ω are the normalized wave number and frequency of the pulse by $k_{pe} = \omega_{pe}/c$ and ω_{pe}. The slow part, enter the argument of the l-th harmonic amplitude $S_l^{(n)}$, depends on the stretched space and time slowly, which are considered as

$$X_m = \varepsilon^m x, \quad T_m = \varepsilon^m t \tag{25}$$

It must be noted that $m = 0$ corresponds to the fast carrier space/time scale, while $m \geq 1$ corresponds to the slower envelope scales. Assuming variable independence, time and space differentiation are obtained as follow

$$\frac{\partial}{\partial t} = \frac{\partial}{\partial t_0} + \varepsilon\frac{\partial}{\partial T_1} + \varepsilon^2\frac{\partial}{\partial T_2} + \cdots \tag{26}$$

$$\frac{\partial}{\partial x} = \frac{\partial}{\partial x_0} + \varepsilon\frac{\partial}{\partial X_1} + \varepsilon^2\frac{\partial}{\partial X_2} + \cdots \tag{27}$$

For obtaining the set of reduced equations, using Eqs. (23)–(24), all parameters are defined as

$$S = S^{(0)} + \varepsilon S_0^{(1)} + \varepsilon S_1^{(1)}e^{i\psi} + \varepsilon^2 S_0^{(2)} + \varepsilon^2 S_1^{(2)}e^{i\psi} + \varepsilon^2 S_2^{(2)}e^{2i\psi} + \varepsilon^3 S_0^{(3)} + \varepsilon^3 S_1^{(3)}e^{i\psi} + \varepsilon^3 S_2^{(3)}e^{2i\psi} + \varepsilon^3 S_3^{(3)}e^{3i\psi} + c.c \tag{28}$$

For all state variables, the reality condition as $S_{-l}^{(n)} = S_l^{(n)*}$ is true. By replacing Eqs. (26)–(28) into Eqs. (7)–(10) and (22) and collect the terms of the same order in ε for l-th harmonic amplitudes, $S_l^{(n)}$, the set of reduced equations are obtained which must be solved separately.

In the first order ($n = 1$) and for zeroth harmonic ($l = 0$), equations convert to

$$n_0^{(1)} = u_0^{(1)} = A_0^{(1)} = 0 \tag{29}$$

For the first harmonic of the first order ($n = 1$, $l = 1$) the following relations are get

$$-\omega n_1^{(1)} + k u_1^{(1)} = 0 \tag{30}$$

$$-k^2\phi_1^{(1)} = n_1^{(1)} \tag{31}$$

$$(\omega^2 - k^2)A_1^{(1)} = A_1^{(1)} \tag{32}$$

$$\omega u_1^{(1)} + k\phi_1^{(1)} = 0 \tag{33}$$

from Eq. (32) linear dispersion relation for the propagation of electromagnetic wave into plasma is obtained as $\omega^2 - k^2 = 1$, indeed from other equations, we have

$$n_1^{(1)} = u_1^{(1)} = \phi_1^{(1)} = 0 \tag{34}$$

It is obvious, there are not any perturbation in electron density, parallel velocity, and scalar potential in the first order.

For the second order and the zeroth harmonic amplitude, it is obtained

$$n_0^{(2)} = u_0^{(2)} = A_0^{(2)} = \frac{\partial\phi_0^{(1)}}{\partial X_1} = 0 \tag{35}$$

For the second order and first harmonics ($n = 2$, $l = 1$) equations convert to

$$n_1^{(2)} = u_1^{(2)} = \phi_1^{(2)} = 0 \tag{36}$$

$$\frac{\partial A_1^{(1)}}{\partial T_1} + v_g\frac{\partial A_1^{(1)}}{\partial X_1} = 0 \tag{37}$$

Here $v_g = \frac{k}{\omega}$, is the group velocity and this relation indicates, up to second order of ε, wave packet moves with constant group velocity.

Proceeding in the perturbation analysis, for ($n = 2$, $l = 2$) we have

$$u_2^{(2)} = \frac{\omega}{k}n_2^{(2)} \tag{38}$$

$$-4k^2\phi_2^{(2)} = n_2^{(2)} \tag{39}$$

$$\omega u_2^{(2)} = \frac{k}{2}\left(A_1^{(1)}\right)^2 - k\phi_2^{(2)} + k\frac{c_{se}^2}{c^2}n_2^{(2)} \tag{40}$$

Then we could derive second harmonics in the density perturbation as a result of nonlinear self-interaction of wave envelope as

$$n_2^{(2)} = \frac{2k^2}{\left(4\omega^2 - 1 - 4k^2\frac{c_{se}^2}{c^2}\right)}\left(A_1^{(1)}\right)^2 \tag{41}$$

Finally, for the third order perturbation following relations are obtained.

In the zeroth harmonics ($n = 3$, $l = 0$)

$$\phi_0^{(2)} = \frac{\left|A_1^{(1)}\right|^2}{2} \tag{42}$$

This relation indicates that zeroth scalar potential is generated by nonlinear self-interaction of envelope.

For the first harmonic ($n = 3$, $l = 1$)

$$n_1^{(3)} = u_1^{(3)} = \phi_1^{(3)} = 0 \tag{43}$$

And wave equation converts to

$$i\left(\frac{\partial A_1^{(1)}}{\partial T_2} + v_g \frac{\partial A_1^{(1)}}{\partial X_2}\right) + \frac{1}{2\omega}\left(\frac{\partial^2 A_1^{(1)}}{\partial X_1^2} - \frac{\partial^2 A_1^{(1)}}{\partial T_1^2}\right) + \frac{3}{4\omega}\left|A_1^{(1)}\right|^2 A_1^{(1)} - \frac{1}{2\omega} n_2^{(2)} A_1^{(1)*} = 0 \tag{44}$$

Using by Eqs. (37) and (41) in Eq. (44), the following equation is obtained

$$i\left(\frac{\partial A_1^{(1)}}{\partial T_2} + v_g \frac{\partial A_1^{(1)}}{\partial X_2}\right) + P\frac{\partial^2 A_1^{(1)}}{\partial X_1^2} + Q\left|A_1^{(1)}\right|^2 A_1^{(1)} = 0 \tag{45}$$

where

$$P = \frac{1}{2\omega^3} \tag{46}$$

and

$$Q = \frac{3}{4\omega} - \frac{k^2}{\omega\left(4\omega^2 - 1 - 4k^2 \frac{c_{se}^2}{c^2}\right)} \tag{47}$$

are the dispersion and nonlinear coefficients, respectively. By introducing the coordinate transformation $\xi = x - v_g t$ and $\tau = t$, Eq. (45) converts to the nonlinear Schrödinger equation as

$$i\frac{\partial a}{\partial \tau} + P\frac{\partial^2 a}{\partial \xi^2} + Q|a|^2 a = 0 \tag{48}$$

where a represents the slower component of the vector potential as

$$A \cong A_1^{(1)} e^{i(kx-\omega t)} + c.c = a e^{i(kx-\omega t)} + c.c \tag{49}$$

Equation (48) indicates the wave envelope modulation with the effects of dispersion and nonlinearity terms. It could be predicted different types of envelop excitation in the propagation of laser in plasma. We consider the solution as follow for the nonlinear Schrödinger equation.

$$a = R\exp(i\Theta) \tag{50}$$

then Eq. (48) converts to KdV equation as

$$\frac{\partial^2 R}{\partial \xi^2} + \frac{Q}{P}R^3 - \frac{E}{P}R = 0 \tag{51}$$

where E is the positive constant value, Θ indicates the phase correction, and $\partial\Theta/\partial x = v(x,t)$ [35].

The first integral of Eq. (51) is

$$\frac{1}{2}\left(\frac{\partial R}{\partial \xi}\right)^2 + \frac{Q}{P}R^4 - \frac{E}{P}R^2 = \text{constant} \tag{52}$$

which can be assumed as energy equation for pseudo-particle. So, the Sagdeev potential is as follow

$$V(R) = \frac{Q}{P}R^4 - \frac{E}{P}R^2 \tag{53}$$

In the present paper we assume $v = v_0$ as a constant.

If in the above equation, $(Q/P) > 0$ and $(E/P) > 0$, the Sagdeev potential has a minimum and bright soliton is formed [36].

In this case R is satisfied the following boundary conditions in the ξ-space

$$\lim_{\xi \to \pm\infty} R(\xi) = 0 \tag{54}$$

and Eq. (51) has the following bright soliton answer as

$$R(\xi) = R_0 \operatorname{sech}\left(\frac{\xi - v_0 \tau}{L}\right) \tag{55}$$

where $R_0 = \sqrt{2E/Q}$, $L = \sqrt{P/E}$ is the amplitude and width of the pulse respectively. v_0 represents the bright soliton envelope group velocity. The amplitude and width of soliton are independent of velocity and $LR_0 = (2P/Q)^{1/2} = \text{constant}$.

On the other side, if in Eq. (51) $(Q/P) < 0$ and $(E/P) < 0$, bright soliton will not be formed in the plasma medium. In this case, we consider R with the following boundary condition

$$\lim_{\xi \to \pm\infty} R(\xi) = |R_0| \tag{56}$$

in which

$$R(\xi) \equiv R_0 + R_1(\xi) \tag{57}$$

where $\lim_{\xi \to \pm\infty} R_1(\xi) = 0$

As a result, dark soliton will be formed which intensity of wave packet is zero at the center of pulse and reaches a nonzero value at the boundary. The dark soliton can be express as follow

$$R(\xi) = \left|R_0 \tanh\left(\frac{\xi - v_0 \tau}{L}\right)\right| \tag{58}$$

where $R_0 = \sqrt{2E/Q}$, $L = \sqrt{P/E}$ are the amplitude and width of the pulse respectively. The amplitude of the pulse does not depend on velocity of the pulse as was discussed in the case of bright type soliton.

Result and discussion

The coupling between the transverse electromagnetic wave and plasma wave could lead to form the EM soliton. In this paper distribution function of electron is assumed to be mixed. The influence of nonextensive and nonthermal parameters on the structure of bright and dark solitons are discussed.

Equation (51) looks like as energy equation of a pseudo-particle. First term is kinetic energy, while two other terms refer to potential energy or Sagdeev potential. For determining soliton solution, Sagdeev potential should have at least one maximum or one minimum. If $(Q/P) > 0$ and $(E/P) > 0$, Sagdeev potential has a minimum and the bright soliton would be arisen. While for $(Q/P) < 0$ and $(E/P) < 0$ Sagdeev potential has two symmetric maximum on the two sides of $R = 0$ [36]. The sort of soliton (bright or dark) depends on sign of Q and P, consequently. It is obvious, P is positive, and the magnitude of its decreases as frequency increase. The dispersion term, P, is independent of value of nonextensive and nonthermal parameters. While, sign and value of nonlinear term, Q, depends on the fast varying frequency, population of nonthermal electrons and nonextensive parameters.

In Figs. 1, 2 and 3, the variation of nonlinear term, Q in Eq. (47), is plotted versus ω for $kT_e = 1$ MeV and $\alpha = 0.2$ for two different region of nonextensive parameter. As it is mentioned, there are two different regions for q. The nonextensive parameter can vary in $0.6 < q < 1$ or $q > 1$. Since the kinetic energy must be positive value, it is necessary that the nonextensive parameter to be more than 0.8 for $\alpha = 0.2$ [3]. In Fig. 1, q is 0.8, 0.9 and 1. In the limit $q = 1$, distribution function tends to the nonthermal distribution function. It is shown, there are singularity in the nonlinear terms for all value of nonextensive parameters. The nonlinear term, Q, contains two nonlinear terms. The first term, $Q_1 = 3/4\omega$ stands on the relativistic nonlinearity. On the other hand, $Q_2 = k^2/\omega\left(4\omega^2 - 1 - 4k^2\frac{c_{se}^2}{c^2}\right)$ indicates the electron density perturbation due to ponderomotive force and pressure. The relativistic nonlinearity diminishes as frequency increases, and it is invariable for all magnitude of q and α. The value and signs of the second term, Q_2, depends tightly on frequency, nonextensive and nonthermal parameters of plasma. The second nonlinear term, Q_2 is positive and less than relativistic nonlinearity in $\omega < \omega_1$. While, it is positive and is more than relativistic nonlinearity term, in the range $\omega_1 < \omega < \omega_2$, where $\omega_1 = \sqrt{(3\beta^2 + 0.25)/(3\beta^2 - 2)}$, $\omega_2 = 0.5\sqrt{(1 - 4\beta^2)/(1 - \beta^2)}$ and

$$\beta^2 = \frac{1}{3}\frac{kT_e}{m_e c^2}\frac{2(35q^2 - 46q + 60\alpha + 15)}{(7q - 5)[12\alpha + (5q - 3)(3q - 1)]} \tag{59}$$

Therefore, the nonlinear term in $\omega_1 < \omega < \omega_2$ is negative and the dark EM soliton could be developed. In the steady nonthermal parameter, by increasing q, its region be wider. It is noted, out of mentioned region, Q_2 is negative, so the nonlinear term is positive. As a result, the bright EM soliton could be performed for all value of nonextensive parameter.

By increasing of nonextensive parameter, more than 1, feature of mixed distribution function is substituted. In the region $q > 1$, raising of nonextensive parameter makes reducing of probability of high energy electron's existence. In $\alpha = 0.2$ and $q = 1.1$, there is singularity in the nonlinear term. Whereas, for $q \geq 1.2$ in the same nonlinear parameter, Q declines as frequency increases. In Fig. 2, the variation of Q is shown versus ω for $kT_e = 1$ MeV, $\alpha = 0.2$ and $q = 1.2$, 1.3 and 1.4. As it is shown, for $q = 1.2$, and ω approximately more than 2.2, the nonlinear term is negative. It indicates, the dark EM soliton is formed in plasma. But, for $q = 1.3$ and $q = 1.4$, the nonlinear term is positive for all frequency, so only bright EM soliton is performed. This figure, denotes, for $\alpha = 0.2$, by increasing

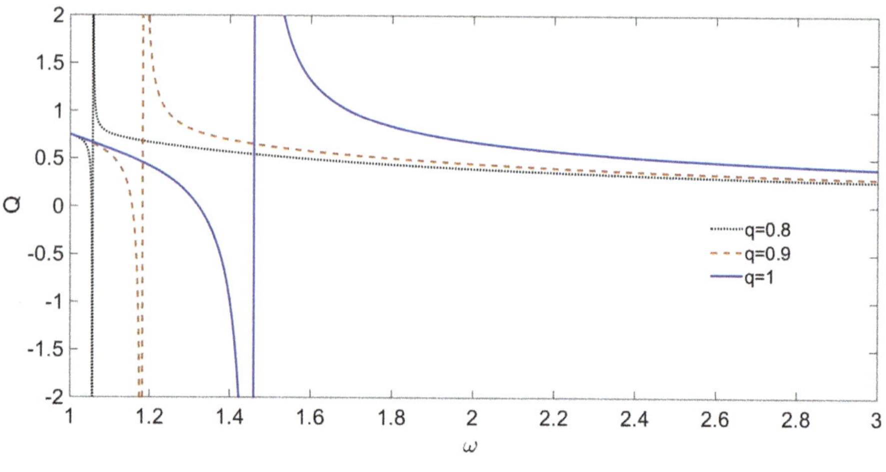

Fig. 1 Variation of Q is plotted versus ω for $kT_e = 1$ MeV, $\alpha = 0.2$ and $q = 0.8$, 0.9 and 1

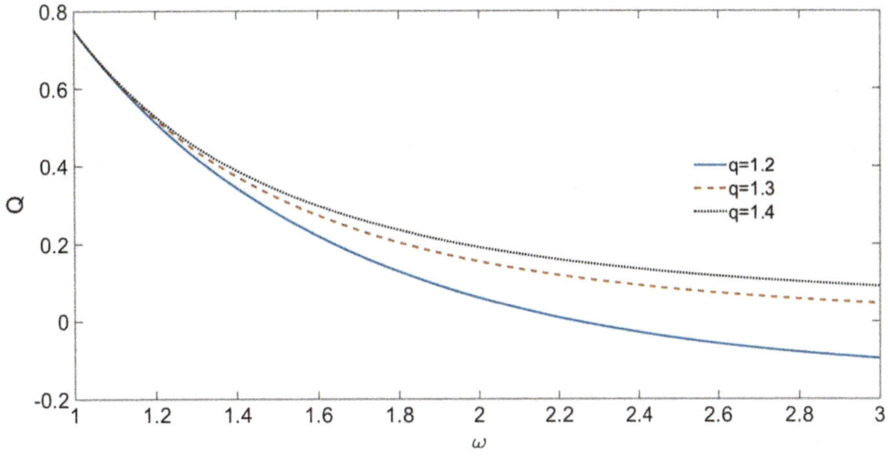

Fig. 2 Variation of Q is plotted versus ω for $kT_e = 1$ MeV, $\alpha = 0.2$ and $q = 1.2$, 1.3 and 1.4

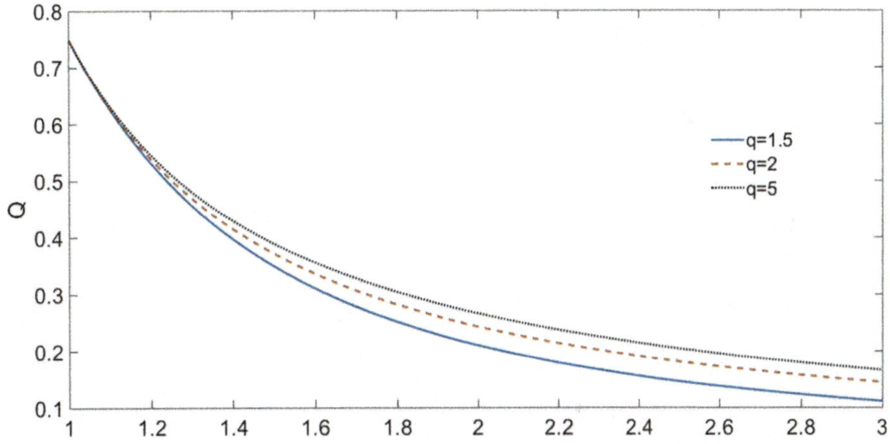

Fig. 3 Variation of Q is plotted versus ω for $kT_e = 1$ MeV, $\alpha = 0.2$ and $q = 1.5$, 2 and 5

of the nonextensive parameter from 1.2 to 1.3, dark soliton converts to the bright soliton.

In Fig. 3, the variation of Q is plotted versus ω for $kT_e = 1$ MeV, $\alpha = 0.2$ and $q = 1.5$, 2 and 5. The second nonlinear term, in the plasma with $\alpha = 0.2$ and $q \geq 1.3$ is positive and less than relativistic term, for all amount of frequency. So, the bright soliton is organized in this case. By increasing frequency, Q_2 increases in low frequency and then reduces in higher magnitude of frequency. Checking of the second nonlinearity term specifies that it decreases as nonextensive parameter raises. Therefore, as q increases, the influence of pressure term diminishes. a result, in the region $q > 1.3$, nonlinear term, Q, increases by enhancement of q for steady value of α.

In this paper, the distribution function of electrons are assumed to obey Cairns–Tsallis model. In mixed model, there are two flexible parameters to vary, nonthermal and nonextensive. In Fig. 4, the variation of Q is shown versus ω for $kT_e = 1$ MeV for steady nonextensive parameter. The nonthermal parameter is 0, 0.1 and 0.2 for $q = 0.9$. In the mixed model, when nonthermal parameter tends to zero, the density of electrons convert to nonextensive distribution function. It indicates, for pure nonextensive distribution with $q = 0.9$, the nonlinear term decrease as frequency arises. The nonlinear term is positive, so bright soliton is formed. The positive nonlinearity requires the relativistic nonlinearity to be more than perturbation of density ($Q_1 > Q_2$). By increasing nonthermal parameter, population of high energy electrons grow. It represents, there are singularity for $q = 0.9$, $\alpha = 0.1$ and $\alpha = 0.2$. Therefore, accretion of the nonthermal parameter results to form the dark soliton in the limited region of frequency. By increasing of nonthermal parameter, the allowed region of frequency to perform of dark EM soliton improves and it is shifted to higher frequency.

In Fig. 5, the variation of Q is plotted versus ω for $kT_e = 1$ MeV and state value of nonextensive parameter. The nonthermal parameter is 0, 0.1 and 0.2 for $q = 1.2$. It is shown, by increasing frequency, the nonlinear term falls. So, by increasing of nonthermal parameter, the nonlinear term reduces. For $\alpha = 0$ and 0.1, the nonlinear term is positive for all value of frequency. Therefore, the effects of relativistic nonlinear term is more than perturbation density. As the nonthermal parameter and population of high energy particle enhance, the influence of density perturbation, Q_2, raises. It could be result that the influence of pressure decreases, while the nonlinear term increases. In

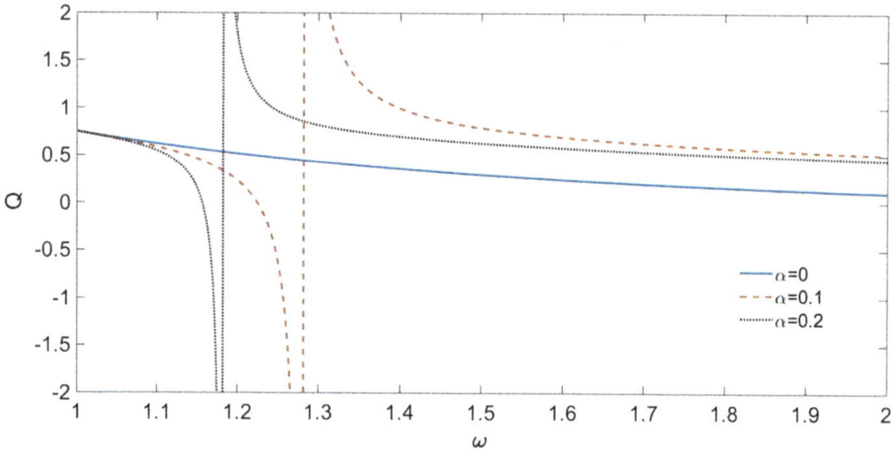

Fig. 4 Variation of Q is plotted versus ω for $kT_e = 1$ MeV, $q = 0.9$, $\alpha = 0$, 0.1 and 0.2

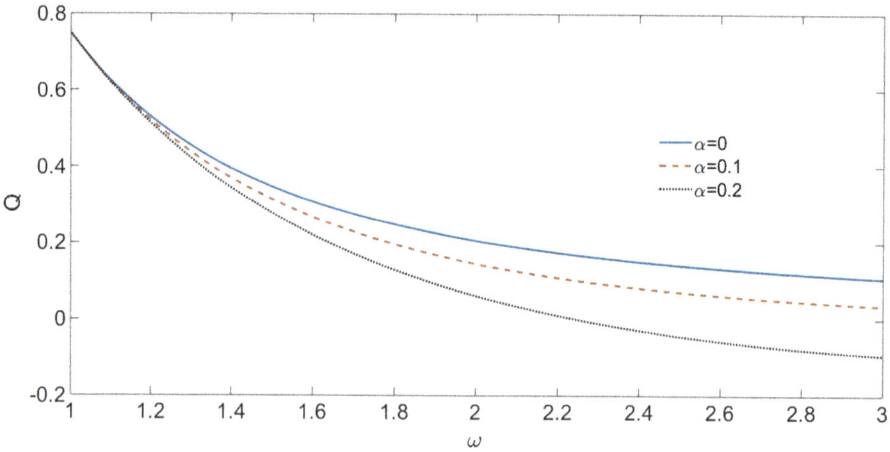

Fig. 5 Variation of Q is plotted versus ω for $kT_e = 1$ MeV, $q = 1.2$, $\alpha = 0$, 0.1 and 0.2

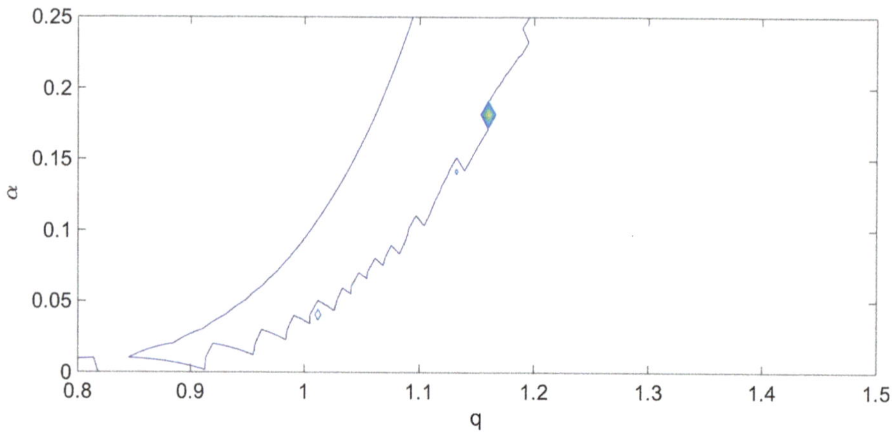

Fig. 6 Variation of P/Q is plotted versus α and q for $\omega = 2$

$q = 1.2$ for $\alpha = 0$ and 0.1, only bright soliton could be formed. By increasing nonthermal parameter the amplitude of the bright soliton amplifies. As the frequency increases, the nonlinear term becomes negative. So, the bright soliton converts to dark soliton in $\alpha = 0.2$.

The contour of P/Q versus q and α is shown in Fig. 6 for $\omega = 1$ and $kT_e = 1$ MeV. The coefficient of P/Q is positive except in the narrow region, between two lines. Then it is shown, for most of value of nonthermal and nonextensive parameters in the plasma with mixed distribution function, the bright soliton is formed. In Fig. 7 the variation of P/Q versus q and α is plotted for $\omega = 1$ and $kT_e = 1$ MeV. In Fig. 7a, $0.8 < q < 1$ and $0.1 < \alpha < 0.2$, while in Fig. 7b $1.3 < q < 1.5$ and $0 < \alpha < 0.2$. As it is shown, in Fig. 6 for both of these regions the factor of P/Q is positive and the bright soliton is created in plasma.

Fig. 7 Variation of P/Q is plotted versus α and q for $\omega = 2$

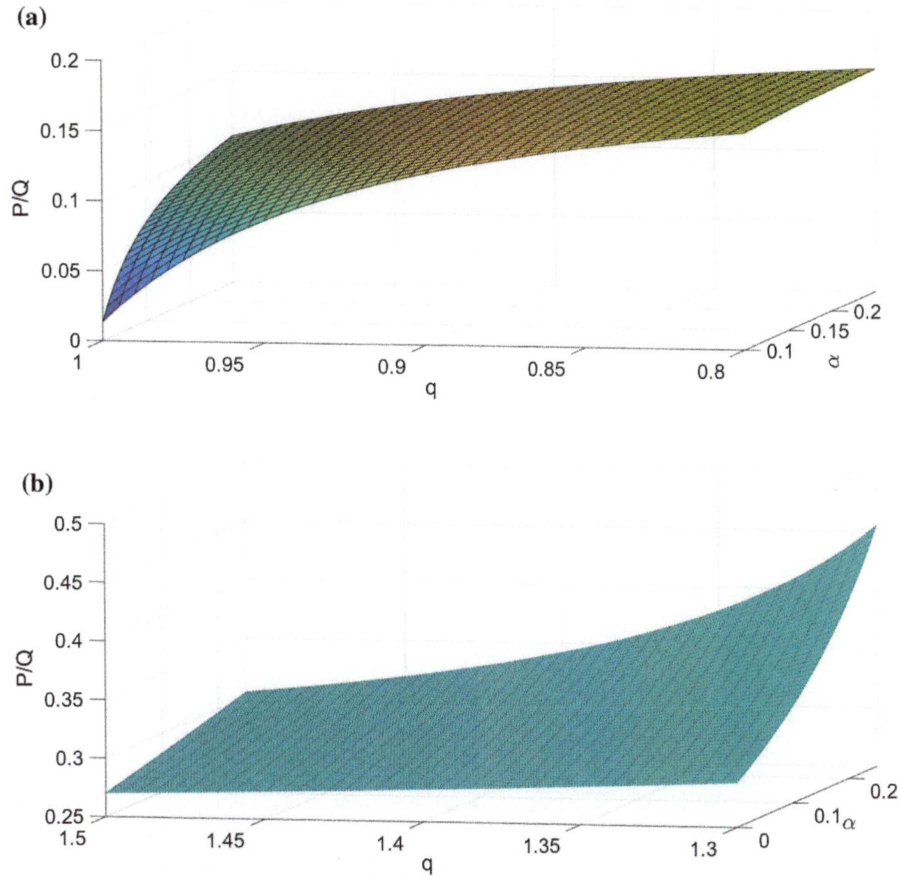

The amplitude and width of the bright soliton is independent of the velocity and $LR_0 = (2P/Q)^{1/2}$ = constant [11]. As it is shown in Fig. 7, by increasing of the nonextensive parameter, P/Q decreases, so LR_0 reduces. While, growth of the nonthermal parameter has contrary effect on P/Q. In both regions of q, increasing of the nonthermal electrons lead to raise of P/Q and LR_0.

Conclusion

In the present paper, the role of mixed electron on properties of the relativistic electromagnetic soliton is investigated. Results show that two types of soliton, bright and dark, may be formed in the interaction of laser pulse and plasma, with mixed distribution of function. In the propagation of laser pulse into plasma, relativistic effect and perturbation of electron density modify the nonlinear term. If nonlinear term is positive, the bright soliton establishes, otherwise the dark soliton forms. Parameter of nonextensive and nonthermal of electrons and frequency of the pulse, adapt the sign and value of the nonlinear term. The nonextensive parameter is a real number, and it must be more than 0.6, while the nonthermal parameter varies in the range 0 till 0.25. It is shown by variation of nonextensive and nonthermal parameters, the sign and magnitude of the nonlinear term changes, so the properties of soliton vary. The effects of the nonthermal and nonextensive parameters are contradictory. Growing of the nonthermal electrons, increase P/Q, so the amplitude of soliton increases. While rising of the nonextensive electrons decrease P/Q, and the amplitude of soliton decreases.

References

1. Heidari, E., Aslaninejad, M., Eshraghi, H., Rajaee, L.: Standing electromagnetic solitons in hot ultra-relativistic electron–positron plasmas. Phys. Plasmas **21**, 032305 (2014)
2. Kominis, Y.: Bright, dark, antidark, and kink solitons in media with periodically alternating sign of nonlinearity. Phys. Rev. A **87**, 063849 (2013)
3. Williams, G., Kourakis, I., Verheest, F., Hellberg, M.A.: Re-examining the Cairns–Tsallis model for ion acoustic solitons. Phys. Rev. E **88**, 023103 (2013)
4. Haas, F., Mahmood, Sh: Nonlinear ion-acoustic solitons in a magnetized quantum plasma with arbitrary degeneracy of electrons. Phys. Rev. E **94**, 033212 (2016)
5. Sánchez-Arriaga, G., Siminos, E., Saxena, V., Kourakis, I.: Relativistic breather-type solitary waves with linear polarization in cold plasmas. Phys. Rev. E **91**, 033102 (2015)
6. D. Farina, S. V. Bulanov: Dynamics of relativistic solitons. Plasma Phys. Controll. Fusion **47**, A73 (2005)

7. Kozlov, V.A., Litvak, A.G., Suvorov, E.V.: Envelope solitons of relativistic strong electromagnetic waves. Sov. Phys. JETP **49**, 75 (1979)
8. Saxena, V., Das, A., Sen, A., Kaw, P.: Fluid simulation studies of the dynamical behavior of one-dimensional relativistic electromagnetic solitons. Phys. Plasmas **13**, 032309 (2006)
9. Saxena, V., Kaurakis, I., Sanchez-Arriaga, G., Siminos, E.: Interaction of spatially overlapping standing electromagnetic solitons in plasmas. Phys. Lett. A **377**, 473 (2013)
10. Farina, D., Bulanov, S.V.: Relativistic electromagnetic solitons in the electron-ion plasma. Phys. Rev. Lett. **86**, 5289 (2001)
11. Hadzievski, L., Mancic, A., Skoric, M.M.: Dynamics of weakly relativistic electromagnetic solitons in laser plasmas. Astron. Obs. Belgrade **82**, 101 (2007)
12. Mikaberidze, G., Berezhiani, V.I.: Standing electromagnetic solitons in degenerate relativistic plasmas. Phys. Lett. A **379**, 2730 (2015)
13. Esirkepov, T., Nishihara, K., Bulanov, S.V., Pegoraro, F.: Three-dimensional relativistic electromagnetic subcycle solitons. Phys. Rev. Lett. **89**, 275002 (2002)
14. T.Z. Esirkepov, F.F. Kamenets, S.V. Bulanov, N.M. Naumova: Low-frequency relativistic electromagnetic solitons in collisionless plasmas. J. Exp. Theor. Lett. **68**, 33 (1998)
15. Kuehl, H.H., Zhang, C.Y.: One-dimensional, weakly nonlinear electromagnetic solitary waves in a plasma. Phys. Rev. E **48**, 1316 (1993)
16. J. Borhanian, I. Kourakis, S. Sobhanian: Electromagnetic envelope solitons in magnetized plasmas. Phys. Lett. A **373**, 3667 (2009)
17. Rios, L.A., Galvao, R.M.O.: Self-modulation of linearly polarized electromagnetic waves in non-Maxwellian plasmas. Phys. Plasmas **17**, 042116 (2010)
18. Qiu, H.B., Song, H.Y., Liu, ShB: Nonlinear Raman forward scattering driven by a short laser pulse in a collisional transversely magnetized plasma with nonextensive distribution. Phys. Plasmas **22**, 092128 (2015)
19. Futaana, Y., Machida, S., Saito, Y., Matsuoka, A., Hayakawa, H.: Moon-related nonthermal ions observed by Nozomi: species, sources, and generation mechanisms. J. Geophys. Res. **108**, 1025 (2003)
20. Bostrom, R.: Observations of weak double layers on auroral field lines. IEEE Trans. Plasma Sci. **20**, 756 (1992)
21. Dovner, P.O., Eriksson, A.I., Bostrom, R., Holback, B.: Freja multiprobe observations of electrostatic solitary structures. Geophys. Res. Lett. **21**, 1827 (1994)
22. Cairns, R.A., Mamun, A.A., Bingham, R., Bostrom, R., Dendy, R.O., Nairn, C.M.C., Shukla, P.K.: Electrostatic solitary structures in non-thermal plasmas. Geophys. Res. Lett. **22**, 2709 (1995)
23. Verheest, F., Hellberg, M.A.: Compressive and rarefactive solitary waves in nonthermal two-component plasmas. Phys. Plasmas **17**, 102312 (2010)
24. Baluku, T.K., Hellberg, M.A.: Ion acoustic solitary waves in an electron–positron-ion plasma with non-thermal electrons. Plasma Phys. Controll. Fusion **53**, 095007 (2011)
25. Rostampooran, Sh, Dorranian, D.: Role of nonthermal electron on the dynamics of relativistic electromagnetic soliton in the interaction of laser-plasma. Phys. Plasmas **23**, 083121 (2016)
26. Latora, V., Rapisarda, A., Tsallis, C.: Non-Gaussian equilibrium in a long-range Hamiltonian system. Phys. Rev. E **64**, 056134 (2001)
27. Latora, V., Rapisarda, A., Tsallis, C.: Fingerprints of nonextensive thermodynamics in a long-range Hamiltonian system. Phys. A **305**, 129 (2002)
28. Douglas, P., Bergamini, S., Renzoni, F.: Tunable Tsallis distributions in dissipative optical lattices. Phys. Rev. Lett. **96**, 110601 (2006)
29. Gougam, L.A., Tribeche, M.: Weak ion-acoustic double layers in a plasma with a q-nonextensive electron velocity distribution. Astrophys. Space Sci. **331**, 181 (2011)
30. Pakzad, H.R.: Effect of q-nonextensive distribution of electrons on electron acoustic solitons. Astrophys. Space Sci. **333**, 247 (2011)
31. Tribeche, M., Amour, R., Shukla, P.K.: Ion acoustic solitary waves in a plasma with nonthermal electrons featuring Tsallis distribution. Phys. Rev. E **85**, 037401 (2012)
32. Amour, R., Tribeche, M., Shukla, P.: Electron acoustic solitary waves in a plasma with nonthermal electrons featuring Tsallis distribution. Astrophys. Space Sci. **338**, 287 (2012)
33. Verheest, F., Pillay, S.R.: Large amplitude dust-acoustic solitary waves and double layers in nonthermal plasmas. Phys. Plasmas **15**, 013703 (2008)
34. Taniuti, T., Yajima, N.: Perturbation method for a nonlinear wave modulation. J. Math. Phys. **10**, 1369 (1969)
35. Fedele, R., Schamel, H.: Solitary waves in the Madelung's fluid: connection between the nonlinear Schrodinger equation and the Korteweg-de Vries. Phys. J. B **27**, 313 (2002)
36. Poornakala, S., Das, A., Sen, A., Kaw, P.K.: Laser envelope solitons in cold overdense plasmas. Phys. Plasmas **9**, 1820 (2002)

Permissions

All chapters in this book were first published in JTAP, by Springer International Publishing AG.; hereby published with permission under the Creative Commons Attribution License or equivalent. Every chapter published in this book has been scrutinized by our experts. Their significance has been extensively debated. The topics covered herein carry significant findings which will fuel the growth of the discipline. They may even be implemented as practical applications or may be referred to as a beginning point for another development.

The contributors of this book come from diverse backgrounds, making this book a truly international effort. This book will bring forth new frontiers with its revolutionizing research information and detailed analysis of the nascent developments around the world.

We would like to thank all the contributing authors for lending their expertise to make the book truly unique. They have played a crucial role in the development of this book. Without their invaluable contributions this book wouldn't have been possible. They have made vital efforts to compile up to date information on the varied aspects of this subject to make this book a valuable addition to the collection of many professionals and students.

This book was conceptualized with the vision of imparting up-to-date information and advanced data in this field. To ensure the same, a matchless editorial board was set up. Every individual on the board went through rigorous rounds of assessment to prove their worth. After which they invested a large part of their time researching and compiling the most relevant data for our readers.

The editorial board has been involved in producing this book since its inception. They have spent rigorous hours researching and exploring the diverse topics which have resulted in the successful publishing of this book. They have passed on their knowledge of decades through this book. To expedite this challenging task, the publisher supported the team at every step. A small team of assistant editors was also appointed to further simplify the editing procedure and attain best results for the readers.

Apart from the editorial board, the designing team has also invested a significant amount of their time in understanding the subject and creating the most relevant covers. They scrutinized every image to scout for the most suitable representation of the subject and create an appropriate cover for the book.

The publishing team has been an ardent support to the editorial, designing and production team. Their endless efforts to recruit the best for this project, has resulted in the accomplishment of this book. They are a veteran in the field of academics and their pool of knowledge is as vast as their experience in printing. Their expertise and guidance has proved useful at every step. Their uncompromising quality standards have made this book an exceptional effort. Their encouragement from time to time has been an inspiration for everyone.

The publisher and the editorial board hope that this book will prove to be a valuable piece of knowledge for researchers, students, practitioners and scholars across the globe.

List of Contributors

A. Asgar and M. Ansari
Department of Mathematics, North-Eastern Hill University, Shillong 793022, Meghalaya, India

Mahsa Fakharpour
Department of Physics, Science and Research Branch, Islamic Azad University, Tehran, Iran

Hadi Savaloni
Department of Physics, University of Tehran, North-Kargar Street, Tehran, Iran

Mohsen Emami-Razavi
Plasma Physics Research Center, Science and Research Branch, Islamic Azad University, Tehran, Iran

Somayeh Asgary
Department of Physics, West Tehran Branch, Islamic Azad University, Tehran, Iran

Saeideh Jurablu, Majid Farahmandjou and Tahereh Pormirjaafari Firoozabadi
Department of Physics, Varamin Pishva Branch, Islamic Azad University, Varamin, Iran

Hassan Ghalami Bavil Olyaee
Islamic Azad University-South Tehran Branch, Iranshaher Shomali, Tehran, Iran

Peter J. S. Foot and Vincent Montgomery
Kingston University London, Penrhyn Road, Kingston upon vThames, Surrey, UK

Somaye Hosseini
Plasma Physics Research Center, Science and Research Branch, Islamic Azad University, Tehran, Iran

Hadi Savaloni
Department of Physics, University of Tehran, North-Kargar Street, Tehran, Iran

Mehran Gholipour-Shahraki
Physics Department, Faculty of Science, Arak University, Arak 38156-8-8349, Iran

Meraj Rajaee and Seyed Mohamad Bagher Ghorashi
Department of Laser and Photonics, University of Kashan, Kashan, Iran

S. Mukherjee
Department of Physics, The University of Burdwan, Golapbag, Burdwan, West Bengal 713 104, India

Papri Dasgupta, Asok Poddar and Chandan Mazumdar
Saha Institute of Nuclear Physics, 1/AF Bidhannagar, Kolkata, West Bengal 700 064, India

B. Rezaei and A. Dashtimoghadam
Physics Department, Razi University, Kermanshah 67149, Iran

H. Tokmehdashi and A. A. Rajabi
Department of Physics, Shahrood University of Technology, Shahrood, Iran

M. Hamzavi
Department of Physics, University of Zanjan, Zanjan, Iran

Ali M. Khalaf
Physics Department, Faculty of Science, Al-Azhar University, Cairo, Egypt

Mahmoud M. Taha
Mathematics and Theoretical Physics Department, Nuclear Research Center, Atomic Energy Authority, P.No. 13759, Cairo, Egypt

Morteza Raeisi
Department of Physics, Faculty of Science, Shahrekord University, Shahrekord, Iran

Esmaeil Shahriari and Mohammad Moradi
Department of Physics, Faculty of Science, Shahrekord University, Shahrekord, Iran
Photonic Research Group, Shahrekord University, Shahrekord, Iran

Jalaj Jain
Universidad de Talca, Talca, Chile
Comisión Chilena de Energía Nuclear, Santiago, Chile

Ram Prakash and Udit Narayan Pal
Microwave Tubes Division, CSIR-Central Electronics and Engineering Research Institute, Pilani 333031, India

Gheesa Lal Vyas and Nilanjan Halder
Manipal University Jaipur, Jaipur 303007, India

Malay Bikas Chowdhuri and Ranjana Manchanda
Institute for Plasma Research, Bhat, Gandhinagar 382428, India

Yaduvendra Choyal
School of Physics, Devi Ahilya Vishwavidyalaya, Indore 452017, India

P. Singh
Department of Physics, Addis Ababa University, P.O.Box 1176, Addis Ababa, Ethiopia

Gezahegn Assefa
Department of Physics, Addis Ababa University, P.O.Box 1176, Addis Ababa, Ethiopia
Department of Physics, Wollo University, P.O.Box 1145, Dessie, Ethiopia

Satish Chand
Department of Physics, Mewar University, Chittorgarh, Rajasthan, India

R. P. Singh
Department of Physics, S. S. V. Degree College (Affiliated to C. C. S. University, Meerut U. P.), Hapur, India

A. Govindan
Department of Physics, M. M. H. College (C. C. S. University-Meerut), Ghaziabad, India

F. Teimoori, K. Khojier and N. Z. Dehnavi
Department of Physics, Faculty of Science, Central Tehran Branch, Islamic Azad University, P.O. Box 13185-786, Tehran, Iran

Pierre Gaillard
Universitéde Bourgogne, 9 Av. Alain Savary, Dijon, France

Mickaël Gastineau
IMCCE, Observatoire de Paris, PSL Research University, CNRS, Sorbonne Universités UPMC Univ, Paris 06 Univ. Lille 77 Av. Denfert-Rochereau, 75014 Paris, France

Mehran Balarastaghi
Department of Electrical Engineering, Science and Research Branch, Islamic Azad University, Tehran, Iran

Vahid Ahmadi
Faculty of Electrical and Computer Engineering, Tarbiat Modares University, Tehran, Iran

Lalita Devi and Hitendra K. Malik
Plasma Waves and Particle Acceleration Laboratory, Department of Physics, Indian Institute of Technology Delhi, New Delhi 110016, India

M. Feshanjerdi, A. A. Masoudi and M. Khorrami
Department of Physics, Alzahra University, 1993891167 Tehran, Iran

Dinesh Varshney and S. Shriya
Materials Science Laboratory, School of Physics, Vigyan Bhavan, Devi Ahilya University, Khandwa Road Campus, Indore 452001, India

M. Varshney
Department of Physics, M. B. Khalsa College, Indore 452002, India

N. Singh
Department of Physics, Ranchi College, Ranchi, Jharkhand 834008, India

R. Khenata
Laboratoire de Physique Quantique et de Modélisation Mathématique (LPQ3M), Département de Technologie, Universitéde Mascara, 29000 Mascara, Algeria

S. Farhad Masoudi
Department of Physics, K.N. Toosi University of Technology, P.O. Box 15875-4416, Tehran, Iran

Mansor Khoramabadi
Department of Physics, Boroujerd Branch, Islamic Azad University, Boroujerd, Iran

M. Ghasemi
Department of Atomic and Molecular Physics, Faculty of Basic Sciences, University of Mazandaran, 47416-95447 Babolsar, Iran

F. Sohbatzadeh and S. Mirzanejhad
Department of Atomic and Molecular Physics, Faculty of Basic Sciences, University of Mazandaran, 47416-95447 Babolsar, Iran
Nano and Biotechnology Research Group, Faculty of Basic Sciences, University of Mazandaran, 47416-95447 Babolsar, Iran

Shabnam Rostampooran and Sharooz Saviz
Laser Laboratory, Plasma Physics Research Center, Science and Research Branch, Islamic Azad University, Tehran, Iran

Index